美国数学会经典影印系列

出版者的话

近年来，我国的科学技术取得了长足进步，特别是在数学等自然科学基础领域不断涌现出一流的研究成果。与此同时，国内的科研队伍与国外的交流合作也越来越密切，越来越多的科研工作者可以熟练地阅读英文文献，并在国际顶级期刊发表英文学术文章，在国外出版社出版英文学术著作。

然而，在国内阅读海外原版英文图书仍不是非常便捷。一方面，这些原版图书主要集中在科技、教育比较发达的大中城市的大型综合图书馆以及科研院所的资料室中，普通读者借阅不甚容易；另一方面，原版书价格昂贵，动辄上百美元，购买也很不方便。这极大地限制了科技工作者对于国外先进科学技术知识的获取，间接阻碍了我国科技的发展。

高等教育出版社本着植根教育、弘扬学术的宗旨服务我国广大科技和教育工作者，同美国数学会（American Mathematical Society）合作，在征求海内外众多专家学者意见的基础上，精选该学会近年出版的数百种专业著作，组织出版了"美国数学会经典影印系列"丛书。美国数学会创建于1888年，是国际上极具影响力的专业学术组织，目前拥有近30000会员和580余个机构成员，出版图书3500多种，冯·诺依曼、莱夫谢茨、陶哲轩等世界级数学大家都是其作者。本影印系列涵盖了代数、几何、分析、方程、拓扑、概率、动力系统等所有主要数学分支以及新近发展的数学主题。

我们希望这套书的出版，能够对国内的科研工作者、教育工作者以及青年学生起到重要的学术引领作用，也希望今后能有更多的海外优秀英文著作被介绍到中国。

高等教育出版社

2016 年 12 月

AMS
AMERICAN
MATHEMATICAL
SOCIETY
美国数学会经典影印系列

Axiomatic Geometry

公理化几何学

John M. Lee

中国教育出版传媒集团
高等教育出版社·北京

Contents

Preface

Why study geometry? Those who have progressed far enough in their mathematical education to read this book can probably come up with lots of answers to that question:

- *Geometry is useful.* It's hard to find a branch of mathematics that has more practical applications than geometry. A precise understanding of geometric relationships is prerequisite for making progress in architecture, astronomy, computer graphics, engineering, mapmaking, medical imaging, physics, robotics, sewing, or surveying, among many other fields.

- *Geometry is beautiful.* Because geometry is primarily about spatial relationships, the subject comes with plenty of illustrations, many of which have an austere beauty in their own right. On a deeper level, the study of geometry uncovers surprising and unexpected relationships among shapes, the contemplation of which can inspire an exquisitely satisfying sense of beauty. And, of course, to some degree, geometric relationships underlie almost all visual arts.

- *Geometry comes naturally.* Along with counting and arithmetic, geometry is one of the earliest areas of intellectual inquiry to have been systematically pursued by human societies. Similarly, children start to learn about geometry (naming shapes) as early as two years old, about the same time they start learning about numbers. Almost every culture has developed some detailed understanding of geometrical relationships.

- *Geometry is logical.* As will be explored in some detail in this book, very early in Western history geometry became the paradigm for logical thought and analysis, and students have learned the rudiments of logic and proof in geometry courses for more than two millennia.

All of these are excellent reasons to devote serious study to geometry. But there is a more profound consideration that animates this book: the story of geometry is the story of mathematics itself. There is no better way to understand what modern mathematics is, how it is done, and why it is the way it is than by undertaking a thorough study of the roots of geometry.

Mathematics occupies a unique position among the fields of human intellectual inquiry. It shares many of the characteristics of science: like scientists, mathematicians strive for precision, make assertions that are testable and refutable, and seek to discover universal laws. But there is at least one way in which mathematics is markedly different from most sciences, and indeed from virtually every other subject: in mathematics, it is possible to have a degree of certainty about the truth of an assertion that is usually impossible to attain in any other field. When mathematicians assert that the area of a unit circle is exactly π, we know this is true as surely as we know the facts of our direct experience, such as the fact that I am currently sitting at a desk looking at a computer screen. Moreover, we know that we could calculate π to far greater precision than anyone will ever be able to measure actual physical areas.

Of course, this is not to claim that mathematical knowledge is absolute. No human knowledge is ever 100% certain—we all make mistakes, or we might be hallucinating or dreaming. But mathematical knowledge is as certain as anything in the human experience. By contrast, the assertions of science are always approximate and provisional: Newton's law of universal gravitation was true enough to pass the experimental tests of his time, but it was eventually superseded by Einstein's general theory of relativity. Einstein's theory, in turn, has withstood most of the tests it has been subjected to, but physicists are always ready to admit that there might be a more accurate theory.

How did we get here? What is it about mathematics that gives it such certainty and thus sets it apart from the natural sciences, social sciences, humanities, and arts? The answer is, above all, the concept of *proof*. Ever since ancient times, mathematicians have realized that it is often possible to demonstrate the truth of a mathematical statement by giving a logical argument that is so convincing and so universal that it leaves essentially no doubt. At first, these arguments, when they were found, were mostly ad hoc and depended on the reader's willingness to accept other seemingly simpler truths: if we grant that such-and-such is true, then the Pythagorean theorem logically follows. As time progressed, more proofs were found and the arguments became more sophisticated. But the development of mathematical knowledge since ancient times has been much more than just amassing ever more convincing arguments for an ever larger list of mathematical facts.

There have been two decisive turning points in the history of mathematics, which together are primarily responsible for leading mathematics to the special position it occupies today. The first was the appearance, around 300 BCE, of Euclid's *Elements*. In this monumental work, Euclid organized essentially all of the mathematical knowledge that had been developed in the Western world up to that time (mainly geometry and number theory) into a systematic logical structure. Only a few simple, seemingly self-evident facts (known as *postulates* or *axioms*) were stated without proof, and all other mathematical statements (known as *propositions* or *theorems*) were proved in a strict logical sequence, with the proofs of the simplest theorems based only on the postulates and with proofs of more complicated theorems based on theorems that had already been proved. This structure, known today as the *axiomatic method*, was so effective and so convincing that it became the model for justifying all further mathematical inquiry.

But Euclid's Eden was not without its snake. Beginning soon after Euclid's time, mathematicians raised objections to one of Euclid's geometric postulates (the famous fifth postulate, which we describe in Chapter 1) on the ground that it was too complicated to be considered truly self-evident in the way that his other postulates were. A consensus developed among mathematicians who made a serious study of geometry that the fifth postulate was really a theorem masquerading as a postulate and that Euclid had simply failed to find the proper proof of it. So for about two thousand years, the study of geometry was largely dominated by attempts to find a proof of Euclid's fifth postulate based only on the other four postulates. Many proofs were offered, but all were eventually found to be fatally flawed, usually because they implicitly assumed some other fact that was also not justified by the first four postulates and thus required a proof of its own.

The second decisive turning point in mathematics history occurred in the first half of the nineteenth century, when an insight occurred simultaneously and independently to three different mathematicians (Nikolai Lobachevsky, János Bolyai, and Carl Friedrich Gauss). There was a good reason nobody had succeeded in proving that the fifth postulate followed from the other four: such a proof is logically impossible! The insight that struck these three mathematicians was that it is possible to construct an entirely self-consistent kind of geometry, now called *non-Euclidean geometry*, in which the first four postulates are true but the fifth is false. This geometry did not seem to describe the physical world we live in, but the fact that it is logically just as consistent as Euclidean geometry forced mathematicians to undertake a radical reevaluation of the very nature of mathematical truth and the meaning of the axiomatic method.

Once that insight had been achieved, things progressed rapidly. A new paradigm of the axiomatic method arose, in which the axioms were no longer thought of as statements of self-evident truths from which reliable conclusions could be drawn, but rather as arbitrary statements that were to be *accepted* as truths in a given mathematical context, so that the theorems following from them must be true if the axioms are true.

Paradoxically, this act of unmooring the axiomatic method from any preconceived notions of absolute truth eventually allowed mathematics to achieve previously undreamed of levels of certainty. The reason for this paradox is not hard to discern: once mathematicians realized that axioms are more or less arbitrary assumptions rather than self-evident truths, it became clear that proofs based on the axioms can use *only* those facts that have been explicitly stated in the axioms or previously proved, and the steps of such proofs must be based on clear and incontrovertible principles of logic. Thus it is no longer permissible to base arguments on geometric or numerical intuition about how things behave in the "real world." If it happens that the axioms are deemed to be a useful and accurate description of something, such as a scientific phenomenon or a class of mathematical objects, then every conclusion proved within that axiomatic system is exactly as certain and as accurate as the axioms themselves. For example, we can all agree that Euclid's geometric postulates are extremely accurate descriptions of the geometry of a building on the scale at which humans experience it, so the conclusions of Euclidean geometry can be relied upon to describe that geometry as accurately as we might wish. On the other hand, it might turn out to be the case that the axioms of non-Euclidean geometry are much more accurate descriptions of the geometry of the universe as a whole (on a scale at which galaxies can be treated as uniformly scattered dust), which would mean that the theorems of non-Euclidean geometry could be treated as highly accurate descriptions of the cosmos.

If mathematics occupies a unique position among fields of human inquiry, then geometry can be said to occupy a similarly unique position among subfields of mathematics. Geometry was the first subject to which Euclid applied his logical analysis, and it has been taught to students for more than two millennia as a model of logical thought. It was the subject that engendered the breakthrough in the nineteenth century that led to our modern conception of the axiomatic method (and thus of the very nature of mathematics). Until very recently, virtually every high-school student in the United States took a geometry course based on some version of the axiomatic method, and that was the only time in the lives of most people when they learned about rigorous logical deduction.

The primary purpose of this book is to tell this story: not exactly the historical story, because it is not a book about the history of mathematics, but rather the intellectual story of how one might begin with Euclid's understanding of the axiomatic method in geometry and progress to our modern understanding of it. Since the axiomatic method underlies the way modern mathematics is universally done, understanding it is crucial to understanding what mathematics is, how we read and evaluate mathematical arguments, and why mathematics has achieved the level of certainty it has.

It should be emphasized, though, that while the axiomatic method is our only tool for ensuring that our mathematical knowledge is sound and for communicating that soundness to others, it is not the only tool, and probably not even the most important one, for *discovering* or *understanding* mathematics. For those purposes we rely at least as much on examples, diagrams, scientific applications, intuition, and experience as we do on proofs. But in the end, especially as our mathematics becomes increasingly abstract, we cannot justifiably claim to be sure of the truth of any mathematical statement unless we have found a proof of it.

This book has been developed as a textbook for a course called *Geometry for Teachers*, and it is aimed primarily at undergraduate students who plan to teach geometry in a North American high-school setting. However, it is emphatically *not* a book about how to teach geometry; that is something that aspiring teachers will have to learn from education courses and from hands-on practice. What it offers, instead, is an opportunity to understand on a deep level how mathematics works and how it came to be the way it is, while at the same time developing most of the concrete geometric relationships that secondary teachers will need to know in the classroom. It is not only for future teachers, though: it should also provide something of interest to anyone who wishes to understand Euclidean and non-Euclidean geometry better and to develop skills for doing proofs within an axiomatic system.

In recent years, sadly, the traditional proof-oriented geometry course has sometimes been replaced by different kinds of courses that minimize the importance of the axiomatic method in geometry. But regardless of what kinds of curriculum and pedagogical methods teachers plan to use, it is of central importance that they attain a deep understanding of the underlying mathematical ideas of the subject and a mathematical way of thinking about them—a version of what the mathematics education specialist Liping Ma [**Ma99**] calls "profound understanding of fundamental mathematics." It is my hope that this book will be a vehicle that can help to take them there.

Many books that treat axiomatic geometry rigorously (such as [**Gre08**], [**Moi90**], [**Ven05**]) pass rather quickly through Euclidean geometry, with a major goal of developing non-Euclidean geometry and its relationship to Euclidean geometry. In this book, by contrast, the primary focus is on Euclidean geometry, because that is the subject that future secondary geometry teachers will need to understand most deeply. I do make a serious excursion into hyperbolic geometry at the end of the book, but it is meant to round out the study of Euclidean geometry and place it in its proper perspective, not to be the main focus of the book.

Organization

The book is organized into twenty chapters. The first chapter introduces Euclid. It seeks to familiarize the reader with what Euclid did and did not accomplish and to explain why the axiomatic method underwent such a radical revision in the nineteenth century. The first book of Euclid's *Elements* should be read in parallel with this chapter. It is a good idea for an aspiring geometry teacher to have a complete copy of the *Elements* in his or her library, and I strongly recommend the edition [**Euc02**], which is a carefully edited edition containing all thirteen volumes of the *Elements* in a single book. But for those who are not ready to pay for a hard copy of the book, it is easy to find translations of the *Elements* on the Internet. An excellent source is Dominic Joyce's online edition [**Euc98**], which includes interactive diagrams.

The second chapter of this book constitutes a general introduction to the modern axiomatic method, using a "toy" axiomatic system as a laboratory for experimenting with the concepts and methods of proof. This system, called *incidence geometry*, contains only a few axioms that describe the intersections of points and lines. It is a complete axiomatic system, but it describes only a very small part of geometry so as to make it easier to concentrate on the logical features of the system. It is not my invention; it is adapted from the axiomatic system for Euclidean geometry introduced by David Hilbert at the turn of the last century (see Appendix A), and many other authors (e.g., [**Gre08**], [**Moi90**], [**Ven05**]) have also used a similar device to introduce the principles of the axiomatic method. The axioms I use are close to those of Hilbert but are modified slightly so that each axiom focuses on only one simple fact, so as to make it easier to analyze how different interpretations do or do not satisfy all of the axioms. At the end of the chapter, I walk students through the process of constructing proofs in the context of incidence geometry.

The heart of the book is a modern axiomatic treatment of Euclidean geometry, which occupies Chapters 3 through 16. I have chosen a set of axioms for Euclidean geometry that is based roughly on the SMSG axiom system developed for high-school courses in the 1960s (see Appendix C), which in turn was based on the system proposed by George Birkhoff [**Bir32**] in 1932 (see Appendix B). In choosing the axioms for this book, I've endeavored to keep the postulates closely parallel to those that are typically used in high-school courses, while maintaining a much higher standard of rigor than is typical in such courses. In particular, my postulates differ from the SMSG ones in several important ways:

- I restrict my postulates to plane geometry only.
- I omit the redundant ruler placement and supplement postulates. (The first is a theorem in Chapter 3, and the second is a theorem in Chapter 4.)
- I rephrase the plane separation postulate to refer only to the elementary concept of intersections between line segments and lines, instead of building convexity into the postulate. (Convexity of half-planes is proved in Chapter 3.)
- I replace the three SMSG postulates about angle measures with a single "protractor postulate," much closer in spirit to Birkhoff's original angle measure postulate, more closely parallel to the ruler postulate, and, I think, more intuitive. (Theorems equivalent to the three SMSG angle measure postulates are proved in Chapter 4.)

- I replace the SMSG postulate about the area of a triangle with a postulate about the area of a unit square, which is much more fundamental.

After the treatment of Euclidean geometry comes a transitional chapter (Chapter 17), which summarizes the most important postulates that have been found to be equivalent to the Euclidean parallel postulate. In so doing, it sets the stage for the study of hyperbolic geometry by introducing such important concepts as the angle defect of a polygon. Then Chapters 18 and 19 treat hyperbolic geometry, culminating in the classification of parallel lines into asymptotically parallel and ultraparallel lines.

Chapter 20 is a look forward at some of the directions in which the study of geometry can be continued. I hope it will whet the reader's appetite for further advanced study in geometry.

After Chapter 20 come a number of appendices meant to supplement the main text. The first four appendices (A through D) are reference lists of axioms for the axiomatic systems of Hilbert, Birkhoff, SMSG, and this book. The next two appendices (E and F) give brief descriptions of the conventions of mathematical language and proofs; they can serve either as introductions to these subjects for students who have not been introduced to rigorous proofs before or as review for those who have. Appendices G and H give a very brief summary of background material on sets, functions, and the real number system, which is presupposed by the axiom system used in this book. Finally, Appendix I outlines an alternative approach to the axioms based on the ideas of transformations and rigid motions; it ends with a collection of challenging exercises that might be used as starting points for independent projects.

There is ample material here for a full-year course at a reasonable pace. For shorter courses, there are various things that can be omitted, depending on the tastes of the instructor and the needs of the students. Of course, if your interest is solely in Euclidean geometry, you can always stop after Chapter 16 or 17. On the other hand, if you want to move more quickly through the Euclidean material and spend more time on non-Euclidean geometry, there are various Euclidean topics that are not used in an essential way later in the book and can safely be skipped, such as the material on nonconvex polygons at the end of Chapter 8 and some or all of Chapters 14, 15, or 16. Don't worry if the course seems to move slowly at first—it has been my experience that it takes a rather long time to get through the first six chapters, but things tend to move more quickly after that.

I adhere to some typographical conventions that I hope will make the book easier to use. Mathematical terms are typeset in ***bold italics*** when they are officially introduced, to reflect the fact that definitions are just as important as theorems and proofs but fit better into the flow of paragraphs rather than being called out with special headings. The symbol □ is used to mark the ends of proofs, and it is also used at the end of the statement of a corollary that follows so easily that it is not necessary to write down a proof. The symbol // marks the ends of numbered examples.

The exercises at the ends of the chapters are essential for mastering the type of thinking that leads to a deep understanding of mathematical concepts. Most of them are proofs. Almost all of them can be done by using techniques very similar to ones used in the proofs in the book, although a few of the exercises in the later chapters might require a bit more ingenuity.

Prerequisites

Because this book is written as a textbook for an advanced undergraduate course, I expect readers to be conversant with most of the subjects that are typically treated at the beginning and intermediate undergraduate levels. In particular, readers should be comfortable at least with one-variable calculus, vector algebra in the plane, mathematical induction, and elementary set theory. It would also be good to have already seen the least upper bound principle and the notions of injective and surjective functions.

Students who have had a course that required them to construct rigorous mathematical proofs will have a distinct advantage in reading the book and doing the exercises. But for students whose experience with proofs is limited, a careful study of Appendices E and F should provide sufficient opportunities to deepen their understanding of what proofs are and how to read and write them.

Acknowledgments

I am deeply indebted to my colleague John Palmieri, who has taught from two different draft versions of this book and contributed immeasurably to improving it. I owe just as much thanks to my students, many of whom—especially Natalie Hobson, Dan Riness, and Gina Bremer—have generously and patiently provided detailed suggestions for helping the book to serve students better. Of course, I owe enormous gratitude to a long list of authors who have written rigorous texts about Euclidean and non-Euclidean geometry that have educated and inspired me, especially Marvin Greenberg [**Gre08**], Robin Hartshorne [**Har00**], George Martin [**Mar96**], Richard Millman and George Parker [**MP91**], Edwin Moise [**Moi90**], and Gerard Venema [**Ven05**].

Additional Resources

I plan to post some supplementary materials, as well as a list of corrections, on the website www.ams.org/bookpages/amstext-21. For the sake of future readers, I encourage all readers to make notes of any mistakes you find and any passages you think could use improvement, whether major or trivial, and send them to me at the email address posted on the website above. Happy reading!

John M. Lee
Seattle, WA
November 27, 2012

Chapter 1

Euclid

The story of axiomatic geometry begins with Euclid, the most famous mathematician in history. We know essentially nothing about Euclid's life, save that he was a Greek who lived and worked in Alexandria, Egypt, around 300 BCE. His best known work is the *Elements* [**Euc02**], a thirteen-volume treatise that organized and systematized essentially all of the knowledge of geometry and number theory that had been developed in the Western world up to that time.

It is believed that most of the mathematical results of the *Elements* were known well before Euclid's time. Euclid's principal achievement was not the discovery of new mathematical facts, but something much more profound: he was apparently the first mathematician to find a way to organize virtually all known mathematical knowledge into a single coherent, logical system, beginning with a list of definitions and a small number of assumptions (called *postulates*) and progressing logically to prove every other result from the postulates and the previously proved results. The *Elements* provided the Western world with a model of deductive mathematical reasoning whose essential features we still emulate today.

A brief remark is in order regarding the authorship of the *Elements*. Scholars of Greek mathematics are convinced that some of the text that has come down to us as the *Elements* was not in fact written by Euclid but instead was added by later authors. For some portions of the text, this conclusion is well founded—for example, there are passages that appear in earlier Greek manuscripts as marginal notes but that are part of the main text in later editions; it is reasonable to conclude that these passages were added by scholars after Euclid's time and were later incorporated into Euclid's text when the manuscript was recopied. For other passages, the authorship is less clear—some scholars even speculate that the definitions might have been among the later additions. We will probably never know exactly what Euclid's original version of the *Elements* looked like.

Since our purpose here is primarily to study the logical development of geometry and not its historical development, let us simply agree to use the name Euclid to refer to the writer or writers of the text that has been passed down to us as the *Elements* and leave it to the historians to explore the subtleties of multiple authorship.

Reading Euclid

Before going any further, you should take some time now to glance at Book I of the *Elements*, which contains most of Euclid's elementary results about plane geometry. As we discuss each of the various parts of the text—definitions, postulates, common notions, and propositions—you should go back and read through that part carefully. Be sure to observe how the propositions build logically one upon another, each proof relying only on definitions, postulates, common notions, and previously proved propositions.

Here are some remarks about the various components of Book I.

Definitions

If you study Euclid's definitions carefully, you will see that they can be divided into two rather different categories. Many of the definitions (including the first nine) are *descriptive definitions*, meaning that they are meant to convey to the reader an intuitive sense of what Euclid is talking about. For example, Euclid defines a *point* as "that which has no part," a *line* as "breadthless length," and a *straight line* as "a line which lies evenly with the points on itself." (Here and throughout this book, our quotations from Euclid are taken from the well-known 1908 English translation of the *Elements* by T. L. Heath, based on the edition [**Euc02**] edited by Dana Densmore.) These descriptions serve to guide the reader's thinking about these concepts but are not sufficiently precise to be used to justify steps in logical arguments because they typically define new terms in terms of other terms that have not been previously defined. For example, Euclid never explains what is meant by "breadthless length" or by "lies evenly with the points on itself"; the reader is expected to interpret these definitions in light of experience and common knowledge. Indeed, in all the books of the *Elements*, Euclid never refers to the first nine definitions, or to any other descriptive definitions, to justify steps in his proofs.

Contrasted with the descriptive definitions are the *logical definitions*. These are definitions that describe a precise mathematical condition that must be satisfied in order for an object to be considered an example of the defined term. The first logical definition in the *Elements* is Definition 10: "When a straight line standing on a straight line makes the adjacent angles equal to one another, each of the equal angles is *right*, and the straight line standing on the other is called a *perpendicular* to that on which it stands." This describes angles in a particular type of geometric configuration (Fig. 1.1) and tells us that we are entitled to call an angle a *right angle* if and only if it occurs in a configuration of that type. (See Appendix E for a discussion about the use of "if and only if" in definitions.) Some other terms for which Euclid provides logical definitions are *circle*, *isosceles triangle*, and *parallel*.

Fig. 1.1. Euclid's definition of right angles.

Postulates

It is in the postulates that the great genius of Euclid's achievement becomes evident. Although mathematicians before Euclid had provided proofs of some isolated geometric facts (for example, the Pythagorean theorem was probably proved at least two hundred years before Euclid's time), it was apparently Euclid who first conceived the idea of arranging all the proofs in a strict logical sequence. Euclid realized that not every geometric fact can be proved, because every proof must rely on some prior geometric knowledge; thus any attempt to prove everything is doomed to circularity. He knew, therefore, that it was necessary to begin by accepting some facts without proof. He chose to begin by postulating five simple geometric statements:

- **Euclid's Postulate 1:** *To draw a straight line from any point to any point.*
- **Euclid's Postulate 2:** *To produce a finite straight line continuously in a straight line.*
- **Euclid's Postulate 3:** *To describe a circle with any center and distance.*
- **Euclid's Postulate 4:** *That all right angles are equal to one another.*
- **Euclid's Postulate 5:** *That, if a straight line falling on two straight lines make the interior angles on the same side less than two right angles, the two straight lines, if produced indefinitely, meet on that side on which are the angles less than the two right angles.*

The first three postulates are *constructions* and should be read as if they began with the words "It is possible." For example, Postulate 1 asserts that "[It is possible] to draw a straight line from any point to any point." (For Euclid, the term *straight line* could refer to a portion of a line with finite length—what we would call a *line segment*.) The first three postulates are generally understood as describing in abstract, idealized terms what we do concretely with the two classical geometric construction tools: a *straightedge* (a kind of idealized ruler that is unmarked but indefinitely extendible) and a *compass* (two arms connected by a hinge, with a sharp spike on the end of one arm and a drawing implement on the end of the other). With a straightedge, we can align the edge with two given points and draw a straight line segment connecting them (Postulate 1); and given a previously drawn straight line segment, we can align the straightedge with it and extend (or "produce") it in either direction to form a longer line segment (Postulate 2). With a compass, we can place the spike at any predetermined point in the plane, place the drawing tip at any other predetermined point, and draw a complete circle whose center is the first point and whose circumference passes through the second point. The statement of Postulate 3 does not precisely specify what Euclid meant by "any center and distance"; but the way he uses this postulate, for example in Propositions I.1 and I.2, makes it clear that it is applicable only when the center and one point on the circumference are already given. (In this book, we follow the traditional convention for referring to Euclid's propositions by number: "Proposition I.2" means Proposition 2 in Book I of the *Elements*.)

The last two postulates are different: instead of asserting that certain geometric configurations can be constructed, they describe relationships that must hold whenever a given geometric configuration exists. Postulate 4 is simple: it says that whenever two right angles have been constructed, those two angles are equal to each other. To interpret this, we must address Euclid's use of the word *equal*. In modern mathematical usage, "*A* equals *B*" just means the *A* and *B* are two different names for the same mathematical object (which could be a number, an angle, a triangle, a polynomial, or whatever). But Euclid uses the word differently: when he says that two geometric objects are equal, he means essentially that they have the *same size*. In modern terminology, when Euclid says two angles are equal, we would say they have the same degree measure; when he says two lines (i.e., line segments) are equal, we would say they have the same length; and when he says two figures such as triangles or parallelograms are equal, we would say they have the same area. Thus Postulate 4 is actually asserting that all right angles are the same size.

It is important to understand why Postulate 4 is needed. Euclid's definition of a right angle applies only to an angle that appears in a certain configuration (one of the two adjacent angles formed when a straight line meets another straight line in such a way as to make equal adjacent angles); it does not allow us to conclude that a right angle appearing in one part of the plane bears any necessary relationship with right angles appearing elsewhere. Thus Postulate 4 can be thought of as an assertion of a certain type of "uniformity" in the plane: right angles have the same size wherever they appear.

Postulate 5 says, in more modern terms, that if one straight line crosses two other straight lines in such a way that the interior angles on one side have degree measures adding up to less than 180° ("less than two right angles"), then those two straight lines must meet on that same side of the first line (Fig. 1.2). Intuitively, it says that if two lines start out

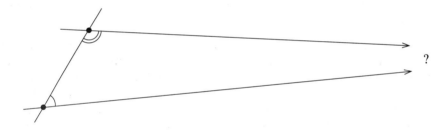

Fig. 1.2. Euclid's Postulate 5.

"pointing toward each other," they will eventually meet. Because it is used primarily to prove properties of parallel lines (for example, in Proposition I.29 to prove that parallel lines always make equal corresponding angles with a transversal), Euclid's fifth postulate is often called the "parallel postulate." We will have much more to say about it later in this chapter.

Common Notions

Following his five postulates, Euclid states five "common notions," which are also meant to be self-evident facts that are to be accepted without proof:

- **Common Notion 1:** *Things which are equal to the same thing are also equal to one another.*
- **Common Notion 2:** *If equals be added to equals, the wholes are equal.*
- **Common Notion 3:** *If equals be subtracted from equals, the remainders are equal.*
- **Common Notion 4:** *Things which coincide with one another are equal to one another.*
- **Common Notion 5:** *The whole is greater than the part.*

Whereas the five postulates express facts about geometric configurations, the common notions express facts about *magnitudes*. For Euclid, magnitudes are objects that can be compared, added, and subtracted, provided they are of the "same kind." In Book I, the kinds of magnitudes that Euclid considers are (lengths of) line segments, (measures of) angles, and (areas of) triangles and quadrilaterals. For example, a line segment (which Euclid calls a "finite straight line") can be equal to, greater than, or less than another line segment; two line segments can be added together to form a longer line segment; and a shorter line segment can be subtracted from a longer one.

It is interesting to observe that although Euclid compares, adds, and subtracts geometric magnitudes of the same kind, he never uses *numbers* to measure geometric magnitudes. This might strike one as curious, because human societies had been using numbers to measure things since prehistoric times. But there is a simple explanation for its omission from Euclid's axiomatic treatment of geometry: to the ancient Greeks, *numbers* meant whole numbers, or at best ratios of whole numbers (what we now call *rational numbers*). However, the followers of Pythagoras had discovered long before Euclid that the relationship between the diagonal of a square and its side length cannot be expressed as a ratio of whole numbers. In modern terms, we would say that the ratio of the length of the diagonal of a square to its side length is equal to $\sqrt{2}$; but there is no rational number whose square is 2. Here is a proof that this is so.

Theorem 1.1 (Irrationality of $\sqrt{2}$). *There is no rational number whose square is 2.*

Proof. Assume the contrary: that is, assume that there are integers p and q with $q \neq 0$ such that $2 = (p/q)^2$. After canceling out common factors, we can assume that p/q is in lowest terms, meaning that p and q have no common prime factors. Multiplying the equation through by q^2, we obtain

$$2q^2 = p^2. \tag{1.1}$$

Because p^2 is equal to the even number $2q^2$, it follows that p itself is even; thus there is some integer k such that $p = 2k$. Inserting this into (1.1) yields

$$2q^2 = (2k)^2 = 4k^2.$$

We can divide this equation through by 2 and obtain $q^2 = 2k^2$, which shows that q is also even. But this means that p and q have 2 as a common prime factor, contradicting our assumption that p/q is in lowest terms. Thus our original assumption must have been false. ☐

This is one of the oldest examples of what we now call a *proof by contradiction* or *indirect proof*, in which we assume that a result is false and show that this assumption leads to a contradiction. A version of this argument appears in the *Elements* as Proposition VIII.8 (although it is a bit hard to recognize as such because of the archaic terminology Euclid used). For a more thorough discussion of proofs by contradiction, see Appendix F. For details of the properties of numbers that were used in the proof, see Appendix H.

This fact had the consequence that, for the Greeks, there was no "number" that could represent the length of the diagonal of a square whose sides have length 1. Thus it was not possible to assign a numeric length to every line segment.

Euclid's way around this difficulty was simply to avoid using numbers to measure magnitudes. Instead, he only compares, adds, and subtracts magnitudes of the same kind.

(In later books, he also compares ratios of such magnitudes.) As mentioned above, it is clear from Euclid's use of the word "equal" that he always interprets it to mean "the same size"; any claim that two geometric figures are equal is ultimately justified by showing that one can be moved so that it coincides with the other or that the two objects can be decomposed into pieces that are equal for the same reason. His use of the phrases "greater than" and "less than" is always based on Common Notion 5: if one geometric object (such as a line segment or an angle) is part of another or is equal (in size) to part of another, then the first is less than the second.

Having laid out his definitions and assumptions, Euclid is now ready to start proving things.

Propositions

Euclid refers to every mathematical statement that he proves as a *proposition*. This is somewhat different from the usual practice in modern mathematical writing, where a result to be proved might be called a *theorem* (an important result, usually one that requires a relatively lengthy or difficult proof); a *proposition* (an interesting result that requires proof but is usually not important enough to be called a theorem); a *corollary* (an interesting result that follows from a previous theorem with little or no extra effort); or a *lemma* (a preliminary result that is not particularly interesting in its own right but is needed to prove another theorem or proposition).

Even though Euclid's results are all called propositions, the first thing one notices when looking through them is that, like the postulates, they are of two distinct types. Some propositions (such as I.1, I.2, and I.3) describe constructions of certain geometric configurations. (Traditionally, scholars of Euclid call these propositions *problems*. For clarity, we will call them *construction problems*.) These are usually stated in the infinitive ("to construct an equilateral triangle on a given finite straight line"), but like the first three postulates, they should be read as asserting the possibility of making the indicated constructions: "[It is possible] to construct an equilateral triangle on a given finite straight line."

Other propositions (traditionally called *theorems*) assert that certain relationships always hold in geometric configurations of a given type. Some examples are Propositions I.4 (the side-angle-side congruence theorem) and I.5 (the base angles of an isosceles triangle are equal). They do not assert the constructibility of anything. Instead, they apply only when a configuration of the given type has already been constructed, and they allow us to conclude that certain relationships always hold in that situation.

For both the construction problems and the theorems, Euclid's propositions and proofs follow a predictable pattern. Most propositions have six discernible parts. Here is how the parts were described by the Greek mathematician Proclus [**Pro70**]:

(1) **Enunciation**: Stating in general form the construction problem to be solved or the theorem to be proved. Example from Proposition I.1: "On a given finite straight line to construct an equilateral triangle."

(2) **Setting out**: Choosing a specific (but arbitrary) instance of the general situation and giving names to its constituent points and lines. Example: "Let AB be the given finite straight line."

(3) **Specification:** Announcing what has to be constructed or proved in this specific instance. Example: "Thus it is required to construct an equilateral triangle on the straight line AB."

(4) **Construction:** Adding points, lines, and circles as needed. For construction problems, this is where the main construction algorithm is described. For theorems, this part, if present, describes any auxiliary objects that need to be added to the figure to complete the proof; if none are needed, it might be omitted.

(5) **Proof:** Arguing logically that the given construction does indeed solve the given problem or that the given relationships do indeed hold.

(6) **Conclusion:** Restating what has been proved.

A word about the conclusions of Euclid's proofs is in order. Euclid and the classical mathematicians who followed him believed that a proof was not complete unless it ended with a precise statement of what had been shown to be true. For construction problems, this statement always ended with a phrase meaning "which was to be done" (translated into Latin as *quod erat faciendum*, or q.e.f.). For theorems, it ended with "which was to be demonstrated" (*quod erat demonstrandum*, or q.e.d.), which explains the origin of our traditional proof-ending abbreviation. In Heath's translation of Proposition I.1, the conclusion reads "Therefore the triangle ABC is equilateral; and it has been constructed on the given finite straight line AB. Being what it was required to do." Because this last step is so formulaic, after the first few propositions Heath abbreviates it: "Therefore etc. q.e.f.," or "Therefore etc. q.e.d."

We leave it to you to read Euclid's propositions in detail, but it is worth focusing briefly on the first three because they tell us something important about Euclid's conception of straightedge and compass constructions. Here are the statements of Euclid's first three propositions:

Euclid's Proposition I.1. *On a given finite straight line to construct an equilateral triangle.*

Euclid's Proposition I.2. *To place a straight line equal to a given straight line with one end at a given point.*

Euclid's Proposition I.3. *Given two unequal straight lines, to cut off from the greater a straight line equal to the less.*

One might well wonder why Euclid chose to start where he did. The construction of an equilateral triangle is undoubtedly useful, but is it really more useful than other fundamental constructions such as bisecting an angle, bisecting a line segment, or constructing a perpendicular? The second proposition is even more perplexing: all it allows us to do is to construct a copy of a given line segment with one end at a certain predetermined point, but we have no control over which direction the line segment points. Why should this be of any use whatsoever?

The mystery is solved by the third proposition. If you look closely at the way Postulate 3 is used in the first two propositions, it becomes clear that Euclid has a very specific interpretation in mind when he writes about "describing a circle with any center and distance." In Proposition I.1, he describes the circle with center A and distance AB and the circle with center B and distance BA; and in Proposition I.2, he describes circles with center B and distance BC and with center D and distance DG. In every case, the center is a point that has already been located, and the "distance" is actually a segment that has already been drawn with the given center as one of its endpoints. Nowhere in these two propositions does he describe what we routinely do with a physical compass: open the compass to the length of a given line segment and then pick it up and draw a circle with that radius somewhere else. Traditionally, this restriction is expressed by saying that Euclid's hypothetical compass is a "collapsing compass"—as soon as you pick it up off the page, it collapses, so you cannot put it down and reproduce the same radius somewhere else.

The purpose of Proposition I.3 is precisely to simulate a noncollapsing compass. After Proposition I.3 is proved, if you have a point O that you want to be the center of a circle and a segment AB somewhere else whose length you want to use for the radius, you can draw a segment from O to some other point E (Postulate 1), extend it if necessary so that it's longer than AB (Postulate 2), use Proposition I.3 to locate C on that extended segment so that $OC = AB$, and then draw the circle with center O and radius OC (Postulate 3).

Obviously, the Greeks must have known how to make compasses that held their separation when picked up, so it is interesting to speculate about why Euclid's postulate described only a collapsing compass. It is easy to imagine that an early draft of the *Elements* might have contained a stronger version of Postulate 3 that allowed a noncollapsing compass, and then Euclid discovered that by using the constructions embodied in the first three propositions he could get away with a weaker postulate. If so, he was probably very proud of himself (and rightly so).

After Euclid

Euclid's *Elements* became the universal geometry textbook, studied by most educated Westerners for two thousand years. Even so, beginning already in ancient times, scholars worked hard to improve upon Euclid's treatment of geometry.

The focus of attention for most of those two thousand years was Euclid's fifth postulate, which usually strikes people as being the most problematic of the five. Whereas Postulates 1 through 4 express possibilities and properties that are truly self-evident to anyone who has thought about our everyday experience with geometric relationships, Postulate 5 is of a different order altogether. Most noticeably, its statement is dramatically longer than those of the other four postulates. More importantly, it expresses an assumption about geometric configurations that cannot fairly be said to be self-evident in the same way as the other four postulates. Although it is certainly plausible to expect that two lines that start out pointing toward each other will eventually meet, it stretches credulity to argue that this conclusion is self-evident. If the sum of the two interior angle measures in Fig. 1.2 were, say, 179.999999999999999999999999998°, then the point where the two lines intersect would be farther away than the most distant known galaxies in the universe! Can anyone really say that the existence of such an intersection point is self-evident? The fifth postulate has the appearance of something that ought to be proved instead of being accepted as a postulate.

There is reason to believe that Euclid himself was less than fully comfortable with his fifth postulate: he did not invoke it in any proofs until Proposition I.29, even though some of the earlier proofs could have been simplified by using it.

For centuries, mathematicians who studied Euclid considered the fifth postulate to be the weakest link in Euclid's tightly argued chain of reasoning. Many mathematicians tried and failed to come up with proofs that the fifth postulate follows logically from the other four, or at least to replace it with a more truly self-evident postulate. This quest, in fact, has motivated much of the development of geometry since Euclid.

The earliest attempt to prove the fifth postulate that has survived to modern times was by Proclus (412–485 CE), a Greek philosopher and mathematician who lived in Asia Minor during the time of the early Byzantine empire and wrote an important commentary on Euclid's *Elements* [**Pro70**]. (This commentary, by the way, contains most of the scant biographical information we have about Euclid, and even this must be considered essentially as legendary because it was written at least 700 years after the time of Euclid.) In this commentary, Proclus opined that Postulate 5 did not have the self-evident nature of a postulate and thus should be proved, and then he proceeded to offer a proof of it. Unfortunately, like so many later attempts, Proclus's proof was based on an unstated and unproved assumption. Although Euclid *defined* parallel lines to be lines in the same plane that do not meet, no matter how far they are extended, Proclus tacitly assumed also that parallel lines are everywhere *equidistant*, meaning the same distance apart (see Fig. 1.3). We will see in Chapter 17 that this assumption is actually equivalent to assuming Euclid's fifth postulate.

Fig. 1.3. Proclus's assumption.

After the fall of the Roman Empire, the study of geometry moved for the most part to the Islamic world. Although the original Greek text of Euclid's *Elements* was lost until the Renaissance, translations into Arabic were widely studied throughout the Islamic empire and eventually made their way back to Europe to be translated into Latin and other languages.

During the years 1000–1300, several important Islamic mathematicians took up the study of the fifth postulate. Most notable among them was the Persian scholar and poet Omar Khayyam (1048–1123), who criticized previous attempts to prove the fifth postulate and then offered a proof of his own. His proof was incorrect because, like that of Proclus, it relied on the unproved assumption that parallel lines are everywhere equidistant.

With the advent of the Renaissance, Western Europeans again began to tackle the problem of the fifth postulate. One of the most important attempts was made by the Italian mathematician Giovanni Saccheri (1667–1733). Saccheri set out to prove the fifth postulate by assuming that it was false and showing that this assumption led to a contradiction. His arguments were carefully constructed and quite rigorous for their day. In the process, he proved a great many strange and counterintuitive theorems that follow from the assumption that the fifth postulate is false, such as that rectangles cannot exist and that the interior angle measures of triangles always add up to less than 180°. In the end, though, he could not find a contradiction that measured up to the standards of rigor he had set for himself. Instead, he punted: having shown that his assumption implied that there must exist parallel lines that approach closer and closer to each other but never meet, he claimed that this result is "repugnant to the nature of the straight line" and therefore his original assumption must have been false.

Saccheri is remembered today, not for his failed attempt to prove Euclid's fifth postulate, but because in attempting to do so he managed to prove a great many results that we now recognize as theorems in a mysterious new system of geometry that we now call *non-Euclidean geometry*. Because of the preconceptions built into the cultural context within which he worked, he could only see them as steps along the way to his hoped-for contradiction, which never came.

The next participant in our story played a minor role, but a memorable one. In 1795, the Scottish mathematician John Playfair (1748–1819) published an edition of the first six books of Euclid's *Elements* [**Pla95**], which he had edited to correct some of what were then perceived as flaws in the original. One of Playfair's modifications was to replace Euclid's fifth postulate with the following alternative postulate.

Playfair's Postulate. *Two straight lines cannot be drawn through the same point, parallel to the same straight line, without coinciding with one another.*

In other words, given a line and a point not on that line, there can be at most one line through the given point and parallel to the given line. Playfair showed that this alternative postulate leads to the same conclusions as Euclid's fifth postulate. This postulate has a notable advantage over Euclid's fifth postulate, because it focuses attention on the uniqueness of parallel lines, which (as later generations were to learn) is the crux of the issue. Most modern treatments of Euclidean geometry incorporate some version of Playfair's postulate instead of the fifth postulate originally stated by Euclid.

The Discovery of Non-Euclidean Geometry

The next event in the history of geometry was the most profound development in mathematics since the time of Euclid. In the 1820s, a revolutionary idea occurred independently and more or less simultaneously to three different mathematicians: perhaps the reason the fifth postulate had turned out to be so hard to prove was that there is a completely consistent theory of geometry in which Euclid's first four postulates are true but the fifth postulate is *false*. If this speculation turned out to be justified, it would mean that proving the fifth postulate from the other four would be a logical impossibility.

In 1829, the Russian mathematician Nikolai Lobachevsky (1792–1856) published a paper laying out the foundations of what we now call *non-Euclidean geometry*, in which the fifth postulate is assumed to be false, and proving a good number of theorems about it. Meanwhile in Hungary, János Bolyai (1802–1860), the young son of an eminent Hungarian mathematician, spent the years 1820–1823 writing a manuscript that accomplished much the same thing; his paper was eventually published in 1832 as an appendix to a textbook written by his father. When the great German mathematician Carl Friedrich Gauss (1777–1855)—a friend of Bolyai's father—read Bolyai's paper, he remarked that it duplicated investigations of his own that he had carried out privately but never published. Although Bolyai and Lobachevsky deservedly received the credit for having invented non-Euclidean geometry based on their published works, in view of the creativity and depth of Gauss's other contributions to mathematics, there is no reason to doubt that Gauss had indeed had the same insight as Lobachevsky and Bolyai.

In a sense, the principal contribution of these mathematicians was more a change of attitude than anything else: while Omar Khayyam, Giovanni Saccheri, and others had also proved theorems of non-Euclidean geometry, it was Lobachevsky and Bolyai (and presumably also Gauss) who first recognized them as such. However, even after this groundbreaking work, there was still no *proof* that non-Euclidean geometry was consistent (i.e., that it could never lead to a contradiction). The *coup de grâce* for attempts to prove the fifth postulate was provided in 1868 by another Italian mathematician, Eugenio Beltrami (1835–1900), who proved for the first time that non-Euclidean geometry was just as consistent as Euclidean geometry. Thus the ancient question of whether Euclid's fifth postulate followed from the other four was finally settled.

The versions of non-Euclidean geometry studied by Lobachevsky, Bolyai, Gauss, and Beltrami were all essentially equivalent to each other. This geometry is now called *hyperbolic geometry*. Its most salient feature is that Playfair's postulate is false: in hyperbolic geometry it is always possible for two or more distinct straight lines to be drawn through the same point, both parallel to a given straight line. As a consequence, many aspects of Euclid's theory of parallel lines (such as the result in Proposition I.29 about the equality of corresponding angles made by a transversal to two parallel lines) are not valid in hyperbolic geometry. In fact, as we will see in Chapter 19, the phenomenon of parallel lines approaching each other asymptotically but never meeting—which Saccheri declared to be "repugnant to the nature of the straight line"—does indeed occur in hyperbolic geometry.

One might also wonder if the Euclidean theory of parallel lines could fail in the opposite way: instead of having two or more parallels through the same point, might it be possible to construct a consistent theory in which there are *no* parallels to a given line through a given point? It is easy to imagine a type of geometry in which there are no parallel lines: the geometry of a sphere. If you move as straight as possible on the surface of a sphere, you will follow a path known as a *great circle*—a circle whose center coincides with the center of the sphere. It can be visualized as the place where the sphere intersects a plane that passes through the center of the sphere. If we reinterpret the term "line" to mean a great circle on the sphere, then indeed there are no parallel "lines," because any two great circles must intersect each other. But this does not seem to have much relevance for Euclid's geometry, because line segments cannot be extended arbitrarily far—in spherical geometry, no line can be longer than the circumference of the sphere. This would seem to contradict Postulate 2, which had always been interpreted to mean that a line segment can be extended arbitrarily far in both directions.

However, after the discovery of hyperbolic geometry, another German mathematician, Bernhard Riemann (1826–1866), realized that Euclid's second postulate could be reinterpreted in such a way that it does hold on the surface of a sphere. Basically, he argued that Euclid's second postulate only requires that any line segment can be extended to a longer one in both directions, but it does not specifically say that we can extend it to any length we wish. With this reinterpretation, spherical geometry can be seen to be a consistent geometry in which no lines are parallel to each other. Of course, a number of Euclid's proofs break down in this situation, because many of the implicit geometric assumptions he used in his proofs do not hold on the sphere; see our discussion of Euclid's Proposition I.16 below for an example. This alternative form of non-Euclidean geometry is sometimes called *elliptic geometry*. (Because of its association with Riemann, it is sometimes erroneously referred to as *Riemannian geometry*, but this term is now universally used to refer to an entirely different type of geometry, which is a branch of differential geometry.)

Perhaps the most convincing confirmation that Euclid's is not the only possible consistent theory of geometry came from Einstein's general theory of relativity around the turn of the twentieth century. If we are to believe, like Euclid, that the postulates reflect self-evident truths about the geometry of the world we live in, then Euclid's statements about "straight lines" should translate into true statements about the behavior of light rays in the real world. (After all, we commonly judge the "straightness" of something by sighting along it, so what physical phenomenon could possibly qualify as a better model of "straight lines" than light rays?) Thus the closest thing in the physical world to a geometric triangle would be a three-sided figure whose sides are formed by the paths of light rays.

Yet Einstein's theory tells us that in the presence of gravitational fields, space itself is "warped," and this affects the paths along which light rays travel. One of the most dramatic confirmations of Einstein's theory comes from the phenomenon known as *gravitational lensing*: this occurs when we observe a distant object but there is a massive galaxy cluster directly between us and the object. Einstein's theory predicts that the light rays from the distant object should be able to follow two (or more) different paths to reach our eyes because of the distortion of space around the galaxy cluster.

Fig. 1.4. A gravitational lens (photograph by W. N. Colley, E. Turner, J. A. Tyson, and NASA).

This phenomenon has indeed been observed; Fig. 1.4 shows a photograph taken by the Hubble Space Telescope, in which a distant loop-shaped galaxy (circled) appears twice in the same photographic image because its light rays have traveled around both sides of the large galaxy cluster in the middle of the photo. Fig. 1.5 shows a schematic view of the same situation. The light coming from a certain point A in the middle of the loop-shaped galaxy follows two paths to our eyes and along the way makes two different dots (B and C) on the photographic plate. As a result, the three points A, B, and C form a triangle whose interior angle measures add up to a number slightly greater than $180°$. (Although it doesn't look like a triangle in this diagram, remember that the edges are paths of light rays. What could be straighter than that?) Yet Euclidean geometry predicts that every triangle has interior angle measures that add up to *exactly* $180°$. We can see why Euclid's arguments fail in this situation by examining the figure: in this case there are two distinct line segments connecting the point A to the observer's eye, which contradicts Euclid's intended meaning of his first postulate. We have no choice but to conclude that the geometry of the physical world we live in does not exactly follow Euclid's rules.

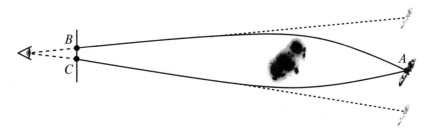

Fig. 1.5. A triangle whose angle sum is greater than $180°$.

Gaps in Euclid's Arguments

As a result of the non-Euclidean breakthroughs of Lobachevsky, Bolyai, and Gauss, mathematicians were forced to undertake a far-reaching reexamination of the very foundations of their subject. Euclid and everyone who followed him had regarded postulates as self-evident truths about the real world, from which reliable conclusions could be drawn. But once it was discovered that two or three conflicting systems of postulates worked equally well as logical foundations for geometry, mathematicians had to face an uncomfortable question: what exactly are we doing when we accept some postulates and use them to prove theorems? It became clear that the system of postulates one uses is in some sense an arbitrary choice; once the postulates have been chosen, as long as they don't lead to any contradictions, one can proceed to prove whatever theorems follow from them.

Thus was born the notion of a mathematical theory as an *axiomatic system*—a sequence of theorems based on a particular set of assumptions called *postulates* or *axioms* (these two words are used synonymously in modern mathematics). We will give a more precise definition of axiomatic systems in the next chapter.

Of course, the axioms we choose are not *completely* arbitrary, because the only axiomatic systems that are worth studying are those that describe something useful or interesting—an aspect of the physical world, or a class of mathematical structures that have proved useful in other contexts, for example. But from a strictly logical point of view, we may adopt any consistent system of axioms that we like, and the resulting theorems will constitute a valid mathematical theory.

The catch is that one must scrupulously ensure that the proofs of the theorems do not use *anything* other than what has been assumed in the postulates. If the axioms represent arbitrary assumptions instead of self-evident facts about the real world, then nothing except the axioms is relevant to proofs within the system. Reasoning based on intuition about the behavior of straight lines or properties that are evident from diagrams or common experience in the real world will no longer be justifiable within the axiomatic system.

Looking back at Euclid with these newfound insights, mathematicians realized that Euclid had used many properties of lines and circles that were not strictly justified by his postulates. Let us examine a few of those properties, as a way of motivating the more careful axiomatic system that we will develop later in the book. We will discuss some of the most problematic of Euclid's proofs in the order in which they occur in Book I. As always, we refer to the edition [**Euc02**].

While reading these analyses of Euclid's arguments, you should bear in mind that we are judging the incompleteness of these proofs based on criteria that would have been utterly irrelevant in Euclid's time. For the ancient Greeks, geometric proofs were meant to be convincing arguments about the geometry of the physical world, so basing geometric conclusions on facts that were obvious from diagrams would never have struck them as an invalid form of reasoning. Thus these observations should not be seen as criticisms of Euclid; rather, they are meant to help point the way toward the development of a new axiomatic system that lives up to our modern (post-Euclidean) conception of rigor.

Euclid's Proposition I.1. *On a given finite straight line to construct an equilateral triangle.*

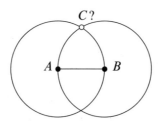

Fig. 1.6. Euclid's proof of Proposition I.1.

Analysis. In Euclid's proof of this, his very first proposition, he draws two circles, one centered at each endpoint of the given line segment AB (see Fig. 1.6). (In geometric diagrams in this book, we will typically draw selected points as small black dots to emphasize their locations; this is merely a convenience and is not meant to suggest that points take up any area in the plane.) He then proceeds to mention "the point C, in which the circles cut one another." It seems obvious from the diagram that there is a point where the circles intersect, but which of Euclid's postulates justifies the fact that such a point always exists? Notice that Postulate 5 asserts the existence of point where two *lines* intersect under certain circumstances; but nowhere does Euclid give us any justification for asserting the existence of a point where two *circles* intersect.

Euclid's Proposition I.3. *Given two unequal straight lines, to cut off from the greater a straight line equal to the less.*

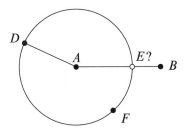

Fig. 1.7. Euclid's proof of Proposition I.3.

Analysis. In his third proof, Euclid implicitly uses another unjustified property of circles, although this one is a little more subtle. Starting with a line segment AD that he has just constructed, which shares an endpoint with a longer line segment AB, he draws a circle DEF with center A and passing through D (justified by Postulate 3). Although he does not say so explicitly, it is evident from his drawing that he means for E to be a point that is simultaneously on the circle DEF and also on the line segment AB. But once again, there is nothing in his list of postulates (or in the two previously proved propositions) that justifies the claim that a circle must intersect a line. (The same unjustified step also occurs twice in the proof of Proposition I.2, but it is a little easier to see in Proposition I.3.)

Euclid's Proposition I.4. *If two triangles have the two sides equal to two sides respectively, and have the angles contained by the equal straight lines equal, they will also have the base equal to the base, the triangle will be equal to the triangle, and the remaining angles will be equal to the remaining angles respectively, namely those which the equal sides subtend.*

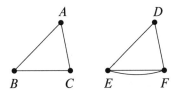

Fig. 1.8. Euclid's proof of Proposition I.4.

Analysis. This is Euclid's proof of the well-known *side-angle-side congruence theorem* (SAS). He begins with two triangles, ABC and DEF, such that $AB = DE$, $AC = DF$, and angle BAC is equal to angle EDF. (For the time being, we are adopting Euclid's convention that "equal" means "the same size.") He then says that triangle ABC should be "applied to triangle DEF." (Some revised translations use "superposed upon" or "superimposed upon" in place of "applied to.") The idea is that we should imagine triangle ABC being moved over on top of triangle DEF in such a way that A lands on D and the segment AB points in the same direction as DE, so that the moved-over copy of ABC occupies the same position in the plane as DEF. (Although Euclid does not explicitly mention it, he also evidently intends for C to be placed on the same side of the line DE as F, to ensure that the moved-over copy of ABC will coincide with DEF instead of being a mirror image of it.) This technique has become known as the *method of superposition.*

This is an intuitively appealing argument, because we have all had the experience of moving cutouts of geometric figures around to make them coincide. However, there is nothing in Euclid's postulates that justifies the claim that a geometric figure can be moved, much less that its geometric properties such as side lengths and angle measures will remain unchanged after the move. Of course, Propositions I.2 and I.3 describe ways of constructing "copies" of line segments at other positions in the plane, but they say nothing about copying angles or triangles. (In fact, he does prove later, in Proposition I.23, that it is possible to construct a copy of an *angle* at a different location; but that proof depends on Proposition I.4!)

This is one of the most serious gaps in Euclid's proofs. In fact, many scholars have inferred that Euclid himself was uncomfortable with the method of superposition, because he used it in only three proofs in the entire thirteen books of the *Elements* (Propositions I.4, I.8, and III.23), despite the fact that he could have simplified many other proofs by using it.

There is another important gap in Euclid's reasoning in this proof: having argued that triangle ABC can be moved so that A coincides with D, B coincides with E, and C coincides with F, he then concludes that the line segments BC and EF will also coincide and hence be equal (in size). Now, Postulate 1 says that it is possible to construct a straight line (segment) from any point to any other point, but it does not say that it is possible to construct *only one* such line segment. Thus the postulates provide no justification for concluding that the segments BC and EF will necessarily coincide, even though they have the same endpoints. Euclid evidently meant the reader to understand that there is a *unique* line segment from one point to another point. In a modern axiomatic system, this would have to be stated explicitly.

Euclid's Proposition I.10. *To bisect a given finite straight line.*

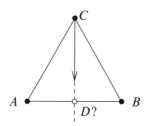

Fig. 1.9. Euclid's proof of Proposition I.10.

Analysis. In the proof of this proposition, Euclid uses another subtle property of intersections that is not justified by the postulates. Given a line segment AB, he constructs an equilateral triangle ABC with AB as one of its sides (which is justified by Proposition I.1) and then constructs the bisector of angle ACB (justified by Proposition I.9, which he has just proved). So far, so good. But his diagram shows the angle bisector intersecting the segment AB at a point D, and he proceeds to prove that AB is bisected (or cut in half) at this very point D. Once again, there is nothing in the postulates that justifies Euclid's assertion that there must be such an intersection point.

Euclid's Proposition I.12. *To a given infinite straight line, from a given point which is not on it, to draw a perpendicular straight line.*

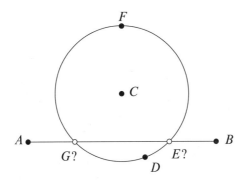

Fig. 1.10. Euclid's proof of Proposition I.12.

Analysis. In this proof, Euclid starts with a line AB and a point C not on that line. He then says, "Let a point D be taken at random on the other side of the straight line AB, and with center C and distance CD let the circle EFG be described." He is stipulating that the circle should be drawn with D on its circumference, which is exactly what Postulate 3 allows one to do. However, he is also implicitly assuming that such a circle will intersect AB in two points, which he calls E and G. Obviously it is the fact that C and D are on opposite sides of AB that is supposed to guarantee the existence of the intersection points; but which of his postulates or previous propositions justifies this conclusion? For that matter, what is "on the other side" supposed to mean? Euclid's definitions and postulates do not mention "sides" of lines at all, but he regularly refers to them in his proofs. It is clear from the diagrams what he means, but it is not justified by the postulates.

Euclid's Proposition I.16. *In any triangle, if one of the sides be produced, the exterior angle is greater than either of the interior and opposite angles.*

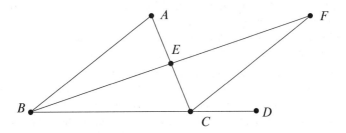

Fig. 1.11. Euclid's proof of Proposition I.16.

Analysis. Nowadays, this result is called the *exterior angle inequality*. Its proof is one of the most subtle and clever in the *Elements*. It is worth reading it over once or twice to absorb the full impact.

It is not easy to see where the gaps are, but there are at least two. After constructing the point E that bisects AC (using Proposition I.10), Euclid then extends BE past E and uses Proposition I.3 to choose a point F on that line such that EF is the same length as BE. Here is where the first problem arises: although Postulate 3 guarantees that a line segment can be extended to form a longer line segment containing the original one, it does not explicitly say that we can make the extended line segment as long as we wish. As we mentioned above, if we were working on the surface of a sphere, this might not be possible because great circles have a built-in maximum length.

The second problem arises toward the end of the proof, when Euclid claims that angle ECD is greater than angle ECF. This is supposed to be justified by Common Notion 5 (the whole is greater than the part). However, in order to claim that angle ECF is "part of" angle ECD, we need to know that F lies in the interior of angle ECD. This seems evident from the diagram, but once again, there is nothing in the axioms or previous propositions that justifies the claim. To see how this could fail, consider once again the surface of a sphere. In Fig. 1.12, we have illustrated an analogous configuration, with A at the north pole and B and C both on the equator. If B and C are far enough apart, it is entirely possible for the point F to end up south of the equator, in which case it is no longer in the interior of angle ECD. (Fig. 1.13 illustrates the same configuration after it has been "unwrapped" onto a plane.)

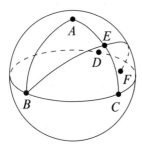

Fig. 1.12. Euclid's proof fails on a sphere.

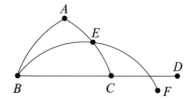

Fig. 1.13. The same diagram "unwrapped."

Some of these objections to Euclid's arguments might seem to be of little practical consequence, because, after all, nobody questions the truth of the theorems he proved. However, if one makes a practice of relying on relationships that seem obvious in diagrams, it is possible to go wildly astray. We end this section by presenting a famous fallacious "proof" of a false "theorem," which vividly illustrates the danger.

The argument below is every bit as rigorous as Euclid's proofs, with each step justified by Euclid's postulates, common notions, or propositions; and yet the theorem being proved is one that everybody knows to be false. This proof was first published in 1892 in a recreational mathematics book by W. W. Rouse Ball [**Bal87**, p. 48]. Exercise 1D asks you to identify the incorrect step(s) in the proof.

Fake Theorem. *Every triangle has at least two equal sides.*

Fake Proof. Let ABC be any triangle, and let AD be the bisector of angle A (Proposition I.9). We consider several cases.

Suppose first that when AD is extended (Postulate 2), it meets BC perpendicularly. Let O be the point where these segments meet (Fig. 1.14(a)). Then angles AOB and AOC are both right angles by definition of "perpendicular." Thus the triangles AOB and

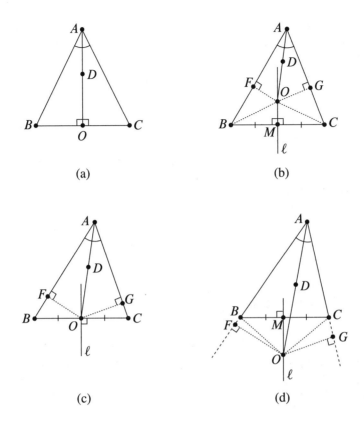

Fig. 1.14. The "proof" that every triangle has two equal sides.

AOC have two pairs of equal angles and share the common side AO, so it follows from Proposition I.26 that the sides AB and AC are equal.

In all of the remaining cases, we assume that the extension of AD is not perpendicular to BC. Let BC be bisected at M (Proposition I.10), let ℓ be the line perpendicular to BC at M (Proposition I.11), and let AD be extended if necessary (Postulate 2) so that it meets ℓ at O. There are now three possible cases, depending on the location of O.

CASE 1: *O lies inside triangle ABC (Fig. 1.14(b)).* Draw BO and CO (Postulate 1). Note that the triangles BMO and CMO have two pairs of equal corresponding sides (MO is common and $BM = CM$), and the included angles BMO and CMO are both right, so the remaining sides BO and CO are also equal by Proposition I.4. Now draw OF perpendicular to AB and OG perpendicular to AC (Proposition I.12). Then triangles AFO and AGO have two pairs of equal corresponding angles and share the side AO, so Proposition I.26 implies that their remaining pairs of corresponding sides are equal: $AF = AG$ and $FO = GO$. Now we can conclude that BFO and CGO are both right triangles in which the hypotenuses BO and CO are equal and the legs FO and GO are also equal. Therefore, the Pythagorean theorem (Proposition I.47) together with Common

Notion 3 implies that the squares on the remaining legs FB and GC are equal, and thus the legs themselves are also equal. Thus we have shown $AF = AG$ and $FB = GC$, so by Common Notion 2, it follows that $AB = AC$.

CASE 2: *O lies on BC* (*Fig.* 1.14(*c*)). Then O must be the point where BC is bisected, because that is where ℓ meets BC. In this case, we argue exactly as in Case 1, except that we can skip the first step involving triangles BMO and CMO, because we already know that $BO = OC$ (because BC is bisected at O). The rest of the proof proceeds exactly as before to yield the conclusion that $AB = AC$.

CASE 3: *O lies outside triangle ABC* (*Fig.* 1.14(*d*)). Again, the proof proceeds exactly as in Case 1, except now there are two changes: first, before drawing OF and OG, we need to extend AB beyond B, extend AC beyond C (Postulate 2), and draw OF and OG perpendicular to the extended line segments (Proposition I.12). Second, in the very last step, having shown that $AF = AG$ and $FB = GC$, we now use Common Notion 3 instead of Common Notion 2 to conclude that $AB = AC$. $\square$

Modern Axiom Systems

We have seen that the discovery of non-Euclidean geometry made it necessary to rethink the foundations of geometry, even Euclidean geometry. In 1899, these efforts culminated in the development by the German mathematician David Hilbert (1862–1943) of the first set of postulates for Euclidean geometry that were sufficient (according to modern standards of rigor) to prove all of the propositions in the *Elements*. (One version of Hilbert's axioms is reproduced in Appendix A.) Following the tradition established by Euclid, Hilbert did not refer to numbers or measurements in his axiom system. In fact, he did not even to refer to comparisons such as "greater than" or "less than"; instead, he introduced new relationships called *congruent* and *between* and added a number of axioms that specify their properties. For example, two line segments are to be thought of as congruent to each other if they have the same length (Euclid would say they are "equal"); and a point B is to be thought of as between A and C if B is an interior point of the segment AC (Euclid would say "AB is part of AC"). But these intuitive ideas were only the motivations for the choice of terms; the only facts about these terms that could legitimately be used in proofs were the facts stated explicitly in the axioms, such as Hilbert's Axiom II.3: *Of any three points on a line there exists no more than one that lies between the other two.*

Although Hilbert's axioms effectively filled in all of the unstated assumptions in Euclid's arguments, they had a distinct disadvantage in comparison with Euclid's postulates: Hilbert's list of axioms was long and complicated and seemed to have lost the elegant simplicity of Euclid's short list of assumptions. One reason for this complexity was the necessity of spelling out all the properties of betweenness and congruence that were needed to justify all of Euclid's assertions regarding comparisons of magnitudes.

In 1932, the American mathematician George D. Birkhoff published a completely different set of axioms for plane geometry using real numbers to measure sizes of line segments and angles. The theoretical foundations of the real numbers had by then been solidly established, and Birkhoff reasoned that since numerical measurements are used ubiquitously in practical applications of geometry (as embodied in the ruler and protractor), there was no longer any good reason to exclude them from the axiomatic foundations of geometry. Using these ideas, he was able to replace Hilbert's long list by only four axioms (see Appendix B).

Once Birkhoff's suggestion started to sink in, high-school text writers soon came around. Beginning with a textbook coauthored by Birkhoff himself [**BB41**], many high-school geometry texts were published in the U.S. that adopted axiom systems based more or less on Birkhoff's axioms. In the 1960s, the School Mathematics Study Group (SMSG), a committee sponsored by the U.S. National Science Foundation, developed an influential system of axioms for high-school courses that used the real numbers in the way that Birkhoff had proposed (see Appendix C). The use of numbers for measuring lengths and angles was embodied in two axioms that the SMSG authors called the *ruler postulate* and the *angle measurement postulate*. In one way or another, the SMSG axioms form the basis for the axiomatic systems used in most high-school geometry texts today. The axioms that will be used in this book are inspired by the SMSG axioms, although they have been modified in various ways: some of the redundant axioms have been eliminated, and some of the others have been rephrased to more closely capture our intuitions about plane geometry.

This concludes our brief survey of the historical events leading to the development of the modern axiomatic method. For a detailed and engaging account of the history of geometry from Euclid to the twentieth century, the book [**Gre08**] is highly recommended.

Exercises

1A. Read all of the definitions in Book I of Euclid's *Elements*, and identify which ones are descriptive and which are logical.

1B. Copy Euclid's proofs of Propositions I.6 and I.10, and identify each of the standard six parts: enunciation, setting out, specification, construction, proof, and conclusion.

1C. Choose several of the propositions in Book I of the *Elements*, and rewrite the statement and proof of each in more modern, idiomatic English. (You are not being asked to change the proofs or to fill in any of the gaps; all you need to do is rephrase Euclid's proofs to make them more understandable to modern readers.) When you do your rewriting, consider the following:

- Be sure to include diagrams, and consider adding additional diagrams if they would help the reader follow the arguments.
- Many terms are used by Euclid without explanation, so make sure you know what he means by them. The following terms, for example, are used frequently by Euclid but seldom in modern mathematical writing, so once you understand what they mean, you should consider replacing them by more commonly understood terms:
  - a *finite straight line*,
  - to *produce* a finite straight line,
  - to *describe* a circle,
  - to *apply* or *superpose* a figure onto another,
  - the *base* and *sides* of an arbitrary triangle,
  - one angle or side is *much greater* than another,
  - a straight line *standing on* a straight line,
  - an angle *subtended* by a side of a triangle.

- In addition, the following terms that Euclid uses without explanation are also used by modern writers, so you don't necessarily need to change them; but make sure that you know what they mean and that the meanings will be clear to your readers:
    - the *base* of an isosceles triangle,
    - to *bisect* a line segment or an angle,
    - *vertical angles,*
    - *adjacent angles,*
    - *exterior angles,*
    - *interior angles.*
- Euclid sometimes writes "I say that [something is true]," which is a phrase you will seldom find in modern mathematical writing. When you see this phrase in Euclid, think about how it fits into the logic of his proof. Is he saying this is a statement that follows from what he has already proved? Or a statement that he thinks is obvious and does not need proof? Or a statement that he claims to be true but has not proved yet? Or something else? How might a modern mathematician express this?
- Finally, after you have rewritten each proof, write a short discussion of the main features of the proof, and try to answer these questions: Why did Euclid construct the proof as he did? Were there any steps that seemed superfluous to you? Were there any steps or justifications that he left out? Why did this proposition appear at this particular place in the *Elements*? What would have been the consequences of trying to prove it earlier or later?

1D. Identify the fallacy that invalidates the proof of the "fake theorem" that says every triangle has two equal sides, and justify your analysis by carefully drawing an example of a nonisosceles triangle in which that step is actually false. [Hint: The problem has to do with drawing conclusions from the diagrams about locations of points. It's not enough just to find a step that is not adequately justified by the axioms; you must find a step that is actually false.]

1E. Find a modern secondary-school geometry textbook that includes some treatment of axioms and proofs, and do the following:
   (a) Read the first few chapters, including at least the chapter that introduces triangle congruence criteria (SAS, ASA, AAS).
   (b) Do the homework exercises in the chapter that introduces triangle congruence criteria.
   (c) Explain whether the axioms used in the book fill in some or all of the gaps in Euclid's reasoning discussed in this chapter.

Chapter 2

Incidence Geometry

Motivated by the advances described in the previous chapter, mathematicians since the early twentieth century have always proved theorems using a modern version of the axiomatic method. The purpose of this chapter is to describe this method and give an example of how it works.

Axiomatic Systems

The first important realization, due to Hilbert, is that it is impossible to precisely define every mathematical term that will be used in a given field. As we saw in Chapter 1, Euclid attempted to give descriptive definitions of words like *point*, *straight line*, etc. But those definitions were not sufficiently precise to be used to justify steps in proofs. Hilbert's insight was that from the point of view of mathematical logic, such definitions have *no meaning*. So, instead, he proposed to eliminate them from the formal mathematical development and simply to leave some terms officially undefined. Such terms are called *primitive terms*—they are given no formal mathematical definitions, but, instead, all of their relevant properties are expressed in the axioms. In Hilbert's axiomatic system, for example, the primitive terms are *point*, *line*, *plane*, *lies on*, *between*, and *congruent*.

Here, then, is the type of system we will be considering. An *axiomatic system* consists of the following elements:

- *primitive terms* (also called *undefined terms*): technical words that will be used in the axioms without formal definitions;
- *defined terms*: other technical terms that are given precise, unambiguous definitions in terms of the primitive terms and other previously defined terms;
- *axioms* (also called *postulates*): mathematical statements about the primitive and defined terms that will be assumed to be true without proof;
- *theorems*: mathematical statements about the primitive and defined terms that can be given rigorous proofs based only on the axioms, definitions, previously proved theorems, and rules of logic.

Each primitive term in an axiomatic system is usually declared to be one of several grammatical types: an *object*, a *relation*, or a *function*.

- A primitive term representing an *object* serves as a noun in our mathematical grammar. Some geometric examples are *point* and *line*.

- A primitive term representing a *relation* serves as a verb connecting two objects of certain types, yielding a statement that is either true or false. A common geometric example is *lies on*: it makes sense to say "a point lies on a line."

- A primitive term representing a *function* serves as a sort of "operator" that can be applied to one or more objects of certain types to produce objects of other types. A common geometric example is *distance*, which can be applied to two points to produce a number, so it makes sense to say "the distance from A to B is 3."

The theorems in an axiomatic system might not all be labeled "theorems"—in a particular exposition of an axiomatic system, some of them might be titled *propositions*, *corollaries*, or *lemmas*, depending on their relationships with the other theorems of the system. But logically speaking, there are no differences among the meanings of these terms; they are all theorems of the axiomatic system.

The Axioms of Incidence Geometry

To illustrate how axiomatic systems work, we will create a "toy" axiomatic system that contains some, but not all, of the features of plane geometry. (It is a genuine axiomatic system; the word "toy" is meant to suggest that it axiomatizes only a very small part of geometry, so we can learn a great deal by playing with this system before we dive into a full-blown axiomatization of Euclidean geometry.) Because it describes how lines and points meet each other, it is called *incidence geometry*. (The word *incidence* descends from Latin roots meaning "falling upon.")

Incidence geometry is based on Hilbert's first three axioms (I.1, I.2, and I.3 in Appendix A). It has three primitive terms: *point*, *line*, and *lies on*. Grammatically, "points" and "lines" are the objects in our system, while "lies on" is a relation that might or might not hold between a point and a line: if A is a point and ℓ is a line, then "A lies on ℓ" is a meaningful mathematical statement in our system; it is either true or false, but not both. The *meanings* of the primitive terms are to be gleaned from the axioms, which we will list below. Intuitively, you can think of points in the same way we thought of them in Euclidean geometry—as indivisible locations in the plane, with no width or area—and you can think of lines as straight one-dimensional figures with no width or depth; but these descriptions are not part of the axiomatic system, and you have to be careful not to assume or use any properties of points and lines other than those that are explicitly stated in the axioms. Hilbert once remarked that one should be able to substitute any other words for the primitive terms in an axiomatic system—such as "tables," "chairs," and "beer mugs" in place of "points," "lines," and "planes"—without changing the validity of any of the proofs in the system.

In addition to the primitive terms, we define the following terms:

- For a line ℓ and a point A, we say that ℓ *contains* A if A lies on ℓ.

- Two lines are said to *intersect* or to *meet* if there is a point that lies on both lines.

- Two lines are ***parallel*** if they do not intersect.

- A collection of points is ***collinear*** if there is a line that contains them all.

The prefix "non" attached to the name of any property negates that property—for example, to say that three points are ***noncollinear*** is to say that it is not true that they are collinear, or equivalently that there is no line that contains them all.

In addition, we will use the terms of mathematical logic with their usual meanings. See Appendix E for a summary of the terminology and conventions of mathematical logic. Note that in modern mathematics (in contrast to Euclid's *Elements*), to say that two objects are ***equal*** is to say that they are the same object, while to say two objects are ***distinct*** or ***different*** is simply to say that they are not equal. When we say three or more objects are distinct, it means that no two of them are equal. In general, a phrase like "*A* and *B* are points" should not be taken to imply that they are distinct points unless explicitly stated. On the other hand, a phrase like "at least two points" means two distinct points and possibly more, while "exactly two points" means two distinct points, no more and no fewer. (We will sometimes explicitly insert the word "distinct" in such phrases to ensure that there is no ambiguity, but it is not strictly necessary.)

In our version of incidence geometry, there are four axioms:

- **Incidence Axiom 1:** *There exist at least three distinct noncollinear points.*

- **Incidence Axiom 2:** *Given any two distinct points, there is at least one line that contains both of them.*

- **Incidence Axiom 3:** *Given any two distinct points, there is at most one line that contains both of them.*

- **Incidence Axiom 4:** *Given any line, there are at least two distinct points that lie on it.*

The second and third axioms could be combined into a single statement: *given any two distinct points, there is a unique line that contains both of them.* But we have separated the existence and uniqueness parts of the statement because they are easier to analyze this way.

Because of Axioms 2 and 3, we can adopt the following notation: if A and B are any two distinct points, we will use the notation $\overleftrightarrow{AB}$ to denote the unique line that contains both A and B. If A and B are points that have already been introduced into the discussion, this notation is meaningful only when we know that A and B are distinct; if they happen to be the same point, there could be many different lines containing that point, so the notation $\overleftrightarrow{AB}$ would not have a definite meaning. If we say "$\overleftrightarrow{AB}$ is a line" or "let $\overleftrightarrow{AB}$ be a line" without having previously introduced A and B, this should be understood as including the assertion that A and B are not equal. Thus such a statement really means "A and B are distinct points and $\overleftrightarrow{AB}$ is the unique line containing them."

Because of the terms we have chosen—*point, line,* and *lies on*—you will probably find yourself visualizing points and lines as marks on paper, much as we did when studying Euclid's propositions. This is not necessarily a bad idea, because we have deliberately chosen the terms to evoke familiar objects of geometric study. But you must develop the habit of thinking critically about each statement to make sure that it is justified only by the axioms, definitions, and previously proved theorems. For example, here are two similar-sounding statements about points and lines, both of which are true in the usual Euclidean setting:

Statement I. *Given any point, there are at least two distinct lines that contain it.*

Statement II. *Given any line, there are at least two distinct points that do not lie on it.*

As we will see later in this chapter, Statement I can be proved from the axioms alone, so it is a theorem of incidence geometry (see Theorem 2.42). However, Statement II cannot be proved from the axioms (see Exercise 2D). The justification for the latter claim depends on some extremely useful tools for deepening our understanding of an axiomatic system: *interpretations* and *models*.

Interpretations and Models of Incidence Geometry

In a certain sense, the theorems of an axiomatic system are not "about" anything; they are just statements that follow logically from the axioms. But the real power of axiomatic systems comes in part from the fact that they can actually tell us things about whole classes of concrete mathematical systems. Here we describe how that works.

An ***interpretation*** of an axiomatic system is an assignment of a mathematical definition for each of its primitive terms. Usually, our interpretations will be constructed in some area of mathematics that we already understand, such as set theory or the real number system. Although these other subjects are ultimately based on their own axiomatic systems, we will simply take them as given for the purposes of constructing interpretations.

An interpretation of an axiomatic system is said to be a ***model*** if each of the axioms is a true statement when the primitive terms are given the stated definitions. Typically, an axiomatic system will have many different models. (The main exception is an axiomatic system that is self-contradictory: for example, if we added a fifth incidence axiom that said *no line contains two distinct points*, then we would have a system that admitted no models.) One source of the power of axiomatic systems is that any theorems we succeed in proving in the axiomatic system become true statements about every model because they all follow logically from the axioms, which are themselves true statements about every model.

Let us explore some models of incidence geometry. We start with a very simple one—in fact, in a certain sense, it is the simplest possible model because every model must have at least three points by Axiom 1.

Example 2.1 (The Three-Point Plane). In this interpretation, we define the term ***point*** to mean any one of the numbers 1, 2, or 3; and we define the term ***line*** to mean any one of the sets $\{1,2\}$, $\{1,3\}$, or $\{2,3\}$. (Here we are using a standard notation for sets: for example, $\{1,2\}$ is the set whose elements are 1 and 2 and nothing else. See Appendix G.) We say a "point" ***lies on*** a "line" if that "point" (i.e., number) is one of the elements in that "line" (i.e., set of two numbers). (In this example, we are placing quotation marks around the primitive terms to emphasize that we have assigned arbitrary meanings to them, which may have nothing to do with our ordinary understanding of the words.)

It is convenient to visualize the three-point plane by means of a diagram like that of Fig. 2.1. In diagrams such as these, dots are meant to represent "points," and line segments are meant to represent "lines"; a "point lies on a line" if and only if the corresponding dot touches the corresponding line segment. It is important to remember, however, that the official definition of the interpretation is the description given above, in terms of numbers and sets of numbers; the diagram is not the interpretation. In particular, looking at the diagram, one might be led to believe that the "line" containing 1 and 2 also contains many other "points" besides the two endpoints; but if you look back at the definition, you will see that this is not the case: each "line" contains exactly two "points." The line segments in the drawing are there merely to remind us which sets of "points" constitute "lines."

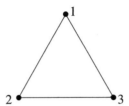

Fig. 2.1. The three-point plane.

We will check that the three-point plane is a model for incidence geometry by showing that each of the four incidence axioms is a true statement about this interpretation. Axiom 1 is true because 1, 2, and 3 are three "points" that are not all contained in any one "line," so they are noncollinear. To prove that Incidence Axiom 2 holds, we have to prove that for each pair of distinct "points," there is a "line" that contains them. In this case, we can prove this by enumerating the possible pairs directly: the "points" 1 and 2 are contained in the "line" $\{1, 2\}$; the "points" 1 and 3 are contained in the "line" $\{1, 3\}$; and the "points" 2 and 3 are contained in the "line" $\{2, 3\}$. Axiom 3 holds because each of these pairs of "points" is contained in only the indicated "line" and no other. Finally, to prove Axiom 4, just note that each "line" contains exactly two "points" by definition.   //

In our proof that the three-point plane is a model, we carried out everything in gory detail. As the models get more complicated, when we need to prove that the axioms hold by enumeration, we will not bother to write out all possible lines and points; but you should be able to see easily how it would be done in each case. (If it's not obvious how to do so, you should pick up your pencil and write down all the possibilities!) Also, from now on, after we have given definitions for "point," "line," and "lies on" for a particular interpretation, we will often dispense with the awkward quotation marks. Just remember that whenever we use these terms in the context of an interpretation, they are to be understood as having the meanings we assign them in that interpretation.

Here are some more models.

Example 2.2 (The *n*-Point Plane). The previous example can be generalized as follows. For any integer $n \geq 3$, we can construct an interpretation by defining a ***point*** to be any one of the numbers $1, 2, 3, \ldots, n$; defining a ***line*** to be any set containing exactly two of those numbers; and defining ***lies on*** to mean "is an element of" as before. Fig. 2.2 illustrates this interpretation when $n = 4$, and Fig. 2.3 illustrates it for $n = 5$.

Remember, in these pictures, the only "points" in the geometry are the ones indicated by dots. In the illustration of the 5-point plane, some pairs of lines, such as $\{1, 3\}$ and $\{2, 4\}$, appear to cross; but in the interpretation those lines do not intersect because there is

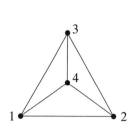

Fig. 2.2. The four-point plane.

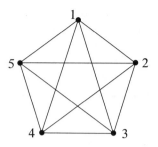

Fig. 2.3. The five-point plane.

no point that lies on both of them. Many other illustrations are possible—for example, Fig. 2.4 shows two other illustrations of the four-point plane. In one of these illustrations, the "lines" are drawn as curves for convenience, but the corresponding sets of points are still "lines" in the model. Although the three illustrations look very different, they all represent exactly the same model because they have the same points, the same lines, and the same "lies on" relation.

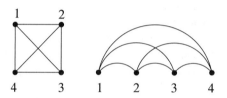

Fig. 2.4. Two more illustrations of the four-point plane.

We can show that for each $n \geq 3$, the n-point plane is a model of incidence geometry. Because 1, 2, and 3 are noncollinear points, Axiom 1 is satisfied. For any pair of distinct points i and j, the line $\{i, j\}$ is the unique line that contains them, so Axioms 2 and 3 are satisfied. Finally, Axiom 4 is satisfied because each line contains two distinct points by definition.   //

Example 2.3 (The Fano Plane). In this interpretation, we define a ***point*** to be one of the numbers $1, 2, 3, 4, 5, 6, 7$ and a ***line*** to be one of the following sets:

$$\{1,2,3\}, \quad \{3,4,5\}, \quad \{5,6,1\}, \quad \{1,7,4\}, \quad \{2,7,5\}, \quad \{3,7,6\}, \quad \{2,4,6\}.$$

(See Fig. 2.5.) As before, ***lies on*** means "is an element of." It is easy to check by enumeration that the Fano plane satisfies all four axioms, so it is a model of incidence geometry. (The verification is not hard, but writing down all the details would take a while. If you are not convinced that the axioms are all satisfied, check them!) Note that we have drawn the "line" $\{2, 4, 6\}$ as a circle for convenience, but it is still to be interpreted as a line in this geometry, no different from any other line.   //

In all of the models we have introduced so far, "points" have been numbers and "lines" have been sets of points. There is nothing about the axioms that requires this. Here is a model that is not of this type.

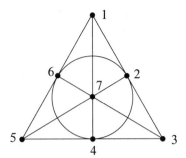

Fig. 2.5. The Fano plane.

Example 2.4 (The Amtrak Model). In this interpretation, we define ***point*** to mean any one of the cities Seattle, Sacramento, or Chicago; and ***line*** to mean any one of the following Amtrak passenger rail lines: the Coast Starlight, the Empire Builder, or the California Zephyr. We define ***lies on*** to mean that the city is one of the stops on that rail line. The route map of these lines is shown in Fig. 2.6. We leave it to the reader to check that this is a model.   ⫽

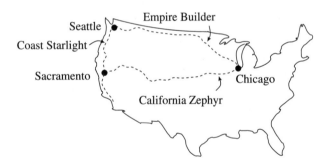

Fig. 2.6. The Amtrak model.

The Amtrak model has a lot in common with the three-point plane: both have exactly three "points" and three "lines," and each "line" contains exactly two "points." In fact, in a certain sense these two models are essentially the same, except for the names given to the points and lines. To describe this relationship more precisely, we introduce the following terminology: suppose we are given two models $\mathcal{A}$ and $\mathcal{B}$ of a given axiomatic system. An ***isomorphism*** between the models is an assignment of one-to-one correspondences between the primitive objects of each type in $\mathcal{A}$ and the objects of the same type in $\mathcal{B}$ in such a way that all of the primitive relations and functions are preserved. For models of incidence geometry, this means a one-to-one correspondence between the points of $\mathcal{A}$ and the points of $\mathcal{B}$ and a one-to-one correspondence between the lines of $\mathcal{A}$ and the lines of $\mathcal{B}$, with the property that a given point of $\mathcal{A}$ lies on a given line of $\mathcal{A}$ if and only if the corresponding point of $\mathcal{B}$ lies on the corresponding line of $\mathcal{B}$. If such a correspondence exists, we say the two models are ***isomorphic***. This means they are "the same" in all of the features that are relevant for this axiomatic system, but the names of the objects might be different. The

Amtrak model is isomorphic to the three-point plane under the correspondence

$1 \leftrightarrow$ Seattle, $\{1,2\} \leftrightarrow$ Coast Starlight,

$2 \leftrightarrow$ Sacramento, $\{1,3\} \leftrightarrow$ Empire Builder,

$3 \leftrightarrow$ Chicago, $\{2,3\} \leftrightarrow$ California Zephyr.

None of the other models we have introduced so far are isomorphic to each other. If two models have different numbers of points, they cannot be isomorphic because there cannot be a one-to-one correspondence between their points. Other than the 3-point plane and the Amtrak model, the only other two models we've seen that have the same number of points are the Fano plane and the 7-point plane; for those, see Exercise 2E.

The variety of models we can construct is limited only by our imagination. Here is another model that is isomorphic to the three-point plane; it illustrates again that lines do not necessarily need to be sets of points.

Example 2.5 (The Three-Equation Model). Define *point* to mean any of the ordered pairs $(1,0)$, $(0,1)$, or $(1,1)$; and *line* to mean any one of the following equations:

$$x = 1, \qquad y = 1, \qquad x + y = 1.$$

(See Fig. 2.7.) We say that a point (a,b) *lies on* a line if the equation is true when we substitute $x = a$ and $y = b$. Thus, for example, the point $(1,0)$ lies on the lines $x = 1$ and $x + y = 1$, but not on $y = 1$. You should convince yourself that this is indeed a model of incidence geometry and that it is isomorphic to both the three-point plane and the Amtrak model.   //

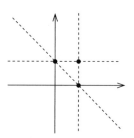

Fig. 2.7. The three-equation model.

Some Nonmodels

Next, let us consider some interpretations that are *not* models.

Example 2.6 (One-Point Geometry). Here is a silly interpretation of incidence geometry: we define *point* to mean the number 1 only; in this interpretation, there are no lines, and so no point lies on any line. Axiom 1 is obviously false in this interpretation, so this is not a model of incidence geometry. However, it is interesting to note that Axioms 2–4 are actually all true. For example, because there are no lines, it is true that for every line (all none of them!), there are at least two points that lie on it. The other axioms are true for similar reasons. (These are examples of statements that are said to be "vacuously true" because their hypotheses are never satisfied; see Appendix E for a discussion of such statements.)   //

Example 2.7 (The Three-Point Line). In this interpretation, we define *point* to mean any one of the numbers 1, 2, or 3, *line* to mean only the set {1, 2, 3}, and *lies on* to mean "is an element of" (see Fig. 2.8). You can check that in this interpretation Axiom 1 is false, but Axioms 2, 3, and 4 are all true; thus this interpretation is not a model of incidence geometry. //

Fig. 2.8. The three-point line.

For the rest of the interpretations in this section, we will leave it to you to figure out why they are not models (see Exercise 2A).

Example 2.8 (Three-Ring Geometry). In this interpretation, a *point* is any one of the numbers 1, 2, 3, 4, 5, 6; a *line* is any one of the sets {1, 2, 5, 6}, {2, 3, 4, 6}, or {1, 3, 4, 5}; and *lies on* means "is an element of" (see Fig. 2.9). //

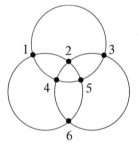

Fig. 2.9. Three-ring geometry.

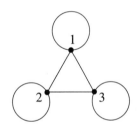

Fig. 2.10. One-two geometry.

Example 2.9 (One-Two Geometry). Here is yet another interpretation that has exactly three points: a *point* is 1, 2, or 3; a *line* is any set consisting of exactly one or two points, namely {1}, {2}, {3}, {1, 2}, {1, 3}, or {2, 3}; and *lies on* means "is an element of" (see Fig. 2.10). //

Example 2.10 (Square Geometry). Define an interpretation of incidence geometry by letting *point* mean any of the vertices of a square, *line* mean any of the sides of that square, and *lies on* mean the point is one of the endpoints of the side (see Fig. 2.11). //

Fig. 2.11. Square geometry.

Consistency and Independence

The interpretations above serve to illustrate some important concepts regarding axiomatic systems. An axiomatic system is said to be **consistent** if it is impossible to derive a contradiction from its axioms, and it is **inconsistent** if it is possible to derive such a contradiction. For example, as we mentioned above, if we add the fifth axiom *no line contains two distinct points* to the axioms of incidence geometry, we would obtain an inconsistent system because the axioms immediately lead to a contradiction (in fact, the new axiom and Axiom 4 are already direct contradictions of one another).

It is usually not too hard to prove that a given axiom system is inconsistent: just show how to derive a contradiction. On the other hand, it might be very difficult to prove that the system is consistent, because you would have to show that nobody, however clever, will ever find a way to derive a contradiction from its axioms. Fortunately, however, there is a handy shortcut: if you can prove that an axiomatic system has a model, then no contradiction can follow from the axioms, because any result that follows logically from the axioms—including any potential contradiction—would have to be true in the model.

Of course, this argument depends on knowing that there is no contradiction in the model. For that, we have to rely on the theoretical context in which the model is constructed. For example, the three-point plane is based on the elementary theory of finite sets. Provided we are sure there can be no contradiction in that theory, then we can conclude that incidence geometry is consistent. For that reason, one often says that demonstrating the existence of a model proves **relative consistency** of an axiomatic system, with relation to the system in which the model is constructed: provided the system we used to construct the model is consistent, it follows that our axiomatic system is also consistent. Our construction of the three-point plane proves that incidence geometry is consistent if set theory is consistent. (Virtually all mathematicians believe that it is.) In the end, we have no way to prove *absolute* consistency of any mathematical system, just as we have no way to prove the absolute truth of any of the theorems of geometry; but if we are willing to accept on faith the consistency of some simple theories, that often allows us to prove the consistency of other more complicated theories by constructing models.

Closely related to consistency is the notion of *independence* of axioms. When we construct an axiomatic system, we would like the list of axioms to be as short as possible—it would be undesirable to assume more axioms than we actually need. For example, there would be no reason to add a fifth axiom to incidence geometry that said *there exist at least two distinct points* because that statement is already provable from the other four. (In fact, it is provable from Axiom 1 alone.)

Thus we make the following definition: given an axiomatic system $\mathcal{A}$, a statement P about the primitive and defined terms of $\mathcal{A}$ is said to be **independent of** $\mathcal{A}$ if it is not possible either to prove or to disprove P using the axioms of $\mathcal{A}$. Like consistency, independence of a statement can be very hard to prove directly; but once again, models provide a shortcut. If there exists a model of $\mathcal{A}$ in which P is a false statement and there also exists another model of $\mathcal{A}$ in which P is a true statement, then P is definitely independent of $\mathcal{A}$. (If P were provable within $\mathcal{A}$, then it would have to be true in every model; while if it were disprovable, it would have to be false in every model.)

For example, consider the statement *every line contains exactly three points*. By examining the models above, we can show that this statement is independent of the axioms of incidence geometry because it is false in the three-point plane but true in the Fano plane.

More interestingly, we can prove that each of the four axioms of incidence geometry is independent of the other three. To do so, we just need to exhibit for each axiom an interpretation in which the other three axioms are true and that axiom is true as well (the three-point plane will do for this) and then exhibit another interpretation in which the other three axioms are true but our chosen axiom is false. In fact, we have already seen such interpretations: Examples 2.7–2.10 describe interpretations in which three of the axioms are true, but in each case a different one is false. (See Exercises 2A and 2B.) Thus each of the four axioms is independent of the other three; no axiom could be removed and still allow us to prove all the same theorems.

Some Infinite Models

Each of the models we have examined so far contains only finitely many points. Next, we will examine models with infinitely many points; these are much closer to our usual understanding of Euclidean geometry.

Our first infinite model should be familiar to you from analytic geometry and calculus. It is named after the seventeenth-century French philosopher and mathematician René Descartes (1596–1650), who was the first to use pairs of numbers to represent points in the plane [**Des54**].

Example 2.11 (The Cartesian Plane). The ***Cartesian plane*** is an interpretation in which a ***point*** is defined to be any ordered pair (x, y) of real numbers; the set of all such ordered pairs is denoted by $\mathbb{R}^2$. A ***line*** is defined to be any set of points (x, y) that is the entire solution set of an equation of one of the following two types:

$$x = c \qquad \text{for some real constant } c \text{ (a } \textbf{\textit{vertical line}}\text{)};$$
$$y = mx + b \qquad \text{for some real constants } m \text{ and } b \text{ (a } \textbf{\textit{nonvertical line}}\text{)}.$$

(A ***horizontal line*** is the special case of a nonvertical line with $m = 0$, or in other words a line defined by an equation of the form $y = b$; but we have no need to treat the horizontal case separately.) We say that a point (x, y) ***lies on*** a line if the ordered pair (x, y) is a solution to the given equation. For a nonvertical line, the constant m in the equation $y = mx + b$ is called the ***slope of the line***. //

Theorem 2.12. *The Cartesian plane is a model of incidence geometry.*

Proof. To see that Axiom 1 is true in this model, consider the three points $(0,0)$, $(0,1)$, and $(1,0)$. There is no vertical line that contains them all, because they do not all have the same x-component. If there were a nonvertical line containing all three points, defined by an equation of the form $y = mx + b$, then plugging in $(x, y) = (0,0)$ would imply $b = 0$, while plugging in $(x, y) = (0,1)$ would imply $b = 1$, which is a contradiction. Thus these three points are noncollinear.

To see that the Cartesian plane satisfies Axiom 2, suppose (x_1, y_1) and (x_2, y_2) are two different points; we have to show that there is a line that contains both of them. This is a standard exercise in analytic geometry—finding the equation of a line through two points. We consider two cases, depending on the relationship between the x-values and leading to vertical and nonvertical lines.

CASE 1: $x_1 = x_2$. If we set $c = x_1 = x_2$, then both points (x_1, y_1) and (x_2, y_2) satisfy the equation $x = c$, so there is a vertical line containing both of them.

CASE 2: $x_1 \neq x_2$. In this case, recalling a little high-school analytic geometry, we can write down the ***two-point formula for a line***:

$$y = mx + b, \qquad \text{where } m = \left(\frac{y_2 - y_1}{x_2 - x_1} \right), \quad b = \left(\frac{x_2 y_1 - x_1 y_2}{x_2 - x_1} \right). \qquad (2.1)$$

A straightforward computation shows that both (x_1, y_1) and (x_2, y_2) are solutions to this equation, so this shows that there exists a line containing our two points.

To prove Axiom 3, suppose that (x_1, y_1) and (x_2, y_2) are two different points and ℓ and ℓ' are two lines that contain both of them. We will show that ℓ and ℓ' are in fact the same line. We have three different cases to consider.

CASE 1: *Both ℓ and ℓ' are vertical.* In this case, the two lines are defined by equations $x = c$ and $x = c'$, respectively. Because the point (x_1, y_1) lies on both lines, we have $x_1 = c$ and $x_1 = c'$. By transitivity, $c = c'$, so both lines are the same. Thus in this case there is only one line containing the two points.

CASE 2: *Both ℓ and ℓ' are nonvertical.* Then they are defined by equations $y = mx + b$ and $y = m'x + b'$, respectively. Because (x_1, y_1) and (x_2, y_2) both lie on ℓ, we conclude that

$$y_1 = mx_1 + b,$$
$$y_2 = mx_2 + b.$$

Subtracting the first equation from the second and simplifying leads to the conclusion that m is given by the same formula as in (2.1). Then we can insert this back into the first equation and solve for b, to conclude that b is also given by (2.1). Exactly the same argument shows that m' and b' are given by the same formulas, so we conclude again that the two lines are actually the same.

CASE 3: *One line is vertical and the other is not.* Then one is described by an equation of the form $x = c$, and the other by $y = mx + b$. The fact that both points are on the line $x = c$ implies that $x_1 = c = x_2$, and then the second equation implies $y_1 = mx_1 + b = mx_2 + b = y_2$. This yields $(x_1, y_1) = (x_2, y_2)$, which contradicts our initial assumption that the points were distinct. Therefore this case cannot occur. This completes the proof of Axiom 3.

Axiom 4 is much easier to prove. Suppose ℓ is any line. If ℓ is a vertical line defined by $x = c$, then it contains the distinct points $(c, 0)$ and $(c, 1)$. If it is nonvertical, defined by $y = mx + b$, then it contains both $(0, b)$ and $(1, m + b)$. $\qquad \square$

Our next interpretation is also familiar.

Example 2.13 (Spherical Geometry). ***Spherical geometry*** is the interpretation in which a ***point*** is defined to be a point on the surface of the unit sphere in $\mathbb{R}^3$ (that is, an ordered triple (x, y, z) of real numbers such that $x^2 + y^2 + z^2 = 1$), and a ***line*** is defined to be a ***great circle***, which is a circle on the surface of the sphere whose center coincides with the center of the sphere. (A great circle can also be thought of as the intersection of the sphere with a plane through the origin in $\mathbb{R}^3$; see Fig. 2.12.) This is not a model of incidence geometry, because Axiom 3 is violated: for example, consider the **north pole** $(0,0,1)$ and the **south pole** $(0,0,-1)$. Any plane that goes through these two points intersects the sphere in a great circle called a **meridian circle**. (On an Earth globe, each half of such a circle would be a curve of constant longitude.) Thus there are infinitely many different "lines" that contain the north and south poles. //

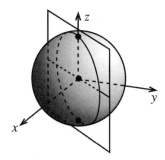

Fig. 2.12. A great circle on the sphere.

Example 2.14 (Single Elliptic Geometry). By making a slight modification to the previous interpretation, we can obtain a model of incidence geometry. **Single elliptic geometry** is the interpretation in which a **point** is defined to be a *pair* of antipodal (diametrically opposite) points on the surface of the unit sphere, and a **line** is a great circle. We say a "point" **lies on** a "line" if the given "point" is a pair of antipodal points that both lie on the given great circle. In this interpretation, two "lines" intersect in exactly one "point" (i.e., pair of antipodal points on the sphere). In fact, all four incidence axioms are satisfied by single elliptic geometry, so it is a model of incidence geometry. The word "single" in the name of this geometry reflects the fact that two "lines" intersect in a single "point." By contrast, the geometry of the sphere is sometimes called **double elliptic geometry**. //

Example 2.15 (The Beltrami–Klein Disk). The **Beltrami–Klein disk** is the interpretation of incidence geometry in which the primitive terms are given the following definitions (see Fig. 2.13): a **point** is an ordered pair $(x,y) \in \mathbb{R}^2$ satisfying the constraint $x^2 + y^2 < 1$ (which forces (x,y) to be strictly inside the unit circle); and a **line** is a **chord** of the circle, i.e., a line segment whose endpoints both lie on the unit circle itself. We define **lies on**, as usual, to mean "is an element of." (Note, however, that the endpoints of the chord are *not* "points" of the model; the only "points" in this geometry that "lie on" the chord are the ones in the interior of the circle.) We will not go through all the details, but a little thought should convince you that the Beltrami–Klein disk is a model of incidence geometry. (This model was introduced by Eugenio Beltrami in 1868 [**Bel68**] as a model of non-Euclidean geometry and later improved by Felix Klein.) //

Example 2.16 (The Poincaré Disk). The **Poincaré disk** is closely related to the Beltrami–Klein disk. In this interpretation, **points** are defined exactly as in the Belrami–Klein model, but lines are defined differently. (See Fig. 2.14.) A **line** in this model is a set of either one of the following two types:

(a) a set consisting of all points (x,y) inside the unit circle that lie on some Cartesian line through $(0,0)$; or

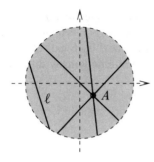

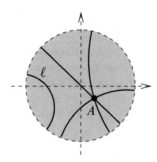

Fig. 2.13. The Beltrami–Klein disk.   **Fig. 2.14.** The Poincaré disk.

(b) a set consisting of all points (x, y) inside the unit circle that lie on some Cartesian circle that intersects the unit circle perpendicularly.

As before, ***lies on*** means "is an element of." Once again, it is possible with a bit of work to prove that the Poincaré disk is a model of incidence geometry. (This model had several sources, but it was named after Henri Poincaré because of his extensive use of it in complex analysis.)   //

Example 2.17 (The Poincaré Half-Plane). Here is yet another model that can be constructed within the Cartesian plane. In the ***Poincaré half-plane model***, a ***point*** is defined to be an ordered pair $(x, y) \in \mathbb{R}^2$ with $y > 0$, and a ***line*** is a set of either one of the following two types:

(a) the set of all points (x, y) with $y > 0$ and satisfying $x = c$ for some constant c (a vertical half-line),

(b) the set of all points (x, y) with $y > 0$ and lying on some Cartesian circle whose center is on the x-axis (a semicircle).

(See Fig. 2.15.) Like the Poincaré disk, the Poincaré half-plane can be shown to be a model of incidence geometry. (This model was also originally introduced by Beltrami in [**Bel68**], and its properties were developed extensively by Poincaré.)   //

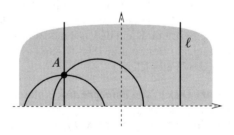

Fig. 2.15. The Poincaré half-plane.

Example 2.18 (The Hyperboloid Model). Finally, here is one more model that, like spherical geometry, is constructed from a surface in three-dimensional space. In this model, a ***point*** is defined to be an ordered triple $(x, y, z) \in \mathbb{R}^3$ that satisfies $z > 0$ and $z^2 = x^2 + y^2 + 1$; the set S of all such points is the "upper sheet" of a surface called a ***two-sheeted hyperboloid***. A ***line*** is defined to be a ***great hyperbola*** on S, which is any set of points where S intersects a plane through the origin (see Fig. 2.16). It is relatively easy to check that Axioms 1 and 4 are satisfied in this interpretation. To see that Axioms 2 and 3 are satisfied, suppose A and B are any two distinct points on S. A little algebra shows that the two points A and B cannot be collinear with the origin and both lie on S, so a standard computation in linear algebra shows that there is a unique plane containing the origin, A, and B. The intersection of this plane with S is the unique "line" containing A and B. Thus Axioms 2 and 3 are satisfied, and the hyperboloid is a model of incidence geometry. //

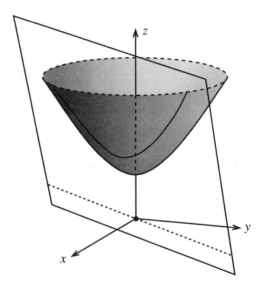

Fig. 2.16. The hyperboloid model.

Parallel Postulates

Because the questions surrounding the existence and uniqueness of parallels turn out to be of such consequence for Euclidean and non-Euclidean geometries, let us explore the phenomenon of parallelism in incidence geometry.

Remember that the definition of ***parallel lines*** in incidence geometry is simply lines that do not intersect. In some models, this can lead to unexpected results. For example, in the four-point plane (see Fig. 2.2), the lines $\{1, 2\}$ and $\{3, 4\}$ are parallel, despite the fact that their representations appear to be perpendicular in the diagram. Similarly, in the five-point plane (Fig. 2.3), the lines $\{1, 3\}$ and $\{2, 4\}$ are parallel.

Here are three potential axioms that we might wish to add to the axioms of incidence geometry.

The Elliptic Parallel Postulate. *For each line ℓ and each point A that does not lie on ℓ, there are no lines that contain A and are parallel to ℓ.*

The Euclidean Parallel Postulate. *For each line ℓ and each point A that does not lie on ℓ, there is one and only one line that contains A and is parallel to ℓ.*

The Hyperbolic Parallel Postulate. *For each line ℓ and each point A that does not lie on ℓ, there are at least two distinct lines that contain A and are parallel to ℓ.*

We will explain the nomenclature of the elliptic and hyperbolic parallel postulates at the end of this section. The postulate we are calling the Euclidean parallel postulate is not, of course, the same as Euclid's fifth postulate (which is also sometimes called the Parallel Postulate). Ours is closer to Playfair's postulate (see p. 10), but we have chosen a slightly stronger statement that emphasizes the similarities and contrasts among the three different parallel postulates.

The elliptic parallel postulate could be stated more simply as *"There are no parallel lines."* But again, the statement given here was chosen to emphasize its relationship with the other parallel postulates.

Let's begin our discussion by looking back at some of the interpretations that we introduced earlier in this chapter and asking which (if any) of the parallel postulates are satisfied in some of those interpretations. Because we are asking this question in the context of the interpretations, not in our axiomatic system, we will not try to construct formal proofs the way we will do for theorems *within* the axiomatic system; instead, we will give informal (but hopefully convincing) arguments for why our conclusions are true.

Example 2.19 (Parallelism in Finite Models of Incidence Geometry).

- **The Three-Point Plane:** In this model, any two lines intersect in one point, so there are no parallel lines. Thus the elliptic parallel postulate holds.

- **The Four-Point Plane:** Look first at the line $\ell = \{1, 2\}$. There are two points not on this line, namely 3 and 4. Consider $A = 3$: there are three lines that contain 3, and by checking them one at a time, we see that exactly one of them (namely $\{3, 4\}$) has no point in common with ℓ, so it is parallel to ℓ. Similarly, considering the point $A = 4$, once again we see that $\{3, 4\}$ is the unique line through A and parallel to ℓ. We can do the same thing with the line $\{3, 4\}$: there are two points (1 and 2) not on it, and in each case $\{1, 2\}$ is the unique line through that point and parallel to ℓ. We can see by symmetry that the other four lines will work the same way: the entire diagram looks the same if we rotate it $120°$, and after one or two such rotations each line will end up in the same position as either $\{1, 2\}$ or $\{3, 4\}$, so the same argument will apply. Thus the Euclidean parallel postulate holds in this model. (This is not quite a rigorous proof, but it is meant to convince you that you could easily write down a complete proof by enumeration if you took the time.)

- **The Five-Point Plane:** In this case also, there are two kinds of lines up to symmetry (sides of the pentagon and diagonals of the pentagon). By choosing one line of each type and checking each of the three points not on it, you can verify that there are always exactly two parallels to the given line through each point. Thus the hyperbolic parallel postulate holds in this model.   //

It is important to note that a model might not satisfy *any* parallel postulate. Here is an example.

Example 2.20. Consider the model of incidence geometry illustrated in Fig. 2.17: the *points* are the numbers $1, 2, 3, 4, 5$; the *lines* are the sets $\{1, 2\}$, $\{2, 3\}$, $\{3, 4\}$, $\{4, 1\}$, $\{1, 5, 3\}$, and $\{2, 5, 4\}$; and *lies on* means "is an element of." (We leave it to you to check that this is indeed a model.) Let ℓ be the line $\{1, 2\}$. Then there is exactly one line that contains the point 3 and is parallel to ℓ, namely the line $\{3, 4\}$. However, if we look instead at the point 5, we see that there is *no* line that contains 5 and is parallel to ℓ, because the lines that contain 5 (namely $\{1, 5, 3\}$ and $\{2, 5, 4\}$) both intersect ℓ. Because the parallel postulates all say that something has to be true for *every* line and *every* point not on that line, none of the parallel postulates are true in this model.   //

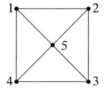

Fig. 2.17. The model of Example 2.20.

In the exercises, you will be asked to explore parallelism in some of our other interpretations of incidence geometry. But before we go any further, let us observe that we already have enough information to conclude that none of the parallel postulates can be proved or disproved from the axioms of incidence geometry.

Theorem 2.21. *Each of the three parallel postulates is independent of the axioms of incidence geometry.*

Proof. The Euclidean parallel postulate is true in the four-point plane and false in the three-point and five-point planes; the hyperbolic parallel postulate is true in the five-point plane and false in the other two; and the elliptic parallel postulate is true in the three-point plane and false in the other two.   □

We can also ask which parallel postulates are satisfied in our infinite models. Let us first see how to tell when two lines are parallel in the Cartesian plane.

Lemma 2.22 (Parallelism in the Cartesian Plane). *In the Cartesian plane, parallel lines have the following properties:*

(a) *Any two distinct vertical lines are parallel.*

(b) *Two distinct nonvertical lines are parallel if and only if they have the same slope.*

(c) *No vertical line is parallel to any nonvertical line.*

Proof. First, suppose ℓ and ℓ' are two distinct vertical lines, described by $x = c$ and $x = c'$, respectively. The assumption that the lines are distinct means that $c \neq c'$. No point (x, y) can lie on both lines, because it would have to satisfy $x = c$ and $x = c'$. This proves (a).

Next, suppose ℓ and ℓ' are distinct nonvertical lines described by equations $y = mx + b$ and $y = m'x + b'$, respectively. Assume first that they have the same slope, which means $m = m'$. Since we are assuming that the lines are distinct, we must have $b \neq b'$. If there were a point (x_0, y_0) where the two lines intersect, then that point would have to satisfy both $y_0 = mx_0 + b$ and $y_0 = m'x_0 + b' = mx_0 + b'$. By algebra, therefore, we obtain $b = b'$, which is a contradiction. Therefore, there can be no such intersection point, so the lines are parallel.

Conversely, assume that the lines have different slopes: $m \neq m'$. We will show that the lines are not parallel by showing there exists a point (x_0, y_0) where they intersect. Consider the following point:

$$(x_0, y_0) = \left(\frac{b' - b}{m - m'}, \; m\frac{b' - b}{m - m'} + b \right). \tag{2.2}$$

By direct substitution, we can see that this point satisfies both equations, so the two lines are not parallel. (In case you wondered, formula (2.2) comes from solving the two equations simultaneously for x_0 and y_0; but the only thing that matters for the proof is that the resulting point is actually a solution of both equations.) This completes the proof of (b).

Finally, to prove (c), assume ℓ is a vertical line defined by the equation $x = c$ and ℓ' is a nonvertical line defined by $y = mx + b$. It is easy to check that the point $(x_0, y_0) = (c, mc + b)$ lies on both lines, so they are not parallel. $\qquad\square$

Theorem 2.23. *The Cartesian plane satisfies the Euclidean parallel postulate.*

Proof. Let ℓ be a line and let A be a point not on ℓ. Write the coordinates of A as (x_0, y_0). Because there are two types of lines, we have two cases to consider.

CASE 1: *ℓ is vertical.* Then ℓ is described by an equation of the form $x = c$ for some real constant c. The assumption that A does not lie on ℓ means that $x_0 \neq c$. Let $c' = x_0$, and let ℓ' be the vertical line defined by the equation $x = c'$. Then A lies on ℓ' (by definition of "lies on" in this model), and Lemma 2.22 shows that ℓ' is parallel to ℓ. This shows that there is at least one line that contains A and is parallel to ℓ.

To prove uniqueness, suppose ℓ'' is another line parallel to ℓ and containing A. Then Lemma 2.22 shows that ℓ'' is also vertical, so it must be described by an equation of the form $x = c''$ for some constant c''. The fact that A lies on ℓ'' implies that $x_0 = c''$; thus $c'' = c'$, which means that ℓ' and ℓ'' are actually the same line.

CASE 2: *ℓ is nonvertical.* In this case, ℓ is described by an equation of the form $y = mx + b$. Before we proceed with the proof, let us reason heuristically for a moment. We are looking for a line ℓ' containing A and parallel to ℓ. Lemma 2.22 tells us that ℓ' will be parallel to ℓ if and only if it is distinct from ℓ, is nonvertical, and has the same slope as ℓ, namely m; thus ℓ' must necessarily have an equation of the form $y = mx + b'$ for some constant b' different from b. We also want it to contain A, which means that (x_0, y_0) must be a solution to the equation, or in other words $y_0 = mx_0 + b'$. By algebra, therefore, we must have $b' = y_0 - mx_0$.

With this heuristic reasoning in hand, we now proceed with the formal proof. Define $b' = y_0 - mx_0$, and let ℓ' be the nonvertical line whose equation is

$$y = mx + b'. \tag{2.3}$$

Then simple algebra shows that (x_0, y_0) solves (2.3), so A lies on ℓ'. Because A does not lie on ℓ, it follows that ℓ and ℓ' must be distinct lines, and because they have the same slope m, they are parallel by Lemma 2.22. This shows that there exists at least one line ℓ' containing A and parallel to ℓ.

To prove that it is unique, suppose ℓ'' is any other line containing A and parallel to ℓ. Lemma 2.22 shows that ℓ'' is nonvertical and has slope m, so it must have an equation of the form $y = mx + b''$ for some constant b''. Because A lies on ℓ'', the same algebraic computation as before shows that $b'' = y_0 - mx_0 = b'$, so in fact ℓ'' is the same line as ℓ. $\qquad\square$

Let us briefly consider our other infinite interpretations.

Example 2.24 (More Parallelism in Incidence Geometry).

- **Spherical Geometry:** It is a straightforward (though not necessarily easy) exercise in analytic geometry to show that any two great circles on the sphere must intersect each other in exactly two antipodal points. (We will not prove it, but you should take some time to convince yourself that it is true, perhaps by drawing diagrams or looking at a globe.) Thus there are no parallel lines in this geometry, so the elliptic parallel postulate is satisfied.

- **Single Elliptic Geometry:** Because two great circles on the sphere always intersect in a pair of antipodal points, two "lines" in single elliptic geometry always intersect in a single "point." Once again, there are no parallel lines, so the elliptic parallel postulate is satisfied.

- **The Beltrami–Klein Disk:** In this model, it is easy to see that given any line and any point not on that line, there are infinitely many parallels to the given line through the given point. (Look back at Fig. 2.13, which shows an example of three parallels to a line ℓ through a point A.) Thus the hyperbolic parallel postulate is satisfied.

- **The Poincaré Disk:** Again, it straightforward (though not quite as easy as in the Beltrami–Klein disk) to see that the hyperbolic parallel postulate is satisfied. (See Fig. 2.14.) In this case, to check all possibilities, one would have to consider separately all combinations of lines of type (a) or lines of type (b) of Example 2.16, together with the point at the center of the circle or a point away from the center. You should take a few minutes to convince yourself that in each case, there are infinitely many parallels to the given line through the given point.

- **The Poincaré Half-Plane:** A similar argument by cases shows that the hyperbolic parallel postulate is satisfied in the Poincaré half-plane and that there are infinitely many parallels through a given point not on a line.

- **The Hyperboloid Model:** Of the models we have introduced, the hyperboloid model is one of the harder ones to visualize, so it is probably not easy to guess which, if any, of the parallel postulates it satisfies. With a bit of analytic geometry, one can prove that this model also satisfies the hyperbolic parallel postulate.   //

The names of the elliptic and hyperbolic parallel postulates are explained, at least in part, by the two surface models we introduced in this chapter: the hyperbolic parallel postulate is satisfied in the hyperboloid model, while the elliptic parallel postulate is satisfied in spherical geometry (and a sphere is a special case of an ellipsoid).

Theorems in Incidence Geometry

So far in our discussion of incidence geometry, we have explored three of its four elements as an axiomatic system: *primitive terms*, *defined terms*, and *axioms*. Now we are ready to introduce the fourth and most important element: *theorems*. Throughout the rest of this chapter, we assume the four incidence axioms, and we will construct our proofs rigorously within that axiomatic system. Before reading this section, it would be a good idea to read through Appendices E and F, which describe the conventions of mathematical logic and typical methods of proof.

Let's begin with something simple.

Theorem 2.25. *Given any point A, there exists another point that is distinct from A.*

Before we prove the theorem, let us take some time to discuss the process of writing such a proof. Usually, there are five stages to writing a proof:

1. Decode what the theorem statement means and what needs to be proved.

2. Choose a template for proving that type of statement.

3. Brainstorm about how the proof should go.

4. Construct the logical steps of the proof.

5. Write the proof in easily readable paragraph form.

Let's walk through each of these stages in turn for Theorem 2.25. We are going to be extremely careful this time through, so that you can see how strictly we need to pay attention to detail in order to ensure that our proofs are airtight.

1. *Decode what the theorem statement means and what needs to be proved.* The statement of Theorem 2.25 is a universal existence statement. In its "top-level" structure, it is a universal statement, with "given any point A" as its universal quantifier. The statement being quantified is an existence statement: "there exists another point that is distinct from A." Thus we might symbolize the whole statement this way, with the understanding that A and B represent points: "$\forall A, \exists B, B \neq A$."

2. *Choose a template for proving that type of statement.* Often, the structure of the theorem statement tells you immediately what kind of template to set up. In this case, we use Template F.12 (see Appendix F) for a universal existence proof. It looks like this:

	Statement	Reason
	Let A be an arbitrary point.	(hypothesis)
	Part 1: Choosing B	
	...	(...)
Goal:	Let $B = \ldots$.	(...)
	Part 2: Proof that $B \neq A$	
	...	(...)
Goal:	$B \neq A$.	(...) $\qquad\square$

We did not include a specific step to prove that B is a point, because it is reasonable to expect that our choice of B will make that obvious. The template does not tell us how we should go about producing the point B, so we leave that to the next step.

3. *Brainstorm about how the proof should go.* At first glance, it might seem that there is hardly anything to prove here. After all, Axiom 1 says there exist at least three distinct points. Why can't we just say "let A be one of them, and let B be another"?

To answer this question, we have to remember that each statement in an axiomatic system means precisely what it says, nothing more and nothing less. Axiom 1 says that there exist three distinct points; it does *not* say we get to choose one of them in advance. (As pointed out in Appendix F, it is useful to imagine that a mischievous gremlin chooses them, with an eye toward making our proof as difficult as possible.) We are given an arbitrary point A; the axiom says there exist three distinct points, but the points chosen by the gremlin might have nothing to do with A. In particular, we cannot assume that one of them is equal to A.

So all we can say is that there exist three distinct noncollinear points; let's call them C, D, and E. Can we just say that C is a point distinct from A? Well, no—just as the axiom does not say we can count on one of the points being equal to A, it also does not say we can count on any particular one of them being unequal to A. It could happen that C is actually the same point as A!

Yes, you will say, but *one of them* must be different from A! Now we're getting somewhere: it is evident that since the two points C and D are different from each other, then one of them must indeed be different from A. But to be thorough, this is something we need to *prove*: Axiom 1 says that there exist three points that are different from *each other*; it does not say anything about being different from a predetermined point A.

Since we do not know which of the three points is different from A, our proof is going to have to be split into cases. This is where we have to be careful to ensure that our cases are known to exhaust all possibilities. We *cannot* take our cases to be "C is different from A," "D is different from A," and "E is different from A," because that would be assuming that at least one of them is different from A, which is exactly what we're trying to prove! Instead, we can take our two cases to be "C is different from A" and "C is equal to A." One of these will certainly be true.

At this point, the rest of the proof comes naturally: if C is different from A, then we can just take $B = C$; and if C is equal to A, we can take $B = D$. In each case, it follows easily from what we already know that B is not equal to A.

4. *Construct the logical steps of the proof.* Now that we have seen the main ideas that will go into the proof, we just need to organize the argument into a two-column proof. There will be two cases, depending on whether $C = A$ or not; and each case will include the two steps of the existence proof: choosing B and proving that $B \neq A$.

Statement	Reason
1. Let A be an arbitrary point.	(hypothesis)
2. There exist distinct points C, D, & E.	(Axiom 1)
3. $C \neq D$.	(definition of *distinct*)
4. Either $C \neq A$ or $C = A$.	(logic)
Case 1:	
5. Assume $C \neq A$.	(hypothesis)
Choosing B:	
6. Let $B = C$.	(defining B)
Proof that $B \neq A$:	
7. $B \neq A$.	(substitute step 6 into step 5)
Case 2:	
8. Assume $C = A$.	(hypothesis)
Choosing B:	
9. Let $B = D$.	(defining B)
Proof that $B \neq A$:	
10. $B \neq A$.	(substitute steps 8 & 9 into step 3)
11. $\exists B$ such that $B \neq A$.	(steps 6, 7, 9, 10; logic) $\square$

5. *Write the proof in easily readable paragraph form.* Finally, having worked out the logic of the proof, we need to write it in a form that is easily understandable by a human reader, meaning in paragraph form. The idea is to take the exact steps that are laid out in the two-column format and express them clearly and fluently in standard English. When you do this, sometimes you will find that multiple steps can easily be compressed into one statement without causing confusion; other times, additional explanations need to be added. As you describe the steps, the justification for every step should be immediately clear to the reader. This does not mean, though, that you have to write out every justification explicitly—if you think a reader is likely to see immediately why a step is true (as when going from step 6 to step 7 above), there is no need to write the reason explicitly in the proof. On the other hand, if you're not sure whether a reason will be clear to your intended audience, it's usually safer to err on the side of including more.

Here is the final form of the proof.

Proof of Theorem 2.25. Let A be an arbitrary point. Axiom 1 guarantees that there exist three distinct points; let's call them C, D, and E. Since they are distinct, we know that $C \neq D$. Either C is distinct from A or it is not. If $C \neq A$, then we can let $B = C$ and conclude that B is a point different from A. Otherwise, if $C = A$, we can let $B = D$, and then the fact that $C \neq D$ ensures that B is different from A. In either case, we have produced a point B that is distinct from A. □

If you look over this written proof, you might decide that introducing the symbol B to represent the chosen point doesn't add anything to the clarity of the proof. It would be perfectly legitimate to shorten the proof as follows.

Another proof of Theorem 2.25. Let A be an arbitrary point. Axiom 1 guarantees that there exist three distinct points; let's call them C, D, and E. Since they are distinct, we know that $C \neq D$. Either C is distinct from A or it is not. If $C \neq A$, then C is a point different from A. Otherwise, if $C = A$, then the fact that $C \neq D$ ensures that D is a point different from A. In either case, we have produced a point that is distinct from A. □

Here is another simple theorem.

Theorem 2.26. *Given any point, there exists a line that contains it.*

Again, let's walk through the steps of constructing a proof.

1. *Decode.* Once again, this is a universal existence statement. To write it symbolically, let's borrow the symbol "$\in$" from set theory to symbolize "lies on," even though lines in incidence geometry need not be sets of points. With the understanding that A represents a point and ℓ represents a line, we can symbolize the theorem statement as "$\forall A$, $\exists \ell$, $A \in \ell$."

2. *Choose a template.* The template for this proof will be just like the preceding one, starting with "let A be an arbitrary point" and then proceeding with the existence proof.

3. *Brainstorm.* After introducing our arbitrary point A, we need to think about how to produce a line containing A. We have only one theorem so far, and it does not say anything about lines, so it is no help. The only axiom that asserts the existence of a line is Axiom 2, but that requires *two points*. Now our previous theorem comes to the rescue: it tell us that there exists a point different from A, and then Axiom 2 says there exists a line containing these two points.

4. *Construct the logical steps.* Next we need to write the steps of the proof in two-column format. From now on, for simplicity, we will often omit the titles of separate parts or cases of short proofs. Here is a two-column proof:

	Statement	Reason
1.	Let A be an arbitrary point.	(hypothesis)
2.	There exists a point B distinct from A.	(Theorem 2.25)
3.	There exists a line ℓ containing A and B.	(Axiom 2)
4.	There exists a line ℓ containing A.	(logic) □

5. *Write the proof.* Here is the final form of the proof:

Proof of Theorem 2.26. Let A be a point. By Theorem 2.25, there is another point B distinct from A, and by Axiom 2, there is at least one line ℓ that contains both A and B. This line ℓ satisfies the desired conclusion. □

The preceding theorem has the following useful corollary. Because the proof (by cases) is so simple, we will just go straight to the final paragraph-style proof.

Corollary 2.27. *If A and B are points (not necessarily distinct), there is a line that contains both of them.*

Proof. Let A and B be any two points. If they are distinct, then Axiom 2 guarantees that there is a line that contains them, while if they are not distinct, then Theorem 2.26 yields the same conclusion.   □

Recall that for any pair of distinct points A and B, we have defined the notation $\overleftrightarrow{AB}$ to mean the unique line containing A and B, whose existence and uniqueness are guaranteed by Axioms 2 and 3 (see p. 25). The next theorem illustrates a common way in which the *uniqueness* statement of Axiom 3 is used.

Theorem 2.28. *If ℓ is a line and A and B are two distinct points on ℓ, then $\overleftrightarrow{AB} = \ell$.*

1. *Decode.* The "top-level" structure of the theorem statement is an implication. Since ℓ, A, and B all appear in the hypothesis and conclusion without being quantified, we can infer that there is an implicit universal quantifier for all three symbols. So we can read the statement as if it started with "For every line ℓ and for all points A and B ... ." The hypothesis of the implication is written as a conjunction ("ℓ is a line and A and B are two distinct points on ℓ"); but since we have now subsumed "ℓ is a line" in the quantifier, we can rewrite the hypothesis simply as "A and B are two distinct points on ℓ." In more detail, this is a conjunction: "A lies on ℓ and B lies on ℓ and $A \neq B$." The conclusion is "$\overleftrightarrow{AB} = \ell$." With the understanding that A and B represent points and ℓ represents a line, we can summarize the statement as follows (again using "$\in$" to denote "lies on"):

$$\forall \ell, \ \forall A, \ \forall B, \ \left(A \in \ell \wedge B \in \ell \wedge A \neq B\right) \Rightarrow \overleftrightarrow{AB} = \ell.$$

2. *Choose a template.* Because the theorem statement is a universal implication, we use that template.

Statement	Reason
Let ℓ be an arbitrary line.	(hypothesis)
Let A, B be arbitrary points.	(hypothesis)
Assume A and B lie on ℓ.	(hypothesis)
Assume $A \neq B$.	(hypothesis)
$\quad\ \ \ldots$	(...)
Goal: $\overleftrightarrow{AB} = \ell$.	(...) □

3. *Brainstorm.* This is essentially an exercise in understanding two things: the definition of the notation $\overleftrightarrow{AB}$ and the meaning of Axiom 3. By definition, $\overleftrightarrow{AB}$ is the (unique) line containing A and B. On the other hand, our hypothesis says that ℓ is also a line containing A and B. Axiom 3 says there is at most one line containing A and B, which really means that if we happen to find two such lines, they must be equal. This is the crux of the proof.

4. *Write the logical steps.* Based on the reasoning in the preceding paragraph, here is a two-column version of the proof.

Statement	Reason
1. Let ℓ be an arbitrary line.	(hypothesis)
2. Let A, B be arbitrary points.	(hypothesis)
3. Assume A and B lie on ℓ.	(hypothesis)
4. Assume $A \neq B$.	(hypothesis)
5. $\overleftrightarrow{AB}$ is a line containing A and B.	(definition of $\overleftrightarrow{AB}$)
6. ℓ is a line containing A and B.	(steps 1 and 3)
7. $\overleftrightarrow{AB} = \ell$.	(steps 5 and 6, Axiom 3) ☐

5. *Write the proof.* Here it is.

Proof of Theorem 2.28. Suppose ℓ is any line and A, B are two distinct points that lie on ℓ. By definition, $\overleftrightarrow{AB}$ is a line containing A and B, and by our hypothesis, ℓ is also a line containing A and B. Thus by Axiom 3, these two line must be equal. ☐

Our next theorem gives a useful criterion for deciding when three points are noncollinear. (Recall that this means there is no line that contains all three of them.)

Theorem 2.29. *If A and B are distinct points and C is any point that does not lie on $\overleftrightarrow{AB}$, then A, B, and C are noncollinear.*

1. *Decode.* Once again, this is a universal implication, with an implicit universal quantifier for A, B, and C. The hypothesis of the implication is a conjunction ("A and B are distinct and C does not lie on $\overleftrightarrow{AB}$"). The conclusion is the negation of "A, B, and C are collinear" and is thus a negated existence statement: "There does not exist a line ℓ that contains A, B, and C." Summarizing,

$$\forall A, B, C, \left(A \neq B \wedge C \notin \overleftrightarrow{AB} \right) \Rightarrow \neg \left(\exists \ell, \, A, B, C \in \ell \right).$$

2. *Choose a template.* We start with a template for a universal implication as before. Until we go through the brainstorming step, there is not much more detail that can be added.

3. *Brainstorm.* The first thing to notice is that there is in fact something to prove here. The hypothesis tells us immediately that A, B, and C do not all lie on the *specific* line $\overleftrightarrow{AB}$; however, in order to prove that A, B, and C are noncollinear, we have to show that there does not exist *any* line that all three points lie on. This is a nonexistence statement, so as explained in Appendix F, the most promising approach will be an indirect proof: we will assume that there *does* exist a line containing all three and derive a contradiction. Suppose ℓ is a line containing A, B, and C. It follows from Theorem 2.28 that $\ell = \overleftrightarrow{AB}$. But then, since C lies on ℓ, it follows by substitution that C lies on $\overleftrightarrow{AB}$, which contradicts our hypothesis.

4. *Construct the logical steps.* Here is the proof in two-column form.

	Statement	Reason
1.	Let A, B, C be points.	(hypothesis)
2.	Assume $A \neq B$.	(hypothesis)
3.	Assume C does not lie on $\overleftrightarrow{AB}$.	(hypothesis)
4.	Assume A, B, C are collinear.	(hypothesis for contradiction)
5.	There is a line ℓ containing A, B, C.	(definition of *collinear*)
6.	A and B are distinct points on ℓ.	(steps 2 and 5)
7.	$\ell = \overleftrightarrow{AB}$.	(Theorem 2.28)
8.	C lies on ℓ.	(step 5, definition of *contains*)
9.	C lies on $\overleftrightarrow{AB}$.	(substituting step 7 into step 8)
10.	Contradiction	(steps 3 and 9)
11.	A, B, C are noncollinear.	(logic) $\square$

5. *Write the proof.* Here is the final version.

Proof of Theorem 2.29. Suppose A and B are distinct points and C is a point not on $\overleftrightarrow{AB}$. Assume for the sake of contradiction that A, B, and C are collinear. This means there is some line ℓ containing all three points. Because the distinct points A and B lie on ℓ, it follows from Theorem 2.28 that $\ell = \overleftrightarrow{AB}$. But since C also lies on ℓ, this implies that C lies on $\overleftrightarrow{AB}$ as well, which is a contradiction. Thus our assumption of collinearity must be false. $\square$

The rest of our theorems about incidence geometry will be left for you to prove as exercises. When brainstorming about the proofs, be sure to be on the lookout for previous theorems that might be useful.

Our next theorem is the converse of the preceding one.

Theorem 2.30. *If A, B, and C are noncollinear points, then A and B are distinct, and C does not lie on $\overleftrightarrow{AB}$.*

You will be asked to prove this theorem shortly. But to get you started, let's think about what needs to be done. This is a universal implication (with an implicit universal quantifier for A, B, and C), whose conclusion is a conjunction. Thus a direct proof would begin by assuming that A, B, and C are noncollinear points, and then there would be two parts: Part 1 would prove that A and B are distinct, and Part 2 would prove that C does not lie on $\overleftrightarrow{AB}$.

If you start thinking about how to set up this proof, things get complicated pretty quickly. How could we prove that A and B are distinct? We don't have an axiom or theorem whose conclusion is that two points are distinct (apart from Axiom 1, which cannot be applied to previously given points). Nor do we have an axiom or theorem whose conclusion is that a point does not lie on a line. For that matter, how would we go about using the hypothesis that A, B, and C are noncollinear? As we observed above, this is a *nonexistence* statement, and it's hard to see how the nonexistence of a line is going to be of much use.

The problem here is that all the parts of the theorem statement—the hypothesis and the two conclusions—are negative statements. The hypothesis is that there does *not* exist a line containing A, B, and C; the first conclusion is that A and B are *not* equal, and the second conclusion is that C does *not* lie on $\overleftrightarrow{AB}$. In situations like this, if you find yourself spinning your wheels when you try to brainstorm about how to set up a proof, it's a good time to try a different strategy. If a direct proof doesn't seem to be working, maybe a contrapositive proof will, or an indirect proof.

For this theorem, it turns out that a contrapositive proof works nicely. The starting assumption will be the negation of the conclusion: either A and B are equal or C lies on $\overleftrightarrow{AB}$ (which includes the assumption that $A \neq B$). Under this assumption, you have to prove that A, B, and C are collinear, meaning that there is a line that contains them all. Because the starting assumption is a disjunction, your proof will have two cases.

Now you should be ready to construct the proof yourself. (See Exercise 2K.)

In the preceding theorem, there is nothing special about the point C: if A, B, and C are noncollinear, the same argument also shows that A and C are distinct, as are B and C; and it also shows that A does not lie on $\overleftrightarrow{BC}$ and B does not lie on $\overleftrightarrow{AC}$. For convenience, let us state this as a corollary. Because it follows immediately from Theorem 2.30 (with the names of the points changed), we do not need to write out the proof.

Corollary 2.31. *If A, B, and C are noncollinear points, then A, B, and C are all distinct. Moreover, A does not lie on $\overleftrightarrow{BC}$, B does not lie on $\overleftrightarrow{AC}$, and C does not lie on $\overleftrightarrow{AB}$.* $\square$

Here are some more theorems in incidence geometry.

Theorem 2.32. *Given a line ℓ and a point A that lies on ℓ, there exists a point B that lies on ℓ and is distinct from A.*

Proof. See Exercise 2L. [Hint: The proof is essentially the same as that of Theorem 2.25, but using Axiom 4 instead of Axiom 1.] $\square$

Theorem 2.33. *Given any line, there exists a point that does not lie on it.*

Proof. See Exercise 2M. $\square$

Theorem 2.34. *Given two distinct points A and B, there exists a point C such that A, B, and C are noncollinear.*

Proof. See Exercise 2N. $\square$

Theorem 2.35. *Given any point A, there exist points B and C such that A, B, and C are noncollinear.*

Proof. See Exercise 2O. $\square$

Theorem 2.36. *Given two distinct points A and B, there exists a line that contains A but not B.*

Proof. See Exercise 2P. $\square$

Theorem 2.37. *Given any point, there exists a line that does not contain it.*

Proof. See Exercise 2Q. $\square$

Theorem 2.38. *If A, B, and C are noncollinear points, then $\overleftrightarrow{AB} \neq \overleftrightarrow{AC}$.*

Proof. See Exercise 2R.   □

Corollary 2.39. *If A, B, and C are noncollinear points, then* $\overleftrightarrow{AB}$, $\overleftrightarrow{AC}$, *and* $\overleftrightarrow{BC}$ *are all distinct.*   □

Theorem 2.40. *If A, B, and C are collinear points and neither B nor C is equal to A, then* $\overleftrightarrow{AB} = \overleftrightarrow{AC}$.

Proof. See Exercise 2S.   □

Theorem 2.41. *Given two distinct, nonparallel lines, there exists a unique point that lies on both of them.*

Proof. See Exercise 2T.   □

Recall Statements I and II that we introduced in Chapter 2 (see p. 26). As you are asked to show in Exercise 2D, Statement II is independent of the axioms of incidence geometry because it is true in some models and false in others, so it cannot be a theorem of incidence geometry. On the other hand, Statement I *is* a theorem of incidence geometry, as you can now show.

Theorem 2.42. *Given any point, there are at least two distinct lines that contain it.*

Proof. See Exercise 2U.   □

Exercises

For Exercises 2A–2J, you do not need to write out formal proofs. Just answer the questions and (if requested) give convincing informal explanations of why your answers are correct.

2A. For each of Examples 2.8, 2.9, and 2.10, figure out which incidence axiom(s) are satisfied and which are not.

2B. Use models to show that each of the incidence axioms is independent of the other three.

2C. Suppose we replace Incidence Axiom 4 with the following:
  • **Incidence Axiom 4′:** *Given any line, there are at least three distinct points that lie on it.*
  What is the smallest number of points in a model for this geometry? More precisely, find a number n such that every model has at least n points and there is at least one model that has only n points, and explain why your answer is correct.

2D. Use models to show that each of the following statements is independent of the axioms of incidence geometry:
  (a) Given any line, there are at least two distinct points that do not lie on it.
  (b) Given any point, there are at least three distinct lines that contain it.
  (c) Given any two distinct points, there is at least one line that does not contain either of them.

2E. Show that the Fano plane and the seven-point plane are not isomorphic to each other. [Remark: It is not enough just to show that a *particular* correspondence is not an isomorphism; you need to demonstrate that there cannot exist *any* isomorphism between the two models.]

2F. Find a model of incidence geometry that has exactly four points but that is not isomorphic to the four-point plane, and explain why it is not isomorphic.

2G. Determine which parallel postulates, if any, are satisfied by the interpretations of Examples 2.3–2.10.

2H. For each of the following interpretations of incidence geometry, decide if it's a model, and decide which parallel postulate(s) (if any) it satisfies. (No proofs necessary.)
   (a) *Point* means an ordinary line in three-dimensional space; *line* means a plane in three-dimensional space; and *lies on* means "is contained in."
   (b) *Point* means an ordinary line containing the origin in three-dimensional space; *line* means an ordinary plane containing the origin in three-dimensional space; *lies on* means "is contained in."

2I. Define a model of incidence geometry with *points* $1,2,3,4,5,6$; *lines* consisting of the sets

$$\{1,2,3\}, \{3,4,5\}, \{5,6,1\}, \{1,4\}, \{2,5\}, \{3,6\}, \{2,6\}, \{4,6\}, \{2,4\};$$

and *lies on* meaning "is an element of" (Fig. 2.18). (You may use without proof the fact that this is indeed a model.) Which, if any, of the parallel postulates is (are) satisfied by this model? Can you figure out what is remarkable about parallelism in this model?

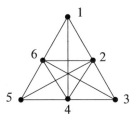

Fig. 2.18. The model of Exercise 2I.

2J. Here is an axiomatic system for an unusual kind of geometry; let's call it *three-two geometry*. The primitive terms are exactly the same as for incidence geometry: *point*, *line*, and *lies on*. As in incidence geometry, we define ℓ *contains* A to mean "A lies on ℓ." There are two axioms:
   • **Axiom 1:** *Every line contains exactly three points.*
   • **Axiom 2:** *Every point lies on exactly two lines.*
   Try to answer as many of the following questions as you can, and justify your answers.
   (a) Is this axiom system consistent?
   (b) Are the two axioms independent of each other?
   (c) What is the minimum number of points in a model for three-two geometry?
   (d) Can you find a model with exactly three points?
   (e) Can you find a model with exactly three lines?

(f) Can you find a model with exactly four points?

(g) Can you find a model with more than four points but still finitely many?

(h) Can you find a model with infinitely many points?

(i) Is there a model of three-two geometry that is also a model of incidence geometry?

(j) Is there a model of three-two geometry that is not a model of incidence geometry?

(k) If a model has finitely many points, is there any necessary relationship between the number of lines and the number of points?

(l) Can there be parallel lines in three-two geometry?

(m) Does this axiom system describe anything useful?

For each of the theorems below, first write a two-column proof, and then translate it into a fluid, clear, and precise paragraph-style proof.

2K. Theorem 2.30.

2L. Theorem 2.32.

2M. Theorem 2.33.

2N. Theorem 2.34.

2O. Theorem 2.35.

2P. Theorem 2.36.

2Q. Theorem 2.37.

2R. Theorem 2.38.

2S. Theorem 2.40.

2T. Theorem 2.41.

2U. Theorem 2.42.

Chapter 3

Axioms for Plane Geometry

Starting in this chapter, we describe a modern, rigorous axiomatic system within which all of Euclid's theorems of plane geometry (and more) can be proved. To begin with, we describe an axiomatic system that does not include any parallel postulate; this system is called ***neutral geometry***. Like Euclid, we will prove as many theorems as we can without relying on a parallel postulate. When the Euclidean parallel postulate is added to the postulates of neutral geometry, we obtain an axiomatic system called ***Euclidean geometry***, and when the hyperbolic parallel postulate is added instead, we obtain ***hyperbolic geometry***. Taken together, this collection of axiomatic systems is called ***plane geometry***.

The postulates we will be using are based roughly on the SMSG postulates proposed for high-school geometry courses in the 1960s (see Appendix C), from which most axiom systems used in high-school geometry textbooks are descended. The SMSG postulates, in turn, were based ultimately on the axiomatic system proposed in 1932 by George D. Birkhoff (Appendix B), the first mathematician to introduce the real number system into the foundations of geometry. The axioms introduced here are closely related to those that are typically used in high-school courses, but they adhere to a higher standard of rigor.

Although Euclid treated the geometry of three-dimensional space in addition to plane geometry and both Hilbert's system and the SMSG system contain postulates that describe the geometry of space, our treatment is limited to plane geometry alone. This is not because three-dimensional geometry is not important (it is), but rather because our main purpose is to explore the way axiomatic systems are used to make mathematics rigorous, and plane geometry serves that purpose perfectly well. Once you have thoroughly understood the approach to plane geometry developed in this book, it should be an easy matter to extend your understanding to the geometry of space.

Because our axiom system uses both set theory and the theory of the real number system as underlying foundations, we assume all of the standard tools of set theory (summarized in Appendix G) and all of the standard properties of the real numbers (summarized in Appendix H) as given.

The proofs in this subject range from very easy to extremely challenging. In order that you might have a bit of forewarning about how difficult a proof is likely to be before you start to tackle it, we use the following rating system for our proofs in plane geometry, borrowed from the movies:

- *Rated G*: simple, straightforward proofs that you should understand thoroughly and be able to reconstruct on your own. Many of these proofs would be quite appropriate to explain to high-school students. All of the geometry proofs in this book are rated G unless otherwise specified.

- *Rated PG*: proofs that are a little less straightforward than the G-rated ones. Some might use slightly more advanced techniques, while others use only elementary techniques but contain unexpected ideas or "tricks." You should understand these thoroughly; but you might not have thought of them yourself, and you might find it a little challenging to reconstruct them on your own unless you have made an effort to commit the main ideas to memory.

- *Rated R*: proofs that are considerably more difficult than most of the proofs in the book because they use more advanced techniques, or they use elementary techniques in highly unconventional ways, or they are unusually long and intricate. You should attempt to follow the logic of these proofs, but you would ordinarily not be expected to be able to reconstruct them without a great deal of effort.

- *Rated X*: proofs that are too advanced or too complex to be included in this book at all.

From now on, we will present all of our proofs in paragraph form, since such proofs are usually the easiest kind to understand. But you are encouraged to continue constructing proofs first in two-column format before writing them in paragraph form, to help you make sure you have correctly worked out the logical steps and their justifications.

Points and Lines

Every axiomatic system must begin with some primitive terms. In our axiomatic approach to geometry, there are four primitive terms: *point*, *line*, *distance* (between points), and *measure* (of an angle). The terms *point* and *line* represent objects, while *distance* and *measure* represent functions: distance is a function of pairs of points, and measure is a function of angles (to be defined later).

Although we do not give the primitive terms formal mathematical definitions within our axiomatic system, here is how you should think about them intuitively:

- *Point*: a precise, indivisible "location" in the plane, without any length, width, depth, area, or volume. Euclid's description of a point as "that which has no part" is as good as any.

- *Line*: a set of points in the plane that forms a "continuous straight path" with no width, no gaps, and no bends, extending infinitely far in both directions.

- *Distance between points*: a nonnegative real number describing how far apart two points are.

- *Measure of an angle*: a real number between 0 and 180 inclusive (representing degrees), describing the size of the opening between two rays with a common endpoint.

It cannot be overemphasized that these are *not* mathematical definitions; they are just intuitive descriptions to help you visualize what the terms mean when they are used in postulates and in further definitions. We will never use any of these descriptions to justify steps in a proof. (It would be impossible to do so while maintaining rigor, because many of the key terms that appear in these descriptions—such as "location," "straight," and "opening"—have themselves not been formally defined!) The only *mathematical* content that can be ascribed to the primitive terms is what the postulates tell us about them.

As a general rule, we use capital letters to denote points, and we use lowercase letters in the range $\ell, m, n, \ldots$ to denote lines. Most of these symbols, however, can also be used to represent other objects, so it is always necessary to say what each particular symbol is meant to represent.

Here is our first postulate.

Postulate 1 (The Set Postulate). *Every line is a set of points, and there is a set of all points called **the plane**.*

This postulate does not contain much geometric information, but it sets the stage for everything we are going to do later. It tells us that we can use all of the tools of set theory (subsets, intersections, unions, etc.) when talking about lines or other collections of points. It rules out models in which lines are anything other than sets of points—such as, for example, the Amtrak model of Chapter 2.

Using the vocabulary of set theory, we can define some commonly used geometric terms. If ℓ is a line and A is a point, we define both *A **lies on** ℓ* and *ℓ **contains** A* to be synonyms for the statement $A \in \ell$. The set postulate thus frees us from having to consider "lies on" as a primitive term.

We define the following terms just as we did in incidence geometry: a set of points is said to be *collinear* if there is a line that contains them all; and two lines ℓ and m are said to *intersect* if there is at least one point that lies on both of them. (In this context, this is just a special case of the set-theoretic definition of what it means for two *sets* to intersect.) If A is a point that lies on both lines ℓ and m, we also say that *ℓ **and** m **meet at** A*. We say that *ℓ **and** m **are parallel**,* denoted by $\ell \parallel m$, if they do not intersect. By this definition, no line is parallel to itself since—as we will see below—every line contains at least one point and thus intersects itself. (Some high-school geometry texts define parallel lines in such a way that every line is parallel to itself. Which definition is chosen is a matter of taste and convenience, and each author is free to choose either definition, as long as he or she uses it consistently. For our purposes, the definition we have chosen is much more useful. Besides, it has an excellent pedigree: it is essentially the same as Euclid's definition.)

The next postulate begins to give the theory a little substance. You will recognize it as Axiom 1 of incidence geometry.

Postulate 2 (The Existence Postulate). *There exist at least three distinct noncollinear points.*

This postulate still does not tell us very much about our geometry, but without it we could not get started because we could not be sure that there exist any points at all!

Our third postulate corresponds to Euclid's first postulate. But ours says more: whereas Euclid assumed only that for any two points, there is a line that contains them, we are assuming in addition that the line is *unique*. (It is clear from Euclid's use of his first postulate in Proposition I.4 that he had in mind that the line should be unique, even though he didn't state it as part of the postulate. Our postulate rectifies this omission.)

> **Postulate 3 (The Unique Line Postulate).** *Given any two distinct points, there is a unique line that contains both of them.*

You will also recognize this as the combination of Axioms 2 and 3 of incidence geometry. (Axiom 4 of incidence geometry is true in our axiomatic system as well, but we do not need to assume it as a separate postulate because it will follow from Postulate 5 below.)

If A and B are two distinct points, we use the notation $\overleftrightarrow{AB}$ to denote the unique line that contains both A and B, just as in incidence geometry (see p. 25). As before, if A and B have not been previously mentioned, then a phrase such as "Let $\overleftrightarrow{AB}$ be a line" is understood to mean "Let A and B be distinct points and let $\overleftrightarrow{AB}$ be the unique line containing them."

Distance

So far, our geometric postulates have not strayed very far from the axioms of incidence geometry. In fact, you can check that every model of incidence geometry described in Chapter 2 in which each line is a set of points (as opposed to, say, a railroad line) is also a model of the axiomatic system we have described so far. Our next two postulates, though, take us in a very different direction, by introducing the real numbers into the study of geometry.

> **Postulate 4 (The Distance Postulate).** *For every pair of points A and B, the distance from A to B is a nonnegative real number determined by A and B.*

We use the notation AB to denote the distance from A to B. This postulate gives us no details about the distance AB other than that it is a nonnegative real number and that it is completely determined by the points A and B. (So, for example, if $A' = A$ and $B' = B$, then $A'B' = AB$; and the distance from A to B will be the same tomorrow as it is today.) The next postulate is what gives distance its geometric meaning. It expresses in rigorous mathematical language our intuitive understanding of what we do with a ruler (Fig. 3.1): we align it with a line along which we wish to determine distances and then read off the distance between two points by subtracting the numbers that appear at the corresponding positions on the ruler. Of course, the postulate says nothing about physical rulers; it is called the "ruler postulate" merely as a way to remember easily what it is about.

If X and Y are sets, recall that a function $f : X \to Y$ is said to be ***injective*** if $f(x_1) = f(x_2)$ implies $x_1 = x_2$; it is said to be ***surjective*** if for every $y \in Y$ there exists $x \in X$

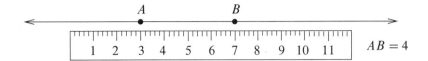

Fig. 3.1. Motivation for the ruler postulate.

such that $f(x) = y$; and it is said to be **bijective** if it is both injective and surjective. (See Appendix G for a brief review of these concepts.)

Postulate 5 (The Ruler Postulate). *For every line ℓ, there is a bijective function $f : \ell \to \mathbb{R}$ with the property that for any two points $A, B \in \ell$, we have*

$$AB = |f(B) - f(A)|. \tag{3.1}$$

If x and y are real numbers, we often refer to the quantity $|y - x|$ as the **distance between x and y**. If ℓ is a line, a function $f : \ell \to \mathbb{R}$ that satisfies equation (3.1) for all $A, B \in \ell$ is said to be **distance-preserving**, because (3.1) says that the distance between the points A and B is the same as that between the numbers $f(A)$ and $f(B)$. A distance-preserving bijective function from ℓ to $\mathbb{R}$ is called a **coordinate function for ℓ**. With this terminology, we can summarize the ruler postulate as follows: *for each line, there exists a coordinate function*. Once a particular coordinate function f has been chosen for a line, then the number $f(A)$ associated with a point A is called the **coordinate of A** with respect to f.

One immediate consequence of the ruler postulate is the following.

Theorem 3.1. *Every line contains infinitely many distinct points.*

Proof. Let ℓ be a line. By the ruler postulate, there exists a coordinate function $f : \ell \to \mathbb{R}$. Because f is surjective, for each positive integer n, there is a point $X_n \in \ell$ such that $f(X_n) = n$. All of the points $\{X_1, X_2, \ldots\}$ must be distinct because f is a well-defined function and assigns different values to them. $\square$

This theorem immediately implies the fourth axiom of incidence geometry.

Corollary 3.2 (Incidence Axiom 4). *Given any line, there are at least two distinct points that lie on it.* $\square$

We have now shown that all four axioms of incidence geometry are true in our axiomatic system. (Incidence Axiom 1 is the existence postulate; Incidence Axioms 2 and 3 follow from the unique line postulate; and we have just proved Incidence Axiom 4.) Therefore, Theorems 2.25–2.42 (including Corollaries 2.27, 2.31, and 2.39) are now theorems of neutral geometry, with exactly the same proofs.

It is important to note that the ruler postulate does not say that there is a *unique* coordinate function, only that there is at least one. In fact, as we will soon show, every line has many different coordinate functions. Because the distance postulate guarantees that the distance between any two points is a well-defined number, one thing that we can conclude from the ruler postulate is that any two coordinate functions for the same line yield the same distances between pairs of points.

The next two lemmas are the mathematical analogues of "sliding the ruler along the line" and "flipping the ruler end over end." The proofs of these lemmas are rated PG, not because they are hard but because they use the concepts of injectivity and surjectivity, which might still be relatively new to some readers.

Lemma 3.3 (Ruler Sliding Lemma). *Suppose ℓ is a line and $f : \ell \to \mathbb{R}$ is a coordinate function for ℓ. Given a real number c, define a new function $f_1 : \ell \to \mathbb{R}$ by $f_1(X) = f(X) + c$ for all $X \in \ell$. Then f_1 is also a coordinate function for ℓ.*

Proof. To show that f_1 is a coordinate function, we must show that it is bijective and distance-preserving. First, to see that it is surjective, suppose y is an arbitrary real number. We need to show that there is a point $P \in \ell$ such that $f_1(P) = y$, which is equivalent to $f(P) + c = y$. Because f is surjective, there is a point $P \in \ell$ such that $f(P) = y - c$, and then it follows that $f_1(P) = y$ as desired.

Next, to see that f_1 is injective, suppose A and B are points on ℓ such that $f_1(A) = f_1(B)$; we need to show that $A = B$. A little algebra gives

$$0 = f_1(A) - f_1(B)$$
$$= (f(A) + c) - (f(B) + c)$$
$$= f(A) - f(B).$$

Thus $f(A) = f(B)$, and the fact that f is injective implies $A = B$. This completes the proof that f_1 is bijective.

Finally, we have to show that f_1 is distance-preserving. If A and B are any two points on ℓ, the definition of f_1 and the fact that f is distance-preserving imply

$$|f_1(B) - f_1(A)| = \left| (f(B) + c) - (f(A) + c) \right|$$
$$= |f(B) - f(A)|$$
$$= AB,$$

which was to be proved.   □

Lemma 3.4 (Ruler Flipping Lemma). *Suppose ℓ is a line and $f : \ell \to \mathbb{R}$ is a coordinate function for ℓ. If we define $f_2 : \ell \to \mathbb{R}$ by $f_2(X) = -f(X)$ for all $X \in \ell$, then f_2 is also a coordinate function for ℓ.*

Proof. Exercise 3A.   □

The preceding two lemmas are the basic ingredients in the proof of the following important strengthening of the ruler postulate.

Theorem 3.5 (Ruler Placement Theorem). *Suppose ℓ is a line and A, B are two distinct points on ℓ. Then there exists a coordinate function $f : \ell \to \mathbb{R}$ such that $f(A) = 0$ and $f(B) > 0$.*

Proof. Given ℓ, A, and B as in the hypothesis, the ruler postulate guarantees that there exists some coordinate function $f_1 : \ell \to \mathbb{R}$. Let $c = -f_1(A)$, and define a new function $f_2 : \ell \to \mathbb{R}$ by $f_2(X) = f_1(X) + c$. Then the ruler sliding lemma guarantees that f_2 is also a coordinate function for ℓ, and it satisfies $f_2(A) = 0$ by direct computation.

Because A and B are distinct points and coordinate functions are injective, it follows that $f_2(B) \neq 0$. Thus we have two cases: either $f_2(B) > 0$ or $f_2(B) < 0$. If $f_2(B) > 0$, then we can set $f = f_2$, and f satisfies the conclusions of the theorem. On the other hand, if $f_2(B) < 0$, then we define a new function $f: \ell \to \mathbb{R}$ by $f(X) = -f_2(X)$. The ruler flipping lemma guarantees that f is a coordinate function. It follows from the definition of f that $f(A) = -f_2(A) = 0$ and $f(B) = -f_2(B) > 0$, so f is the function we seek. $\square$

It is worth noting that the SMSG system (Appendix C) includes the ruler placement theorem as an additional postulate, instead of proving it as a theorem, and many high-school texts follow suit. The main reason seems to be that proving the ruler sliding and flipping lemmas, on which the ruler placement theorem depends, requires a good under-standing of the properties of bijective functions, which high-school students are unlikely to have been introduced to. (Be warned, however, that some texts, such as [**UC02**], state as a postulate an even stronger version of the ruler placement theorem, which claims that it is possible to choose a coordinate function for a line such that any chosen point on the line has coordinate 0 and any other chosen point has coordinate 1. Since another postulate specifies that coordinate functions are distance-preserving, this would lead to the contra-dictory conclusion that the distance between every pair of distinct points is equal to 1. This inconsistency is glossed over without comment in [**UC02**].)

Here are some important properties of distances that can be deduced easily from the ruler postulate.

Theorem 3.6 (Properties of Distances). *If A and B are any two points, their distance has the following properties:*

(a) $AB = BA$.

(b) $AB = 0$ *if and only if* $A = B$.

(c) $AB > 0$ *if and only if* $A \neq B$.

Proof. Let A and B be two points. By Corollary 2.27, there is a line ℓ containing both A and B. By the ruler postulate, there is a coordinate function $f: \ell \to \mathbb{R}$, and it satisfies $AB = |f(B) - f(A)|$. The properties of absolute values guarantee that $BA = |f(A) - f(B)| = |f(B) - f(A)| = AB$, which is (a).

Statement (b) is an equivalence, so we need to prove two implications: if $AB = 0$, then $A = B$; and if $A = B$, then $AB = 0$. Assume first that $AB = 0$. Because f is distance-preserving, this implies $|f(B) - f(A)| = 0$, which in turn implies $f(B) = f(A)$. Because f is injective, this implies $A = B$. Conversely, assume that $A = B$. Then $AB = |f(B) - f(A)| = |f(A) - f(A)| = 0$ because f is a well-defined function.

Finally, we prove (c). If we assume that $AB > 0$, then it follows from (b) that $A \neq B$. Conversely, if $A \neq B$, then (b) implies $AB \neq 0$. Since the distance postulate implies that distances cannot be negative, it follows that AB must in fact be positive. $\square$

Betweenness of Points

The ruler postulate provides the tools we need to address many important issues that Euclid left out of his axiomatic system. The first of these has to do with what it means for a point on a line to be "between" two other points. We already know what "betweenness" means for numbers: if x, y, z are three real numbers, we say that *y is between x and z* if either $x < y < z$ or $x > y > z$. Let us introduce the notation $x * y * z$ to symbolize this relationship. From elementary algebra, it follows that whenever x, y, and z are three distinct real numbers, exactly one of them lies between the other two.

Using the notion of betweenness of numbers, we can define betweenness for points: given points A, B, and C, we say that **B is between A and C** if all three points are distinct and lie on some line ℓ and there is a coordinate function $f : \ell \to \mathbb{R}$ such that $f(A) * f(B) * f(C)$; according to the definition above, this means that either $f(A) < f(B) < f(C)$ or $f(A) > f(B) > f(C)$. (See Fig. 3.2.) We symbolize this relationship with the notation $A * B * C$.

Fig. 3.2. B is between A and C, symbolized by $A * B * C$.

This definition corresponds well to our intuitive idea of what we mean when we say that one point is between two others. There is one awkward thing about it, though: to say B is between A and C means that $f(B)$ is between $f(A)$ and $f(C)$ for *some* coordinate function f; it does not immediately rule out the possibility that a different coordinate function f_1 might put the points in a different order, so that, for example, $f_1(A)$ might be between $f_1(B)$ and $f_1(C)$. By our definition, this would imply that A is also between B and C! We will see below that this cannot happen. But first, we need to establish some basic properties of betweenness.

Theorem 3.7 (Symmetry of Betweenness of Points). *If A, B, C are any three points, then $A * B * C$ if and only if $C * B * A$.*

Proof. Both statements mean the same thing: namely, that A, B, and C are distinct points that lie on some line ℓ, and there is a coordinate function $f : \ell \to \mathbb{R}$ such that either $f(A) < f(B) < f(C)$ or $f(A) > f(B) > f(C)$.   □

The most important fact about betweenness is expressed in the next theorem.

Theorem 3.8 (Betweenness Theorem for Points). *Suppose A, B, and C are points. If $A * B * C$, then $AB + BC = AC$.*

Proof. Assume that A, B, and C are points such that $A * B * C$. This means that A, B, C all lie on some line ℓ and there is some coordinate function $f : \ell \to \mathbb{R}$ such that either $f(A) < f(B) < f(C)$ or $f(A) > f(B) > f(C)$. After interchanging the names of A and C if necessary, we may assume that $f(A) < f(B) < f(C)$. (This is justified because the statement of the theorem means the same thing after A and C are interchanged, by virtue of Theorems 3.6(a) and 3.7.) Then because the absolute value of a positive number is the number itself, we have the following relationships:

$$AB = |f(B) - f(A)| = f(B) - f(A);$$
$$BC = |f(C) - f(B)| = f(C) - f(B);$$
$$AC = |f(C) - f(A)| = f(C) - f(A).$$

Adding the first two equations and subtracting the third, we find that all of the terms on the right-hand side cancel. Thus $AB + BC - AC = 0$, which is equivalent to the conclusion of the theorem.   □

The next theorem was one of the axioms in Hilbert's original axiomatic system for Euclidean geometry. (In later revisions, he found that he could get by with a somewhat weaker axiom; see Axiom II.3 in Appendix A.)

Theorem 3.9 (Hilbert's Betweenness Axiom). *Given three distinct collinear points, exactly one of them lies between the other two.*

Proof. Suppose A, B, and C are distinct collinear points. Let ℓ be the line containing them, and let $f : \ell \to \mathbb{R}$ be a coordinate function for ℓ. Since f is bijective, it follows that $f(A)$, $f(B)$, and $f(C)$ are three distinct real numbers. By algebra, exactly one of these numbers lies between the other two; without loss of generality, let us say that $f(A) * f(B) * f(C)$. (If this is not the case, we can just rename the points so that it is.) Therefore, by definition of betweenness, B is between A and C. Thus we have shown that *at least one* of the three points is between the other two.

Now we need to prove that only one point can be between the other two. Suppose for the sake of contradiction that, say, $A * B * C$ and $A * C * B$ both hold (which could be the case, for example, if some other coordinate function f_1 had the property that $f_1(A) * f_1(C) * f_1(B)$.) Then the betweenness theorem implies

$$AB + BC = AC,$$
$$AC + CB = AB. \tag{3.2}$$

Adding these two equations together, we obtain $AB + AC + 2BC = AB + AC$, from which it follows algebraically that $BC = 0$. On the other hand, since we are assuming B and C are distinct, Theorem 3.6(c) implies that $BC > 0$, which is a contradiction. $\quad\square$

We can now resolve the awkward question raised by our definition of betweenness: whether it is possible to have two coordinate functions f_1 and f_2 for the same line such that $f_1(B)$ is between $f_1(A)$ and $f_1(C)$, while, say, $f_2(A)$ is between $f_2(B)$ and $f_2(C)$. The next lemma shows that no such anomaly is possible.

Corollary 3.10 (Consistency of Betweenness of Points). *Suppose A, B, C are three points on a line ℓ. Then $A * B * C$ if and only if $f(A) * f(B) * f(C)$ for every coordinate function $f : \ell \to \mathbb{R}$.*

Proof. On the one hand, if $f(A) * f(B) * f(C)$ for every coordinate function f, then $A * B * C$ by definition. Conversely, assume that $A * B * C$ and suppose f is any coordinate function for ℓ. The assumption $A * B * C$ means, in part, that A, B, and C are three distinct points, and therefore $f(A)$, $f(B)$, and $f(C)$ are three distinct real numbers. Therefore, one of these numbers is between the other two. If, say, $f(A)$ were between $f(B)$ and $f(C)$, then by definition we would also have $B * A * C$, which contradicts Hilbert's betweenness axiom. The same reasoning shows that $f(C)$ cannot be between $f(A)$ and $f(B)$, so the only remaining possibility is that $f(B)$ is between $f(A)$ and $f(C)$. $\quad\square$

Theorem 3.11 (Partial Converse to the Betweenness Theorem for Points). *If A, B, and C are three distinct collinear points such that $AB + BC = AC$, then $A * B * C$.*

Proof. Suppose A, B, and C are distinct collinear points such that $AB + BC = AC$. Exactly one of the three points lies between the other two. If C were between the other two, then the betweenness of points theorem would imply $AC + CB = AB$. Therefore, both equations in (3.2) hold, and the same argument as in the proof of Theorem 3.9 leads to a contradiction. Similarly, assuming that A is between the other two leads to a contradiction, so the only other possibility is that $A * B * C$. $\square$

Because of this theorem, some authors choose to define betweenness differently: one can define $A * B * C$ to mean that A, B, C are distinct collinear points such that $AB + BC = AC$, and the previous theorem shows that this definition is equivalent to the one we gave. The definition in terms of distances has the advantage that, unlike our definition, it obviously does not depend on any choice of coordinate function for a line. We have chosen the definition we did because it is much more closely connected to our intuitive understanding of betweenness and because it is a definition commonly used in high-school geometry texts.

The previous theorem is only a *partial* converse to the betweenness theorem because it applies only when we know in advance that the three points are collinear. In fact, a stronger version of the converse is true in neutral geometry as well: for *any* three distinct points A, B, C, it is the case that $AB + BC = AC$ if and only if $A * B * C$. The proof will have to wait until later (see Corollary 5.20).

The next theorem gives a useful characterization of the points on a line in terms of betweenness. It says, roughly speaking, that there are five possible locations for a point on the line $\overleftrightarrow{AB}$ (see Fig. 3.3).

Fig. 3.3. The possible locations for a point on $\overleftrightarrow{AB}$.

Theorem 3.12. *Suppose A and B are distinct points. Then*

$$\overleftrightarrow{AB} = \{P : P * A * B \text{ or } P = A \text{ or } A * P * B \text{ or } P = B \text{ or } A * B * P\}. \qquad (3.3)$$

Proof. For brevity, let S denote the set of points on the right-hand side of (3.3). By definition of S, we have

$$P \in S \iff P * A * B \text{ or } P = A \text{ or } A * P * B \text{ or } P = B \text{ or } A * B * P. \qquad (3.4)$$

The theorem statement says that the two sets $\overleftrightarrow{AB}$ and S are equal. Thus we have to prove that each set is contained in the other (see Theorem G.1 in Appendix G and the discussion preceding it).

First we show that $\overleftrightarrow{AB} \subseteq S$. Let P be an arbitrary point of $\overleftrightarrow{AB}$. Then either P is equal to one of the points A or B or it is not. If $P = A$ or $P = B$, then $P \in S$ by (3.4). On the other hand, if P is not equal to A or B, then the three points A, B, and P are distinct (because we are also assuming $A \neq B$), so Hilbert's betweenness axiom guarantees that one of the following holds: $P * A * B$, $A * P * B$, or $A * B * P$. In each case, (3.4) implies again that $P \in S$.

To show the reverse inclusion, assume $P \in S$. This means that it satisfies one of the five conditions in (3.4). If $P = A$ or $P = B$, then $P \in \overleftrightarrow{AB}$ by definition of $\overleftrightarrow{AB}$. On the other hand, if P satisfies one of the three betweenness relations, then P, A, and B are collinear by definition of betweenness. This means there is a line ℓ containing P, A, and B, and Theorem 2.28 guarantees that $\ell = \overleftrightarrow{AB}$. This implies once again that $P \in \overleftrightarrow{AB}$. $\square$

The definition of betweenness applies only to *three* distinct points; but in some circumstances it is useful to extend it to larger sets of points. If $A_1, A_2, \ldots, A_k$ are k distinct points that all lie on a line ℓ, we write $A_1 * A_2 * \cdots * A_k$ to mean that there is some coordinate function $f: \ell \to \mathbb{R}$ such that either $f(A_1) < f(A_2) < \cdots < f(A_k)$ or $f(A_1) > f(A_2) > \cdots > f(A_k)$.

The next lemma is an easy analogue of Theorem 3.7.

Lemma 3.13. *If $A_1, A_2, \ldots, A_k$ are distinct collinear points, then $A_1 * A_2 * \cdots * A_k$ if and only if $A_k * \cdots * A_2 * A_1$.*

Proof. When translated into statements about coordinates, both statements mean the same thing. $\square$

Theorem 3.14. *Given any k distinct collinear points, they can be labeled $A_1, \ldots, A_k$ in some order such that $A_1 * A_2 * \cdots * A_k$.*

Proof. Let ℓ be a line containing all k points, and let $f: \ell \to \mathbb{R}$ be a coordinate function for ℓ. The coordinates of the given points are k distinct real numbers, so they can be arranged in a sequence from smallest to largest. Then we can just label the given points as $A_1, \ldots, A_k$, with the labels chosen so that $f(A_1) < f(A_2) < \cdots < f(A_k)$. $\square$

Theorem 3.15. *Suppose A, B, C are points such that $A * B * C$. If P is any point on $\overleftrightarrow{AB}$, then one and only one of the following relations holds:*

$$P = A, \quad P = B, \quad P = C,$$
$$P * A * B * C, \quad A * P * B * C, \quad A * B * P * C, \quad A * B * C * P. \tag{3.5}$$

Proof. The ruler placement theorem implies that there is a coordinate function $f : \overleftrightarrow{AB} \to \mathbb{R}$ such that $f(A) < f(B)$. Define real numbers a, b, c, p by $a = f(A)$, $b = f(B)$, $c = f(C)$, and $p = f(P)$. The hypothesis $A * B * C$ implies $a < b < c$ (because the other inequality $a > b > c$ would contradict our choice of f). Because p is a real number, it follows that one and only one of the following arrangements must hold: $p = a$, $p = b$, $p = c$, $p < a < b < c$, $a < p < b < c$, $a < b < p < c$, $a < b < c < p$. Each of these corresponds to one and only one of the relations in (3.5). $\square$

Theorem 3.16. *Suppose A, B, C, D are four distinct points. If any of the following pairs of conditions holds, then $A * B * C * D$:*

$$A * B * C \text{ and } B * C * D \qquad or \tag{3.6}$$
$$A * B * C \text{ and } A * C * D \qquad or \tag{3.7}$$
$$A * B * D \text{ and } B * C * D. \tag{3.8}$$

On the other hand, if $A * B * C * D$, then all of the following conditions are true:

$$A * B * C, \quad A * B * D, \quad A * C * D, \quad and \quad B * C * D. \tag{3.9}$$

Proof. Suppose A, B, C, D are distinct points and one of the hypotheses (3.6)–(3.8) holds. In each case, it follows from the first of the two hypotheses and the definition of betweenness that three of the four points are collinear, and then it follows from the second hypothesis that the fourth point also lies on the same line. Let ℓ be a line containing all four points, and let $f: \ell \to \mathbb{R}$ be a coordinate function such that $f(A) < f(B)$. Write $a = f(A)$, $b = f(B)$, $c = f(C)$, and $d = f(D)$.

Assume first that (3.6) holds. Because $A * B * C$ and $a < b$, it follows that $a < b < c$. The assumption $B * C * D$ implies that either $b < c < d$ or $b > c > d$ holds, but only the first is compatible with $a < b < c$. Thus we have $a < b < c < d$, which implies $A * B * C * D$.

Next, assume (3.7) holds. As before, this implies $a < b < c$ and $a < c < d$, which together imply $a < b < c < d$ and therefore $A * B * C * D$. The argument for (3.8) is similar.

To prove the second statement, just note that the assumption $A * B * C * D$ means that there is some coordinate function $f : \overleftrightarrow{AB} \to \mathbb{R}$ such that $f(A) < f(B) < f(C) < f(D)$ (or the other way around), which implies immediately that any three of the four points satisfy the conditions in (3.9). □

Segments

Suppose A and B are two distinct points. We define the **segment from A to B**, denoted by $\overline{AB}$, to be the following set of points:

$$\overline{AB} = \{ P : P = A \text{ or } P = B \text{ or } A * P * B \}.$$

A segment is also sometimes called a **line segment**. In words, the segment $\overline{AB}$ is the set comprising A and B and all points between them. It is obvious from the definition and the symmetry of betweenness that $\overline{AB} = \overline{BA}$. We define the **length of $\overline{AB}$** to be the distance AB. We say two segments $\overline{AB}$ and $\overline{CD}$ are **congruent**, symbolized by $\overline{AB} \cong \overline{CD}$, if they have the same length:

$$\overline{AB} \cong \overline{CD} \qquad \Leftrightarrow \qquad AB = CD.$$

We can extend the definition of parallelism to segments, but it is a bit more complicated than the definition for lines. If $\overline{AB}$ and $\overline{CD}$ are segments, we say that **$\overline{AB}$ and $\overline{CD}$ are parallel** if the lines containing them are parallel: $\overline{AB} \parallel \overline{CD}$ means that $\overleftrightarrow{AB} \parallel \overleftrightarrow{CD}$. Similarly, to say that a segment $\overline{AB}$ is parallel to a line ℓ is to say that $\overleftrightarrow{AB} \parallel \ell$. Note that $\overline{AB} \parallel \overline{CD}$ is a more restrictive condition than simply saying that the two segments do not intersect—for example, two segments can be disjoint without being parallel (see Fig. 3.4).

Parallel   Nonparallel

Fig. 3.4. Parallel and nonparallel segments.

The definition of a segment $\overline{AB}$ includes the stipulation that A and B are distinct points, so every segment has positive length. The statement "$\overline{AB}$ is a segment" should be understood as including the assertion that A and B are distinct, so it is short for "A and B are distinct points and $\overline{AB}$ is the segment from A to B." If A and B are points that have already been introduced into the discussion, it is not meaningful to refer to $\overline{AB}$ until we have verified that A and B are distinct.

The next lemma expresses an important relationship between segments and lines.

Lemma 3.17. *If A and B are two distinct points, then $\overline{AB} \subseteq \overleftrightarrow{AB}$.*

Proof. Let A and B be distinct points. Suppose P is any point on $\overline{AB}$. Then by definition of a segment, there are three cases: $P = A$, $P = B$, or $A * P * B$. In each of these cases, $P \in \overleftrightarrow{AB}$ by Theorem 3.12. □

The next theorem is our analogue of Euclid's second postulate, which asserts that it is possible "to produce a finite straight line continuously in a straight line." Although Euclid did not say precisely what he meant by "produce," the way he actually uses the postulate is by extending a segment to form a longer segment and choosing an arbitrary point on the extended part. Our theorem makes this possibility explicit.

Theorem 3.18 (Segment Extension Theorem). *If $\overline{AB}$ is any segment, there exist points $C, D \in \overleftrightarrow{AB}$ such that $C * A * B$ and $A * B * D$.*

Proof. Given a segment $\overline{AB}$, let $f : \overleftrightarrow{AB} \to \mathbb{R}$ be a coordinate function such that $f(A) = 0$ and $f(B) > 0$, and let c and d be real numbers such that $c < 0$ and $d > f(B)$. Because f is surjective, there are points $C, D \in \overleftrightarrow{AB}$ such that $f(C) = c$ and $f(D) = d$. Because $f(C) < f(A) < f(B) < f(D)$ by construction, it follows that $C * A * B$ and $A * B * D$. □

If $\overline{AB}$ is a segment, we define the ***endpoints of $\overline{AB}$*** to be the points A and B, and the ***interior points of $\overline{AB}$*** to be all points in $\overline{AB}$ other than A and B. The set of all interior points of $\overline{AB}$ is denoted by $\text{Int}\,\overline{AB}$. (Thus $P \in \text{Int}\,\overline{AB}$ if and only if $A * P * B$.)

This definition seems straightforward, but there is a subtlety to it that might not be immediately evident. A segment is just a set of points in the plane, and there might be different ways to express the same set of points that could lead to a different pair of points being considered as the endpoints. To see why this is an issue, consider the analogous case of lines: if we tried to define the "special points" of $\overleftrightarrow{AB}$ to be the points A and B, this definition would not make sense, because someone else might designate the same line as $\overleftrightarrow{CD}$ for two completely different points C, D on the same line. Thus if we want to talk about "the endpoints" of a segment, we need to show that there is a way to characterize them that doesn't depend on the notation we choose to use for the segment.

To this end, suppose S is a set of points in the plane. A point $P \in S$ is called a ***passing point of S*** if there exist distinct points $X, Y \in S$ such that $X * P * Y$; and P is called an ***extreme point of S*** if it is not a passing point.

Lemma 3.19. *Suppose S and T are sets of points in the plane with $S \subseteq T$. Every passing point of S is also a passing point of T.*

Proof. Assume P is a passing point of S. Then there are distinct points $X, Y \in S$ such that $X * P * Y$. Since X and Y are also points of T, this shows that P is also a passing point of T. □

The next proof is rated PG. We won't use this theorem or its corollary in any serious way in the book, but we include it here to justify referring to "the endpoints" of a segment.

Theorem 3.20. *Suppose A and B are distinct points. Then A and B are extreme points of $\overline{AB}$, and every other point of $\overline{AB}$ is a passing point.*

Proof. First we show that A is an extreme point. Suppose for the sake of contradiction that it is a passing point. Then there exist distinct points $X, Y \in \overline{AB}$ such that $X * A * Y$. Then it follows from Theorem 3.15 (together with the fact that $A \neq B$) that B must satisfy one of the following conditions: $B * X * A * Y$, $B = X$, $X * B * A * Y$, $X * A * B * Y$, $B = Y$, or $X * A * Y * B$. The first three conditions imply $B * A * Y$, which contradicts the fact that $Y \in \overline{AB}$; and the last three imply $X * A * B$, which contradicts $X \in \overline{AB}$. Thus A must be an extreme point. A similar argument shows that B is also an extreme point.

Now suppose P is a point of $\overline{AB}$ other than A or B. Then by definition $A * P * B$, so P is a passing point. □

Thanks to Theorem 3.20, the endpoints of a segment are exactly its extreme points, and these depend only on the segment, considered as a set of points in the plane, not on the particular points used to define it. The next corollary makes this precise.

Corollary 3.21 (Consistency of Endpoints of Segments). *Suppose A and B are distinct points and C and D are distinct points, such that $\overline{AB} = \overline{CD}$. Then either $A = C$ and $B = D$, or $A = D$ and $B = C$.*

Proof. The preceding theorem shows that the set of extreme points of $\overline{AB}$ is equal to $\{A, B\}$, and because $\overline{AB} = \overline{CD}$, this set is also equal to $\{C, D\}$. Thus $\{A, B\} = \{C, D\}$, and the result follows. □

Recall that in the *Elements*, Euclid stated five properties of magnitudes that he called "common notions," which he used when comparing line segments, angles, and areas (see Chapter 1). The next theorem gives modern analogues of all five of Euclid's common notions as they apply to segments, in the order in which Euclid stated them. (Later we will see how the common notions also apply to angle measures.)

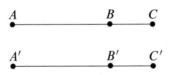

Fig. 3.5. The segment addition and subtraction theorems.

Theorem 3.22 (Euclid's Common Notions for Segments).

(a) **Transitive Property of Congruence:** *Two segments that are both congruent to a third segment are congruent to each other.*

(b) **Segment Addition Theorem:** *Suppose A, B, C, A', B', C' are points such that $A * B * C$ and $A' * B' * C'$ (see Fig. 3.5). If $\overline{AB} \cong \overline{A'B'}$ and $\overline{BC} \cong \overline{B'C'}$, then $\overline{AC} \cong \overline{A'C'}$.*

(c) **Segment Subtraction Theorem:** *Suppose A, B, C, A', B', C' are points such that $A * B * C$ and $A' * B' * C'$. If $\overline{AC} \cong \overline{A'C'}$ and $\overline{AB} \cong \overline{A'B'}$, then $\overline{BC} \cong \overline{B'C'}$.*

(d) **Reflexive Property of Congruence:** *Every segment is congruent to itself.*

(e) **The Whole Segment Is Greater Than the Part:** *If A, B, and C are points such that $A * B * C$, then $AC > AB$.*

Proof. Statements (a) and (d) are immediate consequences of the definition of congruence together with familiar properties of equality. To prove (b), just note that under the given hypotheses, the betweenness theorem for points and the definition of congruence imply

$$AC = AB + BC = A'B' + B'C' = A'C'.$$

The proof of (c) is similar.

Finally, to prove (e), assume that A, B, C are points such that $A * B * C$. By the betweenness theorem for points and algebra, we have $AC - AB = BC$, and by Theorem 3.6, $BC > 0$. By substitution, therefore, $AC - AB > 0$, which implies $AC > AB$. □

The next lemma relates the definition of a segment to coordinate functions.

Lemma 3.23 (Coordinate Representation of a Segment). *Suppose A and B are distinct points and $f : \overleftrightarrow{AB} \to \mathbb{R}$ is a coordinate function for $\overleftrightarrow{AB}$. Then*

$$\overline{AB} = \{P \in \overleftrightarrow{AB} : f(A) \le f(P) \le f(B)\} \qquad \text{if } f(A) < f(B); \qquad (3.10)$$

$$\overline{AB} = \{P \in \overleftrightarrow{AB} : f(A) \ge f(P) \ge f(B)\} \qquad \text{if } f(A) > f(B). \qquad (3.11)$$

Proof. Let A and B be distinct points, and let $f : \overleftrightarrow{AB} \to \mathbb{R}$ be a coordinate function for $\overleftrightarrow{AB}$. First we prove (3.10). For this purpose, assume that $f(A) < f(B)$. To prove that the two given sets are equal, we have to show that each one is contained in the other. We will prove first that $\overline{AB} \subseteq \{P \in \overleftrightarrow{AB} : f(A) \le f(P) \le f(B)\}$, and then we will prove the opposite inclusion.

Let P be a point in $\overline{AB}$. By Lemma 3.17, this implies that $P \in \overleftrightarrow{AB}$. By definition of a segment, the hypothesis means that $P = A$, $P = B$, or $A * P * B$. We consider each case in turn. Assuming $P = A$, the fact that f is a well-defined function implies $f(P) = f(A)$. Together with our hypothesis that $f(A) < f(B)$, this implies $f(A) \le f(P) \le f(B)$. Similarly, $P = B$ implies $f(P) = f(B)$, which in turn implies $f(A) \le f(P) \le f(B)$. Finally, if $A * P * B$, then the consistency theorem for betweenness implies that $f(A) * f(P) * f(B)$, which means either $f(A) < f(P) < f(B)$ or $f(A) > f(P) > f(B)$. Since we have assumed that $f(A) < f(B)$, only the first set of inequalities can occur, and again they imply $f(A) \le f(P) \le f(B)$. This completes the proof that $\overline{AB} \subseteq \{P \in \overleftrightarrow{AB} : f(A) \le f(P) \le f(B)\}$.

Conversely, let P be a point in the right-hand set of (3.10); this means that P lies on $\overleftrightarrow{AB}$ and

$$f(A) \le f(P) \le f(B). \qquad (3.12)$$

There are three possibilities: $f(P) = f(A)$, $f(P) = f(B)$, or $f(P)$ is distinct from both $f(A)$ and $f(B)$. In the first case, injectivity of f implies $P = A$, so certainly $P \in \overline{AB}$. Similarly, if $f(P) = f(B)$, then $P = B$, and again we conclude $P \in \overline{AB}$. Finally, if $f(P)$ is distinct from $f(A)$ and $f(B)$, then injectivity of f implies $f(P) \ne f(A)$ and $f(P) \ne f(B)$. Thus the inequality (3.12) actually implies $f(A) < f(P) < f(B)$, which in turn implies $A * P * B$. Once again, we conclude that $P \in \overline{AB}$. This proves (3.10). The proof of (3.11) is identical, but with the inequalities reversed. □

Theorem 3.24. *If A, B, C are points such that $A * B * C$, then the following set equalities hold:*

(a) $\overline{AB} \cup \overline{BC} = \overline{AC}$.

(b) $\overline{AB} \cap \overline{BC} = \{B\}$.

Proof. Let A, B, C be points such that $A * B * C$, which implies in particular that they are collinear. Let ℓ be the line containing A, B, and C, and let $f : \ell \to \mathbb{R}$ be a coordinate function for ℓ such that $f(A) = 0$ and $f(C) > 0$. (The existence of such a function is guaranteed by the ruler placement theorem.) Then the fact that $A * B * C$ means that $f(A) < f(B) < f(C)$.

First, we prove (a). This is an assertion that two sets are equal, so we need to prove two things: every point in the left-hand set is also in the right-hand set, and vice versa.

Begin by assuming that $P \in \overline{AB} \cup \overline{BC}$. This means that either $P \in \overline{AB}$ or $P \in \overline{BC}$, and in either case $P \in \ell$. In the case that $P \in \overline{AB}$, Lemma 3.23 shows that $f(A) \le f(P) \le f(B)$, and then by transitivity of inequalities it follows that $f(P) < f(C)$. Thus $f(A) \le f(P) < f(C)$, which implies $P \in \overline{AC}$. A similar argument shows that $P \in \overline{BC}$ implies $P \in \overline{AC}$. Thus we have shown that $\overline{AB} \cup \overline{BC} \subseteq \overline{AC}$.

Conversely, assume that $P \in \overline{AC}$. Then $P \in \ell$, and Lemma 3.23 shows that $f(A) \le f(P) \le f(C)$. Now by the trichotomy law for real numbers, there are three cases: $f(P) < f(B)$, $f(P) = f(B)$, or $f(P) > f(B)$. In the first two cases, we have $f(A) \le f(P) \le f(B)$ and thus $P \in \overline{AB}$. In the third case, $f(B) < f(P) \le f(C)$, which implies $P \in \overline{BC}$. Thus $\overline{AC} \subseteq \overline{AB} \cup \overline{BC}$, and we are done.

The proof of (b) is left to you (see Exercise 3E). $\qquad\square$

Corollary 3.25. *If $A * B * C$, then $\overline{AB} \subseteq \overline{AC}$ and $\overline{BC} \subseteq \overline{AC}$.*

Proof. This follows immediately from the preceding theorem and the basic set-theoretic facts that $X \subseteq X \cup Y$ and $Y \subseteq X \cup Y$ for any sets X and Y. $\qquad\square$

Midpoints

If P, A, and B are points, we say that P is **equidistant from A and B** if $PA = PB$. Suppose $\overline{AB}$ is a segment. A point M is said to be a **midpoint of $\overline{AB}$** if M is an interior point of $\overline{AB}$ that is equidistant from A and B. If M is a midpoint of $\overline{AB}$, we often say that **M bisects $\overline{AB}$**, and we also say that a segment or line bisects $\overline{AB}$ if it intersects $\overline{AB}$ only at M.

The next lemma gives some useful alternative characterizations of midpoints.

Lemma 3.26. *Let $\overline{AB}$ be a segment, and let M be a point. The following statements are all equivalent to each other:*

(a) *M is a midpoint of $\overline{AB}$ (i.e., $M \in \operatorname{Int} \overline{AB}$ and $MA = MB$).*

(b) *$M \in \overleftrightarrow{AB}$ and $MA = MB$.*

(c) *$M \in \overline{AB}$ and $AM = \frac{1}{2} AB$.*

Proof. We need to prove the equivalences (a) $\Leftrightarrow$ (b), (b) $\Leftrightarrow$ (c), and (c) $\Leftrightarrow$ (a), which amounts to six implications in all. In situations like this, it is not necessary to prove all six implications, as long as we prove enough to allow us to conclude that any one of the statements implies each of the others. In this case, we will prove (a) $\Rightarrow$ (b) $\Rightarrow$ (c) $\Rightarrow$ (a). (Do you see why this suffices?)

First, we prove (a) $\Rightarrow$ (b). Assume M is a midpoint of $\overline{AB}$. Then by definition $MA = MB$, and $M \in \overline{AB} \subseteq \overleftrightarrow{AB}$, so (b) holds.

Next we prove (b) $\Rightarrow$ (c). Suppose M is a point on $\overleftrightarrow{AB}$ such that $MA = MB$. By Theorem 3.12, one of the following five things must be true: $M * A * B$, $M = A$, $A * M * B$, $M = B$, or $A * B * M$. If $M * A * B$, then the theorem on the whole segment greater than the part implies that $MB > MA$, which contradicts the hypothesis; similarly $A * B * M$ implies $AM > BM$, which is also a contradiction. In each of the remaining cases, $M \in \overline{AB}$, which is one of the conclusions we need to prove. Now if $M = A$ or $M = B$, then one of the distances MA or MB is zero and the other is positive, again contradicting the hypothesis, so the only remaining possibility is $A * M * B$. The betweenness theorem then implies $AM + MB = AB$. Substituting AM for MB in this equation, we obtain $2AM = AB$, which implies $AM = \frac{1}{2}AB$.

Finally, we prove (c) $\Rightarrow$ (a). Suppose M is a point in $\overline{AB}$ such that $AM = \frac{1}{2}AB$. Because $\overline{AB}$ is a segment and thus $AB > 0$, our assumption implies $0 < AM < AB$. If $M = A$, then $AM = 0$, which is a contradiction, and if $M = B$, then $AM = AB$, which is also a contradiction; so M is an interior point of $\overline{AB}$. Then the betweenness theorem yields

$$AM + MB = AB = 2AM.$$

Subtracting AM from both sides, we conclude that $MB = AM$, and therefore M is a midpoint of $\overline{AB}$. $\qquad\square$

In his Proposition I.10, Euclid proved that every segment has a midpoint by showing that it can be constructed using compass and straightedge. For us, the existence and uniqueness of midpoints can be proved more straightforwardly using the ruler postulate.

Theorem 3.27 (Existence and Uniqueness of Midpoints). *Every segment has a unique midpoint.*

Proof. Let $\overline{AB}$ be a segment. We must prove two things: existence ($\overline{AB}$ has *at least one* midpoint) and uniqueness ($\overline{AB}$ has *at most one* midpoint).

To prove existence, let $f : \overleftrightarrow{AB} \to \mathbb{R}$ be a coordinate function for $\overleftrightarrow{AB}$ such that $f(A) = 0$ and $f(B) > 0$; such a function exists by the ruler placement theorem. If we write $b = f(B)$, then the ruler postulate implies that

$$AB = |f(B) - f(A)| = |b - 0| = b.$$

Because f is bijective, there is a unique point $M \in \overleftrightarrow{AB}$ such that $f(M) = \frac{1}{2}b$. We will show that M is a midpoint of $\overline{AB}$. First, because $0 < \frac{1}{2}b < b$, Lemma 3.23 shows that M lies on $\overline{AB}$. Next, the ruler postulate and a little algebra show that

$$AM = |f(M) - f(A)| = \left|\tfrac{1}{2}b - 0\right| = \tfrac{1}{2}b = \tfrac{1}{2}AB.$$

Therefore, it follows from Lemma 3.26 that M is a midpoint of $\overline{AB}$.

To prove uniqueness, suppose that M' is another midpoint of $\overline{AB}$. By Lemma 3.26,

$$AM' = \tfrac{1}{2}AB = \tfrac{1}{2}b. \tag{3.13}$$

Using the same coordinate function f as above, let $m' = f(M')$. Because M' lies on $\overline{AB}$ (by definition of a midpoint), we have $0 \le m' \le b$ by Lemma 3.26, so the ruler postulate gives

$$AM' = \left| f(M') - f(A) \right| = \left| m' - 0 \right| = m'. \tag{3.14}$$

Comparing (3.13) and (3.14), we see that $f(M') = m' = \frac{1}{2}b = f(M)$. Because f is injective, it follows that $M' = M$, and uniqueness is proved. $\qquad\square$

You might wonder why we needed the last paragraph of this proof—didn't we already show in the second paragraph that M is unique when we wrote "there is a *unique* point $M \in \overleftrightarrow{AB}$ such that $f(M) = \frac{1}{2}b$"? The answer is that, although the ruler postulate guarantees that there is a unique point M such that $f(M) = \frac{1}{2}b$, this is *not* precisely the same as saying there is a unique midpoint M, for the simple reason that neither the definition of midpoint nor Lemma 3.26 says anything whatsoever about what $f(M)$ is supposed to be. In order to turn this idea into a proof that M is the only midpoint, we have to show that *every* midpoint must satisfy $f(M) = \frac{1}{2}b$ for this coordinate function, which is essentially what we did in the last paragraph. In general, when you are asked to prove existence and uniqueness of something, the safest course is usually to write two separate proofs, one for existence and one for uniqueness.

It is also important to understand why we need to prove the *existence* of the midpoint of $\overline{AB}$, when we have not bothered to prove the existence of other geometric objects such as the segment $\overline{AB}$ itself. The reason is that just defining an object does not, in itself, ensure that such an object exists. For example, we might have defined "the closest point to A in the interior of $\overline{AB}$" to be the point in $\operatorname{Int}\overline{AB}$ whose distance to A is smaller than that of every other point in the interior of $\overline{AB}$. This is a perfectly unambiguous definition, but unfortunately there is no such point, just as there is no "smallest positive real number." If we wish to talk about "the midpoint" of a segment, we have to prove that it exists.

On the other hand, when we define a *set of points*, as long as the definition stipulates unambiguously what it means for a point to be in that set, then the set exists. Thus the definition of the segment $\overline{AB}$ needs no further elaboration, because it yields a perfectly well-defined set. Of course, if we are not careful, we might inadvertently define a set only to discover that it is empty. Thus in order to be sure that the definition of a segment describes something interesting, we might wish to verify that segments contain plenty of points. The next theorem guarantees that they do.

Theorem 3.28. *Every segment contains infinitely many distinct points.*

Proof. Let $\overline{AB}$ be a segment, and let $f \colon \overleftrightarrow{AB} \to \mathbb{R}$ be a coordinate function such that $f(A) = 0$ and $f(B) > 0$. Let $b = f(B)$. For each positive integer n, there is a point $X_n \in \overleftrightarrow{AB}$ such that $f(X_n) = b/n$. Because f is a well-defined function, the points $\{X_1, X_2, \ldots\}$ are all distinct; and because $0 < b/n \le b$, Lemma 3.23 shows that $X_n \in \overline{AB}$ for each n. $\qquad\square$

Since Euclid gave circles a prominent place in his postulates, we might ask similar questions about them. We define a ***circle*** to be a set consisting of all points whose distance from a given point is equal to a given positive number. The given point is called the ***center of the circle***, and the given distance is called its ***radius***. If O is any point and r is any positive number, the circle with center O and radius r is denoted by $\mathcal{C}(O,r)$; thus

$$\mathcal{C}(O,r) = \{P : OP = r\}. \tag{3.15}$$

Euclid's Postulate 3 asserted that it is possible to describe a circle with any center and any radius. Here is our version of that postulate.

Theorem 3.29 (Euclid's Postulate 3). *Given two distinct points O and A, there exists a circle whose center is O and whose radius is OA.*

Proof. Let O and A be distinct points, and let $r = OA$. Since O and A are distinct, r is a positive number, so $\mathscr{C}(O,r)$ is a well-defined set described by (3.15). $\square$

Exercise 3H shows that every circle, like every segment, contains infinitely many points.

Rays

Given distinct points A and B, we define the ***ray from A through B*** (Fig. 3.6), denoted by $\overrightarrow{AB}$, to be the set

$$\overrightarrow{AB} = \{P : P = A \text{ or } A * P * B \text{ or } P = B \text{ or } A * B * P\}.$$

Intuitively, this is the point A together with all of the points in the same half of $\overleftrightarrow{AB}$ as

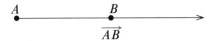

$$\overrightarrow{AB}$$

Fig. 3.6. A ray.

B. Sometimes for brevity we denote a ray by a lowercase letter with an arrow, such as $\vec{a}, \vec{b}, \vec{c}, \vec{r}, \vec{s}, \vec{t}$. In this case, we use the notation $\overleftrightarrow{r}$ to denote the line containing $\vec{r}$. To say "$\vec{r}$ is a ray" is to say that there are two distinct points A and B such that $\vec{r} = \overrightarrow{AB}$. The point A is called the ***endpoint*** or ***starting point*** of $\overrightarrow{AB}$, and any point on the ray other than A is called an ***interior point of*** $\overrightarrow{AB}$. We use the notation $\text{Int}\,\overrightarrow{AB}$ to denote the set of all interior points of $\overrightarrow{AB}$; thus $P \in \text{Int}\,\overrightarrow{AB}$ if and only if $A * P * B$ or $P = B$ or $A * B * P$. Just as with lines and segments, the notation $\overrightarrow{AB}$ has meaning only if A and B are distinct, and a statement such as "$\overrightarrow{AB}$ is a ray" is understood to mean "A and B are distinct points and $\overrightarrow{AB}$ is the ray from A through B."

The endpoint of a ray, like the endpoints of a segment, doesn't depend on what notation we use to denote the ray; we will prove this below (see Corollary 3.41).

Just as we did with segments, we can extend the definition of parallelism to include rays. If $\vec{r}$ is a ray and S is a line or a segment or another ray, to say that $\vec{r}$ ***is parallel to*** S means that the line $\overleftrightarrow{r}$ is parallel to the line containing S.

The next lemma gives a useful criterion for deciding when a point is *not* on a ray.

Lemma 3.30. *Suppose A and B are distinct points and P is a point on the line $\overleftrightarrow{AB}$. Then $P \notin \overrightarrow{AB}$ if and only if $P * A * B$.*

Proof. Let A and B be distinct points and let $P \in \overleftrightarrow{AB}$ be arbitrary. Theorem 3.12 shows that one of the five relations in (3.4) must hold. If $P \notin \overrightarrow{AB}$, then four of the conditions are ruled out by the definition of $\overrightarrow{AB}$, so the only remaining possibility is $P * A * B$.

Conversely, if $P * A * B$, then by definition of betweenness, $P \neq A$ and $P \neq B$, and by Hilbert's betweenness axiom, neither $A * P * B$ nor $A * B * P$ holds; thus $P \notin \overrightarrow{AB}$.   □

The following lemma is a generalization of Lemma 3.17 to rays.

Lemma 3.31. *Suppose A and B are distinct points. Then $\overline{AB} \subseteq \overrightarrow{AB} \subseteq \overleftrightarrow{AB}$* (Fig. 3.7).

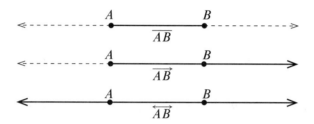

Fig. 3.7. A segment, a ray, and a line.

Proof. First suppose $P \in \overline{AB}$. This means $P = A$, $P = B$, or $A * P * B$, each of which implies $P \in \overrightarrow{AB}$ by definition of a ray. Similarly, if $P \in \overrightarrow{AB}$, then $P = A$, $A * P * B$, $P = B$, or $A * B * P$, each of which implies $P \in \overleftrightarrow{AB}$ by Theorem 3.12.   □

Be sure to observe the distinction between the notations $\overline{AB}$, $\overrightarrow{AB}$, and $\overleftrightarrow{AB}$, which all designate sets of points in the plane, and the notation AB, which designates a real number (the distance from A to B, or the length of the segment $\overline{AB}$).

The next lemma is the analogue for rays of Lemma 3.23.

Lemma 3.32 (Coordinate Representation of a Ray). *Suppose A and B are distinct points and $f : \overleftrightarrow{AB} \to \mathbb{R}$ is a coordinate function for $\overleftrightarrow{AB}$. Then*

$$\overrightarrow{AB} = \{P \in \overleftrightarrow{AB} : f(P) \geq f(A)\} \qquad \text{if } f(A) < f(B); \tag{3.16}$$

$$\overrightarrow{AB} = \{P \in \overleftrightarrow{AB} : f(P) \leq f(A)\} \qquad \text{if } f(A) > f(B). \tag{3.17}$$

Proof. Exercise 3I.   □

The preceding lemma is most useful when used in combination with the ruler placement theorem, which allows us to choose our coordinate function judiciously for a given ray. If A and B are distinct points, let us say that a coordinate function $f : \overleftrightarrow{AB} \to \mathbb{R}$ is *adapted to the ray $\overrightarrow{AB}$* if $f(A) = 0$ and $f(B) > 0$. (If the ray is designated only by a single letter, as in $\vec{a}$, then to say that the coordinate function is adapted to $\vec{a}$ just means that it sends the endpoint of the ray to zero and a point in the interior of the ray to a positive number.) The ruler placement theorem says that for every ray, there is an adapted coordinate function.

Lemma 3.33 (Representation of a Ray in Adapted Coordinates). *Suppose A and B are distinct points and $f : \overleftrightarrow{AB} \to \mathbb{R}$ is a coordinate function adapted to $\overrightarrow{AB}$. If P is any point on $\overleftrightarrow{AB}$, then $P \in \overrightarrow{AB}$ if and only if $f(P) \geq 0$, and $P \in \mathrm{Int}\,\overrightarrow{AB}$ if and only if $f(P) > 0$.*

Proof. Given the hypotheses, it follows from Lemma 3.32 that $P \in \overrightarrow{AB}$ if and only if $f(P) \geq f(A)$, and since $f(A) = 0$, this is equivalent to $f(P) \geq 0$. This proves the first statement.

To prove the second statement, assume first that $P \in \operatorname{Int} \overrightarrow{AB}$. Then $f(P) \geq 0$ by the previous argument, and in addition $P \neq A$, which implies $f(P) \neq f(A)$ because f is injective. It follows that $f(P) > 0$. Conversely, assume that $f(P) > 0$. The argument in the previous paragraph shows that $P \in \overrightarrow{AB}$, and P cannot be equal to A because then $f(P) = f(A) = 0$ would contradict our assumption. Thus $P \in \operatorname{Int} \overrightarrow{AB}$. $\square$

This relationship between coordinate functions and rays allows us to prove some important properties of points on rays. The next lemma is an extremely useful partial converse to Theorem 3.22(e) (the whole segment is greater than the part). Note that the actual converse to that theorem is not true, because $AC > AB$ does not imply $A * B * C$ even if A, B, C are assumed to be collinear (see Fig. 3.8). However, if we assume in addition that B and C are in the "same direction from A" (or, more precisely, that they are interior points of the same ray starting at A), then we can conclude the converse implication (Fig. 3.9).

Fig. 3.8. $AC > AB$ doesn't always imply $A * B * C$ **Fig. 3.9.** ... but it does if B and C are in $\operatorname{Int} \overrightarrow{a}$.

Lemma 3.34 (Ordering Lemma for Points). *Suppose $\overrightarrow{a}$ is a ray starting at a point A, and B and C are interior points of $\overrightarrow{a}$ such that $AC > AB$. Then $A * B * C$.*

Proof. Let $\overrightarrow{a}$ be a ray starting at A, and let B, C be points in $\operatorname{Int} \overrightarrow{a}$ such that $AC > AB$. Let $f : \overrightarrow{a} \to \mathbb{R}$ be a coordinate function adapted to $\overrightarrow{a}$. Then Lemma 3.33 shows that $f(B) > 0$ and $f(C) > 0$. By definition of a coordinate function, therefore,

$$AC = |f(C) - f(A)| = |f(C) - 0| = f(C);$$
$$AB = |f(B) - f(A)| = |f(B) - 0| = f(B).$$

Our hypothesis therefore implies that $f(C) > f(B)$, and because $f(A) = 0$ we have $f(C) > f(B) > f(A)$. Thus $A * B * C$ by definition of betweenness. $\square$

One of the most basic constructions in Euclidean geometry is marking off a segment of a certain predetermined length along a ray. Euclid justifies this with his Proposition I.3, which says that given two segments, it is possible "to cut off from the greater a straight line equal to the less." The next theorem is our analogue of that proposition. There are several differences, however: first, we do not need to assume that we already have a segment of the desired length to use as a model; giving a number that is to be the length of the new segment is sufficient. (Euclid, of course, could not state his proposition this way because he did not use numbers to measure lengths.) Second, whereas Euclid has to start with a segment that is already longer than the model segment, our concept of a ray extending infinitely far in one direction frees us from having to worry about how long the initial segment is. The proof of this theorem is very similar to that of the theorem on existence and uniqueness of midpoints.

Theorem 3.35 (Segment Construction Theorem). *Suppose $\vec{a}$ is a ray starting at a point A and r is a positive real number. Then there exists a unique point C in the interior of $\vec{a}$ such that $AC = r$.*

Proof. Exercise 3J. □

The uniqueness part of this theorem is extremely important in its own right and is often used on its own without the existence part. For this reason, it is worth restating as a separate corollary.

Corollary 3.36 (Unique Point Theorem). *Suppose $\vec{a}$ is a ray starting at a point A, and C and C' are points in Int $\vec{a}$ such that $AC = AC'$. Then $C = C'$.* □

The next corollary is a more literal analogue of Euclid's Proposition I.3.

Corollary 3.37 (Euclid's Segment Cutoff Theorem). *If $\overline{AB}$ and CD are segments with $CD > AB$, there is a unique point E in the interior of $\overline{CD}$ such that $\overline{CE} \cong \overline{AB}$.*

Proof. Exercise 3K. □

Opposite Rays

In the notation $\overrightarrow{AB}$, the point B serves only to indicate which direction along the line $\overleftrightarrow{AB}$ the ray points; if C is a third point on the line $\overleftrightarrow{AB}$, it is possible that $\overrightarrow{AC}$ and $\overrightarrow{AB}$ might be exactly the same ray (Fig. 3.10), or they might be different rays (Fig. 3.11). Two

Fig. 3.10. $\overrightarrow{AB} = \overrightarrow{AC}$.　　　　**Fig. 3.11.** $\overrightarrow{AB}$ and $\overrightarrow{AC}$ are opposite rays.

rays are said to be ***collinear rays*** if there is a line that contains both of them, and they are said to be ***opposite rays*** if they have the same endpoint and their union is a line. The next proof is rated PG, not because the steps are terribly hard, but because it could end up being a lot of work if the parts are proved in a different order from the one given here.

Theorem 3.38 (Rays with the Same Endpoint). *Suppose $\overrightarrow{AB}$ and $\overrightarrow{AC}$ are rays with the same endpoint.*

(a) *If A, B, and C are collinear, then $\overrightarrow{AB}$ and $\overrightarrow{AC}$ are collinear.*

(b) *If $\overrightarrow{AB}$ and $\overrightarrow{AC}$ are collinear, then they are either equal or opposite, but not both.*

(c) *If $\overrightarrow{AB}$ and $\overrightarrow{AC}$ are opposite rays, then $\overrightarrow{AB} \cap \overrightarrow{AC} = \{A\}$ and $\overrightarrow{AB} \cup \overrightarrow{AC} = \overleftrightarrow{AC}$.*

(d) *$\overrightarrow{AB}$ and $\overrightarrow{AC}$ are equal if and only if they have an interior point in common.*

(e) *$\overrightarrow{AB}$ and $\overrightarrow{AC}$ are opposite rays if and only if $C * A * B$.*

Proof. The assumption that $\overrightarrow{AB}$ and $\overrightarrow{AC}$ are rays means, in particular, that neither B nor C is equal to A (but B could be equal to C). Let $f : \overleftrightarrow{AB} \to \mathbb{R}$ be a coordinate function adapted to $\overrightarrow{AB}$. We will use this coordinate function in several of the arguments below.

First we will prove (a). Suppose A, B, and C are collinear, which means there is a line ℓ containing all of them. Since the distinct points A and B both lie on ℓ, it follows from Theorem 2.28 that $\ell = \overleftrightarrow{AB}$; and a similar argument shows that $\ell = \overleftrightarrow{AC}$. Because

$\overrightarrow{AB} \subseteq \overleftrightarrow{AB}$ and $\overrightarrow{AC} \subseteq \overleftrightarrow{AC}$ by Lemma 3.31, it follows that both rays are contained in ℓ and so they are collinear.

Next we will prove (b). Assume $\overrightarrow{AB}$ and $\overrightarrow{AC}$ are collinear. The hypothesis implies that $C \in \overleftrightarrow{AB}$, so we can consider the real number $f(C)$, where f is the coordinate function introduced in the first paragraph. Because $C \neq A$, we see that $f(C) \neq 0$. Thus either $f(C) > 0$ or $f(C) < 0$, but not both. On the one hand, if $f(C) > 0$, Lemma 3.32 shows that

$$\overrightarrow{AC} = \{P \in \overleftrightarrow{AB} : f(P) \geq 0\} = \overrightarrow{AB},$$

so the rays are equal. On the other hand, suppose $f(C) < 0$. We will prove that $\overrightarrow{AB} \cup \overrightarrow{AC} = \overleftrightarrow{AB}$. Lemma 3.32 shows that

$$\overrightarrow{AB} = \{P \in \overleftrightarrow{AB} : f(P) \geq 0\}; \tag{3.18}$$

$$\overrightarrow{AC} = \{P \in \overleftrightarrow{AB} : f(P) \leq 0\}. \tag{3.19}$$

Both $\overrightarrow{AB}$ and $\overrightarrow{AC}$ are subsets of $\overleftrightarrow{AB}$, so $\overrightarrow{AB} \cup \overrightarrow{AC} \subseteq \overleftrightarrow{AB}$. To prove the reverse inclusion, let $P \in \overleftrightarrow{AB}$ be arbitrary. If $f(P) \geq 0$, then (3.18) shows $P \in \overrightarrow{AB}$; while if $f(P) < 0$, then (3.19) shows that $P \in \overrightarrow{AC}$. In either case $P \in \overrightarrow{AB} \cup \overrightarrow{AC}$ as claimed. Thus the union of $\overrightarrow{AB}$ and $\overrightarrow{AC}$ is a line, which means by definition that they are opposite rays.

To prove (c), suppose $\overrightarrow{AB}$ and $\overrightarrow{AC}$ are opposite rays. Arguing as in the preceding paragraph, we conclude that $f(C) < 0$ (since otherwise $\overrightarrow{AB}$ and $\overrightarrow{AC}$ would be equal, not opposite), and therefore Lemma 3.32 shows that $\overrightarrow{AB}$ and $\overrightarrow{AC}$ are described by formulas (3.18) and (3.19), respectively. A point P is in both sets if and only if $f(P) = 0$, which is true if and only if $P = A$; thus $\overrightarrow{AB} \cap \overrightarrow{AC} = \{A\}$. The argument in the preceding paragraph shows that $\overrightarrow{AB} \cup \overrightarrow{AC} = \overleftrightarrow{AC}$.

Next we prove (d). Obviously, if $\overrightarrow{AB}$ and $\overrightarrow{AC}$ are equal, they have lots of interior points in common. Conversely, suppose they have an interior point in common, say P. Since P is not equal to A, it follows that both lines $\overleftrightarrow{AB}$ and $\overleftrightarrow{AC}$ contain the two distinct points P and A, so they must be the same line. Thus $\overrightarrow{AB}$ and $\overrightarrow{AC}$ are collinear, so part (b) shows that they must be equal or opposite. If they were opposite, they would have only A in common by part (c), so in fact they are equal.

Finally, we prove (e). Suppose first that $\overrightarrow{AB}$ and $\overrightarrow{AC}$ are opposite rays. By (c), the only point they have in common is A, so in particular $C \notin \overrightarrow{AB}$. By Lemma 3.30, this implies $C * A * B$. Conversely, if $C * A * B$, then Lemma 3.30 shows that $C \notin \overrightarrow{AB}$, so the two rays cannot be equal and thus are opposite. $\square$

Theorem 3.39 (Opposite Ray Theorem). *Every ray has a unique opposite ray.*

Proof. Let $\overrightarrow{AB}$ be a ray. By the segment extension theorem, there exists a point C such that $C * A * B$. Then $\overrightarrow{AC}$ is opposite to $\overrightarrow{AB}$ by Theorem 3.38(e). This proves existence.

To prove uniqueness, suppose that $\overrightarrow{AC'}$ is another ray that is opposite to $\overrightarrow{AB}$; we will show that $\overrightarrow{AC'}$ is equal to the ray $\overrightarrow{AC}$ we constructed above. Theorem 3.38(e) shows that $C' * A * B$, which means that C' lies on $\overleftrightarrow{AB}$. Note that $\overleftrightarrow{AB} = \overrightarrow{AC} \cup \overrightarrow{AB}$ by Theorem 3.38(c), so C' lies either on $\overrightarrow{AB}$ or on $\overrightarrow{AC}$. Now, the fact that $C' * A * B$ implies that $C' \notin \overrightarrow{AB}$ (by Lemma 3.30), so $C' \in \overrightarrow{AC}$; and since $C' \neq A$, it is an interior point of that ray. Then it follows from Theorem 3.38(d) that $\overrightarrow{AC'} = \overrightarrow{AC}$. $\square$

Now we can prove that the endpoint of a ray is well defined and does not depend on the choice of points used to define the ray. The proof is based on the following theorem.

Theorem 3.40. *Let $\overrightarrow{AB}$ be the ray from A through B. Then A is the only extreme point of $\overrightarrow{AB}$.*

Proof. First we'll show that A is an extreme point. Suppose not. There are points $X, Y \in \overrightarrow{AB}$ such that $X * A * Y$. We will show that $X \notin \overrightarrow{AB}$, which is a contradiction. Since $Y \in \overrightarrow{AB}$ and $Y \neq A$, there are three possible cases: (1) $A * Y * B$, which implies $X * A * Y * B$ and therefore $X * A * B$. (2) $Y = B$, which immediately implies $X * A * B$. (3) $A * B * Y$, which implies $X * A * B * Y$ and therefore $X * A * B$. In each case, we have $X * A * B$, which implies $X \notin \overrightarrow{AB}$, a contradiction.

Next we'll show that every other point of $\overrightarrow{AB}$ is a passing point. If P is any interior point of $\overrightarrow{AB}$, then the segment construction theorem shows that there is a point $Q \in \overrightarrow{AB}$ such that, say, $AQ = AP + 1$, and then the ordering lemma for points shows that $A * P * Q$. Thus P is a passing point. $\qquad\square$

Corollary 3.41 (Consistency of Endpoints of Rays). *If A, B are distinct points and C, D are distinct points such that $\overrightarrow{AB} = \overrightarrow{CD}$, then $A = C$.*

Proof. By Theorem 3.40, each ray has only one extreme point, and that is its endpoint. $\quad\square$

Here is one last theorem about intersections and unions of rays that will be useful.

Theorem 3.42. *Suppose A and B are two distinct points. Then the following set equalities hold:*

(a) $\overrightarrow{AB} \cap \overrightarrow{BA} = \overline{AB}$.
(b) $\overrightarrow{AB} \cup \overrightarrow{BA} = \overleftrightarrow{AB}$.

Proof. Exercise 3L. $\qquad\qquad\qquad\qquad\qquad\qquad\qquad\qquad\qquad\qquad\quad\square$

Plane Separation

We are now ready for our next postulate. This one gives crucial information about how segments and lines intersect.

Postulate 6 (The Plane Separation Postulate). *For any line ℓ, the set of all points not on ℓ is the union of two disjoint subsets called the **sides of** ℓ. If A and B are distinct points not on ℓ, then A and B are on the same side of ℓ if and only if $\overline{AB} \cap \ell = \varnothing$.*

If ℓ is a line and A, B are points not on ℓ, then we say A and B lie on **opposite sides of** ℓ if they do not lie on the same side. Here are some easy consequences of the plane separation postulate.

Theorem 3.43 (Properties of Sides of Lines). *Suppose ℓ is a line.*

(a) *Both sides of ℓ are nonempty sets.*
(b) *If A and B are distinct points not on ℓ, then A and B are on opposite sides of ℓ if and only if $\overline{AB} \cap \ell \neq \varnothing$.*

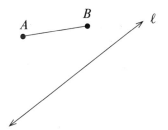

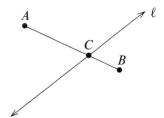

Fig. 3.12. *A* and *B* are on the same side of ℓ.　　**Fig. 3.13.** *A* and *B* are on opposite sides of ℓ.

Proof. We prove (b) first. Given two distinct points A, B that do not lie on ℓ, the plane separation postulate yields two implications: if A and B are on the same side of ℓ, then $\overline{AB} \cap \ell = \varnothing$, and if $\overline{AB} \cap \ell = \varnothing$, then A and B are on the same side. Taking the contrapositives of these two implications shows that $\overline{AB} \cap \ell \neq \varnothing$ if and only if A and B lie on opposite sides of ℓ.

Now we prove (a). By Theorem 2.33, there is a point $A \notin \ell$, and by Theorem 3.1, there is a point $C \in \ell$. By Euclid's segment extension theorem, there is a point $B \in \overleftrightarrow{AC}$ such that $A * C * B$. Because the two lines ℓ and $\overleftrightarrow{AC}$ are distinct, they meet only at C, so neither A nor B lies on ℓ. Since $C \in \overline{AB} \cap \ell$, it follows from (b) that A and B lie on opposite sides of ℓ. Thus both sides are nonempty. $\qquad\square$

The plane separation postulate serves several purposes. First of all, this is the postulate that ensures that we are really talking about the geometry of a plane and not the geometry of three-dimensional (or even higher-dimensional) space. To see why three-dimensional geometry cannot satisfy the plane separation postulate, just imagine a line in space—you will see very quickly that there is no way to divide all the points not on the line into two disjoint subsets such that two points are in the same subset if and only if the line segment between them intersects the line.

More importantly, the plane separation postulate gives an explicit axiomatic justification for a whole raft of "geometrically obvious" facts that Euclid often assumed from diagrams without comment. Many of these follow from the fact that every line has *exactly* two distinct sides. Thus if ℓ is a line and A, B, C are points not on ℓ, we can draw the following conclusions from the plane separation postulate:

- If A and B are on the same side of ℓ and B and C are also on the same side, then A and C are on the same side, so $\overline{AC}$ does not intersect ℓ.

- If A and B are on opposite sides of ℓ and B and C are also on opposite sides, then A and C are on the same side, so $\overline{AC}$ does not intersect ℓ.

- If A and B are on opposite sides while B and C are on the same side, then A and C are on opposite sides and thus $\overline{AC}$ intersects ℓ.

If ℓ is a line, either of the sides of ℓ is called an ***open half-plane***. The union of an open half-plane together with the line that determines it is called a ***closed half-plane***. If ℓ is a line and A is a point not on ℓ, we use the notation $\text{OHP}(\ell, A)$ to denote the open half-plane consisting of all points on the same side of ℓ as A (Fig. 3.14) and $\text{CHP}(\ell, A)$ to denote the closed half-plane consisting of all points that are either on ℓ or on the same side of ℓ as A (Fig. 3.15). It follows from the definitions that $\text{CHP}(\ell, A) = \text{OHP}(\ell, A) \cup \ell$.

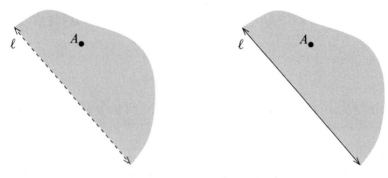

Fig. 3.14. The open half-plane $\text{OHP}(\ell, A)$.   **Fig. 3.15.** The closed half-plane $\text{CHP}(\ell, A)$.

The next two lemmas express the relationship between rays and sides of lines. The first one expresses an important property of rays that seems geometrically "obvious": if a ray starts on a line and goes to one side of the line, it cannot cross over to the other side. We call this the Y-lemma because the drawing that goes with it (Fig. 3.16) is suggestive of the letter Y.

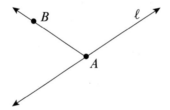

Fig. 3.16. Setup for the Y-lemma.

Lemma 3.44 (The Y-Lemma). *Suppose ℓ is a line, A is a point on ℓ, and B is a point not on ℓ. Then every interior point of $\overrightarrow{AB}$ is on the same side of ℓ as B, and $\overrightarrow{AB} \subseteq \text{CHP}(\ell, B)$.*

Proof. Let ℓ be a line, let A be a point on ℓ, and let B be a point not on ℓ. Suppose P is an arbitrary interior point on $\overrightarrow{AB}$. Since the lines ℓ and $\overleftrightarrow{AB}$ are distinct, they only have the single point A in common by Theorem 2.41, so P does not lie on ℓ. Thus either P lies on the same side of ℓ as B or it lies on the opposite side; assume for the sake of contradiction that it is on the opposite side. This means that there is a point of $\overline{PB}$ that lies on ℓ. But since $\overline{PB}$ is contained in $\overrightarrow{AB}$ and A is the only point of $\ell \cap \overleftrightarrow{AB}$, the point in $\ell \cap \overline{PB}$ must be A itself. As A is not equal to P or B, it must be an interior point of $\overline{PB}$, which means that $P * A * B$. Because P lies on $\overrightarrow{AB}$, this contradicts Lemma 3.30. This proves the first conclusion. The fact that $\overrightarrow{AB} \subseteq \text{CHP}(\ell, B)$ follows from the definition of closed half-planes. $\square$

Because of the Y-lemma, we can make the following definition. Suppose ℓ is a line and $\overrightarrow{r}$ is a ray starting at a point on ℓ. To say that *the ray $\overrightarrow{r}$ lies on a certain side of ℓ*

means that every interior point of $\overrightarrow{r}$ lies on that side. The Y-lemma tells us that if one point of a ray lies on a certain side, then the ray lies on that side. Of course, the endpoint of the ray does not lie on either side of ℓ, but when we say $\overrightarrow{r}$ lies on one side, we mean that its interior points do.

Lemma 3.45 (The X-Lemma). *Suppose $\overrightarrow{OA}$ and $\overrightarrow{OB}$ are opposite rays and ℓ is a line that intersects $\overleftrightarrow{AB}$ only at O (Fig. 3.17). Then $\overrightarrow{OA}$ and $\overrightarrow{OB}$ lie on opposite sides of ℓ.*

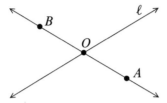

Fig. 3.17. Setup for the X-lemma.

Proof. The hypothesis implies that neither A nor B lies on ℓ. Because $\overrightarrow{OA}$ and $\overrightarrow{OB}$ are opposite rays, it follows from Theorem 3.38(e) that $A * O * B$. This means that $\overline{AB}$ intersects ℓ, so A and B are on opposite sides of ℓ. By the Y-lemma, this implies that $\overrightarrow{OA}$ and $\overrightarrow{OB}$ are on opposite sides of ℓ. $\qquad\square$

The following theorem, which follows easily from the Y-lemma and the X-lemma, shows that every ray can be viewed as the intersection of a line with a closed half-plane.

Theorem 3.46. *Suppose ℓ is a line, A is a point on ℓ, and B is a point not on ℓ (Fig. 3.18). Then*

$$\overrightarrow{AB} = \overleftrightarrow{AB} \cap \mathrm{CHP}(\ell, B). \tag{3.20}$$

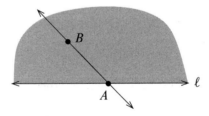

Fig. 3.18. A ray is the intersection of a line with a closed half-plane.

Proof. Suppose first that $Y \in \overrightarrow{AB}$. Then Y is certainly on $\overleftrightarrow{AB}$, so we need only show that it is also in $\mathrm{CHP}(\ell, B)$. Either $Y = A$, which lies on ℓ, or Y is an interior point of $\overrightarrow{AB}$, which lies on the same side of ℓ as B by the Y-lemma. In either case, $Y \in \mathrm{CHP}(\ell, B)$.

Conversely, suppose $Y \notin \overrightarrow{AB}$. Consider two cases: either $Y \in \overleftrightarrow{AB}$ or not. If $Y \in \overleftrightarrow{AB}$, then the fact that $Y \notin \overrightarrow{AB}$ implies that Y is in the ray opposite $\overrightarrow{AB}$ by Theorem 3.38(c). Since $Y \neq A$, the X-lemma implies that Y is on the opposite side of ℓ from B, and thus $Y \notin \mathrm{CHP}(\ell, B)$. On the other hand, if $Y \notin \overleftrightarrow{AB}$, then it is certainly not in $\overleftrightarrow{AB} \cap \mathrm{CHP}(\ell, B)$. $\qquad\square$

Convex Sets

For future use, we introduce an important general concept called *convexity*. If S is a set of points in the plane, we say that S is a ***convex set*** if whenever A and B are distinct points in S, the entire segment $\overline{AB}$ is contained in S. Fig. 3.19 illustrates examples of a convex set and a nonconvex set in the plane.

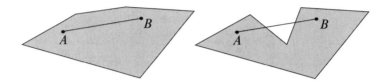

Fig. 3.19. A convex set and a nonconvex set.

One of the most important features of convex sets is the fact that intersections of convex sets are always convex.

Theorem 3.47. *If $S_1,\ldots,S_k$ are convex subsets of the plane, then $S_1 \cap \cdots \cap S_k$ is convex.*

Proof. Suppose A and B are any two distinct points in $S_1 \cap \cdots \cap S_k$. Then A and B both lie in S_i for each $i = 1,\ldots,k$. Since each S_i is convex, this implies that $\overline{AB} \subseteq S_i$ for each i, so $\overline{AB}$ is contained in the intersection. □

The next few theorems show that many of the geometric figures we have introduced so far are convex sets.

Theorem 3.48. *Every line is a convex set.*

Proof. Let ℓ be a line, and let C and D be distinct points on ℓ. Then it follows from Theorem 2.28 that $\ell = \overleftrightarrow{CD}$, and Lemma 3.17 implies that $\overline{CD} \subseteq \ell$. □

Theorem 3.49. *Every segment is a convex set.*

Proof. Let $\overline{AB}$ be a segment, and suppose C and D are two distinct points in $\overline{AB}$. We consider three cases:

CASE 1: *C and D are both endpoints of $\overline{AB}$.* Then $\overline{CD} = \overline{AB}$, which is certainly contained in $\overline{AB}$.

CASE 2: *Only one of the points is an endpoint of $\overline{AB}$.* We may assume without loss of generality that $C = A$ while $D \in \operatorname{Int} \overline{AB}$; this means $A * D * B$, and so $\overline{CD} = \overline{AD} \subseteq \overline{AB}$ by Corollary 3.25.

CASE 3: *Neither C nor D is an endpoint of $\overline{AB}$.* In this case, A, B, C, and D are four distinct collinear points satisfying $A * C * B$ and $A * D * B$. The three points A, C, and D must satisfy either $A * C * D$ or $A * D * C$ (because the other possibility, $D * A * C$, taken together with $A * C * B$, would imply $D * A * B$ by Theorem 3.16, contradicting the fact that $A * D * B$). The argument is the same in both cases, so assume $A * C * D$. Because we also know that $A * D * B$, two applications of Corollary 3.25 yield $\overline{CD} \subseteq \overline{AD} \subseteq \overline{AB}$. □

Theorem 3.50. *Every open or closed half-plane is a convex set.*

Proof. First we prove that open half-planes are convex. Let S be any open half-plane; then $S = \mathrm{OHP}(\ell, A)$ for some line ℓ and some point A not on ℓ. Suppose C and D are any two distinct points in S (Fig. 3.20). We need to show that $\overline{CD} \subseteq S$. We know by assumption

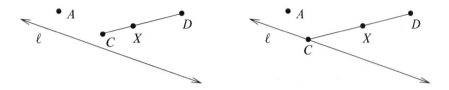

Fig. 3.20. An open half-plane is convex.   **Fig. 3.21.** A closed half-plane is convex.

that C and D themselves lie in S, so we need only show that every interior point of $\overline{CD}$ lies in S. Suppose X is such an interior point. This means that $C * X * D$, and thus by Theorem 3.24, it follows that $\overline{CX} \subseteq \overline{CD}$. Since $\overline{CD}$ does not intersect ℓ, it follows that $\overline{CX}$ does not either. Thus X is on the same side of ℓ as C, and hence also the same side as A; this is the same as saying $X \in S$.

Now we prove that closed half-planes are convex. Let T be a closed half-plane, so $T = \mathrm{CHP}(\ell, A)$ for some ℓ and A as above. Suppose C and D are any two distinct points in T. There are three cases.

CASE 1: C and D both lie in the *open* half-plane $\mathrm{OHP}(\ell, A)$. In this case $\overline{CD} \subseteq \mathrm{OHP}(\ell, A) \subseteq \mathrm{CHP}(\ell, A)$ by the argument in the preceding paragraph.

CASE 2: C and D both lie on ℓ. Then the fact that ℓ is convex implies $\overline{CD} \subseteq \ell \subseteq \mathrm{CHP}(\ell, A)$.

CASE 3: One of the points C or D lies on ℓ and the other lies in $\mathrm{OHP}(\ell, A)$ (see Fig. 3.21). Without loss of generality, say that $C \in \ell$ and $D \in \mathrm{OHP}(\ell, A)$. Suppose X is any point in $\mathrm{Int}\,\overline{CD}$. Then $\overline{XD} \subseteq \overleftrightarrow{CD}$, which meets ℓ only at C. Since $C \notin \overline{XD}$, it follows that $\overline{XD} \cap \ell = \varnothing$, so X is on the same side of ℓ as D, which is the same side as A. This means that $X \in \mathrm{OHP}(\ell, A) \subseteq \mathrm{CHP}(\ell, A)$. □

The next theorem could be proved using an argument similar to that of Theorem 3.49, but it would require checking several cases. The following proof is much easier.

Theorem 3.51. *Every ray is a convex set.*

Proof. Let $\overrightarrow{AB}$ be a ray, and let ℓ be any line that contains A but not B (Theorem 2.36). Then Theorem 3.46 shows that $\overrightarrow{AB}$ is the intersection of the two convex sets $\overleftrightarrow{AB}$ and $\mathrm{CHP}(\ell, B)$, so it is convex. □

Exercises

3A. Prove Lemma 3.4 (the ruler flipping lemma).

3B. Suppose A, B, C, D are four distinct points. Draw pictures of counterexamples that show why none of the following pairs of assumptions is sufficient to conclude that

$A * B * C * D$:

$$A * B * D \text{ and } A * C * D;$$
$$A * B * C \text{ and } A * B * D;$$
$$A * C * D \text{ and } B * C * D.$$

3C. Suppose A, B, C, D are points such that $A * B * C * D$. Prove that $AB + BC + CD = AD$.

3D. Let A, B, C, and D be four distinct collinear points such that $A * C * D$, and suppose that C is not between A and B. Prove that $B * C * D$.

3E. Prove Theorem 3.24(b) (the intersection of segments with a common endpoint).

3F. Prove that every point lies on infinitely many distinct lines. [Hint: Given a point A, let ℓ be a line not containing A, and show that for any two distinct points $B, C \in \ell$, the lines $\overleftrightarrow{AB}$ and $\overleftrightarrow{AC}$ are distinct.]

3G. Let ℓ be a line, let A be a point on ℓ, and let r be a positive real number. Prove that there are exactly two distinct points $B, C \in \ell$ such that $AB = AC = r$. Prove that these points satisfy $B * A * C$.

3H. Prove that every circle contains infinitely many points. [Hint: Use Exercises 3F and 3G.]

3I. Prove Lemma 3.32 (coordinate representations of rays).

3J. Prove Theorem 3.35 (the segment construction theorem). [Hint: Use an adapted coordinate function. Look at the proof of Theorem 3.27 for inspiration.]

3K. Prove Corollary 3.37 (Euclid's segment cutoff theorem).

3L. Prove Theorem 3.42 (on intersections and unions of rays).

Chapter 4

Angles

After lines, segments, and rays, the next most important figures in plane geometry are *angles*. The purpose of this chapter is to give the definition of angles and establish their basic properties.

There are a number of relationships involving angles that appear "obvious" in diagrams and which Euclid took for granted. In order to be confident that our theorems follow logically from the axioms without relying on our geometric intuition, we have to be sure to prove all such relationships. Because of this, many of the proofs in this chapter are rather technical. The payoff for this effort will come in the next chapter when we start discussing triangles, and we can freely use the relationships that have been rigorously proved in this chapter.

Angles

We define an ***angle*** to be a union of two rays with the same endpoint (Fig. 4.1). If the two rays are denoted by $\vec{a}$ and $\vec{b}$, then the angle formed by their union is denoted by $\angle ab$; if they are denoted by $\overrightarrow{OA}$ and $\overrightarrow{OB}$, then the angle is denoted by $\angle AOB$. (Which notation we use will usually depend on whether we need to refer individually to points on the rays.) The common endpoint is called the ***vertex*** of the angle, and the two rays are called its ***sides***.

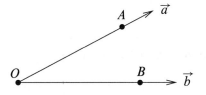

Fig. 4.1. An angle.

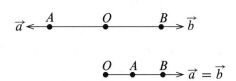

Fig. 4.2. A straight angle and a zero angle.

Our definition of $\angle ab$ allows for the possibility that the rays $\vec{a}$ and $\vec{b}$ are collinear. If they are, Theorem 3.38(b) shows that there are two cases: either $\vec{a}$ and $\vec{b}$ are opposite rays, in which case $\angle ab$ is called a ***straight angle***, or they are the same ray, in which case $\angle ab$ is called a ***zero angle*** (see Fig. 4.2). An angle formed by two noncollinear rays (i.e., an angle that is neither a straight angle nor a zero angle) is called a ***proper angle***. It follows from Theorem 3.38(a) that $\angle AOB$ is a proper angle if and only if the three points A, O, B are noncollinear. For most purposes, we will restrict our attention to proper angles. But in some cases it is useful to allow more general angles, so without further qualification, the term "angle" can mean any of the above types.

It is important to be aware that an angle is a *set of points* in the plane—the union of two rays—not a number of degrees. Thus for two *angles* to be equal means that they are both literally the same set of points. For example, $\angle ab = \angle ba$, because both notations designate exactly the same union of rays. Also, if A' is any interior point of $\overrightarrow{OA}$ and B' is any interior point of $\overrightarrow{OB}$, then $\angle A'OB' = \angle AOB$, thanks to Theorem 3.38(d). As a set of points, a straight angle is the same thing as a line, while a zero angle is just a ray.

Just as we did for endpoints of segments and rays, we must verify that our definition of the vertex of an angle does not depend on the notation we choose to designate the angle. For a proper angle, the next theorem shows how the vertex can be distinguished from all other points.

Theorem 4.1. *If $\angle ab$ is a proper angle, then the common endpoint of $\vec{a}$ and $\vec{b}$ is the only extreme point of $\angle ab$.*

Proof. Suppose $\angle ab$ is a proper angle, and let O denote the common endpoint of $\vec{a}$ and $\vec{b}$. If P is any point in $\angle ab$ other than O, then P is either an interior point of $\vec{a}$ or an interior point of $\vec{b}$; without loss of generality, let us say $P \in \text{Int}\,\vec{a}$. Then it follows from Theorem 3.40 that P is a passing point of $\vec{a}$, and thus by Lemma 3.19 it is also a passing point of $\angle ab$.

To show that O is an extreme point, suppose for contradiction that it is not. Then there are points $X, Y \in \angle ab$ such that $X * O * Y$. Since neither X nor Y is an endpoint of $\vec{a}$ or $\vec{b}$, they must both be interior points of their respective rays. If X and Y were interior points of different rays, say $X \in \text{Int}\,\vec{a}$ and $Y \in \text{Int}\,\vec{b}$, then it would follow from Theorem 3.38(d) that $\vec{a} = \overrightarrow{OX}$ and $\vec{b} = \overrightarrow{OY}$ and from Theorem 3.38(a) that $\vec{a}$ and $\vec{b}$ were collinear, contradicting the assumption that $\angle ab$ is a proper angle. Thus both points must lie in the same ray, say $\vec{a}$. But Theorem 3.40 shows that O is an extreme point of $\vec{a}$, so this is impossible. $\square$

Corollary 4.2 (Consistency of Vertices of Proper Angles). *If $\angle AOB$ and $\angle A'O'B'$ are equal proper angles, then $O = O'$.*

Proof. Theorem 4.1 shows that O is the only extreme point of $\angle AOB$ and O' is the only extreme point of $\angle A'O'B'$, so the fact that $\angle AOB = \angle A'O'B'$ implies $O = O'$. $\square$

Notice that Theorem 4.1 and Corollary 4.2 only apply to vertices of *proper angles*. Because a zero angle is just a ray, its vertex is its endpoint, which is well defined by Corollary 3.41. However, there is no such result concerning the vertex of a straight angle: if $\angle AOB$ and $\angle A'O'B'$ are straight angles with $\angle AOB = \angle A'O'B'$, then O and O' could be any two different points on the line $\overleftrightarrow{AB}$. For convenience, if $\angle AOB$ is a straight angle, we will still sometimes refer to O as its "vertex," but we have to remember that which point is called the vertex depends on which rays we use to designate the straight angle.

In order to speak about the *size* of an angle, we need to introduce the concept of *angle measure*. In our approach to plane geometry, the measure of an angle, like the distance between points, is a primitive term. Its meaning will be captured in the next two postulates. Many features of angle measures are closely analogous to features of distances; our treatment is designed to highlight those analogies. We begin with a postulate about angles that corresponds to the distance postulate.

Postulate 7 (The Angle Measure Postulate). *For every angle $\angle ab$, the measure of $\angle ab$ is a real number in the closed interval $[0, 180]$ determined by $\angle ab$.*

We use the notation $m\angle ab$ to denote the measure of $\angle ab$ (or $m\angle AOB$ if the angle is designated by three points). If this measure is equal to the real number x, we write $m\angle ab = x°$ and say that $\angle ab$ measures ***x degrees***. Technically speaking, an angle measure is just a real number, so $x°$ and x both stand for the same real number; the degree symbol is placed there traditionally as a reminder that we are measuring angles on a scale of 0 to 180. Of course, as you are aware from calculus, it is possible to measure angles using other scales such as radians; if radian measure is desired, one can just modify this postulate and the next by replacing the number 180 with π. Since high-school geometry courses use degrees exclusively, we will stick with them.

The next postulate is analogous to the ruler postulate. In order to state it concisely, we make the following definition. If $\vec{r}$ is a ray starting at a point O and P is a point not on the line $\overleftrightarrow{r}$, the ***half-rotation of rays determined by $\vec{r}$ and P***, denoted by $\mathrm{HR}(\vec{r}, P)$, is the set whose elements are all the rays that start at O and are contained in the closed half-plane $\mathrm{CHP}(\overleftrightarrow{r}, P)$ (see Fig. 4.3). It follows from the definition that $\mathrm{HR}(\vec{r}, P)$ contains the following rays:

- the ray $\vec{r}$;
- the ray opposite $\vec{r}$;
- every ray starting at O and lying on the same side of $\overleftrightarrow{r}$ as P.

If $\vec{a}$ is another ray starting at O and not collinear with $\vec{r}$, sometimes for convenience we will use the notation $\mathrm{HR}\left(\vec{r}, \vec{a}\right)$ to denote the half-rotation determined by $\vec{r}$ and any point on the same side of $\overleftrightarrow{r}$ as the ray $\vec{a}$.

It is important to understand clearly that a half-rotation is a set of *rays*, not a set of points. (If we took the *union* of all the rays in $\mathrm{HR}(\vec{r}, P)$, we would get a set of points, namely $\mathrm{CHP}(\overleftrightarrow{r}, P)$; but that is not what a half-rotation refers to.)

Postulate 8 (The Protractor Postulate). *For every ray $\vec{r}$ and every point P not on $\overleftrightarrow{r}$, there is a bijective function $g: \mathrm{HR}(\vec{r}, P) \to [0, 180]$ that assigns the number 0 to $\vec{r}$ and the number 180 to the ray opposite $\vec{r}$ and such that if $\vec{a}$ and $\vec{b}$ are any two rays in $\mathrm{HR}(\vec{r}, P)$, then*

$$m\angle ab = \left| g\left(\vec{b}\right) - g\left(\vec{a}\right) \right|. \tag{4.1}$$

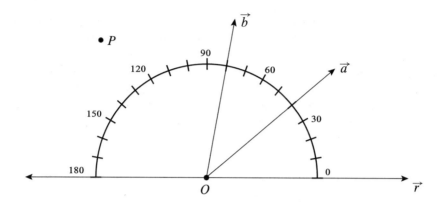

Fig. 4.3. Rays in a half-rotation.

Suppose $\vec{r}$ is a ray and P is a point not on $\overleftrightarrow{r}$. A bijective function $g\colon \mathrm{HR}(\vec{r}, P) \to [0, 180]$ that satisfies (4.1) for all pairs of rays $\vec{a}, \vec{b} \in \mathrm{HR}(\vec{r}, P)$ is called a ***coordinate function for*** $\mathrm{HR}(\vec{r}, P)$. If in addition $g(\vec{r}) = 0$ and $g(\vec{s}) = 180$ (where $\vec{s}$ is the ray opposite $\vec{r}$), we say that the coordinate function ***starts at*** $\vec{r}$. We can summarize the protractor postulate in the following way: *for each ray $\vec{r}$ and each point $P \notin \overleftrightarrow{r}$, there exists a coordinate function for* $\mathrm{HR}(\vec{r}, P)$ *starting at* $\vec{r}$. Once we have decided on a half-rotation and its corresponding coordinate function g, the number $g(\vec{a})$ associated with a particular ray $\vec{a}$ in the half-rotation is called the ***coordinate of the ray*** with respect to the chosen coordinate function.

You might wonder if there is a "protractor placement theorem" analogous to the ruler placement theorem. In fact, most of the choice of where to place our protractor is already taken care of by the choice of a half-rotation $\mathrm{HR}(\vec{r}, P)$. The only flexibility we have is to replace the ray $\vec{r}$ by its opposite ray; if $\vec{s}$ denotes the ray opposite $\vec{r}$, then the half-rotation $\mathrm{HR}(\vec{s}, P)$ is exactly the same set of rays as $\mathrm{HR}(\vec{r}, P)$. The protractor postulate then tells us that there is another coordinate function for this same half-rotation that starts at $\vec{s}$. Intuitively, this corresponds to lifting the protractor off the plane and putting it back upside-down, so that it still lies on the same side of the line $\overleftrightarrow{r}$ but the 0 and 180 marks have changed places. Since the existence of such a coordinate function is already guaranteed by the protractor postulate, there is no need to prove a separate protractor placement theorem.

Given any angle $\angle AOB$, there will typically be many different half-rotations containing both of its rays. If $\angle AOB$ is a proper angle, the half-rotation $\mathrm{HR}\big(\overrightarrow{OA}, B\big)$ defined by the ray $\overrightarrow{OA}$ and the point B is one such half-rotation. (Why can we not use this when $\angle AOB$ is a straight angle or a zero angle?) If $\angle AOB$ is a straight angle or a zero angle, we can instead choose any point C not on $\overleftrightarrow{OA}$ and use $\mathrm{HR}\big(\overrightarrow{OA}, C\big)$. One consequence of the protractor postulate together with the angle measure postulate is that the coordinate functions associated with any two half-rotations will give the same value for $m\angle AOB$. Intuitively, this reflects the fact that no matter where we place our 180° protractor, as long as its center is on the vertex of the angle and its scale intersects both sides of the angle, we will obtain the same value for the angle's measure.

Some simple consequences of the protractor postulate are worth noting. This next theorem is the analogue for angle measures of Theorem 3.6.

Theorem 4.3 (Properties of Angle Measures). *Suppose ∠ab is any angle.*

(a) $m\angle ab = m\angle ba$.

(b) $m\angle ab = 0°$ *if and only if ∠ab is a zero angle.*

(c) $m\angle ab = 180°$ *if and only if ∠ab is a straight angle.*

(d) $0° < m\angle ab < 180°$ *if and only if ∠ab is a proper angle.*

Proof. Let $\text{HR}(\overrightarrow{r}, P)$ be any half-rotation containing both of the rays $\overrightarrow{a}$ and $\overrightarrow{b}$, and let $g: \text{HR}(\overrightarrow{r}, P) \to [0, 180]$ be a coordinate function for it starting at $\overrightarrow{r}$. The protractor postulate says that $m\angle ab$ is given by (4.1). Because this formula gives the same value when $\overrightarrow{a}$ and $\overrightarrow{b}$ are interchanged, it follows that $m\angle ab = m\angle ba$; this proves (a). (This can also be proved simply by noting that the notations ∠ab and ∠ba denote exactly the same angle—i.e., the same set of points—and the measure is completely determined by the angle.)

If ∠ab is a zero angle, then $\overrightarrow{a} = \overrightarrow{b}$, so $g(\overrightarrow{a}) = g(\overrightarrow{b})$ because g is a well-defined function of rays. Thus $m\angle ab = 0°$ by (4.1). Conversely, if $m\angle ab = 0°$, then (4.1) shows that $g(\overrightarrow{a}) = g(\overrightarrow{b})$; since g is injective, this implies that $\overrightarrow{a} = \overrightarrow{b}$ and therefore ∠ab is a zero angle. This proves (b).

To prove (c), suppose first that $m\angle ab = 180°$. Assume without loss of generality that $g(\overrightarrow{a}) \le g(\overrightarrow{b})$. (If this is not the case, just reverse the roles of $\overrightarrow{a}$ and $\overrightarrow{b}$.) Then (4.1) implies

$$180° = m\angle ab = \left| g(\overrightarrow{b}) - g(\overrightarrow{a}) \right| = g(\overrightarrow{b}) - g(\overrightarrow{a}),$$

from which it follows that $g(\overrightarrow{b}) = g(\overrightarrow{a}) + 180$. If $g(\overrightarrow{a})$ were not 0, it would have to be positive, which would imply $g(\overrightarrow{b}) > 180$, contradicting the ruler postulate. Thus $g(\overrightarrow{a}) = 0$ and $g(\overrightarrow{b}) = 180$. Because $g(\overrightarrow{r}) = 0$ and g is injective, it follows that $\overrightarrow{a} = \overrightarrow{r}$. Because g assigns the number 180 only to the ray opposite $\overrightarrow{r}$, it follows that $\overrightarrow{a}$ and $\overrightarrow{b}$ are opposite rays, which means that ∠ab is a straight angle.

Conversely, suppose ∠ab is a straight angle. To show that $m\angle ab = 180°$, it is easiest to work with a specially chosen half-rotation. If we choose any point $P \notin \overleftrightarrow{a}$, then both of the rays $\overrightarrow{a}$ and $\overrightarrow{b}$ are contained in $\text{HR}(\overrightarrow{a}, P)$ by definition. If g is a coordinate function for this half-rotation starting at $\overrightarrow{a}$, then $g(\overrightarrow{a}) = 0$ and $g(\overrightarrow{b}) = 180$, so (4.1) gives $m\angle ab = 180°$.

Finally, (d) follows from the previous two parts, because they show that $0° < m\angle ab < 180°$ if and only if ∠ab is neither a zero angle nor a straight angle, which is the case if and only if it is proper. □

Part (a) of the preceding theorem illustrates an important feature of our approach to angles: in contrast to the treatment of angles in trigonometry and calculus, we don't use positive and negative signs to distinguish between "clockwise" and "counterclockwise" angle measures. (In fact, it is not even clear what meanings could be assigned to those words in the context of our axiomatic system.) Also, we restrict our angle measures to be less than or equal to 180°. For situations in which we need to consider angle measures larger than 180°, we make the following definition. Suppose ∠ab is an angle. We define the **reflex measure of ∠ab** (see Fig. 4.4), denoted by $rm\angle ab$, to be the number

$$rm\angle ab = 360° - m\angle ab.$$

We will sometimes call $m\angle ab$ the **standard measure of ∠ab** to distinguish it from the

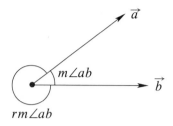

Fig. 4.4. The reflex measure of an angle.

reflex measure. (The word *reflex* is derived from the Latin word for "bent back." Many high-school textbooks define a "reflex angle" to be an angle whose measure is greater than 180°. However, that terminology would be inconsistent with our definitions because the *angle* ∠*ab* is the same union of two rays whether we are interested in its reflex measure or its standard measure. This is why we have adopted the convention that one angle can have two different types of measure.)

In addition to straight angles, zero angles, and proper angles, there are some other special types of angles that are defined in terms of their measures. We say that an angle ∠*ab* is a ***right angle*** if $m\angle ab = 90°$, it is an ***acute angle*** if $0° < m\angle ab < 90°$, and it is a an ***obtuse angle*** if $90° < m\angle ab < 180°$. Two angles ∠*ab* and ∠*cd* are said to be ***congruent***, written $\angle ab \cong \angle cd$, if they have the same measure:

$$\angle ab \cong \angle cd \quad \Leftrightarrow \quad m\angle ab = m\angle cd.$$

Note that we say two *angles are congruent* if their *measures are equal*. To say that two *angles are equal* is to say that they are exactly the same angle.

The next theorem is our version of Euclid's Postulate 4.

Theorem 4.4 (Euclid's Postulate 4). *All right angles are congruent.*

Proof. Any two right angles have the same measure, which means by definition that they are congruent. ☐

The next theorem is one of the most important consequences of the protractor postulate. It is analogous to Euclid's Proposition I.23 but is proved in a very different way.

Theorem 4.5 (Angle Construction Theorem). *Let O be a point, let $\vec{a}$ be a ray starting at O, and let x be a real number such that $0 < x < 180$. On each side of $\overleftrightarrow{a}$, there is a unique ray $\vec{b}$ starting at O such that $m\angle ab = x$.*

Proof. Let O, $\vec{a}$, and x be as in the hypothesis. Choose an arbitrary side of $\overleftrightarrow{a}$ and let P be a point on the chosen side. Let g be a coordinate function for $\mathrm{HR}(\vec{a}, P)$ starting at $\vec{a}$. First we will prove the existence of $\vec{b}$. Because the protractor postulate says that g is surjective, there is a ray $\vec{b} \in \mathrm{HR}(\vec{a}, P)$ such that $g(\vec{b}) = x$. Because g is a coordinate function,

$$m\angle ab = \left| g(\vec{b}) - g(\vec{a}) \right| = |x - 0| = x,$$

as desired.

To prove uniqueness, suppose $\vec{b}'$ is any other ray on the same side of $\overleftrightarrow{a}$ such that $m\angle ab' = x$. Then $\vec{b}' \in \mathrm{HR}(\vec{a}, P)$ by definition. If we set $x' = g(\vec{b}')$, then the fact that

g is a coordinate function means that $x = m\angle ab' = |x' - 0| = x'$. Since g is injective, this implies that $\overrightarrow{b}' = \overrightarrow{b}$. $\qquad\square$

Just as with the segment construction theorem, the uniqueness part of this theorem is worth restating separately.

Corollary 4.6 (Unique Ray Theorem). *Let $\overrightarrow{a}$ be a ray starting at a point O. If $\overrightarrow{b}$ and $\overrightarrow{b}'$ are rays starting at O and lying on the same side of $\overleftrightarrow{a}$ such that $m\angle ab = m\angle ab'$, then $\overrightarrow{b}$ and $\overrightarrow{b}'$ are the same ray.* $\qquad\square$

Betweenness of Rays

The idea of betweenness can be extended to rays. Suppose $\overrightarrow{a}$, $\overrightarrow{b}$, and $\overrightarrow{c}$ are three rays sharing a common endpoint. We say that $\overrightarrow{b}$ ***is between*** $\overrightarrow{a}$ ***and*** $\overrightarrow{c}$ if no two of the rays are collinear and there is some half-rotation containing all three rays, such that the corresponding coordinate $g\left(\overrightarrow{b}\right)$ is between the coordinates $g(\overrightarrow{a})$ and $g(\overrightarrow{c})$. The idea is illustrated in Fig. 4.5. The statement "$\overrightarrow{b}$ is between $\overrightarrow{a}$ and $\overrightarrow{c}$" is symbolized by $\overrightarrow{a} * \overrightarrow{b} * \overrightarrow{c}$.

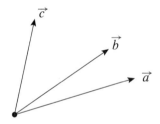

Fig. 4.5. $\overrightarrow{b}$ is between rays $\overrightarrow{a}$ and $\overrightarrow{c}$, symbolized by $\overrightarrow{a} * \overrightarrow{b} * \overrightarrow{c}$.

This definition is closely analogous to the definition of betweenness of points. As we develop properties of betweenness of rays, you will repeatedly see evidence of this analogy. Many of the results in this section are directly analogous to results in Chapter 3.

Theorem 4.7 (Symmetry of Betweenness of Rays). *If $\overrightarrow{a}$, $\overrightarrow{b}$, and $\overrightarrow{c}$ are rays with a common endpoint, then $\overrightarrow{a} * \overrightarrow{b} * \overrightarrow{c}$ if and only if $\overrightarrow{c} * \overrightarrow{b} * \overrightarrow{a}$.*

Proof. Essentially the same as the proof of Theorem 3.7. $\qquad\square$

Theorem 4.8 (Betweenness Theorem for Rays). *If $\overrightarrow{a}$, $\overrightarrow{b}$, and $\overrightarrow{c}$ are rays such that $\overrightarrow{a} * \overrightarrow{b} * \overrightarrow{c}$, then $m\angle ab + m\angle bc = m\angle ac$.*

Proof. Given that $\overrightarrow{a} * \overrightarrow{b} * \overrightarrow{c}$, choose some half-rotation $\text{HR}(\overrightarrow{r}, P)$ containing all three rays and let g be a corresponding coordinate function. Then the hypothesis means that either $g(\overrightarrow{a}) < g\left(\overrightarrow{b}\right) < g(\overrightarrow{c})$ or $g(\overrightarrow{a}) > g\left(\overrightarrow{b}\right) > g(\overrightarrow{c})$; after interchanging the names of $\overrightarrow{a}$ and $\overrightarrow{c}$ if necessary, we may as well assume that the first set of inequalities holds. The rest of the proof proceeds exactly like the proof of the betweenness theorem for points: write down the formulas for $m\angle ab$, $m\angle bc$, and $m\angle ac$ in terms of g, add the first two and subtract the third, and observe that the terms involving g all cancel. $\qquad\square$

The proofs of the next three results are entirely analogous to those of Theorem 3.9, Corollary 3.10, and Theorem 3.22. The first theorem is an analogue for rays of Hilbert's betweenness axiom.

Theorem 4.9. *If $\vec{a}$, $\vec{b}$, and $\vec{c}$ are rays with a common endpoint, no two of which are collinear and all lying in a single half-rotation, then exactly one of them lies between the other two.*

Proof. Exercise 4A. □

Corollary 4.10 (Consistency of Betweenness of Rays). *Let $\vec{a}$, $\vec{b}$, and $\vec{c}$ be distinct rays with a common endpoint, no two of which are collinear, and all lying in a single half-rotation. If $\mathrm{HR}(\vec{r}, P)$ is any half-rotation containing all three rays and g is a corresponding coordinate function, then $\vec{a} * \vec{b} * \vec{c}$ if and only if $g(\vec{a}) * g(\vec{b}) * g(\vec{c})$.*

Proof. Exercise 4B. □

Theorem 4.11 (Euclid's Common Notions for Angles).

(a) **Transitive Property of Congruence:** *Two angles that are both congruent to a third angle are congruent to each other.*

(b) **Angle Addition Theorem:** *Suppose $\vec{a}, \vec{b}, \vec{c}, \vec{a}', \vec{b}', \vec{c}'$ are rays such that $\vec{a} * \vec{b} * \vec{c}$ and $\vec{a}' * \vec{b}' * \vec{c}'$. If $\angle ab \cong \angle a'b'$ and $\angle bc \cong \angle b'c'$, then $\angle ac \cong \angle a'c'$.*

(c) **Angle Subtraction Theorem:** *Suppose $\vec{a}, \vec{b}, \vec{c}, \vec{a}', \vec{b}', \vec{c}'$ are rays such that $\vec{a} * \vec{b} * \vec{c}$ and $\vec{a}' * \vec{b}' * \vec{c}'$. If $\angle ac \cong \angle a'c'$ and $\angle ab \cong \angle a'b'$, then $\angle bc \cong \angle b'c'$.*

(d) **Reflexive Property of Congruence:** *Every angle is congruent to itself.*

(e) **The Whole Angle Is Greater Than the Part:** *If $\vec{a}$, $\vec{b}$, and $\vec{c}$ are rays such that $\vec{a} * \vec{b} * \vec{c}$, then $m\angle ac > m\angle ab$.*

Proof. Exercise 4C. □

There is a close relationship between betweenness of rays and plane separation. The remainder of this chapter is devoted primarily to exploring that relationship.

As a first step in that direction, we introduce the following important concept. Two angles are said to be ***adjacent angles*** if they share one common ray and their other rays lie on opposite sides of the line determined by the common ray. Fig. 4.6 shows several ways that two angles $\angle ab$ and $\angle bc$ can be adjacent:

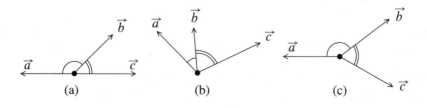

Fig. 4.6. Three pairs of adjacent angles.

Observe that just having a common side is not sufficient for two angles to be adjacent: in Fig. 4.7, the two angles $\angle ab$ and $\angle bc$ share the common side $\overrightarrow{b}$, but they are not adjacent because $\overrightarrow{a}$ and $\overrightarrow{c}$ are not on opposite sides of $\overleftrightarrow{b}$.

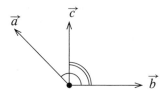

Fig. 4.7. $\angle ab$ and $\angle bc$ are nonadjacent angles sharing a common side.

The next theorem shows that one way to form adjacent angles is to place the common ray between the other two. This corresponds to the situation in Fig. 4.6(b).

Theorem 4.12. *If $\overrightarrow{a}$, $\overrightarrow{b}$, and $\overrightarrow{c}$ are rays such that $\overrightarrow{a} * \overrightarrow{b} * \overrightarrow{c}$, then $\angle ab$ and $\angle bc$ are adjacent angles.*

Proof. By the hypothesis, the angles $\angle ab$ and $\angle bc$ share the common side $\overrightarrow{b}$, so we need only show that $\overrightarrow{a}$ and $\overrightarrow{c}$ are on opposite sides of $\overleftrightarrow{b}$. Since the definition of betweenness of rays includes the stipulation that no two of the rays $\overrightarrow{a}$, $\overrightarrow{b}$, and $\overrightarrow{c}$ are collinear, it follows that neither $\overrightarrow{a}$ nor $\overrightarrow{c}$ is contained in $\overleftrightarrow{b}$. By the Y-lemma, therefore, $\overrightarrow{a}$ and $\overrightarrow{c}$ are either on opposite sides of $\overleftrightarrow{b}$ or on the same side. Assume for the sake of contradiction that they are on the same side. It then follows from the definition of a half-rotation that $\overrightarrow{a}$, $\overrightarrow{b}$, and $\overrightarrow{c}$ all lie in the half-rotation HR $\left(\overrightarrow{b}, \overrightarrow{a}\right)$. Let g be a coordinate function for this half-rotation starting at $\overrightarrow{b}$. Corollary 4.10 guarantees that $g\left(\overrightarrow{b}\right)$ is between $g(\overrightarrow{a})$ and $g(\overrightarrow{c})$; but since $g\left(\overrightarrow{b}\right) = 0$ while $g(\overrightarrow{a})$ and $g(\overrightarrow{c})$ are both positive, this is impossible. Thus our assumption is false, and $\overrightarrow{a}$ and $\overrightarrow{c}$ must be on opposite sides of $\overleftrightarrow{b}$. ∎

As Fig. 4.6 suggests, the converse of the preceding theorem is false: two angles can be adjacent without any of the rays being between the other two. In Fig. 4.6(a), none of the three rays is between the other two because two of the rays are collinear; while in Fig. 4.6(c), the same is true because there is no half-rotation that contains all three rays. (We will prove a partial converse below; see Lemma 4.26.)

Because none of the rays is between the other two, the betweenness theorem for rays does not apply in cases (a) and (c) of Fig. 4.6. However, for case (a), there is a simple substitute for the betweenness theorem. (We'll address case (c) later in this chapter.) First we need to introduce some more terminology.

Two proper angles $\angle ab$ and $\angle cd$ are said to be ***supplementary*** if $m\angle ab + m\angle cd = 180°$ and ***complementary*** if $m\angle ab + m\angle cd = 90°$.

Theorem 4.13. *Supplements of congruent angles are congruent, and complements of congruent angles are congruent.*

Proof. Suppose $\angle ab$ is supplementary to $\angle cd$ and $\angle a'b'$ is supplementary to $\angle c'd'$, and assume that $\angle cd \cong \angle c'd'$. Then

$$m\angle ab = 180° - m\angle cd = 180° - m\angle c'd' = m\angle a'b'.$$

Therefore, $\angle ab \cong \angle a'b'$. The same argument works for complements of congruent angles, with 180° replaced by 90°. □

Two angles form a ***linear pair*** if they are both proper, they share a common side, and their nonshared sides are opposite rays. (It then follows from the X-lemma that their nonshared sides are on opposite sides of the line through the shared side, so they are adjacent.) For example, in Fig. 4.6(a), if we are given that $\vec{a}$ and $\vec{c}$ are opposite rays, then $\angle ab$ and $\angle bc$ form a linear pair. The next theorem is the analogue of Euclid's Proposition I.13, but our proof is different, being based on the protractor postulate.

Theorem 4.14 (Linear Pair Theorem). *If two angles form a linear pair, then they are supplementary.*

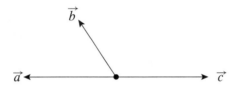

Fig. 4.8. A linear pair.

Proof. Suppose $\angle ab$ and $\angle bc$ form a linear pair with common side $\vec{b}$ (see Fig. 4.8). By definition, the rays $\vec{a}$, $\vec{b}$, and $\vec{c}$ all lie in the half-rotation $\mathrm{HR}(\vec{a}, \vec{b})$. By the protractor postulate, there is a coordinate function g for this half-rotation satisfying $g(\vec{a}) = 0$ and $g(\vec{c}) = 180$. Set $x = g(\vec{b})$. Because the definition of adjacent angles implies that $\vec{a}$ and $\vec{c}$ are not collinear with $\vec{b}$, it follows that $\vec{b}$ is not equal to either $\vec{a}$ or $\vec{c}$, and therefore $0 < x < 180$. Thus

$$m\angle ab = |x - 0| = x,$$
$$m\angle bc = |180 - x| = 180 - x.$$

Adding these two equations yields the result. □

Euclid's definition of a *right angle*, translated into modern terminology, is any angle that forms a linear pair with another angle that is congruent to it. For us, the statement that any such angle is a right angle is the following corollary to the linear pair theorem.

Corollary 4.15. *If two angles in a linear pair are congruent, then they are both right angles.*

Proof. If two congruent angles form a linear pair, their measures add up to 180° by the linear pair theorem. Since the measures are equal, they must both be 90°. □

Note that the converse to the linear pair theorem is certainly not true, because two angles can be supplementary without being adjacent. However, if we add the hypothesis that the angles are adjacent as well as supplementary, then we obtain the following partial converse. It is Euclid's Proposition I.14, with essentially the same proof (except that we express angle comparisons in terms of degree measures). Notice how the unique ray theorem is used in the proof; this technique will reappear frequently.

Theorem 4.16 (Partial Converse to the Linear Pair Theorem). *If two adjacent angles are supplementary, then they form a linear pair.*

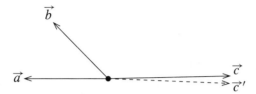

Fig. 4.9. Proof of Theorem 4.16.

Proof. Suppose $\angle ab$ and $\angle bc$ are adjacent supplementary angles that share the common side $\vec{b}$. Let $\vec{c}'$ be the ray opposite $\vec{a}$. Then $\angle ab$ and $\angle bc'$ do in fact form a linear pair by definition. Thus $m\angle ab + m\angle bc' = 180°$ by the linear pair theorem. By hypothesis, we also have the equation $m\angle ab + m\angle bc = 180°$. Subtracting these two equations, we conclude that $m\angle bc' = m\angle bc$.

We noted above that $\vec{a}$ and $\vec{c}'$ lie on opposite sides of $\overleftrightarrow{b}$; and because $\angle ab$ and $\angle bc$ are adjacent by hypothesis, $\vec{a}$ and $\vec{c}$ also lie on opposite sides of $\overleftrightarrow{b}$. It follows therefore that $\vec{c}$ and $\vec{c}'$ are on the same side of $\overleftrightarrow{b}$. Because $m\angle bc' = m\angle bc$, the unique ray theorem shows that $\vec{c} = \vec{c}'$. Therefore $\vec{c}$ is opposite $\vec{a}$, which means that $\angle ab$ and $\angle bc$ form a linear pair. $\square$

Two angles are said to form a pair of ***vertical angles*** if they have the same vertex, they are both proper, and their sides form two distinct pairs of opposite rays. Thus if $\angle ab$ and $\angle cd$ are two proper angles with the same vertex, to say they are vertical angles means either that $\vec{a}$ and $\vec{c}$ are opposite and $\vec{b}$ and $\vec{d}$ are opposite or that $\vec{a}$ and $\vec{d}$ are opposite and $\vec{b}$ and $\vec{c}$ are opposite (see Fig. 4.10).

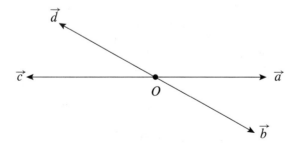

Fig. 4.10. $\angle ab$ and $\angle cd$ are vertical angles.

The next theorem is Euclid's Proposition I.15.

Theorem 4.17 (Vertical Angles Theorem). *Vertical angles are congruent.*

Proof. Exercise 4D. $\square$

Theorem 4.18 (Partial Converse to the Vertical Angles Theorem). *Suppose $\vec{a}$ and $\vec{c}$ are opposite rays starting at a point O, and $\vec{b}$ and $\vec{d}$ are rays starting at O and lying on opposite sides of $\overleftrightarrow{a}$. If $\angle ab \cong \angle cd$, then $\vec{b}$ and $\vec{d}$ are opposite rays.*

Proof. Exercise 4E.   □

The linear pair theorem has a useful generalization to three angles. Suppose $\vec{a}$, $\vec{b}$, $\vec{c}$, and $\vec{d}$ are four distinct rays with the same endpoint. We say the angles $\angle ab$, $\angle bc$, and $\angle cd$ form a ***linear triple*** if all three of the following conditions are satisfied (Fig. 4.11):

   (i) $\vec{a}$ and $\vec{d}$ are opposite rays,
  (ii) $\vec{a} * \vec{b} * \vec{c}$,
 (iii) $\vec{b} * \vec{c} * \vec{d}$.

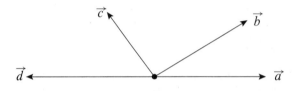

Fig. 4.11. A linear triple.

Theorem 4.19 (Linear Triple Theorem). *If $\angle ab$, $\angle bc$, and $\angle cd$ form a linear triple, then their measures add up to $180°$.*

Proof. Exercise 4F.   □

The Interior of an Angle

For points, the statement that B is between A and C is equivalent (essentially by defini-tion) to the statement that B is an interior point of $\overline{AC}$. This is how we will usually use betweenness of points in geometric proofs: instead of using a coordinate function for a line (which we will rarely mention from now on), we will conclude that a point B is between two other points A and C because it is in the interior of $\overline{AC}$, either because we chose it that way or because we can prove it from other properties of the given configuration.

Betweenness of *rays* is also related to a notion of "interior" (of an angle), which in turn is closely related to plane separation. Developing this relationship rigorously requires some work. Once we have done the work, we will relegate coordinate functions for rays to the background, just as we have done with coordinate functions for lines.

Suppose $\angle AOC$ is a proper angle. We define the ***interior of*** $\angle AOC$, denoted by $\mathrm{Int}\,\angle AOC$, to be the set of all points that are on the same side of $\overleftrightarrow{OA}$ as the ray $\overrightarrow{OC}$ and on the same side of $\overleftrightarrow{OC}$ as $\overrightarrow{OA}$ (see Fig. 4.12). In other words, a point B is in the interior of $\angle AOC$ if and only if it satisfies both of the following conditions:

   • B and C are on the same side of $\overleftrightarrow{OA}$, and
   • B and A are on the same side of $\overleftrightarrow{OC}$.

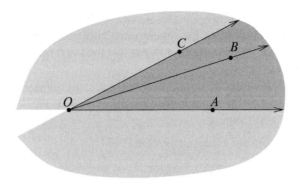

Fig. 4.12. *B* is in the interior of $\angle AOC$.

A point is said to be in the **exterior of an angle** if it is neither on the angle nor in its interior. We do not define the interior or exterior of a zero angle or a straight angle. (For a zero angle, there would be no points in the interior, and for a straight angle there is no way to decide which side of the angle should be considered as its "interior.")

Theorem 4.20. *The interior of a proper angle is a convex set.*

Proof. If $\angle AOB$ is a proper angle, then by definition Int $\angle AOB$ is the intersection of the open half-planes OHP$\left(\overleftrightarrow{OA}, B\right)$ and OHP$\left(\overleftrightarrow{OB}, A\right)$, each of which is convex by Theorem 3.50. The theorem then follows from Theorem 3.47. ☐

From the Y-lemma, we can conclude that if B is in the interior of $\angle AOC$, then every interior point on the ray $\overrightarrow{OB}$ is also in the interior of $\angle AOC$. In that situation, we say that **the ray $\overrightarrow{OB}$ lies in the interior of $\angle AOC$**. From Fig. 4.12, it appears that for a ray $\overrightarrow{OB}$, being in the interior of $\angle AOC$ should be closely related to being between $\overrightarrow{OA}$ and $\overrightarrow{OC}$. In fact, we will show that the two concepts are equivalent. The complete proof will take several steps, but one implication (being in the interior implies betweenness) is easy.

Lemma 4.21. *Suppose $\angle AOC$ is a proper angle and $\overrightarrow{OB}$ is a ray that lies in the interior of $\angle AOC$. Then $\overrightarrow{OA} * \overrightarrow{OB} * \overrightarrow{OC}$.*

Proof. The hypothesis means that B and C are on the same side of $\overleftrightarrow{OA}$ and that B and A are on the same side of $\overleftrightarrow{OC}$. In particular, all three rays lie in HR$\left(\overleftrightarrow{OA}, C\right)$. By Theorem 4.9, one of the rays must lie between the other two. If $\overrightarrow{OA}$ were between $\overrightarrow{OB}$ and $\overrightarrow{OC}$, then Theorem 4.12 would imply that B and C are on opposite sides of $\overleftrightarrow{OA}$, contradicting the hypothesis. Similarly, the assumption that $\overrightarrow{OC}$ is between $\overrightarrow{OA}$ and $\overrightarrow{OB}$ also leads to a contradiction. The only remaining possibility is that $\overrightarrow{OB}$ is between the other two. ☐

Suppose $\vec{a}$, $\vec{b}$, and $\vec{c}$ are three distinct rays with the same endpoint. They form three different angles: $\angle ab$, $\angle bc$, and $\angle ac$. What can we say about the relationships among the measures of these three angles? It depends on how the rays are positioned with respect to each other. There are three possibilities, illustrated in Fig. 4.6:

(a) If two of the rays are collinear (Fig. 4.6(a)), then they must be opposite (because they are distinct), so the linear pair theorem shows that the two angles formed with the remaining ray have measures adding up to 180°.

(b) If no two of the rays are collinear and one ray lies in the interior of the angle formed by the other two (Fig. 4.6(b)), then that ray is between the other two by Lemma 4.21, so the betweenness theorem for rays shows that the largest angle measure is the sum of the two smaller ones.

(c) The only other possibility is that no two of the rays are collinear and none of the rays lies in the interior of the angle formed by the other two (Fig. 4.6(c)). This case is covered by the next theorem.

The proof of the following theorem is rated PG.

Theorem 4.22 (The 360 Theorem). *Suppose $\vec{a}$, $\vec{b}$, and $\vec{c}$ are three distinct rays with the same endpoint, such that no two of the rays are collinear and none of the rays lies in the interior of the angle formed by the other two. Then*

$$m\angle ab + m\angle bc + m\angle ac = 360°. \tag{4.2}$$

Proof. Let O be the common endpoint of $\vec{a}$, $\vec{b}$, and $\vec{c}$, and choose points $A \in \text{Int}\,\vec{a}$, $B \in \text{Int}\,\vec{b}$, and $C \in \text{Int}\,\vec{c}$. The fact that no two of the rays are collinear implies that neither A nor B lies on $\overleftrightarrow{OC}$, and similar statements hold for $\overleftrightarrow{OA}$ and $\overleftrightarrow{OB}$. Consider the following possibilities:

- A and B are on the same side of $\overleftrightarrow{OC}$,
- A and C are on the same side of $\overleftrightarrow{OB}$,
- B and C are on the same side of $\overleftrightarrow{OA}$.

If any two of these statements were true, then one of the rays would be in the interior of the angle formed by the other two; therefore at least two of them must be false. Without loss of generality, let us assume that the first two are false, so that A and B are on opposite sides of $\overleftrightarrow{OC}$ and A and C are on opposite sides of $\overleftrightarrow{OB}$. The situation is illustrated in Fig. 4.13.

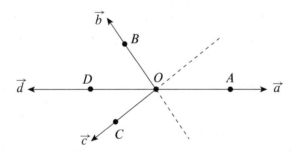

Fig. 4.13. Proof of the 360 theorem.

Let D be a point such that $D * O * A$, and let $\vec{d} = \overrightarrow{OD}$, which is the ray opposite $\vec{a}$. It follows from the X-lemma that D is on the opposite side of both lines $\overleftrightarrow{OB}$ and $\overleftrightarrow{OC}$ from A, so D is on the same side of $\overleftrightarrow{OB}$ as C and on the same side of $\overleftrightarrow{OC}$ as B; in other words, $\vec{d}$ is in the interior of $\angle bc$. It follows from Lemma 4.21 that $\vec{b} * \vec{d} * \vec{c}$. The betweenness theorem for rays implies

$$m\angle bd + m\angle cd = m\angle bc, \tag{4.3}$$

while the linear pair theorem tells us that

$$m\angle ab + m\angle bd = 180°,$$
$$m\angle ac + m\angle cd = 180°.$$

Adding these last two equations together and substituting (4.3) for $m\angle bd + m\angle cd$, we get (4.2). □

Now we are ready to prove the equivalence of the concepts of betweenness and interior for angles. The next lemma has another PG-rated proof.

Lemma 4.23 (Interior Lemma). *Suppose $\vec{a}$, $\vec{b}$, and $\vec{c}$ are three rays with the same endpoint, no two of which are collinear. Then $\vec{b}$ lies in the interior of $\angle ac$ if and only if $\vec{a} * \vec{b} * \vec{c}$.*

Proof. One of the implications—being in the interior implies betweenness—is exactly what Lemma 4.21 says.

Conversely, suppose $\vec{a} * \vec{b} * \vec{c}$, and assume for the sake of contradiction that $\vec{b}$ is not in the interior of $\angle ac$. It is also not the case that $\vec{a}$ is in the interior of $\angle bc$, for that would imply $\vec{b} * \vec{a} * \vec{c}$, which is impossible by Theorem 4.9. For similar reasons, $\vec{c}$ is not in the interior of $\angle ab$. Therefore the hypotheses of the 360 theorem are satisfied, and we have

$$m\angle ab + m\angle bc + m\angle ac = 360°.$$

On the other hand, because we are assuming $\vec{a} * \vec{b} * \vec{c}$, the betweenness theorem implies

$$m\angle ab + m\angle bc = m\angle ac.$$

Subtracting these two equations, we obtain $m\angle ac = 360° - m\angle ac$, from which we conclude that $m\angle ac = 180°$ and thus $\angle ac$ is a straight angle. Since we are assuming that $\vec{a}$ and $\vec{c}$ are noncollinear, this is a contradiction, which proves that our assumption must be false and therefore $\vec{b}$ is in the interior of $\angle ac$. □

Thanks to this lemma, we can now use positions of points with respect to lines to determine when one ray is between two others. The interior lemma shows that a possible alternative definition for betweenness of rays would be to define $\vec{a} * \vec{b} * \vec{c}$ to mean that $\vec{b}$ is in the interior of $\angle ac$. Many authors do in fact adopt this definition. We have chosen the definition we did because it is more closely analogous to the definition of betweenness of points and because it corresponds to our intuitive concept of what it means for one ray to be between two others. From now on, we will use both characterizations of betweenness interchangeably. For example, it is often useful to use the 360 theorem in the following alternative form, which follows immediately from the 360 theorem and the interior lemma.

Corollary 4.24 (Restatement of the 360 Theorem in Terms of Betweenness). *Suppose $\vec{a}$, $\vec{b}$, and $\vec{c}$ are three distinct rays with the same endpoint, such that no two of the rays are collinear and none of the rays lies between the other two. Then*

$$m\angle ab + m\angle bc + m\angle ac = 360°.$$ □

The interior lemma and the 360 theorem are the keys to proving some very useful criteria for deciding when one ray is between two others. To begin, we have the following important converse to the "whole angle greater than the part" theorem. Like its counterpart for segments, this lemma will be extremely useful throughout the book.

Lemma 4.25 (Ordering Lemma for Rays). *Suppose $\vec{a}$, $\vec{b}$, and $\vec{c}$ are rays with the same endpoint, such that $\vec{b}$ and $\vec{c}$ are on the same side of $\overleftrightarrow{a}$ and $m\angle ab < m\angle ac$ (Fig. 4.14). Then $\vec{a} * \vec{b} * \vec{c}$.*

Fig. 4.14. The ordering lemma for rays.

Proof. First we wish to apply Theorem 4.9, which says that exactly one of the rays is between the other two. To do so, we need to verify that all three rays lie in some half-rotation and no two of them are collinear.

The hypotheses guarantee that $\vec{a}$, $\vec{b}$, and $\vec{c}$ all lie in $\mathrm{HR}(\vec{a}, \vec{c})$. They also guarantee that neither $\vec{b}$ nor $\vec{c}$ is collinear with $\vec{a}$ (because then that ray would lie *on* $\overleftrightarrow{a}$, not on one side of it). To show that $\vec{b}$ and $\vec{c}$ cannot be collinear either, suppose for contradiction that they are. Then they are either equal or opposite by Theorem 3.38(b). If they are equal, then $m\angle ab = m\angle ac$, which contradicts our hypothesis. If they are opposite, then the X-lemma shows that they are on opposite sides of $\overleftrightarrow{a}$, which also contradicts our hypothesis. Thus no two of the rays are collinear.

Therefore, the hypotheses of Theorem 4.9 are satisfied, and we conclude that exactly one of the three rays lies between the other two. If $\vec{b} * \vec{a} * \vec{c}$, then $\vec{b}$ and $\vec{c}$ are on opposite sides of $\overleftrightarrow{a}$, contradicting our hypothesis. On the other hand, if $\vec{a} * \vec{c} * \vec{b}$, then the betweenness theorem for rays implies $m\angle ab = m\angle ac + m\angle bc$. Since no two of the rays are collinear, $\angle bc$ is a proper angle, which implies $m\angle bc > 0$ and therefore $m\angle ab > m\angle ac$, again contradicting our hypothesis. Thus the only remaining possibility is $\vec{a} * \vec{b} * \vec{c}$, which is what we wanted to prove. $\square$

The next criterion for betweenness is a partial converse to Theorem 4.12. The proof of the second part is rated PG.

Lemma 4.26 (Adjacency Lemma). *Suppose $\angle ab$ and $\angle bc$ are adjacent angles sharing the common side $\vec{b}$. If either of the following conditions holds, then $\vec{a} * \vec{b} * \vec{c}$.*

(a) $\vec{a}$, $\vec{b}$, and $\vec{c}$ all lie in a single half-rotation.

(b) $m\angle ab + m\angle bc < 180°$.

Proof. First assume that (a) holds. By Theorem 4.9, exactly one of the rays lies between the other two. If $\vec{a}$ is the one between the other two, then the interior lemma implies that $\vec{a}$ lies in the interior of $\angle bc$, which means in particular that $\vec{a}$ and $\vec{c}$ lie on the same side of $\overleftrightarrow{b}$. This contradicts our hypothesis, so this case is impossible. The same argument with $\vec{a}$ and $\vec{c}$ reversed shows that $\vec{c}$ cannot be between the other two either. The only remaining possibility is $\vec{a} * \vec{b} * \vec{c}$.

Now assume (b). By the Y-lemma, there are exactly three possibilities for the location of $\vec{c}$ with respect to $\overleftrightarrow{a}$ (see Fig. 4.15):

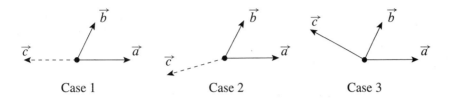

Case 1   Case 2   Case 3

Fig. 4.15. The proof of Lemma 4.26(b).

CASE 1: $\vec{c}$ *is contained in* $\overleftrightarrow{a}$. In this case, $\vec{c}$ and $\vec{a}$ are opposite rays (because the hypothesis implies that they are on opposite sides of $\overleftrightarrow{b}$), so the linear pair theorem implies that $m\angle ab + m\angle bc = 180°$. This contradicts the hypothesis, so this case cannot occur.

CASE 2: $\vec{c}$ *lies on the opposite side of* $\overleftrightarrow{a}$ *from* $\vec{b}$. As before, we also know that $\vec{c}$ lies on the opposite side of $\overleftrightarrow{b}$ from $\vec{a}$. If any of the three rays were in the interior of the angle formed by the other two, at least one of these statements would be false, so the hypotheses of the 360 theorem are satisfied. Therefore,

$$180° > m\angle ab + m\angle bc = 360° - m\angle ac,$$

and algebra yields $m\angle ac > 180°$, which is a contradiction. Thus this case cannot occur either.

CASE 3: $\vec{c}$ *lies on the same side of* $\overleftrightarrow{a}$ *as* $\vec{b}$. This is the only remaining possibility. It implies, in particular, that $\vec{a}, \vec{b}, \vec{c}$ all lie in HR $(\vec{a}, \vec{b})$. Because $\angle ab$ and $\angle bc$ are adjacent angles, part (a) of this lemma implies that $\vec{a} * \vec{b} * \vec{c}$. $\square$

Finally, our last theorem about betweenness shows that there is also a close relationship between betweenness of rays and betweenness of points.

Theorem 4.27 (Betweenness vs. Betweenness). *Suppose ℓ is a line, O is a point not on ℓ, and A, B, C are three distinct points on ℓ. Then $A * B * C$ if and only if $\overrightarrow{OA} * \overrightarrow{OB} * \overrightarrow{OC}$. (See Fig. 4.16.)*

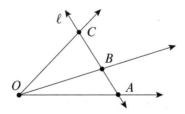

Fig. 4.16. Betweenness vs. betweenness.

Proof. Assume first that $A * B * C$. This means that $B \in \overrightarrow{AC}$, so B and C are on the same side of $\overleftrightarrow{OA}$ by the Y-lemma. The same argument shows that A and B are on the same side of $\overleftrightarrow{OC}$. This implies that $\overrightarrow{OB}$ lies in the interior of $\angle AOC$, so $\overrightarrow{OA} * \overrightarrow{OB} * \overrightarrow{OC}$ by the interior lemma.

Conversely, suppose $\overrightarrow{OA} * \overrightarrow{OB} * \overrightarrow{OC}$. Since A, B, and C are distinct collinear points, exactly one of them is between the other two by Hilbert's betweenness axiom. The interior lemma implies that B and A are on the same side of $\overleftrightarrow{OC}$ and B and C are on the same side of $\overleftrightarrow{OA}$. This means $\overline{BA}$ does not intersect $\overleftrightarrow{OC}$, so C cannot be between B and A; and $\overline{BC}$ does not intersect $\overleftrightarrow{OA}$, so A cannot be between B and C. The only remaining possibility is that B is between A and C.   □

In this chapter, we have seen quite a few characterizations of what it means for one ray to be between two others. Using these various characterizations thoughtfully will free us from having to work with the definition of betweenness in terms of coordinate functions; henceforth, we will never again need to refer to coordinate functions for rays.

Suppose $\vec{a}$, $\vec{b}$, and $\vec{c}$ are three rays with the same endpoint, no two of which are collinear. From the results of this chapter, we see that any one of the following conditions implies $\vec{a} * \vec{b} * \vec{c}$:

- $\vec{b}$ lies in the interior of $\angle ac$ (interior lemma).
- $\vec{b}$ and $\vec{c}$ are on the same side of $\overleftrightarrow{a}$, and $m\angle ab < m\angle ac$ (ordering lemma for rays).
- $\angle ab$ and $\angle bc$ are adjacent, and $\vec{a}, \vec{b}, \vec{c}$ lie in a half-rotation (adjacency lemma).
- $\angle ab$ and $\angle bc$ are adjacent, and $m\angle ab + m\angle bc < 180°$ (adjacency lemma).
- There are points $A \in \mathrm{Int}\,\vec{a}$, $B \in \mathrm{Int}\,\vec{b}$, and $C \in \mathrm{Int}\,\vec{c}$ such that $A * B * C$ (betweenness vs. betweenness).

All of these characterizations will be useful to us in different circumstances. It will be a good idea to remember the conditions on this list and be able to choose the one that is most relevant to the problem at hand whenever you need to prove that one ray is between two others.

Conversely, if we know that $\vec{a} * \vec{b} * \vec{c}$, then the following are all true:

- No two of the rays are collinear (definition of betweenness).
- All three rays lie in some half-rotation (definition of betweenness).
- $m\angle ab + m\angle bc = m\angle ac$ (betweenness theorem for rays).
- $m\angle ac > m\angle ab$ and $m\angle ac > m\angle bc$ (whole angle is greater than the part).
- $\angle ab$ and $\angle bc$ are adjacent (Theorem 4.12).
- $\vec{b}$ lies in the interior of $\angle ac$ (interior lemma).

Angle Bisectors and Perpendiculars

Suppose $\angle ab$ is a proper angle. A ray $\vec{r}$ is called an **angle bisector of $\angle ab$** if $\vec{r}$ is between $\vec{a}$ and $\vec{b}$ and $m\angle ar = m\angle rb$. The next theorem is our analogue of Euclid's Proposition I.9, but we give a different proof based on the angle construction theorem.

Theorem 4.28 (Existence and Uniqueness of Angle Bisectors). *Every proper angle has a unique angle bisector.*

Proof. Let $\angle ab$ be a proper angle. The angle construction theorem shows that there is a ray $\vec{r}$ on the same side of $\overleftrightarrow{a}$ as $\vec{b}$ such that $m\angle ar = \frac{1}{2}m\angle ab$. The ordering lemma for rays implies that $\vec{a} * \vec{r} * \vec{b}$, and then the betweenness theorem for rays implies

$$m\angle rb = m\angle ab - m\angle ar = m\angle ab - \tfrac{1}{2}m\angle ab = \tfrac{1}{2}m\angle ab,$$

so $m\angle ar = m\angle rb$ by transitivity. Thus $\vec{r}$ is an angle bisector of $\angle ab$.

To prove that $\vec{r}$ is the unique bisector, suppose $\vec{r}'$ is another one, which means that $\vec{a} * \vec{r}' * \vec{b}$ and $m\angle ar' = m\angle r'b$. The betweenness theorem for rays implies that

$$m\angle ab = m\angle ar' + m\angle r'b = 2m\angle ar',$$

so $m\angle ar' = \frac{1}{2}m\angle ab = m\angle ar$. The interior lemma implies that $\vec{r}'$ is on the same side of $\overleftrightarrow{a}$ as $\vec{r}$, and so the unique ray theorem guarantees that $\vec{r}' = \vec{r}$. ☐

The restriction to proper angles is necessary in this theorem. For example, if $\angle ab$ is a straight angle, there are two different rays that one might consider to be its "bisectors." Instead of talking about bisectors in this case, we use the following terminology. Two distinct lines ℓ and m are said to be ***perpendicular*** if they intersect at a point O and one of the rays in ℓ starting at O makes an angle of $90°$ with one of the rays in m starting at O. In this case, we write $\ell \perp m$, and we say ***ℓ is perpendicular to m at O***.

Theorem 4.29 (Four Right Angles Theorem). *If ℓ and m are perpendicular lines, then ℓ and m form four right angles.*

Proof. Exercise 4H. ☐

The following theorem is Euclid's Proposition I.11.

Theorem 4.30 (Constructing a Perpendicular). *Let ℓ be a line and let P be a point on ℓ. Then there exists a unique line m that is perpendicular to ℓ at P.*

Proof. Exercise 4I. ☐

Exercises

4A. Prove Theorem 4.9 (one ray must be between the other two).

4B. Prove Corollary 4.10 (consistency of betweenness of rays).

4C. Prove Theorem 4.11 (Euclid's common notions for angles).

4D. Prove Theorem 4.17 (the vertical angles theorem).

4E. Prove Theorem 4.18 (the partial converse to the vertical angles theorem).

4F. Prove Theorem 4.19 (the linear triple theorem).

4G. Suppose $\vec{a}$, $\vec{b}$, $\vec{c}$, and $\vec{d}$ are four distinct rays with the same endpoint, such that $\vec{a}$ and $\vec{d}$ are opposite rays and $\vec{a} * \vec{b} * \vec{c}$. Show that $\angle ab$, $\angle bc$, and $\angle cd$ form a linear triple. (In other words, condition (iii) in the definition of a linear triple is redundant.) [Hint: Use the interior lemma.]

4H. Prove Theorem 4.29 (the four right angles theorem).

4I. Prove Theorem 4.30 (construction of perpendiculars).

Chapter 5

Triangles

Many important theorems of Euclidean geometry are theorems about the geometric properties of triangles. We begin by introducing the relevant terminology.

Definitions

We define a ***triangle*** to be a union of three segments $\overline{AB}$, $\overline{BC}$, and $\overline{AC}$ formed by three noncollinear points A, B, and C. The points A, B, C are called the ***vertices*** (singular: ***vertex***) of the triangle, and the segments $\overline{AB}$, $\overline{BC}$, $\overline{AC}$ are called its ***sides*** or ***edges***. (We will most often use the term "sides," but "edges" is sometimes preferred when there is a possibility of confusion with sides of lines.) The triangle whose vertices are A, B, and C is denoted by $\triangle ABC$. Note that, in contrast to Euclid, we define a triangle to include only the points on the line segments, not the points in the "interior."

The next proof is rated PG.

Theorem 5.1 (Consistency of Triangle Vertices). *If $\triangle ABC$ is a triangle, the only extreme points of $\triangle ABC$ are A, B, and C. Thus if $\triangle ABC = \triangle A'B'C'$, then the sets $\{A, B, C\}$ and $\{A', B', C'\}$ are equal.*

Proof. Exercise 5A. $\qquad\qquad\qquad\qquad\qquad\qquad\qquad\qquad\qquad\qquad\qquad\qquad\qquad$ □

If $\triangle ABC$ is a triangle, the ***angles of $\triangle ABC$*** are the three angles $\angle BAC$, $\angle ABC$, and $\angle BCA$. When there is only one angle under discussion at a given vertex (so there can be no confusion about which angle is meant), we will often label an angle by its vertex alone instead of using the "official" notation, so the three angles of $\triangle ABC$ can also be referred to as $\angle A$, $\angle B$, and $\angle C$, respectively. At other times, for clarity, we might label some of the angles in a drawing with numbers and refer to them as $\angle 1$, $\angle 2$, etc. For example, in Fig. 5.1, the notations $\angle A$ and $\angle 1$ both refer to $\angle CAB$; $\angle B$ and $\angle 2$ refer to $\angle ABC$; and $\angle C$ and $\angle 3$ refer to $\angle BCA$. These notations are just a convenient shorthand, and there should always be enough information given so that the reader can replace each such abbreviation with the full notation without any trouble.

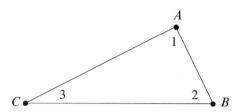

Fig. 5.1. The angles of a triangle.

Caution is required, however, when using diagrams in proofs. It is acceptable to refer to a diagram to establish the meaning of an abbreviation such as ∠1; but each symbol that is so defined should be mentioned in the text of the proof, using a phrase such as "Let ∠1 denote the angle shown in Fig. 5.1." Similarly, diagrams can sometimes be used to simplify the statements of hypotheses, as in "Assume that the three angles marked in the diagram are right angles." (Such assumptions must always be stated, not left for the reader to infer from the diagram.) However, a diagram can *never* be used to justify a conclusion. Even if it is "obvious" from a diagram that two lines, rays, or segments intersect or that a certain angle is acute or that three points are collinear, if such facts are not part of the hypothesis, then they must always be proved according to the conventions outlined in Appendix F. Most of the theorems of the preceding two chapters were designed to give us tools for justifying such conclusions.

This is not to say that diagrams have no place in rigorous geometric proofs. In fact, well-chosen diagrams are essential for making geometric proofs easy to follow. If you don't believe it, try reading some of Euclid's proofs, or the proofs in this chapter, while covering up the diagrams that go with them. As soon as a proof refers to more than one or two points, lines, rays, or segments, you will probably find yourself drawing a diagram to help you remember how all the elements are related to each other. The goal of good mathematical writing is communication, and in geometry more than any other subject, diagrams play an essential role.[1]

Let us continue with our definitions. If ∠1 and ∠2 are two angles of a triangle, a side of the same triangle is said to be *included between ∠1 and ∠2* if its endpoints are the vertices of those angles; thus, for example, in △ABC, the side $\overline{AB}$ is included between ∠A and ∠B. Similarly, an angle of a triangle is said to be *included between two sides* if those sides meet at the vertex of the angle; thus ∠B is included between $\overline{AB}$ and $\overline{BC}$. We say *a side is opposite an angle* (or *an angle is opposite a side*) if the vertex of the angle is not one of the endpoints of the side; thus, for example, ∠C is opposite $\overline{AB}$.

Triangles can be classified by the relationships among their side lengths:

- An *equilateral triangle* is one in which all three sides have the same length.

- An *isosceles triangle* is one in which at least two sides have the same length.

- A *scalene triangle* is one in which all three side lengths are different.

---

[1] It is interesting to note that a group of logicians [**ADM09**] have recently developed a rigorous axiom system for Euclidean geometry that includes some axioms designed specifically for drawing conclusions from diagrams. But of course all such conclusions must be explicitly justified by the axioms and cannot rely on geometric intuition.

They can also be classified by their angle measures:

- An *equiangular triangle* is one in which all three angles have the same measure.
- An *acute triangle* is one in which all three angles are acute.
- A *right triangle* is one that contains a right angle.
- An *obtuse triangle* is one that contains an obtuse angle.

Note that our definition of isosceles triangles is different from Euclid's, which stipulates that an isosceles triangle has *only* two congruent sides; we consider an equilateral triangle to be a special case of an isosceles triangle.

In an isosceles triangle, the *base* is the side that is not congruent to the other two, if there is one (if the triangle is equilateral, any side can be considered the base), and the *base angles* are the two angles that include the base. In a right triangle, the side opposite the right angle is called the *hypotenuse*, and the other two sides are called the *legs*.

Intersections of Lines and Triangles

Before we study geometric properties of triangles, we need to establish some "obvious" results about conditions under which lines and triangles intersect. The first one is a simple application of the plane separation postulate.

Theorem 5.2 (Pasch's Theorem). *Suppose* $\triangle ABC$ *is a triangle and* ℓ *is a line that does not contain any of the points A, B, or C. If* ℓ *intersects one of the sides of* $\triangle ABC$, *then it also intersects another side (see Fig. 5.2).*

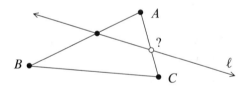

Fig. 5.2. Pasch's Theorem.

Proof. Assume without loss of generality that ℓ intersects $\overline{AB}$. If it also intersects $\overline{BC}$, we are done, so assume it does not do so. Then B and C are on the same side of ℓ, while A and B are on opposite sides. It follows that A and C are on opposite sides, and thus $\overline{AC}$ must intersect ℓ. □

Corollary 5.3. *If* $\triangle ABC$ *is a triangle and* ℓ *is a line that does not contain any of the points A, B, or C, then either* ℓ *intersects exactly two sides of* $\triangle ABC$ *or it intersects none of them.*

Proof. Exercise 5B. □

Pasch's theorem is named after Moritz Pasch, a nineteenth-century German mathematician who was one of the first to point out many of the unstated assumptions in Euclid's *Elements*. He proposed that the statement of Theorem 5.2 be added to Euclid's geometry as an additional axiom; it is in fact included (often under the name *Pasch's Axiom*) in some modern axiom systems, including Hilbert's (see Axiom II.4 in Appendix A). In most systems, however, some version of the plane separation postulate is used instead.

The next theorem is related to Pasch's theorem, but it is slightly different—whereas Pasch's theorem says that a line that intersects one *side* of a triangle must intersect one of the other sides, our next theorem applies to a ray that intersects a *vertex*, provided it lies between the two sides that meet at that vertex.

Theorem 5.4 (The Crossbar Theorem). *Suppose* $\triangle ABC$ *is a triangle and* $\overrightarrow{AD}$ *is a ray between* $\overrightarrow{AB}$ *and* $\overrightarrow{AC}$. *Then the interior of* $\overrightarrow{AD}$ *intersects the interior of* $\overline{BC}$ *(Fig. 5.3).*

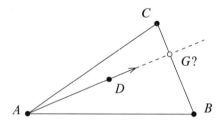

Fig. 5.3. The crossbar theorem.

Proof. Theorem 4.12 shows that B and C are on opposite sides of $\overleftrightarrow{AD}$. This means that there is a point G where the interior of $\overline{BC}$ meets $\overleftrightarrow{AD}$. We just need to show that $G \in \text{Int}\,\overrightarrow{AD}$, and not on its opposite ray. The assumption that $\overrightarrow{AB} * \overrightarrow{AD} * \overrightarrow{AC}$ means that $\overrightarrow{AD}$ is in the interior of $\angle BAC$, so D is on the same side of $\overleftrightarrow{AB}$ as C; and C and G are on the same side of $\overleftrightarrow{AB}$ by the Y-lemma applied to the ray $\overrightarrow{BC}$. It follows that D and G are on the same side of $\overleftrightarrow{AB}$. By Theorem 3.46, $\text{Int}\,\overrightarrow{AD}$ is the only part of $\overleftrightarrow{AD}$ on that side of $\overleftrightarrow{AB}$, so $G \in \text{Int}\,\overrightarrow{AD}$ as claimed.   □

Euclid would never have thought to prove theorems such as these last two, because the existence of a second intersection point would have been obvious from the picture. For example, in his proof of Proposition I.10, Euclid implicitly uses the crossbar theorem without acknowledging it. But our modern understanding of the axiomatic method requires that every such statement be justified.

Students sometimes have a hard time remembering the distinction between the crossbar theorem and the betweenness vs. betweenness theorem because the diagrams that go with both theorems are very similar. But there is a big difference in the circumstances under which the two theorems can be applied: in order to apply the betweenness vs. betweenness theorem, we have to know in advance that all three rays intersect the line ℓ. The crossbar theorem, in contrast, applies when we do not yet know that $\overrightarrow{AD}$ intersects $\overleftrightarrow{BC}$, provided we know that it lies between $\overrightarrow{AB}$ and $\overrightarrow{AC}$.

Congruent Triangles

Two triangles are said to be ***congruent*** if there is a one-to-one correspondence between their vertices such that all three pairs of corresponding angles and all three pairs of corresponding sides are congruent. The notation $\triangle ABC \cong \triangle DEF$ means that these two triangles are congruent under the specific correspondence $A \leftrightarrow D$, $B \leftrightarrow E$, and $C \leftrightarrow F$; thus it means specifically that

$$\overline{AB} \cong \overline{DE}, \quad \overline{BC} \cong \overline{EF}, \quad \overline{CA} \cong \overline{FD}, \quad \angle A \cong \angle D, \quad \angle B \cong \angle E, \quad \angle C \cong \angle F.$$

(See Fig. 5.4.) In diagrams like this, we will frequently follow the common convention of using matching marks to indicate which sides and angles are assumed to be congruent to each other. Note that the order in which the vertices are listed in the notation $\triangle ABC \cong \triangle DEF$ is significant.

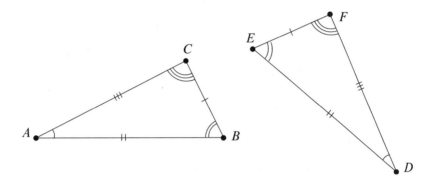

Fig. 5.4. Congruent triangles: $\triangle ABC \cong \triangle DEF$.

Theorem 5.5 (Transitive Property of Congruence of Triangles). *Two triangles that are both congruent to a third triangle are congruent to each other.*

Proof. This is an immediate consequence of the transitive properties of congruence of segments and angles. $\qquad\square$

In order to prove that two triangles are congruent, in principle one must prove that six different "parts" of one triangle (three sides and three angles) are congruent to the six corresponding parts of the other. Conversely, if we know that two triangles are congruent, then we can conclude from the definition that all six pairs of corresponding parts are congruent.

We know from our experience with physical objects that it is usually not necessary to check all six congruences in order to conclude that two triangles are congruent. If you make a triangle out of three sticks, there is only one triangle that can be made—any two people using three sticks of the same lengths will make congruent triangles. This observation is expressed mathematically in Euclid's Proposition I.8, nowadays called the *SSS* (or *side-side-side*) *congruence theorem*. Another such result is one of Euclid's earliest results, Proposition I.4. Called the *SAS* (*side-angle-side*) *congruence theorem*, it asserts that two triangles are congruent if there is a correspondence between their vertices such that two pairs of corresponding sides and the included angles are congruent.

Euclid's proofs of Propositions I.4 and I.8 used a highly unusual technique, called the *method of superposition*, which we described in Chapter 1. As we noted there, this argument is not justified by Euclid's postulates, so we have to look for some other way to prove SAS and SSS congruence. In fact, it is not possible to prove either SAS or SSS congruence from the neutral geometry postulates we have introduced so far: in the next chapter, we will describe an interpretation of our axiomatic system (called *taxicab geometry*) in which

Postulates 1–8 are true, as are all five of Euclid's postulates, but the SAS and SSS congruence theorems are actually false. This shows that these theorems do not follow from the other postulates.

There are two ways to remedy this situation. One way is to try to mimic Euclid's method of superposition—this would require introducing one or more new postulates that assert the existence of certain kinds of "motions" of the plane, or more precisely bijective functions from the plane to itself (called *transformations of the plane*) that preserve all geometric properties such as distances and angle measures. In recent years, this transformational approach has been adopted by a few high-school geometry textbooks. It has an appealing naturalness to it and leads to an interesting study of transformations for their own sake, in addition to justifying Euclid's original arguments for SAS and SSS. Its chief disadvantage is that if one is striving for rigor, one must work rather hard right at the beginning to prove a list of properties of transformations before one can use them to prove anything very interesting. For reference, the basic features of this approach are described in Appendix I.

On the other hand, a simpler approach is just to adopt the SAS congruence criterion itself as a postulate. In one form or another, this is the approach adopted by Hilbert, Birkhoff, SMSG, and most high-school texts, and it is the approach we adopt here. Having just studied the preceding two chapters of this book, which consist largely of proofs of facts about lines, segments, rays, and angles that most people would consider intuitively obvious, you will probably be relieved to know that adopting SAS as a postulate will allow us very quickly to start proving interesting geometric theorems.

With this motivation in mind, we introduce our next postulate.

Postulate 9 (The SAS Postulate). *If there is a correspondence between the vertices of two triangles such that two sides and the included angle of one triangle are congruent to the corresponding sides and angle of the other triangle, then the triangles are congruent under that correspondence.*

All of the other congruence criteria for triangles can be deduced from the SAS postulate combined with our other postulates. Here is the first example. It is the first part of Euclid's Proposition I.26, with the same proof. (The other part is AAS congruence, which is Theorem 5.22 below.)

Theorem 5.6 (ASA Congruence). *If there is a correspondence between the vertices of two triangles such that two angles and the included side of one triangle are congruent to the corresponding angles and side of the other triangle, then the triangles are congruent under that correspondence.*

Proof. Suppose $\triangle ABC$ and $\triangle DEF$ are triangles such that $\angle A \cong \angle D$, $\angle B \cong \angle E$, and $\overline{AB} \cong \overline{DE}$. If $\overline{AC} \cong \overline{DF}$, then $\triangle ABC \cong \triangle DEF$ by SAS, and we are done. So assume for the sake of contradiction that $\overline{AC} \not\cong \overline{DF}$. By the trichotomy law, one of these sides must be longer than the other; without loss of generality, let us say that $AC > DF$. By Euclid's segment cutoff theorem (Corollary 3.37), there is an interior point $C' \in \overline{AC}$ such that $\overline{AC'} \cong \overline{DF}$ (Fig. 5.5). The two triangles $\triangle ABC'$ and $\triangle DEF$ satisfy the hypotheses of the SAS postulate, so they are congruent. It follows from the definition of congruence that $\angle ABC' \cong \angle E$. Since $A * C' * C$ by construction, the betweenness vs. betweenness

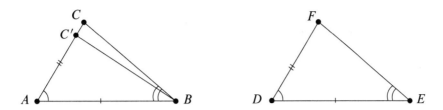

Fig. 5.5. The ASA congruence theorem.

theorem implies that $\overrightarrow{BA} * \overrightarrow{BC'} * \overrightarrow{BC}$, and therefore $m\angle ABC > m\angle ABC'$ by Theorem 4.11(e) (the whole angle is greater than the part). Thus by substitution, $m\angle ABC > m\angle E$. This contradicts our hypothesis that these two angles are congruent and rules out the possibility that $\overline{AC}$ and $\overline{DF}$ are not congruent. ◻

The next theorem corresponds to Euclid's Proposition I.5.

Theorem 5.7 (Isosceles Triangle Theorem). *If two sides of a triangle are congruent to each other, then the angles opposite those sides are congruent.*

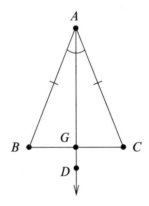

Fig. 5.6. The isosceles triangle theorem.

Proof. Let $\triangle ABC$ be a triangle in which $\overline{AB} \cong \overline{AC}$ (Fig. 5.6). We need to show that $\angle B \cong \angle C$. Let $\overrightarrow{AD}$ be the bisector of $\angle BAC$. Because $\overrightarrow{AD}$ is between $\overrightarrow{AB}$ and $\overrightarrow{AC}$ by definition of an angle bisector, the crossbar theorem implies that there is a point G where $\overrightarrow{AD}$ intersects the interior of $\overline{BC}$. Observe that $\angle BAG \cong \angle CAG$ by definition of angle bisector, $\overline{AB} \cong \overline{AC}$ by hypothesis, and $\overline{AG}$ is congruent to itself; thus $\triangle BAG \cong \triangle CAG$ by SAS. It follows from the definition of congruence that $\angle B \cong \angle C$. ◻

This proof is simple, short, and easy for most people to follow. This is in sharp contrast to Euclid's elaborate proof, which required extending two lines, drawing two additional line segments, and applying SAS twice. Because that proof has been a stumbling block for geometry students at least since the Middle Ages, it was given the epithet *pons asinorum*, Latin for "bridge of fools" (literally "bridge of asses"), reflecting the fact that only serious and hard-working students succeeded in crossing it (and perhaps also reflecting the fact that the diagram included with Euclid's proof resembles a bridge).

There is another, even simpler, proof of the isosceles triangle theorem, apparently discovered by Pappus of Alexandria about 600 years after Euclid's time. It is so simple that it sometimes requires a second look to see what's going on.

Pappus's Proof of the Isosceles Triangle Theorem. As before, suppose $\triangle ABC$ is a triangle in which $\overline{AB} \cong \overline{AC}$. Under the correspondence $A \leftrightarrow A$, $B \leftrightarrow C$, and $C \leftrightarrow B$, the triangles $\triangle ABC$ and $\triangle ACB$ satisfy the hypotheses of the SAS postulate (because $\overline{AB} \cong \overline{AC}$, $\overline{AC} \cong \overline{AB}$, and $\angle BAC \cong \angle CAB$), and therefore $\triangle ABC \cong \triangle ACB$. Thus the corresponding angles $\angle B$ and $\angle C$ are congruent.   □

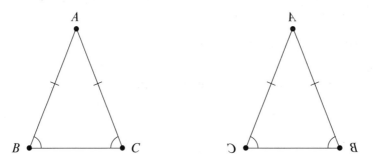

Fig. 5.7. Pappus's proof of the isosceles triangle theorem.

This proof might be a little surprising if you haven't seen it before. To understand how it works, it helps to visualize two copies of $\triangle ABC$, one a mirror image of the other. (See Fig. 5.7; I am indebted to the beautiful geometry text by Harold Jacobs [**Jac03**] for the idea of illustrating the proof this way.) Intuitively, the crux of the idea is that $\triangle ABC$ is congruent to its mirror image. During the centuries since Pappus's proof was first written down, some mathematicians questioned whether it is legitimate to apply SAS to two triangles that happen to be the same triangle. But now that we have a more precise understanding of the meanings of mathematical statements, it is clear that our SAS postulate does indeed apply here, because its statement does not stipulate that the two triangles in question are distinct.

The next theorem is Euclid's Proposition I.6.

Theorem 5.8 (Converse to the Isosceles Triangle Theorem). *If two angles of a triangle are congruent to each other, then the sides opposite those angles are congruent.*

Proof. Exercise 5D.   □

Corollary 5.9. *A triangle is equilateral if and only if it is equiangular.*

Proof. If $\triangle ABC$ is equilateral, then the isosceles triangle theorem shows that any two of its angles are congruent to each other. Conversely, if $\triangle ABC$ is equiangular, then the converse to the isosceles triangle theorem shows that any two sides are congruent to each other.   □

Theorem 5.10 (Triangle Copying Theorem). *Suppose* $\triangle ABC$ *is a triangle and* $\overline{DE}$ *is a segment congruent to* $\overline{AB}$. *On each side of* $\overleftrightarrow{DE}$, *there is a point* F *such that* $\triangle DEF \cong \triangle ABC$.

Proof. Exercise 5E. □

The point F whose existence is asserted in the previous theorem is actually unique, once we decide on a specific side of $\overleftrightarrow{DE}$. Like the unique point and unique ray theorems, this uniqueness statement is often very important in its own right, so it is worth stating as a separate theorem. This is Euclid's Proposition I.7, but our proof is different.

Theorem 5.11 (Unique Triangle Theorem). *Suppose* $\overline{DE}$ *is a segment and* F *and* F' *are points on the same side of* $\overleftrightarrow{DE}$ *such that* $\triangle DEF \cong \triangle DEF'$ *(Fig. 5.8). Then* $F = F'$.

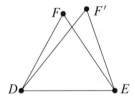

Fig. 5.8. The unique triangle theorem.

Proof. Because $\angle FDE \cong \angle F'DE$ and both are on the same side of $\overleftrightarrow{DE}$, the unique ray theorem implies that $\overrightarrow{DF} = \overrightarrow{DF'}$. Then because $\overline{DF} \cong \overline{DF'}$ and both lie on the same ray starting at D, the unique point theorem implies that $F = F'$. □

Next comes the most natural-seeming congruence theorem of all, the SSS congruence theorem. Somewhat surprisingly, proving it takes considerably more work than proving ASA. It is Euclid's Proposition I.8, but we give a different proof that avoids the unjustified use of superposition.

Theorem 5.12 (SSS Congruence). *If there is a correspondence between the vertices of two triangles such that all three sides of one triangle are congruent to the corresponding sides of the other triangle, then the triangles are congruent under that correspondence.*

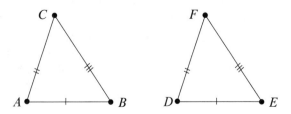

Fig. 5.9. The SSS congruence theorem.

Proof. Suppose $\triangle ABC$ and $\triangle DEF$ are triangles such that $\overline{AB} \cong \overline{DE}$, $\overline{AC} \cong \overline{DF}$, and $\overline{BC} \cong \overline{EF}$ (Fig. 5.9). By Theorem 5.10, we can choose a point C' on the opposite side of $\overleftrightarrow{AB}$ from C such that $\triangle ABC' \cong \triangle DEF$. If A, C, and C' are noncollinear, then because $\overline{AC} \cong \overline{AC'}$, it follows from the isosceles triangle theorem applied to $\triangle ACC'$ that $\angle ACC' \cong \angle AC'C$. Similarly, if B, C, and C' are noncollinear, from $\overline{BC} \cong \overline{BC'}$ it follows that $\angle BCC' \cong \angle BC'C$. We wish to use these angle congruences to show that $\angle ACB \cong \angle AC'B$.

Because C and C' are on opposite sides of $\overleftrightarrow{AB}$, the segment $\overline{CC'}$ intersects $\overleftrightarrow{AB}$ at a point Q. Since Q could lie anywhere on $\overleftrightarrow{AB}$, we have to consider three separate possibilities, depending on how Q is situated relative to A and B (Fig. 5.10).

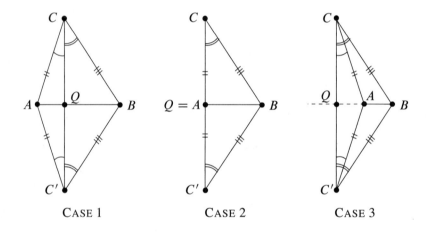

CASE 1   CASE 2   CASE 3

Fig. 5.10. Some possible positions for Q.

CASE 1: *Q is between A and B.* In this case, the betweenness vs. betweenness theorem implies that $\overrightarrow{CC'}$ (which is the same ray as $\overrightarrow{CQ}$) is between $\overrightarrow{CA}$ and $\overrightarrow{CB}$, and likewise $\overrightarrow{C'C}$ is between $\overrightarrow{C'A}$ and $\overrightarrow{C'B}$. Thus it follows from the angle addition theorem (Theorem 4.11(b)) that $m\angle ACB = m\angle AC'B$.

CASE 2: *Q is equal to A or B.* If $Q = A$, then $C * A * C'$, while B does not lie on $\overleftrightarrow{CC'}$ (because otherwise A, B, C would be collinear). It follows that $\angle ACB = \angle BCC' \cong \angle BC'C = \angle AC'B$. Similarly, if $Q = B$, then $\angle ACB = \angle ACC' \cong \angle AC'C = \angle AC'B$.

CASE 3: *Q is not in $\overline{AB}$.* In this case there are two possibilities: either $Q * A * B$ or $A * B * Q$. The argument is the same in both cases except with A and B reversed, so we will just consider the case $Q * A * B$. Now betweenness vs. betweenness tells us that $\overrightarrow{CC'} * \overrightarrow{CA} * \overrightarrow{CB}$ and $\overrightarrow{C'C} * \overrightarrow{C'A} * \overrightarrow{C'B}$, and the angle subtraction theorem gives $m\angle ACB = m\angle AC'B$.

In all three cases, we have proved that $\angle ACB \cong \angle AC'B$, and so $\triangle ABC \cong \triangle ABC'$ by SAS. Since $\triangle ABC' \cong \triangle DEF$ by construction, it follows from transitivity of congruence that $\triangle ABC \cong \triangle DEF$. $\qquad\square$

Inequalities

In this section, we will prove several important inequalities about sides and angles of triangles. The first of these, the exterior angle inequality, will turn out to be the crucial ingredient in a number of important proofs to follow.

In order to state the exterior angle inequality, we need to introduce some more definitions. An angle that forms a linear pair with one of the angles of a triangle is called an ***exterior angle*** of the triangle. To emphasize the distinction, the *angles of the triangle* are often called its ***interior angles***. Thus in $\triangle ABC$, for example, the interior angles are $\angle BAC$, $\angle ABC$, and $\angle ACB$; if P and Q are points such that $A * B * P$ and $C * B * Q$, then $\angle ABQ$ and $\angle CBP$ are two of the exterior angles of $\triangle ABC$ (Fig. 5.11). Given an exterior angle, the interior angle that forms a linear pair with it is called its ***adjacent interior angle***, while the other two interior angles are called its ***remote interior angles***.

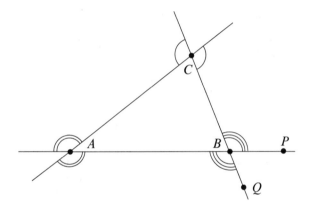

Fig. 5.11. Exterior angles of $\triangle ABC$.

Our next theorem is Euclid's Proposition I.16, which we examined in detail in Chapter 1. It is, surprisingly, one of the most important theorems in neutral geometry—not so much because its result is particularly interesting or significant in its own right, but because its proof involves some unexpected tricks, and it will be the key step in the proofs of a number of significant theorems later. This proof is rated PG.

Theorem 5.13 (Exterior Angle Inequality). *The measure of an exterior angle of a triangle is strictly greater than the measure of either remote interior angle.*

Proof. Let $\triangle ABC$ be a triangle, and let $\angle CBP$ be an exterior angle at B, with $A * B * P$ as in Fig. 5.11. We need to show that

$$m\angle CBP > m\angle BCA \quad \text{and} \quad m\angle CBP > m\angle BAC. \tag{5.1}$$

We begin by proving the first inequality.

Let M be the midpoint of $\overline{BC}$, and let X be the point on the ray opposite $\overrightarrow{MA}$ such that $\overline{MX} \cong \overline{AM}$ (Fig. 5.12). Since $\angle BMX \cong \angle CMA$ by the vertical angles theorem, the SAS postulate implies that $\triangle BMX \cong \triangle CMA$. By definition of congruence, then, we conclude that $\angle MBX \cong \angle MCA$.

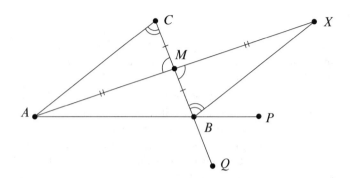

Fig. 5.12. Proof of the exterior angle inequality.

Next we need to show that $\overrightarrow{BX}$ is between $\overrightarrow{BM}$ and $\overrightarrow{BP}$. Because $A * B * P$, the points A and P are on opposite sides of $\overleftrightarrow{BM}$, and A and X are on opposite sides of the same line by the X-lemma. Therefore X and P are on the same side of $\overleftrightarrow{BM}$. On the other hand, X and M are on the same side of $\overleftrightarrow{BP}$ by the Y-lemma applied to the ray $\overrightarrow{AX}$. Therefore X is in the interior of $\angle MBP$, which implies $\overrightarrow{BM} * \overrightarrow{BX} * \overrightarrow{BP}$ by the interior lemma. Therefore $m\angle MBP > m\angle MBX$ by Theorem 4.11(e) (the whole angle is greater than the part). Since $m\angle MBX = m\angle MCA$, it follows by substitution that $m\angle MBP > m\angle MCA$, which is the same as $m\angle CBP > m\angle BCA$. This completes the proof of the first inequality of (5.1).

To prove the second inequality, just note that the same argument (with the roles of A and C reversed) implies that $m\angle ABQ > m\angle BAC$, where Q is any point such that $C * B * Q$. Because $m\angle CBP = m\angle ABQ$ by the vertical angles theorem, substitution yields the second inequality of (5.1). $\qquad\square$

Our proof of the exterior angle inequality is essentially the same as Euclid's, with only one difference: Euclid simply asserted that $m\angle MBP > m\angle MBX$, because it was obvious from the diagram that the latter is "part of" the former. In contrast, our proof gives a detailed justification for this inequality based on properties of betweenness of rays.

The next corollary corresponds to Euclid's Proposition I.17. Our proof uses essentially the same idea as Euclid's, but expressed in more algebraic language.

Corollary 5.14. *The sum of the measures of any two angles of a triangle is less than* $180°$.

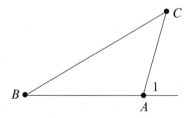

Fig. 5.13. Proof of Corollary 5.14.

Proof. Let $\triangle ABC$ be a triangle. We will show that $m\angle A + m\angle B < 180°$; the argument for any other pair of angles is exactly the same. Let $\angle 1$ denote either of the exterior angles adjacent to $\angle A$ (Fig. 5.13). The exterior angle inequality implies

$$m\angle 1 > m\angle B, \tag{5.2}$$

and the linear pair theorem implies

$$m\angle 1 = 180° - m\angle A. \tag{5.3}$$

Substituting (5.3) into (5.2), we obtain $180° - m\angle A > m\angle B$, which is equivalent to the inequality we want to prove. $\qquad\square$

Corollary 5.15. *Every triangle has at least two acute angles.*

Proof. Suppose $\triangle ABC$ is a triangle, and assume for the sake of contradiction that it has no more than one acute angle. Without loss of generality, we may assume that neither $\angle A$ nor $\angle B$ is acute, so $m\angle A \geq 90°$ and $m\angle B \geq 90°$. Therefore, $m\angle A + m\angle B \geq 90° + 90° = 180°$, contradicting Corollary 5.14. $\qquad\square$

The next theorem combines Euclid's Propositions I.18 and I.19, with the same proofs. It says roughly that the largest side of a triangle is opposite the largest angle, and the smallest side is opposite the smallest angle.

Theorem 5.16 (Scalene Inequality). *Let $\triangle ABC$ be a triangle. Then $AC > BC$ if and only if $m\angle B > m\angle A$.*

Proof. Assume first that $AC > BC$ (Fig. 5.14). By Euclid's segment cutoff theorem

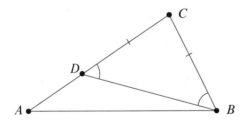

Fig. 5.14. Proof of the scalene inequality.

(Corollary 3.37), we can choose a point D in the interior of $\overline{AC}$ such that $\overline{CD} \cong \overline{CB}$. By the isosceles triangle theorem, $\angle CBD \cong \angle CDB$. Now, the betweenness vs. betweenness theorem implies that $\overrightarrow{BA} * \overrightarrow{BD} * \overrightarrow{BC}$, so $m\angle CBA > m\angle CBD$. On the other hand, the exterior angle inequality implies that $m\angle CDB > m\angle CAB$. Putting these results together, we find

$$m\angle CBA > m\angle CBD = m\angle CDB > m\angle CAB.$$

Conversely, assume that $m\angle B > m\angle A$. By the trichotomy law for real numbers, there are three possibilities for the lengths AC and BC: $AC < BC$, $AC = BC$, or $AC > BC$. If $AC < BC$, then the proof in the preceding paragraph shows that $m\angle A > m\angle B$, which is a contradiction. If $AC = BC$, then the isosceles triangle theorem shows that $m\angle B = m\angle A$, which is also a contradiction. The only remaining possibility is that $AC > BC$. $\qquad\square$

Corollary 5.17. *In any right triangle, the hypotenuse is strictly longer than either leg.*

Proof. By Corollary 5.15, a right triangle has two acute angles, so the right angle is strictly larger than either of the other two angles. Thus the result follows from the scalene inequality. □

Our third major inequality is a version of the familiar adage that "the shortest distance between two points is along a straight line." It is Euclid's Proposition I.20.

Theorem 5.18 (Triangle Inequality). *If A, B, and C are noncollinear points, then $AC < AB + BC$.*

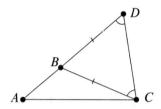

Fig. 5.15. Proof of the triangle inequality.

Proof. Exercise 5F. (Fig. 5.15 suggests the idea of the proof, which uses the scalene inequality and the isosceles triangle theorem.) □

The triangle inequality applies only to three noncollinear points. It is often useful to have a more general version of the inequality, which applies when there are more than three points and when the points are allowed to be collinear or not even assumed to be distinct. Its proof is our first example of a proof by mathematical induction. Now would be a good time to review the discussion of induction proofs in Appendix F.

Theorem 5.19 (General Triangle Inequality).

(a) *If A, B, C are any three points (not necessarily distinct), then*

$$AC \leq AB + BC, \tag{5.4}$$

and equality holds if and only if $A = B$, $B = C$, or $A * B * C$.

(b) *If $n \geq 3$ and $A_1, \ldots, A_n$ are any n points (not necessarily distinct), then*

$$A_1 A_n \leq A_1 A_2 + A_2 A_3 + \cdots + A_{n-1} A_n. \tag{5.5}$$

(See Fig. 5.16.)

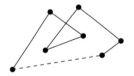

Fig. 5.16. The general triangle inequality.

Proof. Suppose A, B, and C are any three points. There are several cases to consider:

CASE 1: If A, B, C are noncollinear, then the original triangle inequality applies, so (5.4) holds with "less than" instead of "less than or equal to."

CASE 2: If the three points are collinear and distinct and $A * B * C$, then the betweenness theorem for points shows that (5.4) holds with an equal sign.

CASE 3: If the three points are collinear and distinct and $A * C * B$, then Theorem 3.22(e) (the whole segment is greater than the part) implies that

$$AC < AB < AB + BC.$$

(The last inequality follows from the fact that distances between distinct points are positive.) An analogous argument works if $B * A * C$.

CASE 4: If $A = C$ but B is distinct from A and C, then $AB + BC = 2AB > 0 = AC$; thus again the strict inequality holds.

CASE 5: Finally, if $A = B$ or $B = C$, then one of the terms on the right-hand side of (5.4) is zero, while the other term is equal to the left-hand side, so equality holds.

Looking back over these cases, we see that equality holds in Cases 2 and 5, while the strict inequality holds in all other cases. This proves (a).

We will prove (b) by induction on n. The base case $n = 3$ is taken care of by part (a). Let n be an integer greater than or equal to 3, and assume that the inequality is true for every collection of n points; we need to prove the analogous inequality for an arbitrary collection of $n + 1$ points, $A_1, \ldots, A_{n+1}$. We compute

$$A_1 A_{n+1} \leq A_1 A_n + A_n A_{n+1} \leq \left(A_1 A_2 + A_2 A_3 + \cdots + A_{n-1} A_n \right) + A_n A_{n+1},$$

where the first inequality is justified by the base case, and the second by the inductive hypothesis. This completes the induction step and thus the proof. $\square$

Corollary 5.20 (Converse to the Betweenness Theorem for Points). *If A, B, and C are three distinct points and $AB + BC = AC$, then $A * B * C$.*

Proof. Suppose A, B, C are distinct and $AB + BC = AC$. The preceding theorem shows that this equation holds if and only if $A = B$, $B = C$, or $A * B * C$. Since we are assuming A, B, and C are distinct, only the third possibility can occur. $\square$

The exterior angle, scalene, and triangle inequalities all apply to sides or angles within a *single* triangle. By contrast, our final inequality, which corresponds to Euclid's Propositions I.24 and I.25, compares sides and angles in two *different* triangles. The geometric intuition behind this inequality is that if you use two segments of fixed lengths to make triangles and let the included angle vary as if the two segments were connected by a hinge, then the opposite side gets larger as the angle gets larger (Fig. 5.17). We will not use this theorem until Chapter 17, but it will play a very important role when it does appear. Its proof is rated PG. (Our proof of the first part is a little different from Euclid's proof of Proposition I.24, for reasons that will be explained in the proof.)

Theorem 5.21 (Hinge Theorem). *Suppose $\triangle ABC$ and $\triangle DEF$ are two triangles such that $\overline{AB} \cong \overline{DE}$ and $\overline{AC} \cong \overline{DF}$. Then $m\angle A > m\angle D$ if and only if $BC > EF$.*

Proof. First suppose $m\angle A > m\angle D$. Using the triangle copying theorem (Theorem 5.10), make a copy of $\triangle ABC$ by letting G be the point on the same side of $\overleftrightarrow{ED}$ as F such that

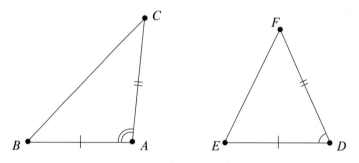

Fig. 5.17. The hinge theorem.

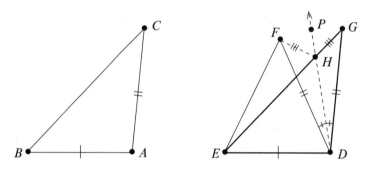

Fig. 5.18. Proof of the hinge theorem.

$\triangle DEG \cong \triangle ABC$ (Fig. 5.18). (In the figure, G is pictured on the same side of $\overleftrightarrow{EF}$ as D, but this need not be the case; the argument works exactly the same regardless of where G lies with respect to $\overleftrightarrow{EF}$. Euclid gives a proof that works only when G and D are on the same side of $\overleftrightarrow{EF}$, which is why we are using a different argument. This proof is a little longer, but it does not require the consideration of separate cases.)

Now we have $m\angle GDE = m\angle A > m\angle FDE$ by hypothesis, so the ordering lemma for rays shows that $\overrightarrow{DF}$ is in the interior of $\angle GDE$. In particular, this means that $\overrightarrow{DG}$ and $\overrightarrow{DF}$ are not collinear and F and E are on the same side of $\overleftrightarrow{DG}$.

Let $\overrightarrow{DP}$ be the bisector of $\angle GDF$. We wish to show that $\overrightarrow{DP}$ contains an interior point of $\overline{GE}$. Since $\overrightarrow{DP}$ is between $\overrightarrow{DG}$ and $\overrightarrow{DF}$, it follows that P is on the same side of $\overleftrightarrow{DG}$ as F, and thus also as E. Note that $m\angle GDP < m\angle GDE$ because

$$m\angle GDP = \tfrac{1}{2}m\angle GDF < m\angle GDF = m\angle GDE - m\angle FDE < m\angle GDE.$$

Therefore, we can apply the ordering lemma for rays once more to conclude that $\overrightarrow{DP}$ is in the interior of $\angle GDE$. Thus by the crossbar theorem, $\overrightarrow{DP}$ meets the interior of $\overline{GE}$ at a point H. It follows from SAS that $\triangle DHF \cong DHG$, and therefore $HF = HG$.

Now consider the distance EF. By the general triangle inequality, we have

$$EF \leq EH + HF = EH + HG.$$

Because $E * H * G$, the last expression on the right is equal to EG, so we conclude that $EF \leq EG = BC$. If EF were equal to BC, then $\triangle ABC$ and $\triangle DEF$ would be congruent by SSS, which is impossible because $\angle A$ is not congruent to $\angle D$; so in fact we must have $EF < BC$ as desired. This completes the proof that $m\angle A > m\angle D$ implies $BC > EF$.

To prove the converse, assume that $BC > EF$. By trichotomy, there are three possibilities: $m\angle A < m\angle D$, $m\angle A = m\angle D$, or $m\angle A > m\angle D$. By the argument in the first part of the proof, the first possibility would imply $BC < EF$, which contradicts our hypothesis. Similarly, the second possibility would imply $BC = EF$ by SAS, which is also a contradiction. The only remaining possibility is $m\angle A > m\angle D$. □

More Congruence Theorems

In this section, we will use the exterior angle inequality to prove some more congruence theorems for triangles. It is remarkable that a theorem about *inequalities* plays a crucial role in the proofs of so many theorems about congruence, which are essentially statements that certain things are *equal*.

The next theorem is the second half of Euclid's Proposition I.26, with the same proof.

Theorem 5.22 (AAS Congruence). *If there is a correspondence between the vertices of two triangles such that two angles and a nonincluded side of one triangle are congruent to the corresponding angles and side of the other triangle, then the triangles are congruent under that correspondence.*

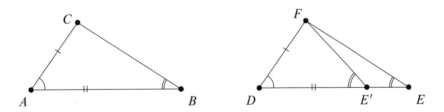

Fig. 5.19. Proof of the AAS congruence theorem.

Proof. Suppose $\triangle ABC$ and $\triangle DEF$ are triangles such that $\angle A \cong \angle D$, $\angle B \cong \angle E$, and $\overline{AC} \cong \overline{DF}$. If $\overline{AB} \cong \overline{DE}$, then the two triangles are congruent by SAS (or ASA), so assume for the sake of contradiction that $AB \neq DE$. Then one of these lengths, say DE, is greater than the other. By Corollary 3.37, there is a point E' in the interior of $\overline{DE}$ such that $\overline{DE'} \cong \overline{AB}$ (Fig. 5.19). Then SAS implies that $\triangle ABC \cong \triangle DE'F$, so in particular $\angle B \cong \angle DE'F$. Since $\angle E$ is also congruent to $\angle B$, transitivity of congruence implies that $\angle DE'F \cong \angle E$. But $\angle DE'F$ is an exterior angle of $\triangle E'EF$, while $\angle E$ is a remote interior angle, so this contradicts the exterior angle inequality. □

Let us pause to make an important observation about rigor. Whenever a postulate or a previously proved theorem is used as justification for a step in a proof, it is absolutely essential to ensure that the hypotheses of the postulate or theorem are satisfied. In the course of the preceding proof, to justify the application of the exterior angle inequality, we asserted without proof that $\angle DE'F$ is an exterior angle of $\triangle EE'F$ and $\angle E$ is a remote interior angle. To support these claims, we would have to verify that the definitions of exterior and remote interior angles are satisfied. Verifying that $\angle DE'F$ is an exterior angle of $\triangle E'EF$ means checking that it forms a linear pair with one of the interior angles of the same triangle. Since we chose E' so that $D * E' * E$, it follows from Theorem 3.38(e) that $\overrightarrow{E'D}$ and $\overrightarrow{E'E}$ are opposite rays; thus because $\angle DE'F$ and $\angle EE'F$ share the common ray $\overrightarrow{E'F}$ and their other rays are opposite rays, they do indeed form a linear pair. Since $\angle E$ is an angle of $\triangle EE'F$ but not the one that forms a linear pair with $\angle DE'F$, it is a remote interior angle.

In this situation, it was reasonable to leave out the justifications for these claims because two things were true: the truth of the claims was immediately obvious from the diagram, *and* the detailed proof of the claims was short, easy, and based on simple facts that we have already used many times before. When both of these conditions hold, it is entirely acceptable to make the assertion and leave it to the reader to fill in the details if he or she is curious. In fact, in most cases a proof will be easier to follow if justifications of simple claims like these are *not* included, because they can cause an argument to get so bogged down in the verification of easy details that it's hard to see the forest for the trees. (By contrast, in the proof of the exterior angle inequality, the verification that $\overrightarrow{BM} * \overrightarrow{BX} * \overrightarrow{BP}$ was anything but obvious and therefore had to be written out.)

When you are reading a proof, if you encounter a claim that you don't see immediately how to prove, you should stop and figure out how to prove it; once you have done so, you can decide whether the author was justified in leaving out the justification according to the criteria mentioned above. When you are *writing* a proof, you are responsible for ensuring that each claim you make is fully justified; once you have worked out the justification for yourself, you can decide whether including the details will make the proof easier or harder to understand. It's a judgment call every time, and there are no rigid rules. At first, you will probably be safer if you err on the side of including more such justifications; but as you gain experience and facility with proof-writing, you will get better at making such choices.

We now have four results that allow us to conclude that two triangles are congruent if they have three congruent pairs of corresponding parts: SAS, ASA, SSS, and AAS. Are there any other such theorems? We would not expect to have an AAA congruence theorem, because two triangles with the same angles might differ by a scale factor. (As we will see in Chapter 12, such triangles are said to be *similar*.)

The only other possible congruence theorem would be what we might call an "ASS congruence theorem"—one that says two triangles are congruent if there is a correspondence between their vertices such that two sides and a nonincluded angle of one triangle are congruent to the corresponding parts of the other. If you experiment for a while, you will probably find that there are typically two noncongruent triangles that can be built with the same two side lengths and a given nonincluded angle. Here is a counterexample that proves that "ASS congruence" is not a theorem in neutral geometry.

Example 5.23 (There Is No ASS Congruence Theorem). Let A and B be two distinct points, and let P be a point not on $\overleftrightarrow{AB}$. By the segment construction theorem, there is a point C on the ray $\overrightarrow{AP}$ such that $\overline{AC} \cong \overline{AB}$. Choose any point $D \in \overleftrightarrow{BC}$ such that $B * C * D$ (see Fig. 5.20). Because $A \notin \overleftrightarrow{BC}$, $\triangle DAB$ and $\triangle DAC$ are both triangles. These triangles have two pairs of congruent sides ($\overline{DA} \cong \overline{DA}$ and $\overline{AB} \cong \overline{AC}$), and the nonincluded angle $\angle D$ is common to both; thus they satisfy the hypotheses of ASS. However, they are not congruent, because $B * C * D$ implies that $BD > CD$. //

It is worth taking careful note of what we did here. Most theorems in geometry are stated as universal implications: "If x satisfies property $P(x)$, then it satisfies property

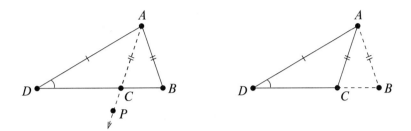

Fig. 5.20. Counterexample to ASS congruence.

$Q(x)$," with the tacit understanding that this is being asserted "for all x." To *disprove* such a statement, it is necessary to exhibit a counterexample—that is, to prove there exists an x that satisfies $P(x)$ but not $Q(x)$. It is not enough just to observe that an x with certain properties would be a counterexample, unless you can *prove* the existence of such an x. This is why we took pains to show how our postulates and previous theorems actually imply the existence of two triangles that satisfy the ASS hypotheses but not the conclusion.

Fig. 5.20 does suggest, though, that the possibilities are limited, and there should be at most two different triangles with congruent sides and nonincluded angle. The next theorem shows that this is indeed the case.

Theorem 5.24 (ASS Alternative Theorem). *Suppose $\triangle ABC$ and $\triangle DEF$ are two triangles such that $\angle A \cong \angle D$, $\overline{AB} \cong \overline{DE}$, $\overline{BC} \cong \overline{EF}$ (the hypotheses of ASS). Then $\angle C$ and $\angle F$ are either congruent or supplementary.*

Proof. Exercise 5G.   □

One consequence of the preceding theorem is the following limited version of ASS congruence, sometimes abbreviated as AsS ("angle-side-longer-side") congruence.

Theorem 5.25 (Angle-Side-Longer-Side Congruence). *Suppose $\triangle ABC$ and $\triangle DEF$ are two triangles such that $\angle A \cong \angle D$, $\overline{AB} \cong \overline{DE}$, $\overline{BC} \cong \overline{EF}$ (the hypotheses of ASS), and assume in addition that $BC \geq AB$. Then $\triangle ABC \cong \triangle DEF$.*

Proof. Exercise 5H.   □

One situation in which the hypotheses of the preceding theorem are automatically satisfied is when the nonincluded angles are right angles. In this case, the two sides are necessarily the hypotenuse and one leg of each triangle, so the following theorem is called the HL ("hypotenuse-leg") congruence theorem.

Theorem 5.26 (HL Congruence). *If the hypotenuse and one leg of a right triangle are congruent to the hypotenuse and one leg of another, then the triangles are congruent under that correspondence.*

Proof. Exercise 5I.   □

Exercises

5A. Prove Theorem 5.1 (consistency of triangle vertices). [Hint: Look at the proof of Theorem 4.1 for inspiration.]

5B. Prove Corollary 5.3 (to Pasch's theorem).

5C. Suppose $\triangle ABC$ is a triangle and ℓ is a line (which might or might not contain one or more vertices). Is it possible for ℓ to intersect exactly one side of $\triangle ABC$? Exactly two? All three? In each case, either give an example or prove that it is impossible.

5D. Prove Theorem 5.8 (the converse to the isosceles triangle theorem). [Hint: One way to proceed is to construct an indirect proof, like Euclid's proof of Proposition I.6. Another is to mimic Pappus's proof of the isosceles triangle theorem.]

5E. Prove Theorem 5.10 (the triangle copying theorem).

5F. Prove Theorem 5.18 (the triangle inequality).

5G. Prove Theorem 5.24 (the ASS alternative theorem). [Hint: Since the conclusion is a disjunction, you will want to follow Template F.8 in Appendix F. Try to construct a diagram that looks like Fig. 5.20.]

5H. Prove Theorem 5.25 (the angle-side-longer-side congruence theorem).

5I. Prove Theorem 5.26 (the HL congruence theorem).

Chapter 6

Models of Neutral Geometry

We have now introduced all of the postulates of neutral geometry (see Appendix D for the statements of the postulates). For easy reference, here is the list:

1. The Set Postulate,
2. The Existence Postulate,
3. The Unique Line Postulate,
4. The Distance Postulate,
5. The Ruler Postulate,
6. The Plane Separation Postulate,
7. The Angle Measure Postulate,
8. The Protractor Postulate,
9. The SAS Postulate.

Eventually, we will add one of the following parallel postulates to this list:

10E. The Euclidean Parallel Postulate,

10H. The Hyperbolic Parallel Postulate.

When we add the Euclidean parallel postulate to the neutral postulates, this axiomatic system becomes ***Euclidean geometry***; when we add the hyperbolic parallel postulate, it becomes ***hyperbolic geometry***. (In the next chapter, we will prove that adding the elliptic parallel postulate would result in an inconsistent axiomatic system.)

In this chapter, we make a short digression from the development of our axiomatic system in order to describe two very different models of neutral geometry and some interpretations that are not models. Throughout this chapter, because we are working with interpretations, we will not be constrained to work within our axiomatic system; instead, we allow ourselves to use whatever standard facts from the rest of mathematics are convenient. Because of this, you will find that the arguments in this chapter have a very different flavor from the rest of our treatment of geometry.

The Cartesian Model

Our first model of neutral geometry is the most familiar one and is in fact the model that modern mathematicians and scientists use for most practical applications of Euclidean geometry, including calculus. It is the *Cartesian plane*, an interpretation we first encountered in the context of incidence geometry (Example 2.11). Now we expand that interpretation so as to make it a model of neutral geometry.

In order to turn the Cartesian plane into an interpretation of neutral geometry, we have to define all of the primitive terms. In Example 2.11, we already gave definitions for *point* and *line* in the Cartesian plane; now we have to define *distance* and *angle measure*.

To describe these definitions, it will be convenient to introduce a little bit of vector notation and terminology. Vectors should be familiar from your study of multivariable calculus as mathematical objects used to model quantities that have direction and magnitude but whose absolute position in the plane is not relevant. There are various ways of representing vectors; for our purposes, the most convenient way is in terms of their x and y components. Thus we officially define a **vector in $\mathbb{R}^2$** to be an ordered pair of real numbers; the two numbers are called its **vector components**, or just its **components**. The vector with components x and y is denoted by $\langle x, y \rangle$ to distinguish it from the *point* (x, y) with the same components.

From a purely formal point of view, there is technically no mathematical difference between a *point* in $\mathbb{R}^2$ and a *vector* in $\mathbb{R}^2$; both are ordered pairs of real numbers. We use different notations primarily to remind ourselves that vectors and points are used for different geometric purposes and that there are different operations that are appropriate to perform on them. We generally visualize a point as a specific location in the plane, while we visualize a vector as an arrow.

Most commonly, vectors arise in the following context. If $A = (x_1, y_1)$ and $B = (x_2, y_2)$ are two points in the Cartesian plane, the **displacement vector from A to B** (Fig. 6.1), denoted by $\mathrm{Vec}(AB)$, is the vector

$$\mathrm{Vec}(AB) = \langle x_2 - x_1, y_2 - y_1 \rangle.$$

(It is tempting to use $\overrightarrow{AB}$ to denote the displacement vector from A to B, but we have already appropriated that notation for rays.) The notion of displacement vectors can be used to explain the relationship between the *point* (x, y) and the *vector* $\langle x, y \rangle$ with the same components: if we let O be the point with coordinates $(0, 0)$ (called the **origin**), then for any point $A = (x, y)$, the displacement vector $\mathrm{Vec}(OA)$ is the vector $\langle x, y \rangle$ whose components are the same as the coordinates of A.

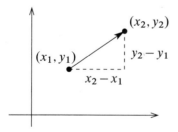

Fig. 6.1. A displacement vector.

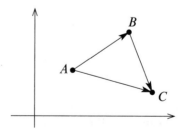

Fig. 6.2. Addition of displacement vectors.

We often denote a vector by a boldface lowercase letter such as $\mathbf{a}$ or $\mathbf{b}$. If $\mathbf{a} = \langle x_1, y_1 \rangle$ and $\mathbf{b} = \langle x_2, y_2 \rangle$ are two vectors, their *vector sum* and *vector difference* are defined by

$$\mathbf{a} + \mathbf{b} = \langle x_1 + x_2, y_1 + y_2 \rangle,$$
$$\mathbf{a} - \mathbf{b} = \langle x_1 - x_2, y_1 - y_2 \rangle.$$

It is straightforward to check from the definitions that displacement vectors satisfy the following identity (see Fig. 6.2): for any three points A, B, C,

$$\text{Vec}(AB) + \text{Vec}(BC) = \text{Vec}(AC). \tag{6.1}$$

If $\mathbf{a} = \langle x_1, y_1 \rangle$ and $\mathbf{b} = \langle x_2, y_2 \rangle$ are vectors, their *dot product*, denoted by $\mathbf{a} \cdot \mathbf{b}$, is the real number obtained by multiplying their corresponding components together and adding:

$$\mathbf{a} \cdot \mathbf{b} = x_1 x_2 + y_1 y_2.$$

The dot product satisfies the following commutative and distributive laws, as you can easily verify from the definition:

$$\mathbf{a} \cdot \mathbf{b} = \mathbf{b} \cdot \mathbf{a};$$
$$(\mathbf{a} \pm \mathbf{b}) \cdot \mathbf{c} = \mathbf{a} \cdot \mathbf{c} \pm \mathbf{b} \cdot \mathbf{c};$$
$$\mathbf{c} \cdot (\mathbf{a} \pm \mathbf{b}) = \mathbf{c} \cdot \mathbf{a} \pm \mathbf{c} \cdot \mathbf{b}.$$

The *length* of $\mathbf{a} = \langle x_1, y_1 \rangle$ is the number

$$\|\mathbf{a}\| = \sqrt{\mathbf{a} \cdot \mathbf{a}} = \sqrt{x_1{}^2 + y_1{}^2}.$$

With these preliminaries out of the way, we are ready to turn the Cartesian plane into a model of neutral geometry.

Example 6.1 (The Cartesian Plane). The *Cartesian plane* is the interpretation of neutral geometry in which the primitive terms have the following meanings:

- A *point* is an ordered pair (x, y) of real numbers.
- A *line* is any set of points (x, y) that is the entire solution set of an equation of one of the following two types:

  $x = c$ for some real constant c (a *vertical line*);

  $y = mx + b$ for some real constants m and b (a *nonvertical line*).

- The *distance* between two points $A = (x_1, y_1)$ and $B = (x_2, y_2)$ is

$$AB = \|\text{Vec}(AB)\| = \sqrt{(x_2 - x_1)^2 + (y_2 - y_1)^2}. \tag{6.2}$$

- The *measure* of an angle $\angle AOB$ is the number

$$m\angle AOB = \arccos\left(\frac{\mathbf{a} \cdot \mathbf{b}}{\|\mathbf{a}\| \, \|\mathbf{b}\|}\right), \tag{6.3}$$

where $\mathbf{a}$ and $\mathbf{b}$ are the displacement vectors $\text{Vec}(OA)$ and $\text{Vec}(OB)$, respectively.   //

The definition of angle measure is motivated by the familiar relation between dot products and angles that you probably learned when you studied vector algebra: $\mathbf{a} \cdot \mathbf{b} = \|\mathbf{a}\| \, \|\mathbf{b}\| \cos\theta$, where θ is supposed to be the measure of the angle between the vectors $\mathbf{a}$ and $\mathbf{b}$. The function $\arccos: [-1, 1] \to [0, 180]$ is the *arccosine* function. In elementary treatments of trigonometry, this function is usually defined as the inverse of the cosine function: if x is a real number in the interval $[-1, 1]$, then $\arccos(x)$ is the unique number $\theta \in [0, 180]$

(in degrees) such that $\cos\theta = x$. However, for the purpose of constructing a model of neutral geometry, in order to avoid circularity, we need a definition of the arccosine function that does not presuppose any knowledge of angles or trigonometric functions. Thus we define it by the following formula:

$$\arccos(x) = 180\left(\frac{\int_x^1 \frac{dt}{\sqrt{1-t^2}}}{\int_{-1}^1 \frac{dt}{\sqrt{1-t^2}}}\right). \tag{6.4}$$

The integral in the numerator represents the arc length of the portion of the unit circle above the x-axis and between the points $(1,0)$ and $\left(x, \sqrt{1-x^2}\right)$ (Fig. 6.3), while the integral in the denominator represents the arc length of the entire upper unit semicircle; thus the fraction expresses the percentage of the semicircle that lies between those two points. The factor of 180 ensures that our angle measures lie in the range from 0 to 180. Using the tools of advanced calculus, one can prove that the arccosine function has all of the properties that one expects from its trigonometric interpretation. (Note that both integrals in (6.4) are improper integrals because the integrands are not defined at $t = \pm 1$, so one of the things that has to be proved is that the integrals converge.) We will not prove all of these properties; but one such property that is easy to prove, and which will be useful in examples, is given in the following lemma.

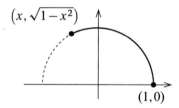

Fig. 6.3. Defining the arccosine function.

Lemma 6.2 (Right Angles in the Cartesian Plane). *Suppose $\angle AOB$ is an angle in the Cartesian plane, and $\mathbf{a} = \mathrm{Vec}(OA)$ and $\mathbf{b} = \mathrm{Vec}(OB)$ are the corresponding displacement vectors. Then $\angle AOB$ is a right angle if and only if $\mathbf{a} \cdot \mathbf{b} = 0$.*

Proof. Let $\angle AOB$, $\mathbf{a}$, and $\mathbf{b}$ be as in the hypothesis, and assume first that $\mathbf{a} \cdot \mathbf{b} = 0$. Then (6.3) shows that $m\angle AOB = \arccos(0)$. By (6.4), this is

$$m\angle AOB = 180\left(\frac{\int_0^1 \frac{dt}{\sqrt{1-t^2}}}{\int_{-1}^1 \frac{dt}{\sqrt{1-t^2}}}\right). \tag{6.5}$$

On the other hand, we can evaluate the definite integral in the denominator by splitting it into two terms:

$$\int_{-1}^0 \frac{dt}{\sqrt{1-t^2}} + \int_0^1 \frac{dt}{\sqrt{1-t^2}}.$$

Applying the change of variables $t = -u$ to the first integral shows that both of these integrals are equal to each other and to the integral in the numerator of (6.5). Therefore, (6.5) reduces to $m\angle AOB = 180 \cdot \frac{1}{2} = 90$.

Conversely, assume that $\angle AOB$ is a right angle. If we define a real number r by $r = \mathbf{a} \cdot \mathbf{b}/(\|\mathbf{a}\| \, \|\mathbf{b}\|)$, then (6.3) shows that $\arccos(r) = 90$, and thus it follows from (6.4) that

$$\int_r^1 \frac{dt}{\sqrt{1-t^2}} = \frac{1}{2} \int_{-1}^1 \frac{dt}{\sqrt{1-t^2}}. \tag{6.6}$$

We already showed in the previous paragraph that this equation holds when $r = 0$. Because the integrand is strictly positive as long as $-1 < t < 1$, it follows from elementary properties of integrals that the left-hand side of (6.6) is less than the right-hand side when $r > 0$ (because we are integrating over a smaller interval) and greater than the right-hand side when $r < 0$; therefore the only value of r that can satisfy (6.6) is $r = 0$. Looking back at the definition of r, we see that this implies $\mathbf{a} \cdot \mathbf{b} = 0$. ☐

The crucial fact about the Cartesian plane is that all of the postulates of neutral geometry are true in this interpretation, so it is a *model* of neutral geometry. To verify this, we need to prove that each of the nine postulates is a true statement in the Cartesian interpretation.

A few of the postulates are quite easy to verify.

Theorem 6.3. *The Cartesian plane satisfies the set postulate, the existence postulate, the unique line postulate, and the distance postulate.*

Proof. The set postulate is satisfied because every line is a set of points by definition, and the collection of all points is the set $\mathbb{R}^2$. Theorem 2.12 showed that the Cartesian plane satisfies the existence postulate (because it satisfies Incidence Axiom 1) and the unique line postulate (because it satisfies Incidence Axioms 2 and 3). The distance postulate is satisfied because the definition of the distance between points A and B in the Cartesian plane always yields a nonnegative real number completely determined by A and B. ☐

The proofs that the other five postulates are satisfied are somewhat more involved. As you read the next few proofs, be sure to bear in mind that these are *not* being carried out within the axiomatic system of neutral geometry. In order to prove that the Cartesian plane satisfies the neutral postulates, we are free to use the full mathematical theory within which the interpretation is constructed—in this case, that means set theory, the real numbers, analytic geometry, vector algebra, and calculus (all of which are based ultimately on set theory).

The next proof is rated R.

Theorem 6.4. *The Cartesian plane satisfies the ruler postulate.*

Proof. Let ℓ be a line in the Cartesian plane. We need to show that there is a coordinate function for ℓ, i.e., a bijective function $f : \ell \to \mathbb{R}$ such that

$$AB = |f(B) - f(A)| \quad \text{for all } A, B \in \ell. \tag{6.7}$$

We will handle the vertical and nonvertical cases separately.

Suppose first that ℓ is a vertical line. This means that there is a certain constant c such that ℓ is the set of all points (x, y) satisfying the equation $x = c$; in other words, ℓ

is the set of all points of the form (c, y) for arbitrary $y \in \mathbb{R}$. Define $f: \ell \to \mathbb{R}$ by setting $f(x, y) = y$ for all $(x, y) \in \ell$. To show that f is a coordinate function, we have to show first that it is bijective. To prove injectivity, suppose (x_1, y_1) and (x_2, y_2) are points on ℓ such that $f(x_1, y_1) = f(x_2, y_2)$. From the definition of f, this means that $y_1 = y_2$, and the fact that both points are on ℓ means that $x_1 = c = x_2$. Thus $(x_1, y_1) = (x_2, y_2)$, so f is injective. To prove surjectivity, let $r \in \mathbb{R}$ be arbitrary, and just note that (c, r) is a point on ℓ such that $f(c, r) = r$.

Finally, we need to show that f satisfies (6.7). If $A = (x_1, y_1)$ and $B = (x_2, y_2)$ are two points on ℓ, then $x_1 = c = x_2$, so the distance formula (6.2) reduces to $AB = \sqrt{(y_2 - y_1)^2} = |y_2 - y_1|$, which is obviously equal to $|f(B) - f(A)|$. (Here we have used the standard algebraic fact that $\sqrt{r^2} = |r|$ for any real number r.)

Now suppose ℓ is nonvertical. Then there are constants m and b such that ℓ is the set of all points (x, y) that satisfy the equation

$$y = mx + b. \tag{6.8}$$

Define $f: \ell \to \mathbb{R}$ by

$$f(x, y) = x\sqrt{1 + m^2}. \tag{6.9}$$

(Although this formula would make sense for any $(x, y) \in \mathbb{R}^2$, we are defining $f(x, y)$ only when (x, y) is a point of ℓ, meaning that it satisfies (6.8).)

To show that f is injective, suppose (x_1, y_1) and (x_2, y_2) are points on ℓ that satisfy $f(x_1, y_1) = f(x_2, y_2)$. From the definition of f, this means that $x_1\sqrt{1 + m^2} = x_2\sqrt{1 + m^2}$, and because $\sqrt{1 + m^2} \neq 0$, this in turn implies that $x_1 = x_2$. Since both points are on ℓ, (6.8) implies

$$y_2 = mx_2 + b = mx_1 + b = y_1. \tag{6.10}$$

Thus we have shown $(x_1, y_1) = (x_2, y_2)$, so f is injective.

To show that f is surjective, suppose r is an arbitrary real number. If we set $x_0 = r/\sqrt{1 + m^2}$ and $y_0 = mx_0 + b$, then (x_0, y_0) is a point on ℓ, and an easy computation shows that $f(x_0, y_0) = r$.

Finally, we need to show that f satisfies (6.7). Suppose A and B are arbitrary points on ℓ. If we write $A = (x_1, y_1)$ and $B = (x_2, y_2)$, then the fact that they are both on ℓ means that $y_1 = mx_1 + b$ and $y_2 = mx_2 + b$. Using formula (6.2) for the distance, we obtain

$$\begin{aligned} AB &= \sqrt{(x_2 - x_1)^2 + (y_2 - y_1)^2} \\ &= \sqrt{(x_2 - x_1)^2 + \big((mx_2 + b) - (mx_1 + b)\big)^2} \\ &= \sqrt{(x_2 - x_1)^2 + m^2(x_2 - x_1)^2} \\ &= \sqrt{(x_2 - x_1)^2(1 + m^2)} \\ &= \sqrt{(x_2 - x_1)^2}\sqrt{1 + m^2} \\ &= |x_2 - x_1|\sqrt{1 + m^2}. \end{aligned}$$

On the other hand, from the definition of f, we obtain

$$
\begin{aligned}
|f(B) - f(A)| &= |f(x_2, y_2) - f(x_1, y_1)| \\
&= \left| x_2 \sqrt{1 + m^2} - x_1 \sqrt{1 + m^2} \right| \\
&= \left| (x_2 - x_1) \sqrt{1 + m^2} \right| \\
&= |x_2 - x_1| \sqrt{1 + m^2}.
\end{aligned}
$$

Comparing these two computations proves the result. $\quad\square$

In case you are wondering where the coordinate function f of (6.9) came from, here is the heuristic reasoning that led to it. Once the equation of the line is expressed in the form (6.8), it is tempting to use x as a coordinate function along the line, because (6.8) shows that x can take any value and y is uniquely determined once x is known. Thus the function $f(x, y) = x$ is a bijective function from ℓ to $\mathbb{R}$, as you can easily check; but unfortunately it does not satisfy (6.7) except in the special case $m = 0$ (which represents a horizontal line). The solution is to rescale the x-coordinate by a constant multiple. A little trial and error leads to the constant $\sqrt{1 + m^2}$ and thus to the coordinate function f we have defined.

The proofs of the remaining postulates are quite a bit longer. Rather than carrying them out in detail, we just sketch the main ideas involved in each.

The following proof is rated R; the missing details, although they would take a while to write out in full, are not beyond our scope.

Theorem 6.5. *The Cartesian plane satisfies the plane separation postulate.*

Sketch of proof. Suppose ℓ is a line. Then it is either a vertical line defined by an equation of the form $x = c$ or a nonvertical line defined by an equation of the form $y = mx + b$. In either case, we define a function $L : \mathbb{R}^2 \to \mathbb{R}$ as follows:

$$
L(x, y) = \begin{cases} x - c & \text{if } \ell \text{ is vertical,} \\ y - mx - b & \text{if } \ell \text{ is nonvertical.} \end{cases}
$$

Then it is immediate that a point (x, y) lies on ℓ if and only if $L(x, y) = 0$.

We define two subsets of the plane by

$$
\begin{aligned}
S_+ &= \{(x, y) : L(x, y) > 0\}, \\
S_- &= \{(x, y) : L(x, y) < 0\}.
\end{aligned}
$$

By the trichotomy law for real numbers, every point not on ℓ is in one and only one of these sets, so the set of points not on ℓ is the union of the two disjoint sets S_+ and S_-.

The other fact that needs to be proved is that for any two distinct points $A, B \notin \ell$, $\overline{AB} \cap \ell = \varnothing$ if and only if A and B lie on the same side of ℓ. As we observed in the proof of Theorem 3.43, this is equivalent to proving that $\overline{AB} \cap \ell \neq \varnothing$ if and only if A and B lie on opposite sides.

The proof relies on two algebraic facts that we leave to you to verify. Let $A = (x_1, y_1)$ and $B = (x_2, y_2)$ be two distinct points that do not lie on ℓ, and for any $t \in \mathbb{R}$, define a point $P_t \in \mathbb{R}^2$ by the formula

$$
P_t = \big(x_1 + t(x_2 - x_1),\ y_1 + t(y_2 - y_1) \big).
$$

Then the following fact can be proved by straightforward algebraic manipulations:

$$L(P_t) = L(A) + t\big(L(B) - L(A)\big) \quad \text{for each } t \in \mathbb{R}. \tag{6.11}$$

In addition, by applying Lemma 3.23 (coordinate representation of a segment) to the coordinate functions constructed in the proof of Theorem 6.4, we can prove the following parametric characterization of the segment $\overline{AB}$:

$$\overline{AB} = \{P_t : 0 \le t \le 1\}. \tag{6.12}$$

(Note that Lemma 3.23 was proved after we had introduced only Postulates 1–5, and we have verified all of those postulates in the Cartesian plane, so the lemma applies here as well.) Granting these facts, we can proceed with the proof.

Assume first that A and B are on opposite sides of ℓ, which means that $L(A)$ and $L(B)$ have opposite signs. Interchanging the names of A and B if necessary, we may assume that $L(A) < 0$ and $L(B) > 0$. Define a real number t by

$$t = \frac{-L(A)}{L(B) - L(A)}.$$

Our assumptions guarantee that the numerator and denominator are both positive and the numerator is strictly less than the denominator, so $0 < t < 1$. Together with (6.12), this implies that $P_t \in \overline{AB}$; and then (6.11) implies that $L(P_t) = 0$, so that $P_t \in \ell$. Therefore $\overline{AB} \cap \ell \ne \varnothing$ as required.

Conversely, assume that $\overline{AB} \cap \ell \ne \varnothing$. Using (6.12), this implies that there is some $t \in [0,1]$ such that $P_t \in \ell$. If t were equal to 0 or 1, then P_t would be equal to A or B, respectively; since we are assuming A and B do not lie on ℓ, this is impossible. Thus $0 < t < 1$. The fact that $P_t \in \ell$ implies that $L(P_t) = 0$, which by (6.11) means $L(A) + t\big(L(B) - L(A)\big) = 0$. Because $t \ne 0$, we can solve this equation for $L(B)$ to obtain

$$L(B) = \frac{t-1}{t} L(A).$$

Because we know $0 < t < 1$, the fraction on the right-hand side has negative numerator and positive denominator, so $L(A)$ and $L(B)$ have opposite signs. This implies that A and B are on opposite sides of ℓ as required. $\qquad\square$

The proof of the angle measure and protractor postulates is the most involved of all, so we just briefly sketch the main ideas. The proof of the following theorem is rated X; one place where it is carried out in full is [**MP91**, Section 5.4].

Theorem 6.6. *The Cartesian plane satisfies the angle measure postulate and the protractor postulate.*

Sketch of proof. With some basic tools of advanced calculus, it can be shown that the arccosine function defined by (6.4) is a bijective function from the interval $[-1, 1]$ to the interval $[0, 180]$. The angle measure postulate follows easily from this fact together with some straightforward properties of the dot product.

To prove the protractor postulate, suppose $\overrightarrow{OA}$ is a ray and P is a point not on $\overrightarrow{OA}$. For any ray $\overrightarrow{OB} \in \mathrm{HR}\big(\overrightarrow{OA}, P\big)$, let

$$g\big(\overrightarrow{OB}\big) = \arccos\left(\frac{\mathbf{a} \cdot \mathbf{b}}{\|\mathbf{a}\|\,\|\mathbf{b}\|}\right),$$

where $\mathbf{a} = \text{Vec}(OA)$ and $\mathbf{b} = \text{Vec}(OB)$ as before. A few basic properties of g follow easily from standard properties of the dot product and properties of the integral: g is bijective, $g\left(\overrightarrow{OA}\right) = 0$, and $g\left(\overrightarrow{OD}\right) = 180$, where $\overrightarrow{OD}$ is the ray opposite $\overrightarrow{OA}$. The proof of the remaining property required of a coordinate function—that $m\angle rs = \left| g\left(\overrightarrow{r}\right) - g\left(\overrightarrow{s}\right) \right|$ for any two rays $\overrightarrow{r}, \overrightarrow{s} \in \text{HR}\left(\overrightarrow{OA}, P\right)$—is considerably more messy, so we leave it to the interested reader to look it up in [**MP91**]. The essential ingredients are the difference formula for the cosine function (which can be translated into a formula involving only the arccosine function) and some nonobvious dot product identities. $\qquad\square$

There is only one postulate of neutral geometry remaining. This proof is rated R.

Theorem 6.7. *The Cartesian plane satisfies the SAS postulate.*

Sketch of proof. Assume $\triangle ABC$ and $\triangle DEF$ are two triangles such that $\overline{AB} \cong \overline{DE}$, $\overline{BC} \cong \overline{EF}$, and $\angle B \cong \angle E$. Define the following displacement vectors (see Fig. 6.4):

$$\begin{aligned} \mathbf{a} &= \text{Vec}(BC), & \mathbf{d} &= \text{Vec}(EF), \\ \mathbf{b} &= \text{Vec}(CA), & \mathbf{e} &= \text{Vec}(FD), \\ \mathbf{c} &= \text{Vec}(BA), & \mathbf{f} &= \text{Vec}(ED). \end{aligned}$$

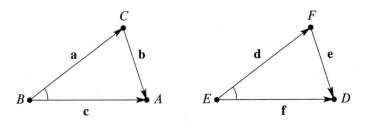

Fig. 6.4. The SAS postulate in the Cartesian plane.

The identity (6.1) yields the following relationships among these vectors:

$$\mathbf{a} + \mathbf{b} = \mathbf{c}, \qquad \mathbf{d} + \mathbf{e} = \mathbf{f}. \tag{6.13}$$

The SAS hypotheses yield the following equations:

$$\|\mathbf{a}\| = \|\mathbf{d}\|, \qquad \|\mathbf{c}\| = \|\mathbf{f}\|, \qquad \arccos\left(\frac{\mathbf{a} \cdot \mathbf{c}}{\|\mathbf{a}\| \, \|\mathbf{c}\|}\right) = \arccos\left(\frac{\mathbf{d} \cdot \mathbf{f}}{\|\mathbf{d}\| \, \|\mathbf{f}\|}\right).$$

Because the arccosine function is injective, the two expressions in parentheses above must be equal. Using this fact together with the first two equations above, we conclude that $\mathbf{a} \cdot \mathbf{c} = \mathbf{d} \cdot \mathbf{f}$, $\mathbf{a} \cdot \mathbf{a} = \mathbf{d} \cdot \mathbf{d}$, and $\mathbf{c} \cdot \mathbf{c} = \mathbf{f} \cdot \mathbf{f}$. Combining these with (6.13), we compute

$$\begin{aligned} \|\mathbf{b}\|^2 = \mathbf{b} \cdot \mathbf{b} &= (\mathbf{c} - \mathbf{a}) \cdot (\mathbf{c} - \mathbf{a}) \\ &= \mathbf{c} \cdot \mathbf{c} - 2\mathbf{c} \cdot \mathbf{a} + \mathbf{a} \cdot \mathbf{a} \\ &= \mathbf{f} \cdot \mathbf{f} - 2\mathbf{f} \cdot \mathbf{d} + \mathbf{d} \cdot \mathbf{d} \\ &= (\mathbf{f} - \mathbf{d}) \cdot (\mathbf{f} - \mathbf{d}) \\ &= \mathbf{e} \cdot \mathbf{e} = \|\mathbf{e}\|^2, \end{aligned}$$

which implies $\|\mathbf{b}\| = \|\mathbf{e}\|$, or $\overline{CA} \cong \overline{FD}$. Similar computations lead to the relations

$$\frac{(-\mathbf{a}) \cdot \mathbf{b}}{\|-\mathbf{a}\| \|\mathbf{b}\|} = \frac{(-\mathbf{d}) \cdot \mathbf{e}}{\|-\mathbf{d}\| \|\mathbf{e}\|}, \qquad \frac{(-\mathbf{b}) \cdot (-\mathbf{c})}{\|-\mathbf{b}\| \|-\mathbf{c}\|} = \frac{(-\mathbf{e}) \cdot (-\mathbf{f})}{\|-\mathbf{e}\| \|-\mathbf{f}\|},$$

which lead in turn to the congruences $\angle C \cong \angle F$ and $\angle A \cong \angle D$. ☐

The following corollary is an immediate consequence of Theorems 6.3–6.7.

Corollary 6.8. *The Cartesian plane is a model of neutral geometry.* ☐

Thus we are assured that all of the theorems we have proved and will prove *within* the axiomatic system are true statements about the Cartesian plane.

Corollary 6.9. *The postulates of neutral geometry are consistent.* ☐

In addition, Theorem 2.23 showed that the Cartesian model satisfies the Euclidean parallel postulate. Thus we have the following corollaries as well.

Corollary 6.10. *The Cartesian plane is a model of Euclidean geometry.* ☐

Corollary 6.11. *The postulates of Euclidean geometry are consistent.* ☐

A noteworthy fact about the Cartesian model is that *every* model of Euclidean geometry is isomorphic to it. Recall from Chapter 2 that an **isomorphism** between two models of an axiomatic system is a bijective correspondence between the objects of the two models that preserves all of the relevant relationships. For models of neutral geometry, this means a one-to-one correspondence between the points of one model and the points of the other, which takes lines to lines, distances to distances, and angle measures to angle measures. An axiomatic system is said to be **categorical** if any two models of it are isomorphic to each other. The basic fact about Euclidean geometry is that it is categorical: every model of Euclidean geometry is isomorphic to the Cartesian plane. We do not yet have the tools to prove this, but in Chapter 13 we will sketch a proof.

The Poincaré Disk Model

Our next interpretation was first introduced as a model of incidence geometry in Chapter 2. Here we expand on that interpretation in order to make it into a model of neutral geometry.

Example 6.12 (The Poincaré Disk). Let $O = (0,0)$ denote the origin in the Cartesian plane, let $\mathcal{C}$ denote the circle in the Cartesian plane with center O and radius 1, and let $\mathbb{D}$ denote the set of all Cartesian points whose Cartesian distance from O is strictly less than 1 (the points "inside" the circle $\mathcal{C}$).

- A *point* is a Cartesian point lying in $\mathbb{D}$, or in other words an ordered pair (x, y) satisfying $x^2 + y^2 < 1$.

- A *line* is a set of either one of the following two types:
  (a) the set of all Cartesian points in $\mathbb{D}$ that lie on some Cartesian line through O or
  (b) the set of all Cartesian points in $\mathbb{D}$ that lie on some Cartesian circle that intersects the circle $\mathcal{C}$ perpendicularly.

- The *distance* between two points A and B in the Poincaré disk model is the number given by the formula

$$AB_{\text{Poincaré}} = \left| \ln \frac{(AB')(BA')}{(AA')(BB')} \right|, \tag{6.14}$$

where ln is the natural logarithm; A' and B' are the points where the Cartesian line or circle that defines the Poincaré line containing A and B meets the unit circle $\mathcal{C}$ (see Fig. 6.5); and AB', BA', AA', and BB' are the ordinary Cartesian distances between these points. (Note that A' and B' are not "points" in the Poincaré disk model; but they are points in the Cartesian model that we are using to define the Poincaré disk model, so the Cartesian distances in (6.14) are all well defined.)

- The **measure** of an angle $\angle AOB$ is the ordinary Cartesian measure of the angle formed by the tangent rays to the curves or lines that make up the angle (see Fig. 6.6). //

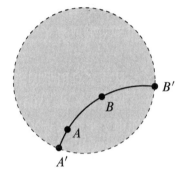
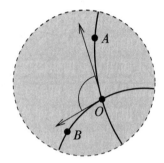

Fig. 6.5. Defining the Poincaré distance.   **Fig. 6.6.** Defining the Poincaré angle measure.

The formula (6.14) for the Poincaré distance is complicated, but it has one feature that is easy to notice: if the Cartesian distance BB' is very small (meaning that B is very close to the circle $\mathcal{C}$), then the denominator in the argument of the logarithm will be very small, and consequently, the Poincaré distance from A to B will be very large. Thus despite appearances, "lines" in this model are infinitely long, and the "plane" extends infinitely far in all directions. The geometry of the Poincaré disk model is illustrated by Fig. 6.7, which shows a family of pentagons that are all congruent to each other in this model. You have probably seen prints by the Dutch artist M. C. Escher that exploit the geometry of the Poincaré disk in a similar way.

One of the most salient features of the Poincaré disk model is that it satisfies the hyperbolic parallel postulate, not the Euclidean one. For example, Fig. 2.14 illustrates a line ℓ and a point $A \notin \ell$ through which there pass many lines parallel to ℓ.

The significance of the Poincaré disk model is due to the following theorem. Its X-rated proof is too involved to include here; but a version of it can be found in [**Moi90**, Chapter 25].

Theorem 6.13. *All of the postulates of neutral geometry are true in the Poincaré disk interpretation, as is the hyperbolic parallel postulate.*

Corollary 6.14. *The postulates of hyperbolic geometry are consistent.*

Proof. The Poincaré disk is a model of hyperbolic geometry. □

Corollary 6.15. *The Euclidean parallel postulate is independent of the axioms of neutral geometry.*

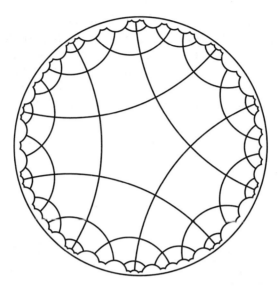

Fig. 6.7. Congruent pentagons in the Poincaré disk.

Proof. We now have seen a model of neutral geometry (the Cartesian plane) in which the Euclidean parallel postulate is true and another model (the Poincaré disk) in which it is false.   □

As we will show in Chapter 17, the Euclidean parallel postulate is equivalent to Euclid's fifth postulate, so it follows also that Euclid's fifth postulate is independent of the neutral postulates. Moreover, because Euclid's first four postulates are true in every model of neutral geometry, this also shows that Euclid's fifth postulate is independent of his other four.

As we discussed in Chapter 2, the only way to prove consistency of an axiomatic system is relative to some other axiomatic system that we assume to be consistent. Our proofs of the consistency of Euclidean geometry and hyperbolic geometry are based on the assumption that the axiomatic underpinnings of set theory (and therefore also those of the real numbers, vector algebra, analytic geometry, and calculus) are themselves consistent, as virtually all mathematicians believe they are.

A remarkable thing about the Poincaré disk model is that it is possible to define a version of it using only the *postulates* of Euclidean geometry, not any particular model. To see how this is done, assume that the postulates of Euclidean geometry are true; thus we assume there is a set of points—which we call the *Euclidean plane*—satisfying all of the axioms of Euclidean geometry. Then in the definition of the Poincaré disk, in place of the origin in the Cartesian plane, we can use any arbitrary point O in the Euclidean plane, and in place of Cartesian lines, circles, distances, and angle measures, we use their Euclidean counterparts. This leads to the conclusion that if the postulates of Euclidean geometry are consistent, then the postulates of hyperbolic geometry are also consistent. It was this observation that finally convinced mathematicians that hyperbolic geometry is no less logically rigorous than its Euclidean counterpart.

There are other models of hyperbolic geometry in addition to the Poincaré disk. The Beltrami–Klein model (Example 2.15), the Poincaré half-plane model (Example 2.17), and the hyperboloid model (Example 2.18) can all be turned into models of neutral geometry by suitably defining distances and angle measures, and as we observed in Chapter 2, they all satisfy the hyperbolic parallel postulate. In fact, in Beltrami's watershed paper [**Bel68**] proving the consistency of hyperbolic geometry, he introduced three of these models (Beltrami–Klein, Poincaré disk, and Poincaré half-plane) and proved that they are all isomorphic to each other. He didn't know about the hyperboloid model, which was developed somewhat later, but it is isomorphic to the other three as well.

The definitions of distances and angle measures in these models are even more involved than the definition of the Poincaré disk model that we have given here, and are more effectively studied in the context of *differential geometry*, a subject that brings sophisticated tools of calculus to bear on geometric questions. See [**dC76**] for an introduction to differential geometry at the advanced undergraduate level.

As you can probably guess by looking at the complexity of the Poincaré disk model, proving Theorem 6.13 would be a challenging task indeed, which is why we have not attempted to write down a proof. This is one reason it took so long for mathematicians to realize that it was possible to construct a consistent model of non-Euclidean geometry and therefore that Euclid's fifth postulate cannot be proved from the other ones. When we study non-Euclidean geometry later in the course, we will often refer to the Poincaré disk model as a source of intuition.

Some Nonmodels

Next, we introduce a few interpretations of neutral geometry that are *not* models, because one or more of the postulates are not satisfied. Looking at such nonmodels is useful because it demonstrates why we need the various postulates: if we can show that there is an interpretation in which one of the postulates is false but some others are true, then we have shown that the postulate that fails is independent of the ones that don't.

To warm up, let us revisit some simple nonmodels that we touched on earlier.

Example 6.16 (One-Point Geometry). *One-point geometry* is the interpretation in which there is exactly one point and there are no lines. We may define distances and angle measures however we like, because there are no pairs of points and no angles to which they might apply. This satisfies all of the postulates of neutral geometry (vacuously) except the existence postulate. //

Example 6.17 (The Three-Point Plane). The *three-point plane* of Example 2.1 satisfies the set postulate, the existence postulate, and the unique line postulate. However, no matter how we define distances, it cannot satisfy the ruler postulate, because there is no bijective function between any line in this interpretation and $\mathbb{R}$, for the simple reason that each line contains only two points, while $\mathbb{R}$ is an infinite set. The same argument shows that *no* interpretation with only finitely many points can be a model of neutral geometry. //

Example 6.18 (Spherical Geometry). *Spherical geometry* is the interpretation described in Example 2.13, in which a *point* is defined to be a point on the surface of the unit sphere in $\mathbb{R}^3$ and a *line* is a great circle on the sphere, i.e., a circle on the surface of the sphere whose center coincides with the center of the sphere. No matter how we define distances and angle measures, this interpretation cannot be turned into a model of neutral geometry because it does not satisfy the unique line postulate.   //

Although the sphere does not satisfy all the postulates of neutral geometry, Riemann argued that it *does* satisfy Euclid's first four postulates, provided we are careful how we interpret them. For example, it is a straightforward exercise in three-dimensional analytic geometry to prove that any two distinct points on the sphere are contained in a great circle, so Euclid's Postulate 1 is satisfied. The trickiest postulate to interpret is Euclid's Postulate 2: if we interpret a "finite straight line" to mean an arc of a great circle and interpret "produce" to mean that we can extend any arc to a longer arc in both directions, then this postulate is also satisfied. The other three postulates are satisfied as well. However, as we mentioned in Chapter 1, Euclid's proof of the exterior angle inequality (Proposition I.16) does not work on the sphere, and in fact that theorem is false in this interpretation. This illustrates the fact that Proposition I.16 actually requires more postulates than Euclid assumed. The postulates of neutral geometry are designed to fill in these gaps.

Example 6.19 (Single Elliptic Geometry). *Single elliptic geometry* is the interpretation of Example 2.14, in which a *point* is defined to be a pair of antipodal points on the surface of the unit sphere and a *line* is the set of all pairs of antipodal points lying on a specific great circle. This interpretation satisfies the set, existence, and unique line postulates, as well as the elliptic parallel postulate, as you can check. However, there is no way to define distances and angle measures in such a way that the other postulates of neutral geometry are true in this interpretation. We do not quite have the tools to prove this yet; but in the next chapter we will show that there exist parallel lines in every model of neutral geometry, which shows that single elliptic geometry cannot be such a model.   //

The next two nonmodels are the most important ones in this chapter.

Example 6.20 (The Rational Plane). In this interpretation, we define a *point* to be an ordered pair (x, y) of *rational* numbers and a *line* to be the set of all rational solutions (x, y) of an equation of one of the following two types:

$$x = c \qquad\qquad \text{for some rational constant } c;$$
$$y = mx + b \qquad\qquad \text{for some rational constants } m \text{ and } b.$$

Distances between points and measures of angles are defined exactly as in the Cartesian plane. (Note that the distance between two points might not be a rational number, but this is not a problem because the distance postulate says that distance is supposed to be a uniquely defined *real* number.)   //

The rational plane satisfies the set, existence, unique line, and distance postulates, as you can easily check. However, the next theorem shows that it is not a model of neutral geometry.

Theorem 6.21. *The rational plane does not satisfy the ruler postulate.*

Proof. Suppose for the sake of contradiction that the ruler postulate were satisfied. Consider the line ℓ defined by the equation $x = 0$, which is the set of all points of the form $(0, y)$ for arbitrary rational numbers y, and let A be the point $(0, 0) \in \ell$. Let $f : \ell \to \mathbb{R}$ be a coordinate function for ℓ, let a be the real number $f(A)$, and let $b = a + \sqrt{2}$. Because f is surjective, there is a point $B \in \ell$ such that $f(B) = b$. Then the fact that f is distance-preserving means that

$$AB = |f(B) - f(A)| = |b - a| = \sqrt{2}.$$

However, if we write the coordinates of B as $(0, y_0)$, the definition of distance in the rational plane implies that this distance can also be computed as

$$AB = \sqrt{(0-0)^2 + (y_0 - 0)^2} = \sqrt{(y_0)^2} = |y_0|.$$

Thus $y_0 = \pm\sqrt{2}$. Since B is a point in the rational plane, y_0 is a rational number, so this is a contradiction. $\square$

One reason the rational plane is significant is that it satisfies all five of Euclid's postulates (or at least reasonable modern interpretations of those postulates) and yet some of Euclid's theorems are not true in it. Thus it demonstrates that some of the steps that Euclid used in his proofs were not justified by his postulates. One such step occurs right away in his very first proof: in Proposition I.1 (constructing an equilateral triangle), Euclid starts with an arbitrary line segment $\overline{AB}$ and draws circles with centers at A and B, both with radius equal to AB. He then mentions "the point C, in which the circles cut one another." However, Euclid's postulates do not guarantee that there exists such a point. For example, in the rational plane, take A and B to be the points $(-1, 0)$ and $(1, 0)$, respectively, so $AB = 2$.

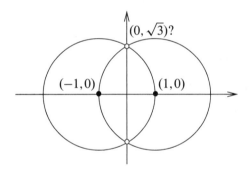

Fig. 6.8. Circles in the rational plane.

Suppose for the sake of contradiction that there is a point $C = (x, y)$ in the rational plane that lies on both of the circles of radius 2 with centers at A and B (see Fig. 6.8). Because $CA = CB = 2$, the Cartesian formula for distance yields the following simultaneous equations for the coordinates x and y:

$$4 = (CA)^2 = (x+1)^2 + y^2,$$
$$4 = (CB)^2 = (x-1)^2 + y^2.$$

Expanding the right-hand sides and subtracting the second equation from the first, we see that most of the terms cancel, yielding $0 = 2x - (-2x)$, which implies $x = 0$. Then substituting this back into the first equation, we obtain $4 = 1 + y^2$, or $y^2 = 3$. This means $y = \pm\sqrt{3}$. But the next lemma (a generalization of Thoerem 1.1) implies that $\sqrt{3}$ is irrational, contradicting the assumption that y is rational.

Lemma 6.22. *Suppose n is a positive integer. If $\sqrt{n}$ is not an integer, then it is irrational.*

Proof. Exercise 6A.   □

Our final nonmodel exhibits another gap in Euclid's logic.

Example 6.23 (Taxicab Geometry). In this interpretation, we define **point**, **line**, and **angle measure** exactly as we did in the Cartesian model. However, distance is defined differently: if $A = (x_1, y_1)$ and $B = (x_2, y_2)$ are two points in the taxicab plane, we define the **taxicab distance** between them by the formula

$$AB = |x_2 - x_1| + |y_2 - y_1|.$$

The reason this is called the "taxicab distance" is because it suggests the distance you would have to travel if you rode a taxicab between two points in a place like Manhattan where streets are laid out on a rectangular grid: to get from (x_1, y_1) to (x_2, y_2), you have to travel due east or west a distance $|x_2 - x_1|$ and then due north or south a distance $|y_2 - y_1|$ or vice versa. (This only describes the *distance* between two points; a *line* in taxicab geometry is a set of points satisfying a linear equation, just as in the Cartesian plane.)   //

It can be shown that this geometry satisfies all of the postulates of neutral geometry *except* the SAS postulate. (See Exercise 6C for a suggestion on proving the ruler postulate.) It follows that the SAS postulate is independent of the other postulates of neutral geometry.

To see why the SAS postulate fails in taxicab geometry, consider the triangles $\triangle ABC$ and $\triangle A'B'C'$ (see Fig. 6.9), where

$$A = (0,0), \qquad B = (2,0), \qquad C = (0,2);$$
$$A' = (0,1), \qquad B' = (1,0), \qquad C' = (-1,0).$$

Using the definition of the taxicab distance, we can compute easily that $AB = A'B' = 2$

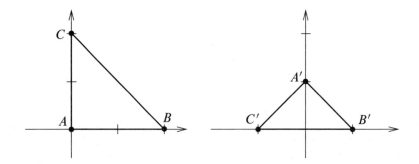

Fig. 6.9. SAS is false in taxicab geometry.

and $AC = A'C' = 2$, so these triangles have two pairs of corresponding sides that are congruent. Moreover, a simple computation based on Lemma 6.2 shows that both $\angle A$ and $\angle A'$ are right angles (remember, angle measures in taxicab geometry are exactly the same as in the Cartesian model). Therefore, these triangles satisfy the hypotheses of the SAS postulate. However, they are not congruent, because $BC = 4$, while $B'C' = 2$. The problem is that in this geometry, the geometric properties of a figure depend on how it is oriented in the plane, because of the special role played by the vertical and horizontal directions.

It can also be shown that taxicab geometry satisfies all five of Euclid's postulates. This demonstrates that Euclid's proof of the SAS congruence theorem (Proposition I.4) is not justified by his postulates, because if it were, it would have to be true in this model.

Independence of the Neutral Geometry Postulates

One desirable characteristic of any axiomatic system is that it should not contain any unnecessary axioms. In particular, any axiom that can be proved as a theorem based on the remaining axioms is redundant and does not need to be included in the system. The way to avoid redundant axioms is to ensure that each axiom is *independent* of the others in the sense we described in Chapter 2, meaning that it can neither be proved nor disproved from the other axioms.

Of course, for pedagogical reasons, it is sometimes desirable to include some redundant axioms—if their proofs would be too long or hard for the intended audience, for example—and this is regularly done in high-school geometry books. But for the purpose of understanding the structure of axiomatic Euclidean geometry, it is useful to know if the postulates we have chosen are independent.

Some of our postulates require previous postulates before they even make sense—for example, the angle measure postulate refers to half-rotations, which are defined in terms of sides of lines, which depend on the plane separation postulate for their very definition. For this reason, we restrict our attention to the question of whether each postulate is independent of the *preceding* ones. If the answer were no, then in a strictly logical sense, that postulate would add nothing new to the axiomatic system and could have been omitted.

The next theorem shows that each of the neutral postulates does add something new to our axiomatic system.

Theorem 6.24. *Each of the postulates of neutral geometry is independent of the ones that preceded it.*

Sketch of proof. Because the Cartesian plane is an interpretation in which all nine neutral postulates are true, to prove that each successive postulate is independent of the previous ones, we need only exhibit a model in which the preceding postulates are true and the new one is false. We consider each of the postulates in turn. For Postulate 1, there are no preceding postulates, so we begin with the second.

2. THE EXISTENCE POSTULATE: One-point geometry (described in Example 2.6) is an interpretation in which the set postulate is true but the existence postulate is false.

3. THE UNIQUE LINE POSTULATE: Square geometry (Example 2.10) is an interpretation in which Postulates 1 and 2 are true but Postulate 3 is false.

4. THE DISTANCE POSTULATE: If we start with the three-point plane and define the distance between any two points to be equal to -1, we get an interpretation in which Postulates 1–3 are true but the distance postulate is false.

5. THE RULER POSTULATE: Again start with the three-point plane, but now define the distance between any two distinct points to be 1, while the distance between any point and itself is zero. This satisfies the first four postulates but not the ruler postulate.

6. THE PLANE SEPARATION POSTULATE: There is a three-dimensional analogue of the Cartesian plane, in which points are ordered triples of real numbers, lines are

sets of points satisfying pairs of independent simultaneous linear equations, and the distance between two points (x_1, y_1, z_1) and (x_2, y_2, z_2) is

$$\sqrt{(x_2 - x_1)^2 + (y_2 - y_1)^2 + (z_2 - z_1)^2}. \tag{6.15}$$

Using an analysis similar to the one we used for the Cartesian plane, it can be shown that this interpretation satisfies Postulates 1–5 but not the plane separation postulate.

7. THE ANGLE MEASURE POSTULATE: If we start with the Cartesian plane but redefine the measures of all angles to be, say, 360°, then the first six postulates hold but the angle measure postulate does not.

8. THE PROTRACTOR POSTULATE: As in the previous example, start with the Cartesian plane, but now redefine all angle measures to be 0°. Then Postulates 1–7 are satisfied, but the protractor postulate is not.

9. THE SAS POSTULATE: As we noted above, taxicab geometry is an interpretation in which Postulates 1–8 are all satisfied, but the SAS postulate is not.

Thus each of the nine postulates of neutral geometry is independent of all of the previous ones.   □

Exercises

6A. Prove Lemma 6.22 (the square root of a positive integer is either an integer or irrational). [Hint: Let n be a positive integer, and suppose $\sqrt{n}$ is rational but not an integer. If $\sqrt{n} = p/q$ in lowest terms, show that q has at least one prime factor that does not divide p, and proceed as in the proof of Theorem 1.1.]

6B. Show that Euclid's Proposition I.3 is false in the rational plane by constructing explicit segments $\overline{AB}$ and $\overline{CD}$ such that $AB > CD$ but such that there is no point $E \in \overline{AB}$ with $\overline{AE} \cong \overline{CD}$.

6C. Show that taxicab geometry satisfies the ruler postulate. [Hint: If ℓ is a line defined by an equation of the form $y = mx + b$, show that the function $f(x, y) = x(1 + |m|)$ is a coordinate function for it. What can you do for vertical lines?]

6D. In taxicab geometry, prove that there exists an equilateral triangle that is not equiangular, and use this to prove that the isosceles triangle theorem is false. [Hint: There is a picture of such a triangle in this chapter. Lemma 6.2 might be helpful for showing that it is not equiangular.]

6E. In taxicab geometry, prove that the triangle inequality (Theorem 5.18) is false by finding three specific noncollinear points A, B, C such that $AC \geq AB + BC$.

6F. In taxicab geometry, show that the SSS congruence theorem is false by finding triangles $\triangle ABC$ and $\triangle A'B'C'$ with $AB = A'B'$, $AC = A'C'$, and $BC = B'C'$ but such that $\triangle ABC \not\cong \triangle A'B'C'$.

Chapter 7

Perpendicular and Parallel Lines

In this chapter, we undertake an in-depth study of properties of perpendicular and parallel lines in neutral geometry.

Perpendicular Lines

First we focus on perpendicular lines. Recall that if ℓ and m are lines, we say by definition that ***ℓ and m are perpendicular***, denoted by $\ell \perp m$, if they intersect at a point O and at least one of the rays in ℓ starting at O makes a right angle with at least one of the rays in m starting at O. The four right angles theorem (Theorem 4.29) shows that in this case, all four angles formed by ℓ and m are right angles.

Sometimes it is also useful to talk about segments or rays being perpendicular to other segments, rays, or lines. In that case, as we did for parallelism, we need a slightly more complicated definition. If S_1 is a segment, ray, or line and S_2 is another segment, ray, or line, we say that ***S_1 and S_2 are perpendicular***, symbolized also by $S_1 \perp S_2$, if S_1 and S_2 intersect and the lines containing them are perpendicular (see Fig. 7.1).

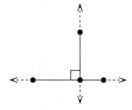

Fig. 7.1. Perpendicular segments.

Given a line ℓ and a point $A \in \ell$, Theorem 4.30 showed that there is a unique line that is perpendicular to ℓ through A. The next theorem (corresponding to Euclid's Proposition

I.12) is complementary to that one: it shows that we can always find a unique perpendicular to ℓ through a point A *not on* ℓ. Constructing such a line is called ***dropping a perpendicular from A to ℓ***. As mentioned in Chapter 1, Euclid's proof of Proposition I.12 relied on facts about intersections of lines and circles that were not justified by his postulates. We will be able to justify those facts in Chapter 14, but the following proof circumvents the problem by avoiding the use of circles altogether. The reason we did not prove this theorem in Chapter 4 is that the proof uses the SAS postulate.

Theorem 7.1 (Dropping a Perpendicular). *Suppose ℓ is a line and A is a point not on ℓ. Then there exists a unique line that contains A and is perpendicular to ℓ.*

Proof. First we will prove existence. Choose three points $B, C, D \in \ell$ such that $B * C * D$. By the angle construction theorem, there is a unique ray on the other side of ℓ from A that makes an angle with $\overrightarrow{CD}$ that is congruent to $\angle ACD$. Choose A' on that ray such that $\overline{CA'} \cong \overline{CA}$, and let $m = \overleftrightarrow{AA'}$. The proof will be completed by showing that $m \perp \ell$.

Now, $\angle ACD \cong \angle A'CD$ by construction. Note also that $\angle ACB$ and $\angle ACD$ form a linear pair, as do $\angle A'CB$ and $\angle A'CD$ (Fig. 7.2). Because $\angle ACB$ and $\angle A'CB$ are supplements of the congruent angles $\angle ACD$ and $\angle A'CD$, they are also congruent to each other.

Because A and A' are on opposite sides of ℓ, there is a point F where $\overline{AA'}$ intersects ℓ. If F is equal to C, then $\angle AFD$ and $\angle A'FD$ form a linear pair. Since they are congruent by construction, they are both right angles. This means that $m \perp \ell$, and we are done.

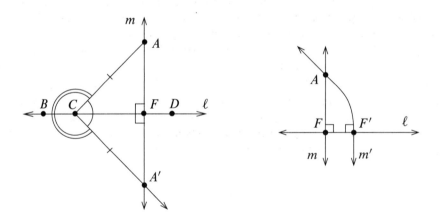

Fig. 7.2. Dropping a perpendicular.   **Fig. 7.3.** Uniqueness of the perpendicular.

On the other hand, suppose $F \neq C$. Then either $F \in \text{Int} \overrightarrow{CD}$ or $F \in \text{Int} \overrightarrow{CB}$; since the argument is the same in both cases except with the roles of B and D reversed, we may assume that $F \in \text{Int} \overrightarrow{CD}$. Since $\angle ACF \cong \angle A'CF$, $\overline{CA} \cong \overline{CA'}$, and $\overline{CF}$ is congruent to itself, we have $\triangle ACF \cong \triangle A'CF$ by SAS. Therefore $\angle AFC \cong \angle A'FC$. Since these angles are congruent and form a linear pair, they are both right angles, and thus $m \perp \ell$. This completes the proof of existence.

Now we have to prove uniqueness. Suppose for the sake of contradiction that m and m' are two distinct lines through A and perpendicular to ℓ. Let F be the point where m meets

ℓ, and let F' be the point where m' meets ℓ (Fig. 7.3). Then $F \neq F'$ because otherwise m would be equal to m', and A, F, and F' are noncollinear because collinearity would imply $A \in \ell$. Thus $\triangle AFF'$ is a triangle. By the four right angles theorem, both $\angle AFF'$ and $\angle AF'F$ are right angles. Thus $\triangle AFF'$ has two right angles, which contradicts Corollary 5.15 (every triangle has at least two acute angles). $\square$

In the situation of the preceding theorem, the point F where m intersects ℓ is called the *foot of the perpendicular from A to ℓ*.

There will be many situations in which it is important to know whether the foot of a perpendicular to a given line lies in a certain ray or segment within that line. There are some simple criteria that can be used to determine this. We begin with rays.

Theorem 7.2. *Suppose $\overrightarrow{AB}$ is a ray, P is a point not on $\overleftrightarrow{AB}$, and F is the foot of the perpendicular from P to $\overleftrightarrow{AB}$ (Fig. 7.4).*

(a) $F = A$ *if and only if $\angle PAB$ is a right angle.*

(b) $F \in \operatorname{Int} \overrightarrow{AB}$ *if and only if $\angle PAB$ is acute.*

(c) F *lies in the interior of the ray opposite $\overrightarrow{AB}$ if and only if $\angle PAB$ is obtuse.*

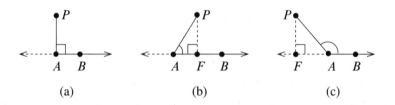

(a)   (b)   (c)

Fig. 7.4. The three cases of Theorem 7.2.

Proof. Let $\overrightarrow{AB}$, P, and F be as in the hypothesis. We begin by proving the "only if" part of each statement.

For (a), suppose $F = A$. This means that $\overleftrightarrow{PA}$ is perpendicular to $\overleftrightarrow{AB}$, so $\angle PAB$ is a right angle by the four right angles theorem.

Next, for (b), suppose $F \in \operatorname{Int} \overrightarrow{AB}$. Because $P \notin \overleftrightarrow{AB}$, the points P, F, and A are noncollinear, so they form a right triangle with right angle at F. It follows from Corollary 5.15 that $\angle PAF$ is acute, and our hypothesis guarantees that $\angle PAF = \angle PAB$.

For (c), assume that F lies in the ray opposite $\overrightarrow{AB}$. Then just as above, $\triangle PFA$ is a right triangle with right angle at F, and thus $\angle PAF$ is acute. In this case, $\angle PAF$ and $\angle PAB$ form a linear pair and are thus supplementary, from which it follows that $\angle PAB$ is obtuse. This completes the proofs of the "only if" parts.

Now, the "if" parts follow from these: suppose first that $\angle PAB$ is a right angle. Then F cannot lie in $\operatorname{Int} \overrightarrow{AB}$, because the argument above would imply that $\angle PAB$ is acute; and F cannot lie in the interior of the ray opposite $\overrightarrow{AB}$, because then $\angle PAB$ would be obtuse; so $F = A$ is the only possibility. Similarly, if $\angle PAB$ is acute, then F cannot be equal to A or lie in the interior of the ray opposite $\overrightarrow{AB}$, so it must lie in $\operatorname{Int} \overrightarrow{AB}$. The argument for (c) is the same. $\square$

The Altitudes of a Triangle

Many applications of the preceding theorem are based on dropping a perpendicular from a vertex of a triangle to the line containing the opposite side. If $\triangle ABC$ is a triangle and F is the foot of the perpendicular from a vertex A to the line $\overleftrightarrow{BC}$ that contains the opposite side, then the segment $\overline{AF}$ is called the ***altitude from A to $\overline{BC}$***.

In general, an altitude to a side might or might not intersect that side (see Fig. 7.5). The next theorem shows how to determine where the foot of an altitude is located.

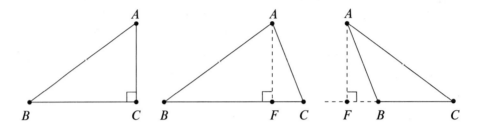

Fig. 7.5. Possible locations for an altitude of a triangle.

Theorem 7.3. *Let $\triangle ABC$ be a triangle, and let F be the foot of the altitude from A to $\overline{BC}$.*

(a) *$F = B$ if and only if $\angle B$ is right, and $F = C$ if and only if $\angle C$ is right.*

(b) *$B * F * C$ if and only if $\angle B$ and $\angle C$ are both acute.*

(c) *$F * B * C$ if and only if $\angle B$ is obtuse, and $B * C * F$ if and only if $\angle C$ is obtuse.*

Proof. Part (a) follows immediately from Theorem 7.2(a).

Next we prove (c). Note that $F * B * C$ if and only if F lies in the interior of the ray opposite $\overrightarrow{BC}$, which by Theorem 7.2(c) is true if and only if $\angle B$ is obtuse. The argument for $\angle C$ is similar.

Finally, to prove (b), suppose $B * F * C$. Then (a) shows that neither B nor C is a right angle, and (c) shows that neither one is obtuse, so they must both be acute. Conversely, suppose $\angle B$ and $\angle C$ are both acute. Then (a) shows that F is not equal to B or C, and (c) shows that it cannot satisfy $F * B * C$ or $B * C * F$. By Theorem 3.12, the only remaining possibility is $B * F * C$. $\qquad\square$

Corollary 7.4. *In any triangle, the altitude to the longest side always intersects the interior of that side.*

Proof. Let $\triangle ABC$ be a triangle, and assume the vertices are labeled so that $\overline{AB}$ is the longest side. Then $\angle C$ is the largest angle by the scalene inequality. At least two of the angles of $\triangle ABC$ are acute; thus if any angle is not acute, it must be $\angle C$. It follows that $\angle A$ and $\angle B$ are both acute, and the result follows from the previous theorem. $\qquad\square$

Corollary 7.5. *In a right triangle, the altitude to the hypotenuse always intersects the interior of the hypotenuse.*

Proof. The hypotenuse is the longest side. $\qquad\square$

In any triangle, a segment joining a vertex to the midpoint of the opposite side is called a **median** of the triangle. Ordinarily, the median and the altitude to a given side are different; but for the base of an isosceles triangle they are not.

Theorem 7.6 (Isosceles Triangle Altitude Theorem). *The altitude to the base of an isosceles triangle is also the median to the base and is contained in the bisector of the angle opposite the base (see Fig. 7.6).*

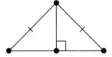

Fig. 7.6. The altitude to the base of an isosceles triangle.

Proof. See Exercise 7A. □

Perpendicular Bisectors

Suppose $\overline{AB}$ is a segment. A **perpendicular bisector of $\overline{AB}$** is a line that is perpendicular to $\overline{AB}$ and contains its midpoint.

Theorem 7.7 (Existence and Uniqueness of a Perpendicular Bisector). *Every segment has a unique perpendicular bisector.*

Proof. Exercise 7B. □

Theorem 7.8 (Perpendicular Bisector Theorem). *If $\overline{AB}$ is any segment, every point on the perpendicular bisector of $\overline{AB}$ is equidistant from A and B.*

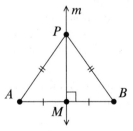

Fig. 7.7. P is on the perpendicular bisector of $\overline{AB}$.

Proof. Suppose $\overline{AB}$ is a segment and m is its perpendicular bisector, and let M denote the midpoint of $\overline{AB}$. Let P be an arbitrary point of m (Fig. 7.7). There are two cases: either $P \in \overleftrightarrow{AB}$ or not. If $P \in \overleftrightarrow{AB}$, then $P = M$ (because M is the only point where the lines m and $\overleftrightarrow{AB}$ intersect), so P is equidistant from A and B in this case. On the other hand, if $P \notin \overleftrightarrow{AB}$, then $\triangle PMA$ and $\triangle PMB$ are congruent triangles by SAS, so it follows that $PA = PB$. □

Theorem 7.9 (Converse to the Perpendicular Bisector Theorem). *Suppose* $\overline{AB}$ *is a segment and* P *is a point that is equidistant from* A *and* B. *Then* P *lies on the perpendicular bisector of* $\overline{AB}$.

Proof. Exercise 7C. $\square$

Theorem 7.10 (Reflection across a Line). *Let* ℓ *be a line and let* A *be a point not on* ℓ. *Then there is a unique point* A', *called the **reflection of** A **across** ℓ, *such that* ℓ *is the perpendicular bisector of* $\overline{AA'}$ *(Fig. 7.8).*

Proof. Exercise 7D. $\square$

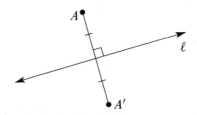

Fig. 7.8. The reflection of a point across a line

The Distance from a Point to a Line

We all understand from our everyday experience that the shortest distance from a point to a line is along a straight path perpendicular to the line. For example, if you want to get to a nearby road, you will have the shortest distance to walk if you follow a path that meets the road at right angles. We can now give a rigorous geometric justification for this fact.

First we start with some general definitions. Suppose S is a set of points in the plane and P is any point (which might or might not lie in S). A point $C \in S$ is said to be a ***closest point to*** P ***in*** S if $PC \leq PX$ for every point $X \in S$. Note that for certain sets, there might be more than one closest point, and for others, there might be none at all, as the following examples show.

If there is a closest point to P in S, we define the ***distance from*** P ***to*** S, symbolized by $d(P, S)$, to be the distance from P to any closest point in S.

Example 7.11 (Closest Points).

(a) If S is any set of points and $P \in S$, then the distance from P to itself is zero, while the distance from P to any other point of S is positive; thus P itself is the unique closest point to P in S, and $d(P, S) = 0$.

(b) If $\mathcal{C}$ is a circle and O is its center, then the distance from O to every point of $\mathcal{C}$ is the same, so every point on $\mathcal{C}$ is closest to O and $d(O, \mathcal{C})$ is equal to the radius of $\mathcal{C}$.

(c) If $\overline{AB}$ is a segment and S is the set of all interior points of $\overline{AB}$, then there is no closest point to A in S.   //

The next simple lemma can be useful in identifying closest points.

Lemma 7.12 (Properties of Closest Points). *Let P be a point and let S be any set of points.*

(a) *If C is a closest point to P in S, then another point $C' \in S$ is also a closest point to P if and only if $PC' = PC$.*

(b) *If C is a point in S such that $PX > PC$ for every point $X \in S$ other than C, then C is the unique closest point to P in S.*

Proof. Exercise 7E. □

The next theorem shows that the closest point on a line is easy to find: it is the foot of the perpendicular from the point to the line. It also says that points farther away from the foot are also farther away from the original point.

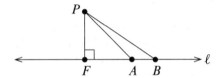

Fig. 7.9. The closest point to *P* on *ℓ*.

Theorem 7.13 (Closest Point on a Line). *Suppose ℓ is a line, P is a point not on ℓ, and F is the foot of the perpendicular from P to ℓ (Fig. 7.9).*

(a) *F is the unique closest point to P on ℓ.*

(b) *If A and B are points on ℓ such that $F * A * B$, then $PB > PA$.*

Proof. Exercise 7F. □

The next theorem shows that segments also have unique closest points. Its proof is rated PG; the only place it will be used in this book is in Exercise 8A.

Theorem 7.14 (Closest Point on a Segment). *Suppose $\overline{AB}$ is a segment and P is any point. Then there is a unique closest point to P in $\overline{AB}$.*

Proof. Exercise 7G. □

If ℓ and *m* are two distinct lines, a point *P* is said to be ***equidistant from ℓ and m*** if $d(P, \ell) = d(P, m)$. This definition is useful for expressing the following property of angle bisectors, analogous to the perpendicular bisector theorem.

Theorem 7.15 (Angle Bisector Theorem). *Suppose $\angle AOB$ is a proper angle and P is a point on the bisector of $\angle AOB$. Then P is equidistant from $\overrightarrow{OA}$ and $\overrightarrow{OB}$. (See Fig. 7.10.)*

Proof. Exercise 7H. □

Like the perpendicular bisector theorem, the angle bisector theorem has an important converse. However, the full converse is not true: a point can be equidistant from the lines containing the sides of an angle without lying on the bisector, because it might not be in the interior of the angle. But if we assume in addition that the point is in the interior, then

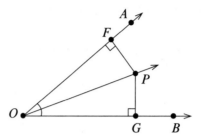

Fig. 7.10. The angle bisector theorem.

it must lie on the bisector, as Theorem 7.17 below will show. The proof will rely on the following lemma, whose proof is rated PG.

Lemma 7.16. *Suppose* $\angle AOB$ *is a proper angle and* P *is a point in* $\mathrm{Int}\,\angle AOB$ *that is equidistant from* $\overleftrightarrow{OA}$ *and* $\overleftrightarrow{OB}$. *Then the feet of the perpendiculars from* P *to* $\overleftrightarrow{OA}$ *and* $\overleftrightarrow{OB}$ *lie in the interiors of the rays* $\overrightarrow{OA}$ *and* $\overrightarrow{OB}$, *respectively.*

Proof. Given $\angle AOB$ and P as in the hypothesis, let F and G be the feet of the perpendiculars from P to $\overleftrightarrow{OA}$ and $\overleftrightarrow{OB}$, respectively, so the fact that P is equidistant from the two lines means that $PF = PG$. Suppose for the sake of contradiction that one of these feet does not lie in the interior of the corresponding ray; without loss of generality, assume $G \notin \mathrm{Int}\,\overrightarrow{OB}$. There are two possibilities: either $G = O$ or G lies in the interior of the ray opposite $\overrightarrow{OB}$.

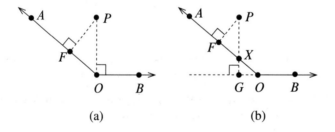

(a)   (b)

Fig. 7.11. The proof of Lemma 7.16.

If $G = O$ (see Fig. 7.11(a)), then F cannot be equal to O, because otherwise $\overleftrightarrow{OA}$ and $\overleftrightarrow{OB}$ would both be perpendicular to $\overleftrightarrow{PO}$ at O, so they would be the same line, contradicting our assumption that $\angle AOB$ is a proper angle. It follows from Theorem 7.13 that $PO > PF$, contradicting the assumption that P is equidistant from the two lines.

On the other hand, suppose G is in the interior of the ray opposite $\overrightarrow{OB}$ (Fig. 7.11(b)). Then G and B are on opposite sides of $\overleftrightarrow{OA}$ by the X-lemma. The assumption that $P \in \mathrm{Int}\,\angle AOB$ implies that P and B are on the same side of $\overleftrightarrow{OA}$, so G and P are on opposite sides of $\overleftrightarrow{OA}$. Thus there is a point X where the interior of $\overline{PG}$ meets $\overleftrightarrow{OA}$. This implies $PX < PG$ (the whole segment is greater than the part). But $PG = PF$ by hypothesis, so X is a point on $\overleftrightarrow{OA}$ that is closer to P than F, which contradicts Theorem 7.13.   ☐

Theorem 7.17 (Partial Converse to the Angle Bisector Theorem). *Suppose* $\angle AOB$ *is a proper angle. If* P *is a point in the interior of* $\angle AOB$ *that is equidistant from* $\overrightarrow{OA}$ *and* $\overleftrightarrow{OB}$, *then* P *lies on the angle bisector of* $\angle AOB$.

Proof. Exercise 7I.   □

Parallel Lines

Now we are ready to start studying parallel lines. Recall that our definition of *parallel lines* is simply lines that do not intersect.

One consequence of the definition of parallelism is immediate.

Lemma 7.18. *Suppose* ℓ *is a line and* S *is a segment, ray, or line that is parallel to* ℓ. *Then all points of* S *lie on the same side of* ℓ.

Proof. Let A be any point of S. We need to show that every other point $B \in S$ is on the same side of ℓ as A. If $B = A$, there is nothing to prove, so suppose B is a point of S that is distinct from A. Then the line containing S is $\overleftrightarrow{AB}$, so our hypothesis implies that $\overleftrightarrow{AB} \parallel \ell$. Because $\overline{AB} \subseteq \overleftrightarrow{AB}$, it follows that $\overline{AB}$ does not intersect ℓ, so A and B are on the same side of ℓ.   □

Many of the criteria for determining when two lines are parallel refer to the following situation: suppose ℓ and ℓ' are lines (not necessarily assumed to be parallel). We say that a line t is a ***transversal for*** ℓ ***and*** ℓ' if t, ℓ, and ℓ' are all distinct and t intersects ℓ and ℓ' at two different points (Fig. 7.12). If t is a transversal for ℓ and ℓ', we often say that ℓ ***and*** ℓ' ***are cut by*** t.

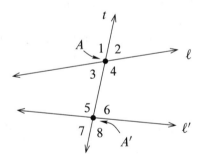

Fig. 7.12. A transversal.

A transversal makes eight different angles at the points of intersection with the lines that it cuts, which can be classified into several different categories. In order to describe these efficiently, we introduce the following terminological shortcut, which is an extension of our convention for saying that "a ray lies on a side of a line": if $\angle AOB$ is an angle and S is one of the sides of the line $\overleftrightarrow{OA}$, we will say that $\angle AOB$ ***lies on side*** S if the ray $\overrightarrow{OB}$ lies on that side (which just means, of course, that its interior points lie on side S).

Suppose t is a transversal that intersects ℓ at A and ℓ' at A'. The ***interior angles*** are the two angles made by t and ℓ on the same side of ℓ as A' and the two angles made by

t and ℓ' on the same side of ℓ' as A. (Note that in general, it is not enough to say "the same side of ℓ' as ℓ" or vice versa—we are not insisting that ℓ and ℓ' are parallel, so not all points of ℓ will necessarily be on the same side of ℓ'.) The *exterior angles* are the other four: the two angles made by t and ℓ on the opposite side of ℓ from A' and the two angles made by t and ℓ' on the opposite side of ℓ' from A. Thus in Fig. 7.12, the interior angles are $\angle 3$, $\angle 4$, $\angle 5$, and $\angle 6$, while the others are exterior angles.

Two angles, one of which has its vertex at A and the other at A', are called

- *alternate interior angles* if they are interior angles lying on opposite sides of the transversal;

- *consecutive interior angles* if they are interior angles lying on the same side of the transversal; and

- *corresponding angles* if one is an interior angle and the other is an exterior angle and both lie on the same side of the transversal.

Thus in Fig. 7.12, we have the following pairs of angles.

- Alternate interior angles: $\angle 3$ and $\angle 6$; $\angle 4$ and $\angle 5$.

- Consecutive interior angles: $\angle 3$ and $\angle 5$; $\angle 4$ and $\angle 6$.

- Corresponding angles: $\angle 1$ and $\angle 5$; $\angle 2$ and $\angle 6$; $\angle 3$ and $\angle 7$; $\angle 4$ and $\angle 8$.

When proving theorems involving angles formed by transversals, it is important to verify that two angles that are claimed to be alternate interior angles (or corresponding angles or consecutive interior angles) do indeed satisfy the definitions. Usually this will follow very easily from the way the angles are defined (as well as being evident from the diagram), and in these cases it can simply be asserted that a pair of angles is as it appears. But occasionally there are situations in which it is not at all straightforward to show that an angle lies on a certain side of a line, so a detailed proof must be given. We will mention these situations when they arise.

The next theorem is one of the most important criteria for proving that two lines are parallel. It is Euclid's Proposition I.27, with the same proof.

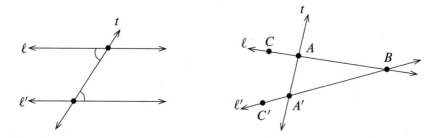

Fig. 7.13. Alternate interior angles theorem.   **Fig. 7.14.** Proof of the theorem.

Theorem 7.19 (Alternate Interior Angles Theorem). *If two lines are cut by a transversal making a pair of congruent alternate interior angles (Fig. 7.13), then they are parallel.*

Proof. Suppose ℓ and ℓ' are lines and t is a transversal that meets them at A and A', respectively. (The assumption that t is a transversal includes the assertion that ℓ and ℓ' are distinct.) We need to show that the existence of a pair of congruent alternate interior angles implies $\ell \parallel \ell'$. We will prove the contrapositive: if ℓ and ℓ' are not parallel, then neither pair of alternate interior angles can be congruent.

Assume ℓ and ℓ' are not parallel. Let B be the point where ℓ meets ℓ' (Fig. 7.14), and choose points C and C' such that $C * A * B$ and $C' * A' * B$. Note that both $\angle CAA'$ and $\angle C'A'A$ are exterior angles for $\triangle AA'B$, so the exterior angle inequality shows that $m\angle CAA' > m\angle AA'B$ and $m\angle C'A'A > m\angle A'AB$. Thus neither pair of alternate interior angles is congruent. $\qquad\square$

Here are several other important criteria for parallelism, all of which follow easily from the preceding theorem. The first of these is Euclid's Proposition I.28.

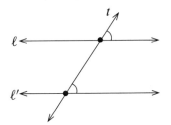

Fig. 7.15. Corresponding angles theorem.

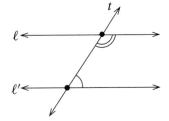

Fig. 7.16. Consecutive interior angles theorem.

Corollary 7.20 (Corresponding Angles Theorem). *If two lines are cut by a transversal making a pair of congruent corresponding angles (Fig. 7.15), then they are parallel.*

Proof. Exercise 7J. $\qquad\square$

Corollary 7.21 (Consecutive Interior Angles Theorem). *If two lines are cut by a transversal making a pair of supplementary consecutive interior angles (Fig. 7.16), then they are parallel.*

Proof. Exercise 7K. $\qquad\square$

Corollary 7.22 (Common Perpendicular Theorem). *If two distinct lines have a common perpendicular (i.e., a line that is perpendicular to both), then they are parallel.*

Proof. Suppose ℓ and ℓ' are distinct lines and t is a line perpendicular to both of them. Then ℓ and ℓ' cannot intersect t at the same point, because there is only one perpendicular to t through a given point. Thus the intersection points are different, which means that t is a transversal for ℓ and ℓ'. Because it makes a pair of congruent corresponding angles, it follows that $\ell \parallel \ell'$. $\qquad\square$

One property of parallel lines that is familiar from our everyday experience is that they are everywhere the same distance apart. To make this idea precise, we make the following definitions. If A and B are points and m is a line, we say that **A and B are equidistant from m** if $d(A,m) = d(B,m)$. If ℓ is another line, we say that **ℓ is equidistant from m** if all points on ℓ are equidistant from m.

We will show that if two lines are equidistant, then they are parallel. In fact, we can prove something much stronger: as long as there are at least *two* distinct points on one line that are the same distance from another line and on the same side of that line, it follows that the two lines are parallel. The next proof is rated PG.

Theorem 7.23 (Two-Point Equidistance Theorem). *Suppose ℓ and m are two distinct lines and there exist two distinct points on ℓ that are on the same side of m and equidistant from m. Then $\ell \parallel m$.*

Proof. Let A and B be distinct points on ℓ that are on the same side of m and equidistant from m, and assume for the sake of contradiction that ℓ and m are not parallel. Let X be the (unique) point where they intersect. Then X cannot be between A and B, because that would force A and B to lie on opposite sides of m. Without loss of generality, assume that $A * B * X$ (Fig. 7.17), which implies $AX > BX$.

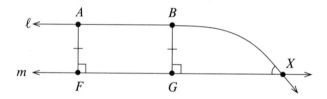

Fig. 7.17. Two points equidistant from a line.

Let F be the foot of the perpendicular from A to m, and let G be the foot of the perpendicular from B to m, so the hypothesis means $AF = BG$. First we need to prove that F, G, and X are all distinct. Assume they are not; then we will show that in fact $F = G = X$, and this will lead to a contradiction. If either F or G were equal to X, then $\overleftrightarrow{AB}$ would be perpendicular to m, which would imply $F = G = X$. On the other hand, $F = G$ would imply $F \in \overleftrightarrow{AB}$ because the unique perpendicular to m through F would contain both A and B, and once again we would have $F = G = X$. In either case, we conclude $AF = AX > BX = BG$, contradicting our assumption that $AF = BG$.

Next we need to show that $F * G * X$. The common perpendicular theorem implies that $\overleftrightarrow{AF} \parallel \overleftrightarrow{BG}$, so A and F lie on the same side of $\overleftrightarrow{BG}$. Since $A * B * X$ implies A and X lie on opposite sides of that line, we conclude that F and X also lie on opposite sides, which in turn implies $F * G * X$ as claimed.

It follows from the foregoing argument that $\overrightarrow{XA} = \overrightarrow{XB}$ and $\overrightarrow{XF} = \overrightarrow{XG}$, and therefore $\angle AXF = \angle BXG$. Therefore, the two right triangles $\triangle AXF$ and $\triangle BXG$ are congruent by AAS. This contradicts the inequality $AX > BX$ and shows that our original assumption that ℓ and m intersect must have been wrong. □

Corollary 7.24 (Equidistance Theorem). *If one of two distinct lines is equidistant from the other, then they are parallel.*

Proof. Suppose ℓ and m are distinct lines and ℓ is equidistant from m. Since the lines are distinct, there can be at most one point that lies on both, so there are infinitely many distinct points that lie on ℓ and not on m. At least two of these points must lie on the same side of *m*. The hypothesis implies that these two points are equidistant from *m*, so Theorem 7.23 implies that $\ell \parallel m$. $\qquad\square$

In high-school geometry, you might have been introduced to some or all of the above theorems and corollaries as "if and only if" statements. For example, in some high-school texts, the name "alternate interior angles theorem" is given to a statement like the following: *if two lines are cut by a transversal, then the two lines are parallel if and only if the transversal makes congruent alternate interior angles*, with similar statements for Corollaries 7.20–7.22 and 7.24. Be sure to observe that we have only proved one implication in each of these theorems—in every case, we have proved *sufficient criteria for parallelism* ("congruent alternate interior angles imply parallel lines") but not *necessary consequences of parallelism* ("parallel lines make congruent alternate interior angles"). The converses of all of these theorems are true in Euclidean geometry, as we will see later; however, they are decidedly false in hyperbolic geometry, so they cannot be proved using the axioms of neutral geometry.

The Existence of Parallel Lines

The next theorem is our version of Euclid's Proposition I.31. It shows that parallel lines exist in abundance. We give a proof that is different from Euclid's, because it allows us to prove a little more than Euclid did (namely, the final statement about a common perpendicular).

Theorem 7.25 (Existence of Parallels). *For every line ℓ and every point A that does not lie on ℓ, there exists a line m such that A lies on m and $m \parallel \ell$. It can be chosen so that ℓ and m have a common perpendicular that contains A.*

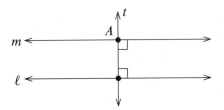

Fig. 7.18. Constructing a parallel.

Proof. Let ℓ be a line and let A be a point not on ℓ. Let t be the line through A and perpendicular to ℓ (which exists and is unique by Theorem 7.1), and then let m be the line through A and perpendicular to t (which exists and is unique by Theorem 4.30); see Fig. 7.18. Because ℓ and m have t as a common perpendicular, they are parallel. $\qquad\square$

It is important to note that the preceding theorem is a theorem in *neutral geometry*, so it is true in every model of neutral geometry. We can now say conclusively that the elliptic parallel postulate is inconsistent with the postulates of neutral geometry, because it contradicts the preceding theorem. Therefore, neither spherical geometry nor single elliptic geometry is a model of neutral geometry, because they both satisfy the elliptic parallel postulate.

On the other hand, it is equally important to note that we have not yet claimed that the parallel through A is *unique*. Indeed, we will see later that as long as we are working in the context of neutral geometry, it need not be. You might be tempted to think that we have actually proved uniqueness, because the lines t and m we constructed in the proof above were the unique ones through A and perpendicular to ℓ and t, respectively. However, this in itself does not prove that m is the unique line through A and parallel to ℓ; it only proves that it is the unique such line *that can be constructed by this procedure*. This is a good illustration of why it is usually necessary to provide a proof of uniqueness that is separate from the proof of existence. It is the uniqueness of the parallel through A that distinguishes Euclidean geometry from non-Euclidean geometry; and it cannot be proved from the axioms of neutral geometry.

Exercises

7A. Prove Theorem 7.6 (the isosceles triangle altitude theorem).

7B. Prove Theorem 7.7 (existence and uniqueness of perpendicular bisectors).

7C. Prove Theorem 7.9 (the converse to the perpendicular bisector theorem).

7D. Prove Theorem 7.10 (existence and uniqueness of a reflected point).

7E. Prove Lemma 7.12 (properties of closest points).

7F. Prove Theorem 7.13 (the closest point on a line).

7G. Prove Theorem 7.14 (the closest point on a segment). [Hint: Consider separately the cases in which $P \in \overleftrightarrow{AB}$ and $P \notin \overleftrightarrow{AB}$, and divide the second case into two subcases depending on whether the foot of the perpendicular from P to $\overleftrightarrow{AB}$ does or does not lie in $\overline{AB}$.]

7H. Prove Theorem 7.15 (the angle bisector theorem). [Hint: The strategy is suggested by Fig. 7.10. Why do F and G lie in the interiors of $\overrightarrow{OA}$ and $\overrightarrow{OB}$, respectively?]

7I. Prove Theorem 7.17 (the partial converse to the angle bisector theorem).

7J. Prove Corollary 7.20 (the corresponding angles theorem).

7K. Prove Corollary 7.21 (the consecutive interior angles theorem).

Chapter 8

Polygons

So far, we have concentrated most of our attention on triangles because they are the basic building blocks for most geometric shapes. Nearly as important as triangles are figures with four or more sides. Roughly speaking, a closed figure formed by three or more segments joined end to end is called a *polygon*. This chapter is devoted to developing some basic properties of polygons in neutral geometry.

Polygons

Here is the official definition. A *polygon* is a union of n distinct segments for some $n \geq 3$, satisfying the following conditions:

(i) There are n distinct points $A_1, \ldots, A_n$ such that the given segments are $\overline{A_1 A_2}$, $\overline{A_2 A_3}$, $\ldots$, $\overline{A_{n-1} A_n}$, $\overline{A_n A_1}$.

(ii) Two of the segments intersect only if they share a common endpoint, in which case the two intersecting segments are noncollinear.

Fig. 8.1 shows several examples of polygons, while Fig. 8.2 shows some examples of unions of segments that are not polygons.

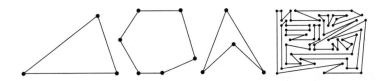

Fig. 8.1. Polygons.

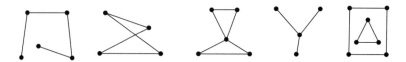

Fig. 8.2. Not polygons.

The n points A_1, ..., A_n are called the **vertices** of the polygon, and the n segments $\overline{A_1A_2}, \ldots, \overline{A_nA_1}$ are called its **sides** or **edges**.

Polygons with three sides are exactly the same as triangles, as you can easily check. In general, we will use the notation $A_1A_2\ldots A_n$ (the vertices simply listed in order) to denote the polygon formed by the union of the n segments $\overline{A_1A_2}, \overline{A_2A_3}, \ldots, \overline{A_nA_1}$. (For triangles, however, we will usually continue to use the notation $\triangle ABC$ for the triangle with vertices A, B, and C to clearly distinguish it from the angle $\angle ABC$ determined by the same three points.) Note that the order in which the vertices are listed is significant: if $ABCD$ is a polygon, then some reorderings of the vertices, such as $BCDA$ and $DCBA$, represent the same polygon (i.e., the union of the same four segments), but other orderings, such as $ACBD$, do not. In fact, $ACBD$ might not represent a polygon at all, because two of the segments might intersect at an interior point (Fig. 8.4). If we want to refer to a generic polygon but do not need to list the vertices explicitly, we often use a script letter like $\mathcal{P}$ to name the polygon.

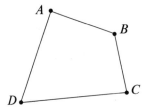

Fig. 8.3. $ABCD$ is a polygon.

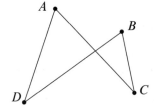

Fig. 8.4. $ACBD$ is not a polygon.

Suppose $\mathcal{P} = A_1 \ldots A_n$ is a polygon with n vertices.

- Two edges that intersect at a vertex are called **adjacent edges**.

- Two vertices that form the endpoints of an edge are called **consecutive vertices**; similarly, three vertices A, B, C are said to be consecutive if $\overline{AB}$ and $\overline{BC}$ are edges of the polygon.

- The **angles of $\mathcal{P}$** are the n angles defined by possible choices of three consecutive vertices: $\angle A_1A_2A_3, \angle A_2A_3A_4, \ldots, \angle A_{n-2}A_{n-1}A_n, \angle A_{n-1}A_nA_1$, and $\angle A_nA_1A_2$. The requirement that adjacent edges be noncollinear ensures that all the angles of a polygon are proper angles.

- Angles at consecutive vertices are called **consecutive angles**.

- Any segment joining two nonconsecutive vertices is called a **diagonal of $\mathcal{P}$**.

Polygons are classified by the number of edges each one has (which is the same as its number of vertices). We have already noted that three-sided polygons are triangles. Some commonly used names for polygons with higher numbers of sides are **quadrilateral** (four sides), *pentagon* (five), *hexagon* (six), *heptagon* (seven), *octagon* (eight), *nonagon* (nine), *decagon* (ten), and *dodecagon* (twelve). Sometimes we will call a polygon with n sides an *n-gon*.

The definition of a polygon requires that each pair of adjacent edges must be non-collinear, to rule out cases like the "pentagon" $ABCDE$ pictured in Fig. 8.5, which has a pair of collinear adjacent edges. This union of line segments is actually a quadrilateral,

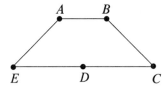

Fig. 8.5. Not a pentagon.

not a pentagon. If we allowed it to be considered as a pentagon just by picking one of the points in the middle of a side and calling it a vertex, there would be nothing stopping us from adding two, three, or a hundred more vertices to the same figure. One of the main reasons why we require properness in the definition of a polygon is so that each polygon will have an unambiguous number of edges and vertices.

Just as we did for triangles, in order to talk about "the vertices" of a polygon, we should first verify that they do not depend on the particular notation chosen to designate the polygon. The PG-rated proof of the next theorem is somewhat more difficult than that of the corresponding theorem for triangles since vertices of polygons might not be extreme points. Since we will not use this result anywhere in the book, we leave the proof as an exercise with hints. (For the kind of polygons on which we will focus most of our attention, called *convex polygons*, there is an easier proof; see Theorem 8.3 below.)

Theorem 8.1 (Consistency of Polygon Vertices). *Suppose $\mathcal{P} = A_1 \ldots A_n$ and $\mathcal{Q} = B_1 \ldots B_m$ are polygons such that $\mathcal{P} = \mathcal{Q}$. Then the two sets of vertices $\{A_1, \ldots, A_n\}$ and $\{B_1, \ldots, B_m\}$ are equal.*

Proof. Exercise 8A. □

Convex Polygons

Some of the most important theorems we will prove about polygons, like theorems we have already seen about triangles, refer to the *interior angles* of the polygon. For a triangle $\triangle ABC$, the interior angles are just the angles of the triangle: $\angle ABC$, $\angle BCA$, and $\angle CAB$. But when defining interior angles for general polygons, there is a complication that did not arise in the case of triangles. To see why, consider the quadrilateral $ABCD$ pictured in Fig. 8.6: the two edges that meet at D determine $\angle CDA$, which is by definition one of the "angles of $ABCD$." However, for vertices that "point inward" like D, we would wish to use the *reflex measure* of $\angle D$ as the "interior angle measure" at that vertex instead of its standard measure. This will be an issue, for example, in defining what we mean by *congruence* of polygons (where we wish to say that the interior angle measures of corresponding angles are equal), in defining *equiangular* polygons (since it is the *interior* angle measures that should be equal), and in defining *interior angle sums* for arbitrary

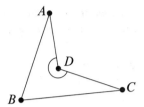

Fig. 8.6. A nonconvex quadrilateral.

polygons. As we will see later in this chapter, it is possible to classify the vertices of an arbitrary polygon into *convex vertices* (which point outward) and *concave vertices* (which point inward), but doing so takes a good deal of work. For most of the purposes we have in mind it is sufficient to focus on polygons that do not have this problem.

For that reason, we will primarily restrict our attention to polygons that "lie on one side of each edge." More precisely, we make the following definitions. Suppose $\mathscr{P}$ is a polygon. Any line that contains an edge of $\mathscr{P}$ is called an ***edge line*** for $\mathscr{P}$. We say that $\mathscr{P}$ is a ***convex polygon*** if for each edge line ℓ of $\mathscr{P}$, the entire polygon $\mathscr{P}$ is contained in one of the closed half-planes determined by ℓ. A polygon that is not convex is called a ***concave polygon***. The polygons in Figs. 8.3 and 8.7 are convex; however, the one in Fig. 8.6 is concave because A and B are on opposite sides of $\overleftrightarrow{CD}$.

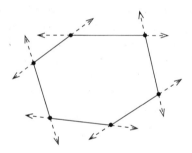

Fig. 8.7. A convex polygon.

Suppose $\mathscr{P}$ is a convex polygon and ℓ is an edge line for $\mathscr{P}$. Then by definition of convexity, there is one and only one side of ℓ that contains points of $\mathscr{P}$. For convenience, we call that the ***$\mathscr{P}$-side of ℓ***. The definition of convexity implies that every point of $\mathscr{P}$ is either on ℓ or on the $\mathscr{P}$-side of ℓ. The next lemma shows exactly which points lie on ℓ.

Lemma 8.2 (Edge-Line Lemma). *If $\mathscr{P}$ is a convex polygon and $\overline{AB}$ is an edge of $\mathscr{P}$, then $\mathscr{P} \cap \overleftrightarrow{AB} = \overline{AB}$.*

Proof. Suppose $\mathscr{P}$ is a polygon and $\overline{AB}$ is an edge of $\mathscr{P}$. Obviously $\overline{AB} \subseteq \mathscr{P} \cap \overleftrightarrow{AB}$, so we need only show that all points of $\mathscr{P} \cap \overleftrightarrow{AB}$ lie on $\overline{AB}$ (Fig. 8.8).

The assumption that $\mathscr{P}$ is convex means that all the points of $\mathscr{P}$ lie in one closed half-plane defined by $\overleftrightarrow{AB}$; let us call that half-plane S. Let C be the vertex other than A that is consecutive with B. Then C cannot lie on $\overleftrightarrow{AB}$ (because otherwise the adjacent edges $\overline{AB}$ and $\overline{BC}$ would be collinear). The hypothesis also implies that $\mathscr{P}$ is contained entirely in

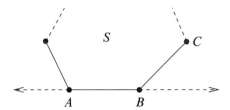

Fig. 8.8. Proof of the edge-line lemma.

one of the closed half-planes determined by $\overleftrightarrow{BC}$, so in particular, no points of $\mathcal{P}$ can lie on the opposite side of $\overleftrightarrow{BC}$ from A. Thus the interior of the ray opposite $\overrightarrow{BA}$ cannot contain any points of $\mathcal{P}$. A similar argument shows that the interior of the ray opposite $\overrightarrow{AB}$ cannot contain any points either; thus all points of $\mathcal{P}$ on the line $\overleftrightarrow{AB}$ must be in $\overline{AB}$. □

The edge-line lemma is a major ingredient of the proof of the following theorem.

Theorem 8.3. *If $\mathcal{P}$ is a convex polygon, then the extreme points of $\mathcal{P}$ are its vertices and no other points.*

Proof. Exercise 8B. □

Despite the terminology, a convex polygon (which is, after all, just a union of segments) is *not* a convex set, because there are many pairs of points on the polygon such that the segment connecting them is not contained in the union of the edges. But the two notions of convexity are closely related, as we will see later in the chapter.

For the special case of convex polygons, we can define several concepts relating to interior angle measures without difficulty. First, if $\mathcal{P}$ is a convex polygon, we simply define the ***interior angle measure*** of each of its angles to be the standard measure of that angle. We say a polygon is ***equilateral*** if all of its side lengths are equal, and it is ***equiangular*** if it is convex and all of its interior angle measures are equal. A ***regular polygon*** is one that is both equilateral and equiangular.

A major focus of our study will be *congruence* of polygons. For convex polygons, the definition of congruence is a straightforward generalization of the definition for triangles. Two convex polygons are said to be ***congruent*** if there is a one-to-one correspondence between their vertices such that consecutive vertices correspond to consecutive vertices and all pairs of corresponding sides and all pairs of corresponding angles are congruent. The notation $A_1 \ldots A_n \cong B_1 \ldots B_n$ means that the polygons $A_1 \ldots A_n$ and $B_1 \ldots B_n$ are congruent under the correspondence $A_1 \leftrightarrow B_1, \ldots, A_n \leftrightarrow B_n$. We are restricting the definition to convex polygons for now because this definition might give undesired congruences for nonconvex polygons, since it doesn't distinguish between convex and concave vertices. For example, each of the polygons pictured in Fig. 8.9 has twenty congruent sides and twenty right angles; thus there is a correspondence between their vertices such that consecutive vertices correspond to consecutive vertices and all pairs of corresponding angles and all pairs of corresponding sides are congruent. But we do not want to consider these polygons to be congruent to each other, for obvious reasons. At the end of the chapter, we will briefly describe how congruence can be defined for nonconvex polygons.

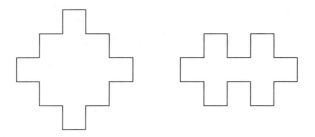

Fig. 8.9. Noncongruent polygons.

Because convexity of polygons is such an important property, it is useful to have several different ways of characterizing it. The next few theorems will describe several such ways.

Theorem 8.4 (Vertex Criterion for Convexity). *A polygon $\mathcal{P}$ is convex if and only if for every edge of $\mathcal{P}$, the vertices of $\mathcal{P}$ that are not on that edge all lie on the same side of the line containing the edge.*

Proof. Let $\mathcal{P}$ be a polygon. First assume that $\mathcal{P}$ is convex, and let $\overline{AB}$ be one of the edges of $\mathcal{P}$. By definition, all of the vertices of $\mathcal{P}$ are contained in one of the closed half-planes defined by $\overleftrightarrow{AB}$. The edge-line lemma shows that no vertices other than A and B lie on $\overleftrightarrow{AB}$, so all of the other vertices actually lie in the corresponding *open* half-plane, which means that they all lie on the same side of $\overleftrightarrow{AB}$.

Conversely, assume that $\mathcal{P}$ has the property that for every edge, the vertices not on that edge all lie on the same side of the line containing the edge. If $\overline{AB}$ is any edge, then it follows from the hypothesis that *all* of the vertices of $\mathcal{P}$ lie in one of the *closed* half-planes defined by $\overleftrightarrow{AB}$; since a closed half-plane is a convex set, it follows that all of the edges of $\mathcal{P}$ lie in that closed half-plane as well. Thus $\mathcal{P}$ is convex. $\qquad\square$

Corollary 8.5. *Every triangle is a convex polygon.*

Proof. Let $\triangle ABC$ be a triangle. For each edge of $\triangle ABC$, the vertices not on that edge lie on one side of the line containing the edge, for the simple reason that there is only one such vertex, and it does not lie *on* the line because the vertices of a triangle are noncollinear. $\qquad\square$

The next characterization relates convexity of a polygon to interiors of angles.

Theorem 8.6 (Angle Criterion for Convexity). *A polygon $\mathcal{P}$ is convex if and only if for each vertex A_i of $\mathcal{P}$, all the vertices of $\mathcal{P}$ are contained in the interior of $\angle A_i$ except A_i itself and the two vertices consecutive with it (see Fig. 8.10).*

Proof. Exercise 8C. $\qquad\square$

To describe the next criterion for convexity, we introduce another definition. Recall that we have defined two segments to be parallel if the lines containing them are parallel. More generally, we say that two segments are ***semiparallel*** if neither of the segments intersects the line containing the other (Fig. 8.11). Thus segments $\overline{AB}$ and $\overline{CD}$ are semiparallel if $\overline{AB} \cap \overleftrightarrow{CD} = \varnothing$ and $\overline{CD} \cap \overleftrightarrow{AB} = \varnothing$. It is easy to check that parallel implies semiparallel

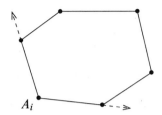

Fig. 8.10. The angle criterion for convexity.

and semiparallel implies disjoint; but neither of these implications is reversible, as Fig. 8.11 illustrates.

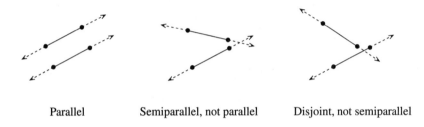

Parallel   Semiparallel, not parallel   Disjoint, not semiparallel

Fig. 8.11. Parallel, semiparallel, and disjoint segments.

Theorem 8.7 (Semiparallel Criterion for Convexity). *A polygon is convex if and only if all pairs of nonadjacent edges are semiparallel.*

Proof. Suppose first that $\mathcal{P}$ is convex. Given any pair of nonadjacent edges of $\mathcal{P}$, say $\overline{AB}$ and $\overline{CD}$, Theorem 8.4 implies that A and B lie on one side of $\overleftrightarrow{CD}$, so $\overline{AB} \cap \overleftrightarrow{CD} = \varnothing$. A similar argument shows that $\overleftrightarrow{CD} \cap \overline{AB} = \varnothing$, so $\overline{AB}$ and $\overline{CD}$ are semiparallel.

Conversely, suppose that all pairs of nonadjacent edges are semiparallel. Choose an arbitrary edge of $\mathcal{P}$, and label the vertices $A_1, \ldots, A_n$ consecutively in such a way that the chosen edge is $\overline{A_1 A_2}$. Because $\mathcal{P}$ is a polygon, A_3 does not lie on $\overleftrightarrow{A_1 A_2}$. Let S denote the side of $\overleftrightarrow{A_1 A_2}$ that contains A_3. If there are more than three vertices, the hypothesis implies that $\overline{A_3 A_4}$ does not meet $\overleftrightarrow{A_1 A_2}$, so A_4 also lies on side S. Continuing in this way, we see that $A_3, A_4, \ldots, A_n$ all lie on side S. The same argument works starting with any other edge, so $\mathcal{P}$ is a convex polygon by the vertex criterion. $\qquad\square$

One way in which convex polygons frequently arise is by splitting other convex polygons into two pieces along a line segment. Some typical situations are illustrated in Fig. 8.12; in each of these pictures, a large convex polygon is split into two smaller convex polygons by a line segment connecting two points on the large polygon.

The next theorem will demonstrate an important and useful fact: the new polygons that result from such splittings are always convex. To streamline the statement of the theorem, let us introduce a few definitions. If $\mathcal{P}$ is a convex polygon, a ***chord of $\mathcal{P}$*** is a line segment whose endpoints both lie on $\mathcal{P}$ but are not both contained in any single edge.

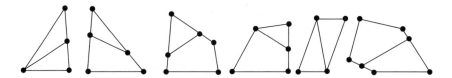

Fig. 8.12. Splitting convex polygons.

Lemma 8.8. *If $\mathcal{P}$ is a convex polygon and $\overline{BC}$ is a chord of $\mathcal{P}$, then* Int $\overline{BC}$ *is disjoint from $\mathcal{P}$.*

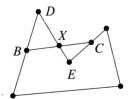

Fig. 8.13. Proof of Lemma 8.8.

Proof. Suppose $\mathcal{P}$ is a convex polygon and $\overline{BC}$ is a chord of $\mathcal{P}$. Assume for the sake of contradiction that there is a point $X \in$ Int $\overline{BC}$ that also lies on $\mathcal{P}$ (Fig. 8.13). Let $\overline{DE}$ be an edge containing X. If either B or C were on $\overleftrightarrow{DE}$, then $\overleftrightarrow{BC}$ and $\overleftrightarrow{DE}$ would have two distinct points in common and thus would be equal. Because the edge-line lemma guarantees that all points of $\mathcal{P}$ lying on the line $\overleftrightarrow{DE}$ are contained in $\overline{DE}$, this would imply that B and C are both in $\overline{DE}$, which contradicts the hypothesis. Thus neither B nor C lies on $\overleftrightarrow{DE}$. Since X is an interior point of $\overline{BC}$ lying on $\overleftrightarrow{DE}$, it follows that B and C are on opposite sides of $\overleftrightarrow{DE}$, contradicting the definition of a convex polygon. Thus there can be no such point X. $\qquad\square$

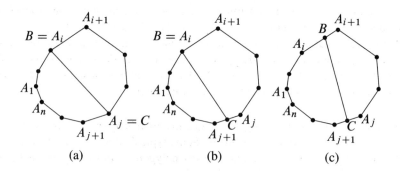

Fig. 8.14. Subpolygons cut off by a chord.

Now suppose $\mathcal{P}$ is a convex polygon and $\overline{BC}$ is a chord of $\mathcal{P}$. We define two more polygons $\mathcal{P}_1$ and $\mathcal{P}_2$, called the ***subpolygons cut off by*** $\overline{BC}$, as follows:

(a) If both B and C are (necessarily nonconsecutive) vertices (Fig. 8.14(a)), we can label the vertices of $\mathcal{P}$ as $A_1, \ldots, A_n$ such that $B = A_i$ and $C = A_j$ with $i < i+1 < j$, and then

$$\mathcal{P}_1 = A_1 \ldots A_{i-1} B C A_{j+1} \ldots A_n \qquad \text{and} \qquad \mathcal{P}_2 = B A_{i+1} \ldots A_{j-1} C.$$

(b) If one of the endpoints of the chord (say B) is a vertex but the other is not (Fig. 8.14(b)), we can label the vertices so that $B = A_i$ and $C \in \operatorname{Int} \overline{A_j A_{j+1}}$ with $i < j < n$, and then

$$\mathcal{P}_1 = A_1 \ldots A_{i-1} B C A_{j+1} \ldots A_n \qquad \text{and} \qquad \mathcal{P}_2 = B A_{i+1} \ldots A_{j-1} A_j C.$$

(c) If neither B nor C is a vertex (Fig. 8.14(c)), we can label the vertices so that $B \in \operatorname{Int} \overline{A_i A_{i+1}}$ and $C \in \operatorname{Int} \overline{A_j A_{j+1}}$ with $i < j < n$, and then

$$\mathcal{P}_1 = A_1 \ldots A_i B C A_{j+1} \ldots A_n \qquad \text{and} \qquad \mathcal{P}_2 = B A_{i+1} \ldots A_j C.$$

It is not immediately obvious that $\mathcal{P}_1$ and $\mathcal{P}_2$ will necessarily be polygons (for example, do the definitions guarantee that three consecutive vertices are necessarily noncollinear?); but the next theorem shows that they are indeed polygons, and in fact they are always convex. This proof is rated PG.

Theorem 8.9 (Polygon Splitting Theorem). *If $\mathcal{P}$ is a convex polygon and $\overline{BC}$ is a chord of $\mathcal{P}$, then the two subpolygons cut off by $\overline{BC}$ are both convex polygons.*

Proof. Let $\mathcal{P} = A_1 \ldots A_n$ be a convex polygon, let $\overline{BC}$ be a chord of $\mathcal{P}$, and let $\mathcal{P}_1$ and $\mathcal{P}_2$ be the subpolygons cut off by $\overline{BC}$ as in the definition above. The proofs that $\mathcal{P}_1$ and $\mathcal{P}_2$ are convex polygons are identical except for notation, so we will consider only $\mathcal{P}_2$.

We need to show that $\mathcal{P}_2$ is in fact a polygon; but first we will show that all pairs of nonadjacent edges of $\mathcal{P}_2$ are semiparallel. Any two nonadjacent edges of $\mathcal{P}_2$ that are already edges (or parts of edges) of $\mathcal{P}$ are automatically semiparallel because $\mathcal{P}$ is convex; so we need only show that if $\overline{A_k A_{k+1}}$ is any edge of $\mathcal{P}_2$ not adjacent to $\overline{BC}$, then $\overline{BC}$ and $\overline{A_k A_{k+1}}$ are semiparallel. Suppose for the sake of contradiction that they are not. One way this can happen is that there is a point $X \in \overline{BC} \cap \overrightarrow{A_k A_{k+1}}$. If $X = B$ or $X = C$, then $X \in \mathcal{P}$, so X actually lies in $\overline{A_k A_{k+1}}$ by the edge-line lemma, but this is a contradiction because $\overline{A_k A_{k+1}}$ is not adjacent to $\overline{BC}$. On the other hand, if $B * X * C$, then either B and C both lie on $\overleftrightarrow{A_k A_{k+1}}$, which again contradicts the edge-line lemma, or B and C lie on opposite sides of $\overleftrightarrow{A_k A_{k+1}}$, which contradicts the definition of a convex polygon.

The other possibility is that there is a point $X \in \overleftrightarrow{BC} \cap \overline{A_k A_{k+1}}$. The argument in the preceding paragraph shows that X cannot lie on $\overline{BC}$, so either $B * C * X$ or $X * B * C$; without loss of generality, assume it is the former. In this case, B and X are two points of $\mathcal{P}$ lying on opposite sides of the line $\overleftrightarrow{A_j A_{j+1}}$ (which contains C), again contradicting the fact that $\mathcal{P}$ is convex. This completes the proof that $\overline{BC}$ and $\overline{A_k A_{k+1}}$ are semiparallel, and thus all pairs of nonadjacent edges are semiparallel.

Now we can verify that $\mathcal{P}_2$ is a polygon. We have just seen that nonadjacent edges are semiparallel, so they are disjoint. Adjacent edges intersect at their endpoints by definition of $\mathcal{P}_2$. To complete the proof, we need only show that no two adjacent edges are collinear, which also implies that they intersect *only* at their endpoints. Of course, if neither of the edges is $\overline{BC}$, then this follows from the fact that $\mathcal{P}$ is a polygon. On the other hand, $\overline{BC}$ cannot be collinear with $\overline{BA_{i+1}}$, because then C would be a point of $\mathcal{P}$ on $\overleftrightarrow{A_i A_{i+1}}$ but not on $\overline{A_i A_{i+1}}$, contradicting the edge-line lemma. Similarly, $\overline{BC}$ cannot be collinear with

$\overline{A_jC}$ (or $\overline{A_{j-1}C}$ if C happens to be equal to A_j). This completes the proof that $\mathcal{P}_2$ is a convex polygon, and the same proof applies to $\mathcal{P}_1$.   □

The Interior of a Convex Polygon

Now we can describe the relationship between convex polygons and convex sets. The key is to consider, instead of the polygon itself, the union of the polygon with its "interior," which is a set of points that we now define.

Suppose $\mathcal{P}$ is a convex polygon and Q is a point in the plane. We say that Q is an ***interior point of*** $\mathcal{P}$ if Q lies in the interior of each of the angles of $\mathcal{P}$. If Q is neither on $\mathcal{P}$ nor in the interior of $\mathcal{P}$, we say it is an ***exterior point of*** $\mathcal{P}$. The set of all interior points is denoted by $\operatorname{Int}\mathcal{P}$ and is called the ***interior of*** $\mathcal{P}$; the set of all exterior points is denoted by $\operatorname{Ext}\mathcal{P}$ and is called the ***exterior of*** $\mathcal{P}$. (Thus every point of the plane is an element of one and only one of the sets $\mathcal{P}$, $\operatorname{Int}\mathcal{P}$, or $\operatorname{Ext}\mathcal{P}$.) The ***region determined by*** $\mathcal{P}$, denoted by $\operatorname{Reg}\mathcal{P}$, is the union of $\mathcal{P}$ and its interior (see Fig. 8.15).

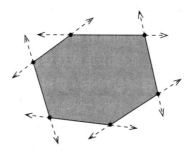

Fig. 8.15. The region determined by a convex polygon.

The following lemma gives useful alternative characterizations of the interior and region of a convex polygon.

Lemma 8.10. *Suppose $\mathcal{P}$ is a convex polygon and Q is a point in the plane.*

(a) *$Q \in \operatorname{Int}\mathcal{P}$ if and only if Q lies on the $\mathcal{P}$-side of every edge line.*

(b) *$Q \in \operatorname{Reg}\mathcal{P}$ if and only if for each edge of $\mathcal{P}$, Q lies in the closed half-plane determined by that edge and the vertices not on that edge.*

Proof. Let $\mathcal{P}$ be a polygon and let Q be any point. To prove (a), assume first that $Q \in \operatorname{Int}\mathcal{P}$. Let $\overline{AB}$ be any edge of $\mathcal{P}$, and let C be the other vertex consecutive with B, so $\angle ABC$ is one of the angles of $\mathcal{P}$. The fact that $Q \in \operatorname{Int}\angle ABC$ means, in particular, that Q is on the same side of $\overleftrightarrow{AB}$ as C, and therefore it is on the $\mathcal{P}$-side of $\overleftrightarrow{AB}$. The same argument applies to each edge.

Conversely, assume that Q lies on the $\mathcal{P}$-side of every edge line. Let $\angle B$ be any one of the angles of $\mathcal{P}$, and let A, C denote the two vertices consecutive with B, so $\angle B = \angle ABC$. The hypothesis implies that Q is on the same side of $\overleftrightarrow{AB}$ as C and on the same side of $\overleftrightarrow{BC}$ as A, which is to say that $Q \in \operatorname{Int}\angle ABC$.

Next, to prove (b), assume that $Q \in \operatorname{Reg}\mathcal{P}$. Then either $Q \in \mathcal{P}$ or $Q \in \operatorname{Int}\mathcal{P}$. If $Q \in \mathcal{P}$, then Q lies in each closed half-plane determined by an edge of $\mathcal{P}$ and the vertices not on that edge by definition of convex polygons. If $Q \in \text{Int}\,\mathcal{P}$, then by part (a), Q lies in every *open* half-plane determined by an edge and the vertices not on that edge, and therefore it also lies in every such *closed* half-plane.

Conversely, suppose Q lies in every closed half-plane determined by an edge of $\mathcal{P}$ and the vertices not on that edge. If Q actually lies in each of the corresponding *open* half-planes, then it is in $\text{Int}\,\mathcal{P} \subseteq \text{Reg}\,\mathcal{P}$ by part (a). On the other hand, if Q lies on the line determined by one of the edges, then the edge-line lemma shows that it actually lies on the edge, so $Q \in \mathcal{P} \subseteq \text{Reg}\,\mathcal{P}$. □

The next lemma expresses the intuitively obvious fact that if a line segment starts in the interior of a convex polygon and ends in its exterior, then it must cross the polygon somewhere.

Lemma 8.11. *Suppose $\mathcal{P}$ is a convex polygon, A is a point in* $\text{Int}\,\mathcal{P}$, *and B is a point in* $\text{Ext}\,\mathcal{P}$. *Then $\overline{AB}$ intersects $\mathcal{P}$.*

Proof. By Lemma 8.10(a), the fact that $A \in \text{Int}\,\mathcal{P}$ means that A is on the $\mathcal{P}$-side of each edge line, while the fact that $B \notin \text{Int}\,\mathcal{P}$ means that there is at least one edge line such that B is not on the $\mathcal{P}$-side of that edge line. This means that $\overline{AB}$ meets at least one edge line, and it might meet more than one (see Fig. 8.16). Since there are only finitely many edge lines, there can be only finitely many such intersections on $\overline{AB}$; let X be the intersection point on $\overline{AB}$ closest to A. Let V and W be the vertices such that $X \in \overleftrightarrow{VW}$. We will complete the proof by showing that X is actually on the edge $\overline{VW}$, which is part of $\mathcal{P}$.

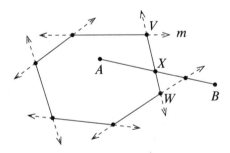

Fig. 8.16. Proof of Lemma 8.11.

Let m denote the line containing the edge adjacent to $\overline{VW}$ at V. If X also lies on m, then X must be equal to V (because adjacent edge lines intersect only at their common vertex). In that case $X \in \mathcal{P}$ and we are done. On the other hand, if X does not lie on m, then the stipulation that X is the intersection point closest to A ensures that $\overline{AX}$ does not meet m, which means that A and X are on the same side of m. Since A is on the $\mathcal{P}$-side of m, so is X. By Theorem 3.46, the portion of $\overleftrightarrow{VW}$ on the $\mathcal{P}$-side of m is exactly the ray $\overrightarrow{VW}$, so we have shown that $X \in \overrightarrow{VW}$. The same argument with W and V reversed shows that $X \in \overrightarrow{WV}$. Because $\overrightarrow{VW} \cap \overrightarrow{WV} = \overline{VW}$, this shows that $X \in \overline{VW} \subseteq \mathcal{P}$ as claimed. □

The next theorem explains, at least in part, the relationship between our two notions of convexity.

Theorem 8.12. *If $\mathcal{P}$ is a convex polygon, then both* Int $\mathcal{P}$ *and* Reg $\mathcal{P}$ *are convex sets.*

Proof. Lemma 8.10 says that a point is in Int $\mathcal{P}$ if and only if it is in each of the open half-planes determined by an edge of $\mathcal{P}$ and the vertices not on that edge. To put it another way, Int $\mathcal{P}$ is the intersection of n open half-planes, one for each edge. Since each open half-plane is convex and an intersection of convex sets is convex, it follows that Int $\mathcal{P}$ is convex. Similarly, Lemma 8.10 shows that Reg $\mathcal{P}$ is the intersection of n closed half-planes, one for each edge.   □

Nonconvex Polygons

Many of the concepts developed in this chapter can be extended to nonconvex polygons. For example, if you look at various examples of polygons, both convex and nonconvex, you will probably convince yourself that *every* polygon divides the plane into an "interior" and an "exterior." However, defining exactly what this means is not straightforward. To determine whether a point is in the interior of a complicated polygon like the one in Fig. 8.17, we have to examine the entire polygon.

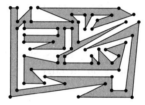

Fig. 8.17. The interior of a complicated polygon.

In the remainder of this chapter, we give definitions of interior and exterior points that work for arbitrary polygons, and we develop some of their basic properties. The proofs of many of the theorems in this section are quite long and complicated (although they can be done using only standard tools of plane separation and interiors of angles), so some of the theorems will be stated without proofs. The most important things to take away from this section are the definitions and the statements of the main theorems. We will not make essential use of this material elsewhere in the book; it is here primarily to acquaint you with some of the theoretical underpinnings of the facts about nonconvex polygons that you are likely to encounter in high-school textbooks.

Our first task is to make sense of the interior and exterior of an arbitrary polygon. It is immediately clear that the definition of interior that we gave for convex polygons will not work in the nonconvex case: for the polygon in Fig. 8.17, you can see that there are points in the interior of the polygon that are not in the interior of, say, the angle at the lower left corner. For the time being, let us temporarily suspend the previous definition of interior, and develop a totally new definition that works in both the convex and nonconvex cases. Later we will show that when the polygon is convex, the two definitions are equivalent.

Here is the key insight: if you start at a point not on the polygon itself and move straight in any direction that misses all the vertices, each time you cross an edge of the polygon you move from the interior to the exterior or vice versa. Thus if you start at an interior point, you will eventually cross the polygon an odd number of times in all, while if you start at an exterior point, you will cross it an even number of times (see Fig. 8.18). The reason we rule out directions passing through vertices is that a ray crossing a vertex might pass from the interior to the exterior or from the exterior to the interior, or it might not, depending on how the angle at that vertex is situated; some of the possibilities are illustrated in Fig. 8.19. Because there are only finitely many vertices, there are always plenty of directions that miss them all.

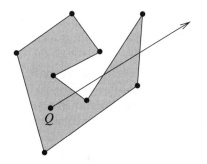

Fig. 8.18. An interior point of a polygon.

Fig. 8.19. Rays containing vertices.

Motivated by these observations, we make the following definitions. Suppose $\mathcal{P}$ is a polygon and $\vec{r}$ is a ray that does not contain any vertices of $\mathcal{P}$. We define the *parity of $\vec{r}$* to be the evenness or oddness of the number of intersections of $\vec{r}$ with $\mathcal{P}$: we say $\vec{r}$ has *even parity* if it has an even number of intersections with $\mathcal{P}$, and it has *odd parity* otherwise. If Q is a point not on $\mathcal{P}$, then we say the point Q has even parity if every ray starting at Q and not containing any vertices of $\mathcal{P}$ has even parity, and it has odd parity if every such ray has odd parity.

A point not on $\mathcal{P}$ that has odd parity is said to be an *interior point of $\mathcal{P}$*, and a point not on $\mathcal{P}$ that has even parity is said to be an *exterior point of $\mathcal{P}$*. The set of all interior points is denoted by $\operatorname{Int}\mathcal{P}$ and is called the *interior of $\mathcal{P}$*; the set of all exterior points is denoted by $\operatorname{Ext}\mathcal{P}$ and is called the *exterior of $\mathcal{P}$*.

The thing that these definitions do not tell us is whether there can be a point $Q \notin \mathcal{P}$ such that some rays starting at Q and missing the vertices have odd parity while others have even parity; such a point would be neither an exterior point nor an interior point of $\mathcal{P}$. The following important theorem shows, among other things, that this cannot happen. To state it, we need another definition. A *polygonal path from P to Q* is a union of finitely many segments that can be arranged in a sequence in such a way that the first endpoint of the first segment is P, the second endpoint of the last segment is Q, and the second endpoint of each segment other than the last is equal to the first endpoint of the next one in the sequence (see Fig. 8.20).

Theorem 8.13 (Jordan Polygon Theorem). *Suppose $\mathcal{P}$ is a polygon.*

(a) *Every point not on $\mathcal{P}$ is in either $\operatorname{Int}\mathcal{P}$ or $\operatorname{Ext}\mathcal{P}$, but not both.*

(b) *If A and B are two points such that $A \in \operatorname{Int}\mathcal{P}$ and $B \in \operatorname{Ext}\mathcal{P}$, then every polygonal path from A to B intersects $\mathcal{P}$.*

(c) *If A and B are two points that both lie in $\operatorname{Int}\mathcal{P}$ or both lie in $\operatorname{Ext}\mathcal{P}$, then there is a polygonal path from A to B that lies entirely in $\operatorname{Int}\mathcal{P}$ or $\operatorname{Ext}\mathcal{P}$, respectively.*

Fig. 8.20. A polygonal path from P to Q.

Proof. We will give a full (R-rated) proof of (a), but we'll just sketch the main ideas in the proofs of the other two statements.

Let $\mathcal{P}$ be a polygon, and let Q be a point not on $\mathcal{P}$. Clearly Q cannot be both an interior point and an exterior point of $\mathcal{P}$, because it is not possible for a ray to have both even and odd parity. In order to show that Q is either an interior point or an exterior point, we have to show that all rays starting at Q and missing the vertices have the same parity.

Thus suppose $\overrightarrow{QC}$ and $\overrightarrow{QD}$ are any two distinct rays starting at Q and not containing any vertices of $\mathcal{P}$. Assume for the time being that $\overrightarrow{QC}$ and $\overrightarrow{QD}$ are not opposite rays (see Fig. 8.21). Notice that because $Q \notin \mathcal{P}$, the number of points in $\angle CQD \cap \mathcal{P}$ is equal to the number of points in $\overrightarrow{QC} \cap \mathcal{P}$ plus the number of points in $\overrightarrow{QD} \cap \mathcal{P}$; thus it suffices to show that $\angle CQD$ must intersect $\mathcal{P}$ an even number of times.

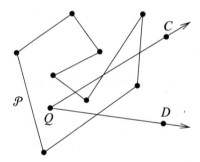

Fig. 8.21. Counting intersections between a polygon and an angle.

One case is easy to dispense with: if all of the vertices of $\mathcal{P}$ are in the interior of $\angle CQD$, then all of the edges are in the interior as well because the interior of a proper angle is a convex set; thus in this case the number of intersections is zero. Since zero is even, the lemma is proved in this case. Henceforth, we assume that at least one vertex of $\mathcal{P}$ is in the exterior of $\angle CQD$.

The crux of the argument is Lemma 8.14 below, which says roughly that the number of intersections of a segment with a proper angle is one when one endpoint of the segment is in the interior of the angle and the other is in the exterior; and otherwise it is zero or two. We will give a precise statement of the lemma and prove it after the end of this proof.

To see why this does the trick, we argue as follows: if we start at a vertex of $\mathcal{P}$ that is exterior to the angle and count up the number of intersection points edge by edge, then the running total number of intersections remains even as long as we stay in the exterior of the angle; it changes from even to odd or odd to even only when we add an odd number, which occurs exactly when one endpoint of the edge is in the interior of the angle and the other is in the exterior. Thus we will have counted an even number of total intersections whenever we are at an exterior vertex and an odd number whenever we are at an interior vertex. When we come back to the original exterior vertex, therefore, the total number of intersections must be even. This proves that $\overrightarrow{QC}$ and $\overrightarrow{QD}$ have the same parity when they are not opposite rays.

If $\overrightarrow{QC}$ and $\overrightarrow{QD}$ are opposite rays, we can choose a third ray $\overrightarrow{QE}$ that is noncollinear with them and does not contain any vertices of $\mathscr{P}$. The argument above shows that both $\overrightarrow{QC}$ and $\overrightarrow{QD}$ have the same parity as $\overrightarrow{QE}$, and thus they have the same parity as each other. This completes the proof of (a).

The X-rated proofs of the other two conclusions, when written out in full detail, are somewhat more complicated than what we have already done, but the underlying ideas are not too hard to explain. To prove (b), the first step is to show that if $\overline{AB}$ is a segment that does not intersect $\mathscr{P}$, then A and B have the same parity. This is easy if one of the rays $\overrightarrow{AB}$ or $\overrightarrow{BA}$ does not contain any vertices of $\mathscr{P}$. To handle the general case, one can show that there is some number $\theta \in (0, 180)$ such that the rays starting at points of $\overline{AB}$, lying on one side of $\overleftrightarrow{AB}$, and making an angle of $\theta°$ with one of the rays in $\overleftrightarrow{AB}$ all miss the vertices of $\mathscr{P}$ and all have the same number of intersections with $\mathscr{P}$ (see Fig. 8.22(a)). Once this is verified, it follows by induction that if two points A and B are connected by a polygonal path that does not intersect $\mathscr{P}$, then they are either both interior points or both exterior points. The contrapositive of this statement is (b).

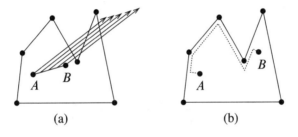

Fig. 8.22. Proof of the Jordan polygon theorem.

To prove (c), one shows that there is a polygonal path that starts by going from A to a point very close to the nearest edge, then follows along parallel to the edges until it gets to the edge nearest to B, then goes straight to B (see Fig. 8.22(b)). The tricky part is showing that by placing the segments close enough to the edges, it is possible to choose them in such a way that they do not intersect $\mathscr{P}$. Once this is verified, it then follows from the argument in the preceding paragraph that the entire polygonal path from A to B lies in either Int $\mathscr{P}$ or Ext $\mathscr{P}$. □

To complete the proof of part (a) of the preceding theorem, we need the following lemma. Its proof is rated R.

Lemma 8.14. *If* $\angle CQD$ *is a proper angle and* $\overline{AB}$ *is a segment such that A and B are not on* $\angle CQD$ *and Q is not on* $\overline{AB}$, *then the number of intersections of* $\overline{AB}$ *with* $\angle CQD$ *is exactly one when one endpoint of* $\overline{AB}$ *is in the interior of the angle and the other is in the exterior; and otherwise it is zero or two.*

Proof. First, we will show that the number of intersection points of $\overline{AB}$ with $\angle CQD$ is always zero, one, or two. If there were three or more intersection points, then at least two of them would have to lie on one of the lines $\overleftrightarrow{QC}$ or $\overleftrightarrow{QD}$; but if that were the case, then the entire segment $\overline{AB}$ would be contained in that line. Since we are assuming that neither A nor B lies on $\angle CQD$, the only way this can happen is if $\overline{AB}$ is contained in the ray opposite $\overrightarrow{QC}$ or the ray opposite $\overrightarrow{QD}$; but this would imply that $\overline{AB}$ contains *no* points on $\angle CQD$, contradicting the assumption that there are at least three. It follows that there cannot be more than two intersection points.

Thus to prove the lemma, it suffices to show that $\overline{AB}$ intersects $\angle CQD$ exactly once if and only if one of the endpoints of $\overline{AB}$ is in the interior of $\angle CQD$ and the other is in the exterior.

Assume first that one endpoint of $\overline{AB}$ is in the interior of the angle and the other is in the exterior. Without loss of generality, say A is in the interior. Then $\overline{AB}$ must intersect at least one of the lines $\overleftrightarrow{QC}$, $\overleftrightarrow{QD}$ (because otherwise B would also be in the interior of the angle), and it might intersect both. Let X be the point of intersection on $\overline{AB}$ closest to A, and for definiteness say that $X \in \overleftrightarrow{QC}$ (see Fig. 8.23). Because $\overline{AX}$ does not meet $\overleftrightarrow{QD}$,

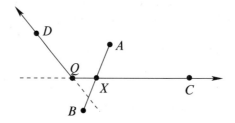

Fig. 8.23. Proof of Lemma 8.14.

it follows that X is on the same side of $\overleftrightarrow{QD}$ as A, which in turn is the same side as C, so X lies in $\overrightarrow{QC}$ (and not its opposite ray). Thus there is at least one intersection between $\overline{AB}$ and $\angle CQD$. We need to show that X is the only such point. On the one hand, in $\overline{AX}$, there are no intersection points with the angle other than X itself, because we chose X to be the closest one. On the other hand, the X-lemma implies that all points of $\overline{XB}$ other than X are on the opposite side of $\overleftrightarrow{QC}$ from A, and thus also on the opposite side from $\overleftrightarrow{QD}$; thus $\overline{XB}$ cannot contain a second intersection point either. This proves that there is exactly one intersection point.

Conversely, assume that $\overline{AB}$ intersects $\angle CQD$ exactly once, and call the intersection point X. Again without loss of generality, we may as well assume that X lies on $\overrightarrow{QC}$. One of the two points A or B, therefore, is on the same side of $\overleftrightarrow{QC}$ as D; renaming the points if necessary, we can assume that

$$A \text{ is on the same side of } \overleftrightarrow{QC} \text{ as } D. \tag{8.1}$$

Thus we are again in the situation illustrated in Fig. 8.23. Now, the Y-lemma guarantees that the entire segment $\overline{XA}$ (except X) lies on the same side of $\overleftrightarrow{QC}$ as A, and hence on the same side as D; thus if $\overline{AX}$ met $\overleftrightarrow{QD}$, the intersection point would have to be a point of $\overrightarrow{QD}$, not its opposite ray. Since are assuming that $\overline{AB}$ meets $\angle CQD$ only once, there can be no such intersection. It follows that

$$A \text{ is on the same side of } \overleftrightarrow{QD} \text{ as } C. \tag{8.2}$$

Then (8.1) and (8.2) imply that A is in the interior of $\angle CQD$. On the other hand, since B is on the opposite side of $\overleftrightarrow{QC}$ from A, it is in the exterior of the angle. $\square$

The Jordan polygon theorem is actually a special case of a much deeper theorem called the *Jordan curve theorem*, which asserts an analogous conclusion for any "simple closed curve" in the plane, not just for polygons. The proof of that theorem is surprisingly hard (see, for example, [**Moi77**]).

In the case of a convex polygon $\mathcal{P}$, we have now given two definitions of what it means for a point to be in the interior of $\mathcal{P}$. The next theorem shows that the two definitions are equivalent. This is another R-rated proof.

Theorem 8.15. *Suppose $\mathcal{P}$ is a convex polygon and Q is any point not on $\mathcal{P}$. Then Q has odd parity if and only if Q is in the interior of each of the angles of $\mathcal{P}$.*

Proof. Let $\mathcal{P}$ be a convex polygon and let Q be an arbitrary point not on $\mathcal{P}$. Assume first that Q lies in the interior of each of the angles of $\mathcal{P}$. We need to show that Q has odd parity. Because all rays that start at Q and miss all the vertices have the same parity, it suffices to show that there is at least one such ray that has odd parity.

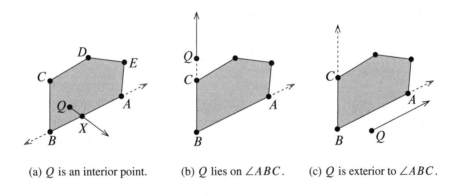

(a) Q is an interior point.   (b) Q lies on $\angle ABC$.   (c) Q is exterior to $\angle ABC$.

Fig. 8.24. Interior and exterior points of a convex polygon.

By Lemma 8.10, the assumption that Q is in the interiors of all the angles implies that Q lies on the $\mathcal{P}$-side of each edge line. Choose an edge $\overline{AB}$, let X be an interior point of $\overline{AB}$, and consider the ray $\overrightarrow{QX}$ (Fig. 8.24(a)). We will show that this ray has no points in common with $\mathcal{P}$ other than X, which implies that it has odd parity. First note that since Q is not on $\overleftrightarrow{AB}$, the ray $\overrightarrow{QX}$ can have no intersections with $\overline{AB}$ other than X. The only part of $\overrightarrow{QX}$ that could possibly intersect other edges of $\mathcal{P}$ is the segment $\overline{QX}$, because the rest of the ray is on the opposite side of $\overleftrightarrow{AB}$ from the rest of $\mathcal{P}$. If $\overline{BC}$ is the edge adjacent to $\overline{AB}$ at B, then our hypothesis shows that Q and A are on the same side of $\overleftrightarrow{BC}$, while the Y-lemma shows that A and X are on the same side; hence Q and X are on the same side, which implies that $\overline{QX} \cap \overline{BC} = \varnothing$. The same argument shows that $\overline{QX}$ cannot intersect the edge adjacent to $\overline{AB}$ at A. Finally, if $\overline{DE}$ is any edge that is not adjacent to $\overline{AB}$, then our hypothesis implies that Q, A, and B are all on the same side of $\overleftrightarrow{DE}$; since half-planes are convex, it follows that X is on that side too, so once again we conclude that $\overline{QX} \cap \overline{DE} = \varnothing$.

To prove the converse, we will actually prove its contrapositive: if it is not true that Q lies in the interior of each angle, then Q does not have odd parity. Assume therefore that for at least one of the angles $\angle ABC$ of $\mathcal{P}$, Q does not lie in its interior; thus either Q lies on the angle or it lies in its exterior.

CASE 1: Q lies on $\angle ABC$. Without loss of generality, let us say $Q \in \overrightarrow{BC}$ (Fig. 8.24(b)). Then Q cannot lie in $\overline{BC}$ because we are assuming $Q \notin \mathcal{P}$. The edge-line lemma shows that the ray opposite $\overrightarrow{QB}$ does not intersect $\mathcal{P}$ at all. That ray therefore has even parity, so Q has even parity.

CASE 2: Q is in the exterior of $\angle ABC$. Then Q is on the opposite side of $\overleftrightarrow{AB}$ from C or on the opposite side of $\overleftrightarrow{BC}$ from A; without loss of generality, let us say it is the former. Then any ray starting at Q and parallel to $\overleftrightarrow{AB}$ remains on the same side of $\overleftrightarrow{AB}$ as Q, and thus does not intersect $\mathcal{P}$ (Fig. 8.24(c)), and once again we conclude that Q has even parity. □

Because of the previous theorem, for convex polygons we can use whichever definition of interior and exterior is more convenient for the problem at hand. For polygons that are not already known to be convex, however, we have no choice but to use the definition in terms of parity.

Using the preceding theorem, we can prove a more accurate statement about intersections of rays with a convex polygon.

Corollary 8.16. *Suppose $\mathcal{P}$ is a convex polygon and $Q \in \operatorname{Int} \mathcal{P}$. Then every ray starting at Q intersects $\mathcal{P}$ exactly once.*

Proof. Let $\overrightarrow{r}$ be a ray starting at Q. If $\overrightarrow{r}$ passes through a vertex of $\mathcal{P}$, then it meets $\mathcal{P}$ at that vertex; while if it misses all the vertices, then it must have odd parity, so it must have at least one intersection point with $\mathcal{P}$. In either case, $\overrightarrow{r}$ intersects $\mathcal{P}$ at least once.

Now we must show that $\overrightarrow{r}$ cannot meet $\mathcal{P}$ more than once. Assume for the sake of contradiction that there are at least two points of intersection; call them R and S. After renaming the points if necessary, we may assume that $QR < QS$, which implies that $Q * R * S$ by the ordering lemma for points. Because $R \in \mathcal{P}$, there is some edge line ℓ containing R. The fact that $Q \in \operatorname{Int} \mathcal{P}$ implies that Q is on the $\mathcal{P}$-side of ℓ by Lemma 8.10. This means in particular that Q does not lie on ℓ, so ℓ and $\overrightarrow{QS}$ are distinct. By the X-lemma, Q and S lie on opposite sides of ℓ. But this is a contradiction, because the fact that $\mathcal{P}$ is convex implies that every point of $\mathcal{P}$ lies either on ℓ or on the $\mathcal{P}$-side of ℓ. □

Convex and Concave Vertices

Suppose $\mathcal{P}$ is a polygon, V is a vertex of $\mathcal{P}$, and $\overrightarrow{r}$ is a ray starting at V and not containing any other vertices of $\mathcal{P}$. Then $\overrightarrow{r}$ is said to be ***inward-pointing*** if $\operatorname{Int} \overrightarrow{r}$ intersects $\mathcal{P}$ an odd number of times, and it is said to be ***outward-pointing*** if $\operatorname{Int} \overrightarrow{r}$ intersects $\mathcal{P}$ an even number of times. The idea is that along an inward-pointing ray, the points between V and the next intersection with $\mathcal{P}$ are all interior points of $\mathcal{P}$. We say that V is a ***convex vertex*** if every inward-pointing ray starting at V lies in the interior of $\angle V$; and V is a ***concave vertex*** if every such ray lies in the exterior of $\angle V$.

Lemma 8.17. *If $\mathcal{P}$ is a convex polygon, then every vertex of $\mathcal{P}$ is a convex vertex.*

Proof. Let $\mathcal{P}$ be a convex polygon, and let B be a vertex of $\mathcal{P}$. Let A and C be the two vertices consecutive with B, so the angle at B is $\angle ABC$. Suppose $\vec{r}$ is any inward-pointing ray starting at $\mathcal{P}$; we need to show that $\vec{r}$ lies in the interior of $\angle ABC$. The assumption means that $\operatorname{Int}\vec{r}$ meets $\mathcal{P}$ an odd number of times, so there is at least one point $W \in \operatorname{Int}\vec{r} \cap \mathcal{P}$. The fact that $\mathcal{P}$ is convex means that W lies in both of the closed half-planes $\operatorname{CHP}\left(\overleftrightarrow{AB},C\right)$ and $\operatorname{CHP}\left(\overleftrightarrow{BC},A\right)$. If W were on $\overleftrightarrow{AB}$, then the edge-line lemma would imply that W is actually in $\overline{AB}$; but this would imply that $\vec{r} = \overrightarrow{AB}$ and thus that $\operatorname{Int}\vec{r}$ contains the vertex B, contradicting the definition of inward-pointing. A similar argument shows that W cannot lie on $\overleftrightarrow{BC}$. Thus W actually lies in the corresponding *open* half-planes, which means it lies in the interior of $\angle ABC$. $\qquad\square$

For nonconvex polygons, we have instead the following lemma. We omit its (R-rated) proof, but it is quite similar to that of Theorem 8.13(a).

Lemma 8.18. *If $\mathcal{P}$ is any polygon, then every vertex of $\mathcal{P}$ is either convex or concave.*

With the definition of concave and convex vertices in hand, we can now extend the definitions of interior angle measures and congruence to arbitrary polygons. If $\mathcal{P}$ is a polygon and V is a vertex of $\mathcal{P}$, we define the ***interior angle measure of*** $\angle V$ to be the standard measure of $\angle V$ if V is a convex vertex and its reflex measure if V is a concave one (see Fig. 8.25). We say two polygons $\mathcal{P}_1$ and $\mathcal{P}_2$ are ***congruent*** if there is a correspondence between their vertices such that consecutive vertices correspond to consecutive vertices, corresponding edges are congruent, and corresponding interior angle measures are equal.

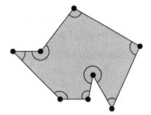

Fig. 8.25. Interior angle measures.

We can now state the main theorem tying together all of the various types of convexity for polygons. The X-rated proof of the following theorem is quite long, so we simply state the theorem without proof. We won't use the theorem in this book, except for the parts we have already proved (Theorem 8.12 and Lemma 8.17).

Theorem 8.19 (Characterizations of Convex Polygons). *If $\mathcal{P}$ is a polygon, the following are equivalent:*

(a) *$\mathcal{P}$ is a convex polygon.*

(b) *$\operatorname{Int}\mathcal{P}$ is a convex set.*

(c) *$\operatorname{Reg}\mathcal{P}$ is a convex set.*

(d) *All vertices of $\mathcal{P}$ are convex.*

In high-school textbooks, you will find various definitions of a *convex polygon*, such as a polygon whose region is a convex set or a polygon whose interior angle measures are all less than 180° or a polygon that lies on one side of the line through each of its edges.

These textbooks typically do *not* give precise definitions of the "interior" or the "interior angle measure" of a general polygon. After having seen how intricate some of the proofs are in the last part of this chapter, you can probably appreciate why not. High-school students generally do not have any trouble accepting notions such as interior points and interior angle measures without proving that they are well defined.

Exercises

8A. Prove Theorem 8.1. [Hint: Suppose S is a set of points in the plane and $Q \in S$. Let us say that Q is a ***strong passing point of S*** if there exist points $X, Y \in S$ such that $X * Q * Y$ and $\overline{XY} \subseteq S$. Say that Q is a ***weak extreme point of S*** if it is a point of S but not a strong passing point. (Although it is not needed for this proof, you can check that every strong passing point of S is also a passing point, while every extreme point of S is also a weak extreme point.) Prove the theorem by showing that if $\mathcal{P} = A_1 \ldots A_n$ is a polygon, then a point of $\mathcal{P}$ is a weak extreme point if and only if it is one of the vertices $A_1, \ldots, A_n$. The hard part is showing that a vertex is a weak extreme point. To prove this, let A_i be a vertex, and assume for contradiction that it is a strong passing point. Let r be a positive number less than the distance from A_i to any edge that does not contain A_i (Theorem 7.14), and show that there exist points $X, Y \in \mathcal{P}$ such that $X * A_i * Y$, $\overline{XY} \subseteq \mathcal{P}$, and $XY < r$, and show that this leads to a contradiction.]

8B. Prove Theorem 8.3 (extreme points of convex polygons).

8C. Prove Theorem 8.6 (the angle criterion for convexity).

8D. Suppose $\triangle ABC$ is a triangle. Show that any point that lies in the interiors of two of its angles is in $\operatorname{Int} \triangle ABC$.

Chapter 9

Quadrilaterals

After triangles, the next most important figures in geometry are quadrilaterals. In this chapter, we explore some properties of quadrilaterals that can be proved in neutral geometry.

Convex Quadrilaterals

Recall that a *quadrilateral* is a polygon with four sides. In addition to all of the standard terminology that we use for general polygons, there are some terms that are specific to quadrilaterals: two sides of a quadrilateral that are not adjacent are called *opposite sides*, and two vertices or angles that are not consecutive are called *opposite vertices* or *opposite angles*. For example, if $ABCD$ is a quadrilateral, then A and C are opposite vertices, as are B and D; and $\overline{AB}$ and $\overline{CD}$ are opposite sides, as are $\overline{BC}$ and $\overline{DA}$. There are only two diagonals: $\overline{AC}$ and $\overline{BD}$. Note that *no* three vertices of a quadrilateral can be collinear, because any three vertices are the endpoints of a pair of adjacent sides (namely, the ones that do not contain the missing vertex).

Here are some special types of quadrilaterals. A quadrilateral is called a

- *trapezoid* if it has at least one pair of parallel sides;
- *parallelogram* if it has two different pairs of parallel sides;
- *rhombus* if all four of its sides are congruent;
- *rectangle* if it has four right angles;
- *square* if all four sides are congruent and all four angles are right angles.

Theorem 9.1. *Every rectangle is a parallelogram.*

Proof. If $ABCD$ is a rectangle, then $\overleftrightarrow{AB}$ and $\overleftrightarrow{CD}$ are parallel by the common perpendicular theorem, as are $\overleftrightarrow{AD}$ and $\overleftrightarrow{BC}$. □

The next two lemmas give some simple but useful properties of convex quadrilaterals, which we will use frequently in the rest of the book.

Lemma 9.2. *In a convex quadrilateral, each pair of opposite vertices lies on opposite sides of the line through the other two vertices.*

Proof. Let $ABCD$ be a convex quadrilateral. The angle criterion for convexity says that C lies in the interior of $\angle A$. This means that $\overrightarrow{AB} * \overrightarrow{AC} * \overrightarrow{AD}$, and then it follows from Theorem 4.12 that B and D lie on opposite sides of $\overleftrightarrow{AC}$. The same argument shows that A and C lie on opposite sides of $\overleftrightarrow{BD}$. □

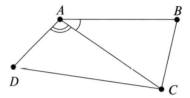

Fig. 9.1. The setup for Lemma 9.3.

Lemma 9.3. *Suppose $ABCD$ is a convex quadrilateral. Then $m\angle BAD = m\angle BAC + m\angle CAD$ (Fig. 9.1), with similar statements for the angles at the other vertices.*

Proof. The angle criterion for convexity guarantees that C lies in the interior of $\angle BAD$, so $\overrightarrow{AB} * \overrightarrow{AC} * \overrightarrow{AD}$. The result then follows from the betweenness theorem for rays. □

In addition to the general criteria for convexity that we proved in Chapter 8, there are two more criteria that apply specifically to quadrilaterals.

Theorem 9.4 (Diagonal Criterion for Convex Quadrilaterals).

(a) *If the diagonals of a quadrilateral intersect, then the quadrilateral is convex.*

(b) *If a quadrilateral is convex, then its diagonals intersect at a point that is in the interiors of both diagonals and of the quadrilateral (see Fig. 9.2).*

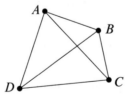

Fig. 9.2. The diagonals of a convex quadrilateral have an interior point in common.

Proof. Exercise 9B. □

Theorem 8.7 showed that a polygon is convex if and only if each pair of nonadjacent sides is semiparallel. For quadrilaterals, it turns out that only one pair of sides needs to be checked.

Theorem 9.5. *If a quadrilateral has at least one pair of semiparallel sides, it is convex.*

Proof. We prove the contrapositive: assuming that $ABCD$ is a nonconvex quadrilateral, we prove that it does not have a pair of semiparallel sides. Thus assume $ABCD$ is nonconvex. By Theorem 8.7, this implies that there is at least one pair of opposite edges that are not semiparallel. Without loss of generality, we can assume that $\overline{AD}$ and $\overline{BC}$ are not semiparallel (Fig. 9.3). We need to show that $\overline{AB}$ and $\overline{CD}$ are not semiparallel either.

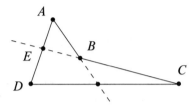

Fig. 9.3. A nonconvex quadrilateral does not have semiparallel sides.

The fact that $\overline{AD}$ and $\overline{BC}$ are not semiparallel implies that one of these two segments must intersect the line containing the other. Again without loss of generality, let us say that $\overline{AD}$ intersects $\overleftrightarrow{BC}$ at a point E. Now, E cannot be a vertex, because then $ABCD$ would have three collinear vertices; so E must be an interior point of $\overline{AD}$. Since B, C, and E are distinct collinear points and $E \notin \overline{BC}$, one of the relations $E * B * C$ or $B * C * E$ must hold. Assume it is the former; the argument for the latter case is nearly identical.

The assumption $E * B * C$ implies that E and C are on opposite sides of $\overleftrightarrow{AB}$. On the other hand, the Y-lemma applied to the ray $\overrightarrow{AD}$ implies that E and D are on the same side of $\overleftrightarrow{AB}$. Therefore C and D are on opposite sides of $\overleftrightarrow{AB}$. Thus $\overline{CD}$ intersects $\overleftrightarrow{AB}$, which implies that $\overline{AB}$ and $\overline{CD}$ are not semiparallel either. □

Corollary 9.6. *Every trapezoid is a convex quadrilateral.*

Proof. A trapezoid has a pair of parallel sides and thus a pair of semiparallel sides. □

Corollary 9.7. *Every parallelogram is a convex quadrilateral.*

Proof. Every parallelogram is a trapezoid. □

Corollary 9.8. *Every rectangle is a convex quadrilateral.*

Proof. Every rectangle is a parallelogram by Theorem 9.1. □

Often, before we can apply the preceding results, it is necessary to verify that we actually have a quadrilateral to begin with. Since the general definition of a polygon includes several conditions that would have to be verified, this could be tedious. Fortunately, for convex quadrilaterals we have the following shortcuts. The first one says roughly that if you connect the four corners of a cross, you automatically get a convex quadrilateral.

Lemma 9.9 (Cross Lemma). *Suppose $\overline{AC}$ and $\overline{BD}$ are noncollinear segments that have an interior point in common. Then $ABCD$ is a convex quadrilateral.*

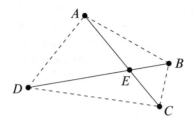

Fig. 9.4. The cross lemma.

Proof. Let E be a point that lies in both Int $\overline{AC}$ and Int $\overline{BD}$, which means $A * E * C$ and $B * E * D$ (Fig. 9.4). Since $\overline{AC}$ and $\overline{BD}$ are noncollinear, the lines $\overleftrightarrow{AC}$ and $\overleftrightarrow{BD}$ are distinct, so they intersect *only* at E. In particular, this implies that A, B, C, and D are distinct points.

The main point is to show that the set $ABCD = \overline{AB} \cup \overline{BC} \cup \overline{CD} \cup \overline{DA}$ is a polygon. By construction, it is the union of four segments determined by four distinct points, so we need only show that consecutive edges are noncollinear and distinct edges intersect only at common endpoints.

To prove that consecutive edges are noncollinear, note that if, say, $\overline{AB}$ and $\overline{BC}$ were collinear, then the lines $\overleftrightarrow{AC}$ and $\overleftrightarrow{BD}$ would intersect at two distinct points (E and B), contradicting the observation in the first paragraph. The same argument shows that no other three points are collinear either.

To prove that distinct edges intersect only at common endpoints, note first that adjacent edges intersect only at their endpoints because no three vertices are collinear. It remains only to show that nonadjacent edges do not intersect. Consider the edges $\overline{AB}$ and $\overline{CD}$. The fact that $A * E * C$ implies that A and C lie on opposite sides of $\overleftrightarrow{BD}$, and then the Y-lemma implies that all interior points of $\overline{BA}$ are on the opposite side of $\overleftrightarrow{BD}$ from all interior points of $\overrightarrow{DC}$. That leaves only B and D as possible intersection points; but since $B \neq D$, we conclude that $\overline{AB}$ and $\overline{CD}$ are disjoint. The same argument shows that $\overline{BC}$ and $\overline{DA}$ are disjoint. Thus $ABCD$ is a quadrilateral, and it is convex because its diagonals intersect.   □

Lemma 9.10 (Trapezoid Lemma). *Suppose A, B, C, and D are four distinct points such that $\overline{AB} \parallel \overline{CD}$ and $\overline{AD} \cap \overline{BC} = \varnothing$. Then $ABCD$ is a trapezoid.*

Proof. Once again, the main point is to prove that $ABCD$ is a polygon. If any three vertices were collinear, then the lines $\overleftrightarrow{AB}$ and $\overleftrightarrow{CD}$ would intersect, contradicting the hypothesis that they are parallel. To prove that distinct edges intersect only at common endpoints, note as before that adjacent edges intersect only at their endpoints because no three vertices are collinear; and the hypotheses imply that opposite edges do not intersect at all. Thus $ABCD$ is a quadrilateral, and because it has a pair of parallel sides it is a trapezoid.   □

The next lemma is really just a special case of the trapezoid lemma, but it is so frequently useful that it is worth stating as a separate lemma.

Lemma 9.11 (Parallelogram Lemma). *Suppose A, B, C, and D are four distinct points such that $\overline{AB} \parallel \overline{CD}$ and $\overline{AD} \parallel \overline{BC}$. Then $ABCD$ is a parallelogram.*

Proof. Given such points A, B, C, and D, the trapezoid lemma implies that $ABCD$ is a trapezoid, and then the hypothesis implies that it is in fact a parallelogram.   □

The next three theorems are generalizations of familiar results about triangles.

Theorem 9.12 (SASAS Congruence). *Suppose $ABCD$ and $EFGH$ are convex quadrilaterals such that $\overline{AB} \cong \overline{EF}$, $\overline{BC} \cong \overline{FG}$, $\overline{CD} \cong \overline{GH}$, $\angle B \cong \angle F$, and $\angle C \cong \angle G$ (Fig. 9.5). Then $ABCD \cong EFGH$.*

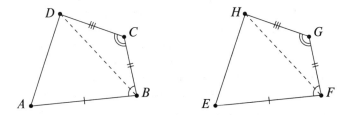

Fig. 9.5. Proof of the SASAS congruence theorem.

Proof. Draw the diagonals $\overline{BD}$ and $\overline{FH}$. Then $\triangle BCD$ and $\triangle FGH$ are congruent by SAS, which implies in particular that $\overline{BD} \cong \overline{FH}$ and $\angle DBC \cong \angle HFG$. Lemma 9.3 shows that $m\angle ABD = m\angle ABC - m\angle DBC = m\angle EFG - m\angle HFG = m\angle EFH$. Now $\triangle ABD$ and $\triangle EFH$ are congruent by SAS, so it follows that $\overline{AD} \cong \overline{EH}$ and $\angle A \cong \angle E$. The two triangle congruences imply $\angle ADB \cong \angle EHF$ and $\angle BDC \cong \angle FHG$; and then Lemma 9.3 implies $\angle ADC \cong \angle EHG$. Since all eight pairs of corresponding parts are congruent, we conclude that $ABCD \cong EFGH$.   □

Theorem 9.13 (AASAS Congruence). *Suppose $ABCD$ and $EFGH$ are convex quadrilaterals such that $\angle A \cong \angle E$, $\angle B \cong \angle F$, $\angle C \cong \angle G$, $\overline{BC} \cong \overline{FG}$, and $\overline{CD} \cong \overline{GH}$. Then $ABCD \cong EFGH$.*

Proof. Exercise 9C.   □

The proof of the next theorem is rated PG.

Theorem 9.14 (Quadrilateral Copying Theorem). *Suppose $ABCD$ is a convex quadrilateral and $\overline{EF}$ is a segment congruent to $\overline{AB}$. On either side of $\overleftrightarrow{EF}$, there are distinct points G and H such that $EFGH \cong ABCD$.*

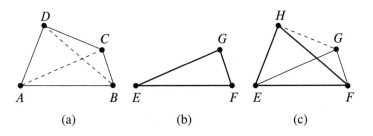

Fig. 9.6. Copying a quadrilateral.

Proof. First, using the triangle copying theorem (Theorem 5.10), let G be the point on the chosen side of $\overleftrightarrow{EF}$ such that $\triangle EFG \cong \triangle ABC$ (see Fig. 9.6(b)). Then let H be the point on the same side of $\overleftrightarrow{EF}$ such that $\triangle EFH \cong \triangle ABD$ (Fig. 9.6(c)).

We wish to use SASAS to prove that $EFGH \cong ABCD$; but before we do so, we need to verify that $EFGH$ is a convex quadrilateral. For this purpose, we will use the cross lemma. The angle criterion for convexity guarantees that C is in the interior of $\angle BAD$, so $m\angle BAD > m\angle BAC$ (the whole angle is greater than the part). Because $\angle FEH \cong \angle BAD$ and $\angle FEG \cong \angle BAC$, substitution implies $m\angle FEH > m\angle FEG$. It follows, first of all, that G and H are distinct points, since otherwise $\angle FEH$ would be equal to $\angle FEG$; and since neither G nor H lies on $\overleftrightarrow{EF}$, all four of the points E, F, G, H are distinct. Moreover, the ordering lemma for rays implies that $\overrightarrow{EF} * \overrightarrow{EG} * \overrightarrow{EH}$. Thus $\overrightarrow{EG}$ must intersect Int $\overline{FH}$ by the crossbar theorem. The same argument shows that $\overrightarrow{FH}$ meets Int $\overline{EG}$. Since these two lines can intersect only at one point, that point must lie in both Int $\overline{FH}$ and Int $\overline{EG}$. Thus $EFGH$ is a convex quadrilateral by the cross lemma.

We have $\overline{AB} \cong \overline{EF}$ by hypothesis, and the congruences $\triangle EFG \cong \triangle ABC$ and $\triangle EFH \cong \triangle ABD$ imply $\overline{FG} \cong \overline{BC}$, $\angle EFG \cong \angle ABC$, $\overline{EH} \cong \overline{AD}$, and $\angle FEH \cong \angle BAD$. Therefore, $EFGH$ is congruent to $ABCD$ by SASAS.   □

Parallelograms

When we think of parallelograms, we typically think of them not only as figures whose opposite sides are parallel, but also as figures whose opposite sides are congruent and whose opposite angles are congruent. We cannot yet prove that every parallelogram has these properties; in fact, just like many familiar properties of parallel lines, they cannot be proved in neutral geometry. However we can prove that having either congruent opposite angles or congruent opposite sides is a *sufficient* condition for a convex quadrilateral to be a parallelogram. (In fact, if we assume congruent opposite sides, we do not even need to assume convexity, as Theorem 9.17 below demonstrates.)

The next proof is rated PG.

Theorem 9.15. *A convex quadrilateral with two pairs of congruent opposite angles is a parallelogram.*

Proof. Suppose $ABCD$ is a convex quadrilateral with $\angle A \cong \angle C$ and $\angle B \cong \angle D$ (see Fig. 9.7). We need to show that $\overleftrightarrow{AB} \parallel \overleftrightarrow{CD}$ and $\overleftrightarrow{AD} \parallel \overleftrightarrow{BC}$. The proofs of both statements are the same, so we only prove the first one.

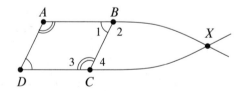

Fig. 9.7. A quadrilateral with congruent opposite angles.

Assume for the sake of contradiction that $\overleftrightarrow{AB}$ and $\overleftrightarrow{CD}$ are not parallel. Then they meet at a point X (see Fig. 9.7). Because $\overleftrightarrow{AB}$ and $\overleftrightarrow{CD}$ are semiparallel, X does not lie on $\overline{AB}$ or $\overline{CD}$. Thus it must be the case that either $A * B * X$ or $X * A * B$. Without loss of generality, we may assume the former. This means that X is on the opposite side of $\overleftrightarrow{BC}$ from A. Since A and D are on the same side of $\overleftrightarrow{BC}$ (because $\overline{AD}$ and $\overline{BC}$ are semiparallel), it follows that D and X are also on opposite sides of $\overleftrightarrow{BC}$, so $D * C * X$. Therefore, the angles marked $\angle 1$ and $\angle 2$ in Fig. 9.7 form a linear pair, as do $\angle 3$ and $\angle 4$. From the linear pair theorem and the hypothesis, we have

$$m\angle 2 + m\angle D = m\angle 2 + m\angle 1 = 180°,$$
$$m\angle 4 + m\angle A = m\angle 4 + m\angle 3 = 180°. \tag{9.1}$$

Adding these two equations yields $m\angle 2 + m\angle 4 + m\angle D + m\angle A = 360°$.

On the other hand, Corollary 5.14 applied to $\triangle XBC$ and $\triangle XAD$ shows that

$$m\angle 2 + m\angle 4 < 180°,$$
$$m\angle A + m\angle D < 180°. \tag{9.2}$$

Adding these two inequalities yields $m\angle 2 + m\angle 4 + m\angle D + m\angle A < 360°$, which contradicts the equation we derived above. $\square$

Corollary 9.16. *Every equiangular quadrilateral is a parallelogram.*

Proof. An equiangular quadrilateral is convex by definition, so it satisfies the hypotheses of the preceding theorem. $\square$

Next we consider congruent opposite sides. This proof is also rated PG.

Theorem 9.17. *A quadrilateral with two pairs of congruent opposite sides is a parallelogram.*

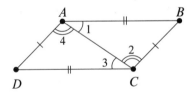

Fig. 9.8. A quadrilateral with congruent opposite sides.

Proof. Suppose $ABCD$ is a quadrilateral with $\overline{AB} \cong \overline{CD}$ and $\overline{AD} \cong \overline{BC}$ (Fig. 9.8). Draw diagonal $\overline{AC}$, and label the angles $\angle 1 = \angle BAC$, $\angle 2 = \angle ACB$, $\angle 3 = \angle DCA$, and $\angle 4 = \angle CAD$ as in the diagram. It follows from SSS that $\triangle ABC \cong \triangle CDA$, and therefore $\angle 1 \cong \angle 3$ and $\angle 2 \cong \angle 4$. The diagram suggests that B and D are on opposite sides of $\overleftrightarrow{AC}$. Provided this is true, $\angle 1$ and $\angle 3$ are congruent alternate interior angles for the transversal $\overleftrightarrow{AC}$ to $\overleftrightarrow{AB}$ and $\overleftrightarrow{CD}$, and therefore $\overleftrightarrow{AB} \parallel \overleftrightarrow{CD}$. Similarly, the fact that $\angle 2$ and $\angle 4$ are congruent implies that $\overleftrightarrow{AD} \parallel \overleftrightarrow{BC}$. This proves that $ABCD$ is a parallelogram.

It remains to show that B and D are indeed on opposite sides of $\overleftrightarrow{AC}$. This is one instance in which a fact that appears to be obvious from the diagram is actually quite tricky to prove. (If we knew in advance that $ABCD$ was a convex quadrilateral, then this would follow immediately from Lemma 9.2; but until we prove that it is a parallelogram, we have no easy way to prove that it is convex.)

Assume for the sake of contradiction that B and D are not on opposite sides of $\overleftrightarrow{AC}$. Since no three vertices of a quadrilateral are collinear, neither B nor D can be on $\overleftrightarrow{AC}$, so they must both be on the same side of the line. There are two cases to consider (Fig. 9.9):

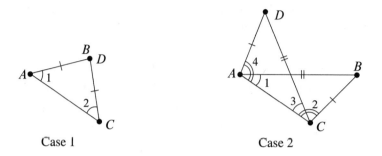

Case 1   Case 2

Fig. 9.9. Proof that B and D cannot be on the same side of $\overleftrightarrow{AC}$.

CASE 1: $AB = BC$. In this case, the hypothesis implies $CD = AB = BC = AD$, and thus $\triangle ACB \cong \triangle ACD$ by SSS. Then the unique triangle theorem guarantees that $B = D$. This contradicts the fact that the vertices of a polygon are all distinct.

CASE 2: $AB \neq BC$. Then one of them is larger; without loss of generality, say $AB > BC$. By the scalene inequality, this implies that $m\angle 2 > m\angle 1$. Using the congruences we proved above, we also have $m\angle 2 > m\angle 3$ and $m\angle 4 > m\angle 1$. The ordering lemma for rays implies that $\overrightarrow{AD} * \overrightarrow{AB} * \overrightarrow{AC}$ and $\overrightarrow{CB} * \overrightarrow{CD} * \overrightarrow{CA}$. Now, the crossbar theorem shows that $\overrightarrow{AB}$ must intersect $\overline{CD}$ and also that $\overrightarrow{CD}$ must intersect $\overline{AB}$. Since these two lines can only have one point in common (otherwise A, B, C, D would all be collinear), the intersection point must lie on both $\overline{AB}$ and $\overline{CD}$. But again, this contradicts the fact that $ABCD$ is a polygon and completes the proof. $\qquad\square$

Corollary 9.18. *Every rhombus is a parallelogram.*

Proof. A rhombus has two pairs of congruent opposite sides. $\qquad\square$

Using these results, we can prove some interesting criteria for identifying special types of quadrilaterals in terms of their diagonals.

Theorem 9.19. *Suppose $ABCD$ is a quadrilateral.*

(a) *If its diagonals bisect each other, then $ABCD$ is a parallelogram.*

(b) *If its diagonals are congruent and bisect each other, then $ABCD$ is equiangular.*

(c) *If its diagonals are perpendicular bisectors of each other, then $ABCD$ is a rhombus.*

(d) *If its diagonals are congruent and are perpendicular bisectors of each other, then $ABCD$ is a regular quadrilateral.*

Proof. Exercise 9D. $\qquad\square$

Corollary 9.20. *There exists a regular quadrilateral.*

Proof. Exercise 9E. $\qquad\square$

In Euclidean geometry, as we will see, an equiangular quadrilateral is just a rectangle, and a regular quadrilateral is a square. But Theorem 9.19 and its corollary did not mention rectangles and squares for a good reason: since we will not be able to prove an angle-sum theorem until we introduce the Euclidean parallel postulate, we do not know that equiangular quadrilaterals have 90° angles. (In fact, in hyperbolic geometry, they do not!)

Exercises

9A. Suppose $ABCD$ is a convex quadrilateral. Prove that any point that lies in the interiors of two nonconsecutive angles is in Int $ABCD$.

9B. Prove Theorem 9.4 (the diagonal criterion for convex quadrilaterals).

9C. Prove Theorem 9.13 (AASAS congruence).

9D. Prove Theorem 9.19 (characterizing special quadrilaterals by their diagonals).

9E. Prove Corollary 9.20 (existence of a regular quadrilateral).

9F. A *kite* is a convex quadrilateral whose four sides consist of two pairs of congruent adjacent sides (Fig. 9.10).
   (a) Show that a quadrilateral is a kite if and only if its diagonals are perpendicular and one of them bisects the other.
   (b) In a nonequilateral kite, show that the two angles between noncongruent sides are congruent to each other.

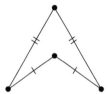

Fig. 9.10. A kite.   **Fig. 9.11.** A dart.

9G. A *dart* is a nonconvex quadrilateral whose four sides consist of two pairs of congruent adjacent sides (Fig. 9.11). Show that the two angles of a dart between noncongruent sides are congruent to each other.

Chapter 10

The Euclidean Parallel Postulate

From this point on in our study of Euclidean geometry, we officially add the Euclidean parallel postulate to our list of axioms. Thus, in addition to the nine axioms of neutral geometry, we assume the following new axiom.

Postulate 10E (The Euclidean Parallel Postulate). *For each line ℓ and each point A that does not lie on ℓ, there is a unique line that contains A and is parallel to ℓ.*

The term ***Euclidean geometry*** refers to the axiomatic system consisting of the primitive terms and axioms of neutral geometry together with the Euclidean parallel postulate. (We call this the "Euclidean parallel postulate" because it enables one to prove all of the theorems that Euclid proved. But, as we know, this is not one of the postulates that Euclid assumed. We will explain the relationship between this postulate and Euclid's fifth postulate later in this chapter and in Chapter 17.)

As we saw in Chapter 7, the axioms of neutral geometry are already sufficient to prove that for every line ℓ and every point $A \notin \ell$, there exists *at least one* line through A and parallel to ℓ. Thus the real content of the Euclidean parallel postulate is the statement that there is *only one* such line. We will see that many familiar properties of Euclidean geometry follow from this postulate.

The first such property is a converse to the alternate interior angles theorem. It is part of Euclid's Proposition I.29, the first proposition in which he makes use of his fifth postulate. Our proof is a little different from his because we are using Postulate 10E in place of Euclid's fifth postulate.

Theorem 10.1 (Converse to the Alternate Interior Angles Theorem). *If two parallel lines are cut by a transversal, then both pairs of alternate interior angles are congruent.*

Proof. Suppose ℓ and ℓ' are parallel lines cut by a transversal t, and let A and A' denote the points where t meets ℓ and ℓ', respectively. Choose either pair of alternate interior angles, and choose points $C \in \ell$ and $D \in \ell'$ such that the chosen angles are $\angle CAA'$ and $\angle AA'D$ (Fig. 10.1).

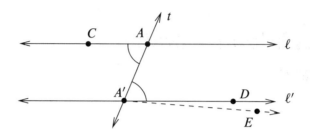

Fig. 10.1. Proof of the converse to the alternate interior angles theorem.

By the angle construction theorem, there is a ray $\overrightarrow{A'E}$ on the same side of t as D that makes an angle with $\overrightarrow{A'A}$ that is congruent to $\angle CAA'$. It follows from the alternate interior angles theorem that $\overleftrightarrow{A'E}$ is parallel to ℓ. By the Euclidean parallel postulate, therefore, $\overleftrightarrow{A'E}$ is in fact equal to ℓ'. Since D and E are on the same side of t, this means that $\overrightarrow{A'E}$ and $\overrightarrow{A'D}$ are the same ray, and therefore $\angle AA'D = \angle AA'E$. Since $\angle AA'E$ is congruent to $\angle CAA'$ by construction, we conclude that $\angle AA'D \cong \angle CAA'$.   □

The next two corollaries follow easily from the preceding theorem; they are the second and third parts of Euclid's Proposition I.29.

Corollary 10.2 (Converse to the Corresponding Angles Theorem). *If two parallel lines are cut by a transversal, then all four pairs of corresponding angles are congruent.*

Proof. Exercise 10A.   □

Corollary 10.3 (Converse to the Consecutive Interior Angles Theorem). *If two parallel lines are cut by a transversal, then both pairs of consecutive interior angles are supplementary.*

Proof. Exercise 10B.   □

Many additional properties of parallel lines are based on the following innocent-looking lemma. It is named after the Greek mathematician Proclus, who first stated it in his fifth-century commentary on Euclid's *Elements* [**Pro70**, p. 371]. (Proclus thought he had given a proof of this lemma that did not rely on Euclid's fifth postulate; but as we remarked in Chapter 1, his proof assumed without justification that parallel lines are equidistant.)

Lemma 10.4 (Proclus's Lemma). *Suppose ℓ and ℓ' are parallel lines. If t is a line that is distinct from ℓ but intersects ℓ, then t also intersects ℓ'.*

Proof. Suppose t is a line that is distinct from ℓ and intersects ℓ at A (Fig. 10.2). If t does not intersect ℓ', then t and ℓ are two distinct lines parallel to ℓ' through A, which contradicts the Euclidean parallel postulate.   □

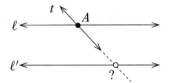

Fig. 10.2. Proclus's lemma.

Theorem 10.5. *Suppose ℓ and ℓ' are parallel lines. Then any line that is perpendicular to one of them is perpendicular to both.*

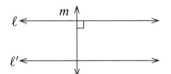

Fig. 10.3. Theorem 10.5.

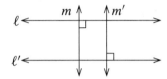

Fig. 10.4. Corollary 10.6.

Proof. Suppose m is perpendicular to one of the lines, say $m \perp \ell$ (see Fig. 10.3). By Proclus's lemma, m also intersects ℓ'. The converse to the corresponding angles theorem shows that each of the angles made by m and ℓ' is congruent to one of the angles made by m and ℓ. Since all of the angles made by m and ℓ are right angles, so are the angles made by m and ℓ', and thus $m \perp \ell'$. □

Corollary 10.6. *Suppose ℓ and ℓ' are parallel lines and m and m' are distinct lines such that $m \perp \ell$ and $m' \perp \ell'$. Then $m \parallel m'$.*

Proof. Under the given hypotheses (Fig. 10.4), Theorem 10.5 shows that m is also perpendicular to ℓ', and therefore m and m' are parallel because they have ℓ' as a common perpendicular. □

Corollary 10.7 (Converse to the Common Perpendicular Theorem). *If two lines are parallel, then they have a common perpendicular.*

Proof. Exercise 10C. □

Theorem 10.8 (Converse to the Equidistance Theorem). *If two lines are parallel, then each one is equidistant from the other.*

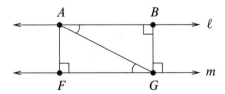

Fig. 10.5. Parallel lines are equidistant.

Proof. Suppose ℓ and m are parallel lines and A and B are any two distinct points on ℓ. Let F and G be the feet of the perpendiculars to m from A and B, respectively, and draw $\overline{AG}$ (Fig. 10.5). Theorem 10.5 shows that $\overleftrightarrow{AF}$ and $\overleftrightarrow{BG}$ are both perpendicular to ℓ, so $ABGF$ is a rectangle and thus convex. Therefore, B and F are on opposite sides of $\overleftrightarrow{AG}$ by Lemma 9.2, which implies that $\angle FGA$ and $\angle BAG$ are alternate interior angles for the transversal $\overleftrightarrow{AG}$. It follows from the converse to the alternate interior angles theorem that $\angle FGA \cong \angle BAG$, and so $\triangle FGA \cong \triangle BAG$ by AAS. This implies $AF = BG$, which means that A and B are equidistant from m. Since A and B were arbitrary points on ℓ, this shows that ℓ is equidistant from m, and the same argument shows that m is equidistant from ℓ. □

Corollary 10.9 (Symmetry of Equidistant Lines). *If ℓ and m are two distinct lines, then ℓ is equidistant from m if and only if m is equidistant from ℓ.*

Proof. By the equidistance theorem and its converse, both statements are equivalent to $\ell \parallel m$. □

As we mentioned in Chapter 7, the alternate interior angles theorem and its corollaries, all of which are valid in neutral geometry, are merely *sufficient criteria for parallelism*; they do not assert any *necessary properties of parallelism*. The converses that we just proved, on the other hand, are properties of lines that are known to be parallel and have been proved only in Euclidean geometry because their proofs depend on the Euclidean parallel postulate.

We end this section with one final consequence of the Euclidean parallel postulate. This is Euclid's Proposition I.30.

Theorem 10.10 (Transitivity of Parallelism). *If ℓ, m, and n are distinct lines such that $\ell \parallel m$ and $m \parallel n$, then $\ell \parallel n$.*

Proof. Exercise 10D. □

Theorem 10.10 seems so "obvious" that you might be tempted to think it should follow immediately from the definition of parallel lines together with the axioms of neutral geometry. But if the Euclidean parallel postulate were false, then for some line m and some point $A \notin m$, there would be two different lines parallel to m and passing through A and therefore not parallel to each other (Fig. 10.6). Thus this theorem cannot be proved without the Euclidean parallel postulate (or something equivalent to it).

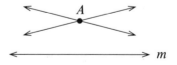

Fig. 10.6. Transitivity of parallelism cannot be proved without the Euclidean parallel postulate.

Angle-Sum Theorems

The next theorem is one of the most important facts in Euclidean geometry. To state it concisely, we introduce the following terminology. If $\mathcal{P}$ is a convex polygon, its **angle sum**, denoted by $\sigma(\mathcal{P})$, is the sum of the measures of its angles. (Later, we will define the angle sum for a more general polygon as the sum of its *interior* angle measures.) For example, for a triangle $\triangle ABC$, the angle sum is just

$$\sigma(\triangle ABC) = m\angle A + m\angle B + m\angle C.$$

The following theorem is the second half of Euclid's Proposition I.32; our proof is a slight modification of his.

Theorem 10.11 (Angle-Sum Theorem for Triangles). *Every triangle has angle sum equal to* $180°$.

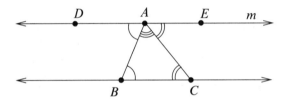

Fig. 10.7. Proof of the angle-sum theorem for triangles.

Proof. The diagram in Fig. 10.7 makes the proof seem nearly obvious; the only work that has to be done is to justify what the diagram seems to be telling us. Let $\triangle ABC$ be a triangle. By Theorem 7.25, there is a line m through A and parallel to $\overleftrightarrow{BC}$. We can choose points D, E on m such that D is on the same side of $\overleftrightarrow{AC}$ as B, and E is on the same side of $\overleftrightarrow{AB}$ as C. Since $\overleftrightarrow{BC}$ and m are parallel, Lemma 7.18 shows that B and C are on the same side of m. Combining these relations and using the interior lemma, we conclude that $\overrightarrow{AD} * \overrightarrow{AB} * \overrightarrow{AC}$ and $\overrightarrow{AB} * \overrightarrow{AC} * \overrightarrow{AE}$. Thus the linear triple theorem shows that

$$m\angle DAB + m\angle BAC + m\angle CAE = 180°. \tag{10.1}$$

On the other hand, it follows from Theorem 4.12 that D and C are on opposite sides of $\overleftrightarrow{AB}$. Therefore, $\angle DAB$ and $\angle ABC$ form a pair of alternate interior angles for the transversal $\overleftrightarrow{AB}$. Similarly, $\angle CAE$ and $\angle ACB$ form a pair of alternate interior angles for the transversal $\overleftrightarrow{AC}$. It follows from the converse to the alternate interior angles theorem that $m\angle DAB = m\angle ABC$ and $m\angle CAE = m\angle ACB$. Substituting these relations into (10.1) completes the proof. $\square$

The next corollary is the other half of Euclid's Proposition I.32.

Corollary 10.12. *In any triangle, the measure of each exterior angle is equal to the sum of the measures of the two remote interior angles.*

Proof. This is an immediate consequence of the angle-sum theorem and the linear pair theorem. $\square$

Theorem 10.13 (60-60-60 Theorem). *A triangle has all of its interior angle measures equal to* $60°$ *if and only if it is equilateral.*

Proof. Exercise 10E.   □

Theorem 10.14 (30-60-90 Theorem). *A triangle has interior angle measures* 30°, 60°, *and* 90° *if and only if it is a right triangle in which the hypotenuse is twice as long as one of the legs.*

Proof. Exercise 10F.   □

Theorem 10.15 (45-45-90 Theorem). *A triangle has interior angle measures* 45°, 45°, *and* 90° *if and only if it is an isosceles right triangle.*

Proof. Exercise 10G.   □

Another important application of the angle-sum theorem is to prove that Euclid's fifth postulate follows from the Euclidean parallel postulate together with the theorems of neutral geometry.

Theorem 10.16 (Euclid's Fifth Postulate). *If ℓ and ℓ′ are two lines cut by a transversal t in such a way that the measures of two consecutive interior angles add up to less than* 180°, *then ℓ and ℓ′ intersect on the same side of t as those two angles.*

Proof. Suppose ℓ, ℓ′, and t satisfy the hypotheses. First note that ℓ and ℓ′ are not parallel, because if they were, the converse to the consecutive interior angles theorem would imply that the two consecutive interior angles would have measures adding up to exactly 180°. Thus there is a point C where ℓ and ℓ′ intersect. It remains only to show that C is on the same side of t as the two interior angles whose measures add up to less than 180°.

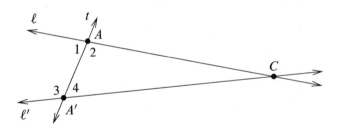

Fig. 10.8. Euclid's fifth postulate.

Let A and A' denote the points where t meets ℓ and ℓ', respectively. Denote the two interior angles at A as $\angle 1$ and $\angle 2$, and denote those at A' as $\angle 3$ and $\angle 4$, with the labels chosen so that $\angle 1$ and $\angle 3$ are the two interior angles whose measures add up to less than 180°. Assume for the sake of contradiction that C is on the opposite side of t from these two angles (Fig. 10.8). Because $\angle 1$ and $\angle 2$ form a linear pair, as do $\angle 3$ and $\angle 4$,

$$m\angle 2 = 180° - m\angle 1,$$
$$m\angle 4 = 180° - m\angle 3.$$

Adding these two equations and using the fact that $m\angle 1 + m\angle 3 < 180°$ yield

$$m\angle 2 + m\angle 4 = 360° - (m\angle 1 + m\angle 3) > 180°.$$

Because $\angle 2$ and $\angle 4$ are two angles of $\triangle AA'C$, this contradicts Corollary 5.14.   □

In fact, as we will show in Chapter 17, the converse is also true: if we assume Euclid's Postulate 5 in addition to the postulates of neutral geometry, the Euclidean parallel postulate follows as a theorem. Thus, in the presence of the axioms of neutral geometry, the Euclidean parallel postulate and Euclid's Postulate 5 are **equivalent**, which means that if either one of them is added to the postulates of neutral geometry as an additional postulate, then the other one can be proved as a theorem. When we are ready to begin our study of non-Euclidean geometry, we will see that there are many other statements that are also equivalent to the Euclidean parallel postulate.

Theorem 10.17 (AAA Construction Theorem). *Suppose $\overline{AB}$ is a segment and α, β, and γ are three positive real numbers whose sum is 180. On each side of $\overleftrightarrow{AB}$, there is a point C such that $\triangle ABC$ has the following angle measures: $m\angle A = \alpha°$, $m\angle B = \beta°$, and $m\angle C = \gamma°$.*

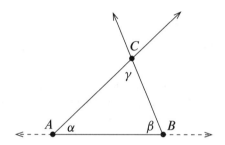

Fig. 10.9. The AAA construction theorem.

Proof. Exercise 10H. □

Now we can prove Euclid's Proposition I.1. (This result is also true in neutral geometry, but it is somewhat harder to prove in that context. Since we will only need it in Euclidean geometry, we prove it only in that case. For a proof that works in neutral geometry, see [**Mar96**, Theorem 20.15].)

Corollary 10.18 (Equilateral Triangle Construction Theorem). *If $\overline{AB}$ is any segment, then on each side of $\overleftrightarrow{AB}$ there is a point C such that $\triangle ABC$ is equilateral.*

Proof. Apply the AAA construction theorem with $\alpha = \beta = \gamma = 60°$. The resulting triangle is equilateral by Corollary 5.9. □

Angle-Sum Theorems for Polygons

The angle-sum theorem for triangles extends easily to convex polygons with any number of vertices. This proof is rated PG because of its use of mathematical induction, which might still be a little unfamiliar to some readers.

Theorem 10.19 (Angle-Sum Theorem for Convex Polygons). *In a convex polygon with n sides, the angle sum is equal to $(n-2) \times 180°$.*

Proof. We will prove the theorem by induction on n. The base case is $n = 3$ (triangles), which was proved in Theorem 10.11. For the inductive step, let n be an integer greater than or equal to 3, and suppose that the theorem is true for all convex polygons with n sides; we need to show that this implies it is also true for every convex polygon with $n + 1$ sides. Suppose $\mathscr{P}_{n+1} = A_1 A_2 \dots A_{n+1}$ is such a polygon; we need to show that its angle sum is $((n+1) - 2) \times 180°$.

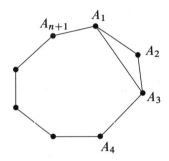

Fig. 10.10. The angle-sum theorem for convex polygons.

Consider the diagonal $\overline{A_1 A_3}$. It is a chord, so the polygon splitting theorem (Theorem 8.9) shows that it cuts $\mathscr{P}_{n+1}$ into a triangle $\triangle A_1 A_2 A_3$ and a convex n-sided polygon $\mathscr{P}_n = A_1 A_3 A_4 \dots A_{n+1}$ (see Fig. 10.10). Theorem 10.11 shows that $\sigma(\triangle A_1 A_2 A_3) = 180°$, and our inductive hypothesis implies that

$$\sigma(\mathscr{P}_n) = (n - 2) \times 180°.$$

By the angle criterion for convexity, A_3 is in the interior of $\angle A_{n+1} A_1 A_2$, and A_1 is in the interior of $\angle A_2 A_3 A_4$. Therefore,

$$m\angle A_{n+1} A_1 A_2 = m\angle A_{n+1} A_1 A_3 + m\angle A_3 A_1 A_2,$$
$$m\angle A_2 A_3 A_4 = m\angle A_2 A_3 A_1 + m\angle A_1 A_3 A_4.$$

Inserting these relations into the formula for the angle sum and simplifying, we find

$$\sigma(\mathscr{P}_{n+1}) = \sigma(\mathscr{P}_n) + \sigma(\triangle A_1 A_2 A_3) = (n - 2) \times 180° + 180° = ((n+1) - 2) \times 180°.$$

This proves the inductive step and thus the theorem. $\qquad\qquad\square$

The next corollary follows immediately from the angle-sum theorem.

Corollary 10.20. *In a regular n-gon, the measure of each angle is $\frac{n-2}{n} \times 180°$.* $\qquad\square$

One useful way to understand intuitively why the angle-sum formula should be true (and to reconstruct the formula if you forget it) is to recast it in terms of exterior angles. Just as for triangles, if $\mathscr{P}$ is a convex polygon, we define an ***exterior angle of $\mathscr{P}$*** to be an angle that forms a linear pair with one of the interior angles of $\mathscr{P}$. If you imagine starting on one of the sides of the polygon and walking all the way around it counterclockwise, then each time you reach a vertex you will have to change the direction you are facing by turning left an amount equal to the exterior angle measure at that vertex. When you have made one complete circuit, your direction will have rotated exactly 360° (see Fig. 10.11). The next corollary makes this intuition rigorous.

Corollary 10.21 (Exterior Angle Sum for a Convex Polygon). *In any convex polygon, the sum of the measures of the exterior angles (one at each vertex) is 360°.*

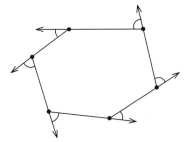

Fig. 10.11. The exterior angle sum for a convex polygon.

Proof. Suppose $\mathcal{P}$ is a convex polygon with n vertices. Then its interior angle sum is $(n-2) \times 180°$. Let $\theta_1, \ldots, \theta_n$ denote the interior angle measures, and let $\varepsilon_1, \ldots, \varepsilon_n$ denote the corresponding exterior angle measures. Since each exterior angle and its corresponding interior angle form a linear pair, they are supplementary, and so

$$
\begin{aligned}
\varepsilon_1 + \cdots + \varepsilon_n &= (180° - \theta_1) + \cdots + (180° - \theta_n) \\
&= (180° + \cdots + 180°) - (\theta_1 + \cdots + \theta_n) \\
&= n \times 180° - (n-2) \times 180° \\
&= 2 \times 180° = 360°. \qquad \square
\end{aligned}
$$

There is also an angle-sum theorem for arbitrary (not necessarily convex) polygons. Recall from Chapter 8 that the *interior angle measure* at a vertex of a polygon is defined to be the standard angle measure at a convex vertex, but the reflex angle measure at a concave one. The proof of this theorem is rated X, so we can only sketch the main ideas.

Theorem 10.22 (Angle-Sum Theorem for General Polygons). *If $\mathcal{P}$ is any polygon with n sides, the sum of its interior angle measures is $(n-2) \times 180°$.*

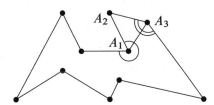

Fig. 10.12. Proving the angle-sum theorem for general polygons.

Sketch of proof. As in the convex case, the proof is by induction on n, with the base case taken care of by Theorem 10.11. The idea of the proof of the inductive step is the same as in the convex case: write $\mathcal{P} = A_1 A_2 A_3 \ldots A_{n+1}$ in such a way that the diagonal $A_1 A_3$ cuts $\mathcal{P}$ into a triangle and an n-sided polygon, and apply the inductive hypothesis. There are two stumbling blocks that make the proof much more difficult than in the convex case. The first is proving that it is always possible to find three consecutive vertices A_1, A_2, A_3 such that the interior of the diagonal $A_1 A_3$ lies in the interior of the polygon. The second is proving that once such vertices have been found, the interior angle measures of the original polygon at A_1 and A_3 are the sums of those of the two smaller polygons. (There are many cases to check, because some interior angle measures are reflex measures, while others are ordinary measures.)   □

Quadrilaterals in Euclidean Geometry

The preceding results lead to some new properties of quadrilaterals that could not be proved in neutral geometry.

Theorem 10.23 (Angle-Sum Theorem for Quadrilaterals). *Every convex quadrilateral has an angle sum of* 360°.

Proof. Apply Theorem 10.19 with $n = 4$.   □

Corollary 10.24. *A quadrilateral is equiangular if and only if it is a rectangle, and it is a regular quadrilateral if and only if it is a square.*

Proof. Let $ABCD$ be a quadrilateral. If it is equiangular, then its angle measures are all $360°/4 = 90°$ by Theorem 10.23, so it is a rectangle. Conversely, if it is a rectangle, then it is convex by Corollary 9.8, and thus equiangular. (Recall that convexity is part of the definition of an equiangular polygon.)

Now suppose $ABCD$ is regular, meaning equiangular and equilateral. Then it is a rectangle by the argument in the preceding paragraph, and because it is also equilateral it is a square. Conversely, if $ABCD$ is a square, then it is an equilateral rectangle by definition, and thus it is also equiangular by the argument in the preceding paragraph.   □

Because the Euclidean parallel postulate allows us to draw many more conclusions about parallel lines, it also allows us to derive many more properties of parallelograms. The properties expressed in the next few theorems are probably already familiar. Parts (a)–(c) of the next theorem are essentially the same as Euclid's Proposition I.34.

Theorem 10.25. *Every parallelogram has the following properties.*

(a) *Each diagonal cuts it into a pair of congruent triangles.*

(b) *Both pairs of opposite sides are congruent.*

(c) *Both pairs of opposite angles are congruent.*

(d) *Its diagonals bisect each other.*

Proof. Exercise 10I.   □

Next, we have some new criteria for a quadrilateral to be a parallelogram or a rectangle. The first one is Euclid's Proposition I.33, with essentially the same proof.

Theorem 10.26. *If a quadrilateral has a pair of congruent and parallel opposite sides, then it is a parallelogram.*

Proof. Suppose $ABCD$ is a polygon in which $\overline{AB}$ and $\overline{CD}$ are congruent and parallel. Draw the diagonal $\overline{AC}$, and label the angles as shown in Fig. 10.13. Because $ABCD$ is a trapezoid, it is convex, and therefore B and D lie on opposite sides of $\overleftrightarrow{AC}$ by Lemma 9.2. It follows that $\angle 1$ and $\angle 3$ are alternate interior angles for the transversal $\overleftrightarrow{AC}$ to

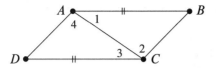

Fig. 10.13. Theorem 10.26.

$\overleftrightarrow{AB}$ and $\overleftrightarrow{CD}$, and thus by the converse to the alternate interior angles theorem they are congruent. Then $\triangle ABC \cong \triangle CDA$ by SAS. This means that $\overline{AD} \cong \overline{BC}$, so $ABCD$ is a parallelogram by Theorem 9.17. □

Theorem 10.27. *If a quadrilateral has a pair of congruent opposite sides that are both perpendicular to a third side, then it is a rectangle.*

Fig. 10.14. Theorem 10.27.

Proof. Suppose $ABCD$ is a quadrilateral in which $\overline{AD}$ and $\overline{BC}$ are congruent to each other and both are perpendicular to $\overline{CD}$ (see Fig. 10.14). Then $\overline{AD} \parallel \overline{BC}$ by the common perpendicular theorem, so the preceding theorem implies that $ABCD$ is a parallelogram. Therefore, $\angle A$ and $\angle B$ are both right angles by Theorem 10.25(c), so $ABCD$ is a rectangle. □

Here is a useful application of Theorem 10.27.

Theorem 10.28 (Rectangle Construction). *Suppose a and b are positive real numbers and $\overline{AB}$ is a segment of length a. On either side of $\overleftrightarrow{AB}$, there exist points C and D such that $ABCD$ is a rectangle with $AB = CD = a$ and $AD = BC = b$.*

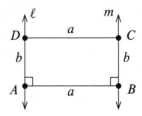

Fig. 10.15. Constructing a rectangle.

Proof. Let ℓ and m be the lines perpendicular to $\overleftrightarrow{AB}$ through A and B, respectively. On the chosen side of $\overleftrightarrow{AB}$, let D be a point on ℓ such that $AD = b$, and let C a point on m such that $BC = b$. Then $\overline{AD} \parallel \overline{BC}$ by the common perpendicular theorem, and since we chose C and D on the same side of $\overleftrightarrow{AB}$, it follows that $\overline{CD} \cap \overline{AB} = \varnothing$. Thus the trapezoid lemma shows that $ABCD$ is a trapezoid, and it follows from Theorem 10.27 that it is a rectangle.   □

The next corollary is Euclid's Proposition I.46. His proof is very similar to our proof of Theorem 10.28, but specialized to the case of a square.

Corollary 10.29 (Square Construction). *If $\overline{AB}$ is any segment, then on each side of $\overleftrightarrow{AB}$ there are points C and D such that $ABCD$ is a square.*

Proof. Just apply the preceding theorem with $a = b = AB$.   □

These properties of parallelograms lead to the following interesting theorem about triangles. In any triangle, a segment connecting the midpoints of two different sides is called a *midsegment* (Fig. 10.16).

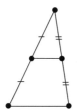

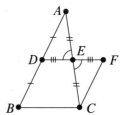

Fig. 10.16. A midsegment.   Fig. 10.17. Proof of the midsegment theorem.

Theorem 10.30 (Midsegment Theorem). *Any midsegment of a triangle is parallel to the third side and is half as long.*

Proof. Let $\triangle ABC$ be a triangle, and let $\overline{DE}$ be one of its midsegments; renaming the vertices if necessary, we may assume that D is the midpoint of $\overline{AB}$ and E is the midpoint of $\overline{AC}$ (Fig. 10.17). Let F be the point on the ray opposite $\overrightarrow{ED}$ such that $EF = ED$. Then $\angle AED$ and $\angle CEF$ are vertical angles and are thus congruent, so $\triangle AED \cong \triangle CEF$ by SAS. It follows that $\overline{AD} \cong \overline{FC}$ and $\angle DAE \cong \angle FCE$.

Now, D and F are on opposite sides of $\overleftrightarrow{AC}$ because $D * E * F$, and therefore $\angle DAE$ and $\angle FCE$ are alternate interior angles for the transversal $\overleftrightarrow{AC}$ to $\overleftrightarrow{AB}$ and $\overleftrightarrow{FC}$. By the alternate interior angles theorem, $\overleftrightarrow{AB} \parallel \overleftrightarrow{FC}$. Moreover, $\overline{DB}$ and $\overline{FC}$ are both congruent to $\overline{AD}$ and thus are congruent to each other. Since $DBCF$ has two sides (namely $\overline{DB}$ and $\overline{CF}$) that are both parallel and congruent, it is a parallelogram by Theorem 10.26. This implies that $\overline{DF}$ and $\overline{BC}$ are both parallel and congruent, so $\overline{DE} \parallel \overline{BC}$. Using the fact that $D * E * F$, we compute $BC = DF = DE + EF = 2DE$, which can be solved for DE to obtain $DE = \frac{1}{2}BC$.   □

Exercises

10A. Prove Corollary 10.2 (the converse to the corresponding angles theorem).

10B. Prove Corollary 10.3 (the converse to the consecutive interior angles theorem).

10C. Prove Corollary 10.7 (the converse to the common perpendicular theorem).

10D. Prove Theorem 10.10 (transitivity of parallelism).

10E. Prove Theorem 10.13 (the 60-60-60 theorem).

10F. Prove Theorem 10.14 (the 30-60-90 theorem).

10G. Prove Theorem 10.15 (the 45-45-90 theorem).

10H. Prove Theorem 10.17 (the AAA construction theorem).

10I. Prove Theorem 10.25 (properties of parallelograms).

10J. Prove that the diagonals of a rectangle are congruent.

10K. Each of the following is a shorthand notation for a possible congruence theorem for convex quadrilaterals: SASSS, SAASS, ASASA, AAASA, SAAAS, ASASS, ASSAS, ASAAS. Try to decide which ones represent valid congruence theorems in Euclidean geometry. When possible, give proofs of those that are valid, and give brief descriptions of counterexamples (no proofs necessary) for those that are not. [Some of these are easy, while some are quite hard.]

10L. In the triangle of Fig. 10.18, $AD = DC = CB$ and $AB = AC$. Find the measure of angle A, and prove your answer correct.

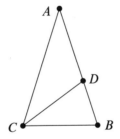

Fig. 10.18. The setup for Exercise 10L.

10M. VARIGNON'S THEOREM: Suppose $ABCD$ is a convex quadrilateral and E, F, G, and H are the midpoints of $\overline{AB}$, $\overline{BC}$, $\overline{CD}$, and $\overline{DA}$, respectively. Prove that $EFGH$ is a parallelogram (see Fig. 10.19).

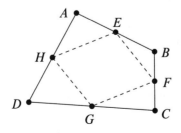

Fig. 10.19. Varignon's theorem.

Chapter 11

Area

In this chapter, we start working with one of the most important concepts in geometry, *area*. The essential properties of area in Euclidean geometry are described by a postulate, which we will explain below. But first, we need some definitions, to pin down the types of sets to which we can assign areas.

Polygonal Regions

The most important types of regions are those whose boundaries are polygons. Recall from Chapter 8 that if $\mathcal{P}$ is a polygon, the ***region determined by*** $\mathcal{P}$ is the set $\operatorname{Reg}\mathcal{P}$ consisting of the union of $\mathcal{P}$ and its interior. Let us call a set of points $\mathcal{R}$ a ***simple polygonal region*** if it is of the form $\mathcal{R} = \operatorname{Reg}\mathcal{P}$ for some polygon $\mathcal{P}$ (see Fig. 11.1). The interior of $\mathcal{P}$ is also called the ***interior of*** $\mathcal{R}$, and the polygon $\mathcal{P}$ itself is called the ***boundary of*** $\mathcal{R}$. A simple polygonal region defined by a triangle is called a ***triangular region***, and ***rectangular regions*** and ***square regions*** are defined similarly. We say that two simple polygonal regions are ***congruent*** if their associated polygons are congruent.

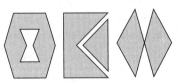

Fig. 11.1. Simple polygonal regions.   **Fig. 11.2.** General polygonal regions.

Simple polygonal regions will suffice for most purposes. But for many applications, it is useful to assign areas to somewhat more general sets, such as regions with holes or regions with disconnected pieces or regions whose boundaries are more complicated than polygons (see Fig. 11.2). For this reason, we make the following definitions. Two simple polygonal regions are said to ***overlap*** if their interiors have nonempty intersection and to be ***nonoverlapping*** if their interiors are disjoint. (Note that nonoverlapping regions are allowed to intersect on their boundaries.) More generally, a collection of finitely many simple polygonal regions is said to be nonoverlapping if no two of them overlap. A set of points $\mathcal{R}$ is called a ***general polygonal region*** if it can be expressed as a union of finitely many nonoverlapping simple polygonal regions. A ***polygonal region*** without qualification means a general polygonal region. The second and third regions in Fig. 11.2 can each be expressed as a union of two nonoverlapping simple polygonal regions in an obvious way. Fig. 11.3 illustrates several different ways to realize the first one as such a union. By means of such decompositions, all questions about areas of general polygonal regions can be reduced to questions about areas of simple polygonal regions.

Fig. 11.3. Some ways of expressing a general polygonal region as a union of simple ones.

There are various approaches to the concept of area in plane geometry. All of these approaches are based on the idea that area is a function that assigns a positive real number to every polygonal region. Thus we make the following definition. An ***area function*** is a function α from the set of all general polygonal regions into the positive real numbers, satisfying the following properties:

(i) (AREA CONGRUENCE PROPERTY) If $\mathcal{R}_1$ and $\mathcal{R}_2$ are congruent simple polygonal regions, then

$$\alpha(\mathcal{R}_1) = \alpha(\mathcal{R}_2).$$

(ii) (AREA ADDITION PROPERTY) If $\mathcal{R}_1, \ldots, \mathcal{R}_n$ are nonoverlapping simple polygonal regions, then

$$\alpha(\mathcal{R}_1 \cup \cdots \cup \mathcal{R}_n) = \alpha(\mathcal{R}_1) + \cdots + \alpha(\mathcal{R}_n).$$

The simplest approach to area is to postulate the existence and uniqueness of a certain area function. This is the approach that most high-school texts adopt, and it is the one we will use here. At the end of the chapter, we will discuss another possible approach.

Intuitively, the area of a region is a positive number describing how much of the plane it occupies—roughly, the number of unit squares it would take to fill up the region. Rigorously, everything we know about it is expressed in the following postulate.

> **Postulate 11E (The Euclidean Area Postulate).** *There exists a unique area function α with the property that $\alpha(\mathcal{R}) = 1$ whenever $\mathcal{R}$ is a square region with sides of length 1.*

Nearly all of the polygonal regions we encounter will be simple ones. (In fact, the majority will be regions determined by triangles or quadrilaterals.) In order to simplify the terminology and notation, if $\mathscr{P}$ is a polygon, instead of speaking of "the area of the region determined by $\mathscr{P}$," we will simply refer to the ***area of $\mathscr{P}$*** and write $\alpha(\mathscr{P})$, with the understanding that this really refers to $\alpha(\text{Reg}\,\mathscr{P})$. For example, if $\triangle ABC$ is a triangle, then $\alpha(\triangle ABC)$ means the area of the triangular region consisting of $\triangle ABC$ together with its interior.

Admissible Decompositions

In order to derive formulas for areas, we will need to carve certain regions into smaller regions and use the area addition property. Suppose $\mathcal{P}$ is a polygon. An **admissible decomposition of $\mathcal{P}$** is a finite collection $\{\mathcal{P}_1, \dots, \mathcal{P}_n\}$ of other polygons such that the regions $\operatorname{Reg}\mathcal{P}_1, \dots, \operatorname{Reg}\mathcal{P}_n$ are nonoverlapping and $\operatorname{Reg}\mathcal{P} = \operatorname{Reg}\mathcal{P}_1 \cup \dots \cup \operatorname{Reg}\mathcal{P}_n$. When this is the case, it follows from the area addition property that $\alpha(\mathcal{P}) = \alpha(\mathcal{P}_1) + \dots + \alpha(\mathcal{P}_n)$.

In order to make practical use of the area addition property, we will need to verify that some common constructions yield admissible decompositions. The following lemma will be our primary tool for constructing admissible decompositions. Its proof is rated PG.

Lemma 11.1 (Convex Decomposition Lemma). *Suppose $\mathcal{P}$ is a convex polygon and $\overline{BC}$ is a chord of $\mathcal{P}$. Then the two convex polygons $\mathcal{P}_1$ and $\mathcal{P}_2$ described in the polygon splitting theorem (Theorem 8.9) form an admissible decomposition of $\mathcal{P}$, and therefore $\alpha(\mathcal{P}) = \alpha(\mathcal{P}_1) + \alpha(\mathcal{P}_2)$.*

Proof. Throughout this proof, we will use the same notation as in the proof of Theorem 8.9, and we will use the results of Lemma 8.10 to characterize the interiors and regions of the various convex polygons. First we have to show that $\operatorname{Reg}\mathcal{P}_1$ and $\operatorname{Reg}\mathcal{P}_2$ are nonoverlapping, which means that their interiors are disjoint. Note that $\mathcal{P}_1$ and $\mathcal{P}_2$ lie in opposite closed half-planes determined by $\overleftrightarrow{BC}$. (If either vertex of $\overline{BC}$, say B, is an interior point of an edge $\overline{A_i A_{i+1}}$ of $\mathcal{P}$, this follows from the fact that $A_i * B * A_{i+1}$; while if both B and C are vertices of $\mathcal{P}$, it follows from the angle criterion for convexity and Theorem 4.12.) Thus it follows from Lemma 8.10 that every point in the interior of $\mathcal{P}_1$ lies on one side of $\overleftrightarrow{BC}$ while every point in the interior of $\mathcal{P}_2$ lies on the other side, so the interiors of the two polygons are disjoint.

Next, we need to prove that $\operatorname{Reg}\mathcal{P} = \operatorname{Reg}\mathcal{P}_1 \cup \operatorname{Reg}\mathcal{P}_2$. First assume Q is a point in $\operatorname{Reg}\mathcal{P}$. If $Q \in \overleftrightarrow{BC}$, then it must lie on $\overline{BC}$ because it is contained in the closed half-planes determined by $\overleftrightarrow{A_i A_{i+1}}$ and C and by $\overleftrightarrow{A_j A_{j+1}}$ and B. Since $\overline{BC}$ is contained in both $\mathcal{P}_1$ and $\mathcal{P}_2$, it follows that Q is contained in $\operatorname{Reg}\mathcal{P}_1 \cup \operatorname{Reg}\mathcal{P}_2$. On the other hand, if $Q \notin \overleftrightarrow{BC}$, then it must lie on one side or the other of this line; without loss of generality, assume it lies on the $\mathcal{P}_2$-side. Then Q also lies in the appropriate closed half-plane determined by each of the edges of $\mathcal{P}_2$ (which are *some* of the closed half-planes determined by edges of $\mathcal{P}$), so it lies in $\operatorname{Reg}\mathcal{P}_2$ and therefore in $\operatorname{Reg}\mathcal{P}_1 \cup \operatorname{Reg}\mathcal{P}_2$.

Conversely, assume $Q \in \operatorname{Reg}\mathcal{P}_1 \cup \operatorname{Reg}\mathcal{P}_2$ and assume for definiteness that Q lies in $\operatorname{Reg}\mathcal{P}_1$. If Q is on one of the edges of $\mathcal{P}_1$ that also lie on $\mathcal{P}$, then clearly $Q \in \mathcal{P} \subseteq \operatorname{Reg}\mathcal{P}$. If Q is on $\overline{BC}$, then Q lies on both $\mathcal{P}_1$ and $\mathcal{P}_2$; because these are convex polygons, it follows that Q lies in the appropriate closed half-plane defined by each of the edges of $\mathcal{P}_1$ and by each of the edges of $\mathcal{P}_2$, which implies that it lies in all of the appropriate closed half-planes defined by edges of $\mathcal{P}$. Therefore, it lies in $\operatorname{Reg}\mathcal{P}$.

Finally, suppose $Q \in \operatorname{Int}\mathcal{P}_1$. In this case, we will prove that $Q \in \operatorname{Int}\mathcal{P}$ and therefore is in $\operatorname{Reg}\mathcal{P}$. For this purpose, it is easiest to use the parity definition of interior points (see Theorem 8.15). For any point $Y \in \operatorname{Int}\overline{BC}$, the ray opposite $\overrightarrow{QY}$ cannot intersect $\overline{BC}$ and cannot intersect any of the vertices of $\mathcal{P}_2$ (Fig. 11.4). For some such choice of Y, this ray

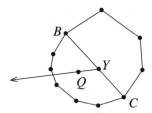

Fig. 11.4. Proof of the convex decomposition lemma.

will not intersect any vertices of $\mathcal{P}$, and then its parity with respect to $\mathcal{P}$ will be the same as its parity with respect to $\mathcal{P}_2$. Therefore $Q \in \operatorname{Int}\mathcal{P} \subseteq \operatorname{Reg}\mathcal{P}$ in this case too. $\square$

Two situations that occur frequently are described in the next two lemmas: in the first, a convex region is carved up into "wedges" by drawing segments outward from a fixed interior point, like slices of a pizza; in the second, a parallelogram is cut into four smaller parallelograms by two segments parallel to its edges. Both of these lemmas are rated PG.

Lemma 11.2 (Pizza Lemma). *Suppose $\mathcal{P}$ is a convex polygon, O is a point in* $\operatorname{Int}\mathcal{P}$, *and* $\{B_1, \ldots, B_m\}$ *are distinct points on* $\mathcal{P}$, *ordered in such a way that for each $i = 1, \ldots, m$, the angle $\angle B_i O B_{i+1}$ is proper and contains none of the B_j's in its interior (where we interpret B_{m+1} to mean B_1). For each $i = 1, \ldots, m$, let $\mathcal{P}_i$ denote the following set:*

$$\mathcal{P}_i = \overline{OB_i} \cup \overline{OB_{i+1}} \cup \left(\mathcal{P} \cap \operatorname{Int}\angle B_i O B_{i+1}\right).$$

(See Fig. 11.5.) Then each $\mathcal{P}_i$ is a convex polygon, and

$$\alpha(\mathcal{P}) = \alpha(\mathcal{P}_1) + \cdots + \alpha(\mathcal{P}_m). \tag{11.1}$$

Proof. We first treat a special case: assume that B_1 and O are collinear with another of the chosen points, say B_k (see Fig. 11.5(a)). Because we are assuming that each of the angles $\angle B_i O B_{i+1}$ is proper, there must be at least two other chosen points, B_2 on one side of $\overleftrightarrow{B_1 B_k}$ and B_{k+1} on the other side, both different from B_1 and B_k. The points $B_2, \ldots, B_{k-1}$ must all lie on one side of $\overleftrightarrow{B_1 B_k}$, and $B_{k+1}, \ldots, B_m$ must lie on the other: if not, there is some i such that B_i lies on one side of $\overleftrightarrow{B_1 B_k}$ and B_{i+1} lies on the other. Because $\angle B_i O B_{i+1}$ is a proper angle, one of the opposite rays $\overrightarrow{OB_1}$ or $\overrightarrow{OB_k}$ must lie in $\operatorname{HR}\left(\overrightarrow{OB_i}, B_{i+1}\right)$; without loss of generality, say it is B_1. This means that $\overrightarrow{OB_i} * \overrightarrow{OB_1} * \overrightarrow{OB_{i+1}}$ by the adjacency lemma, and thus B_1 lies in $\operatorname{Int}\angle B_i O B_{i+1}$, contradicting our hypothesis.

Because $\overline{B_1 B_k}$ is a chord of $\mathcal{P}$, the convex decomposition lemma shows that it cuts $\mathcal{P}$ into two convex subpolygons that form an admissible decomposition of $\mathcal{P}$. Name them $\mathcal{Q}$ and $\mathcal{Q}'$, with the names chosen so that $\mathcal{Q}$ is the one that contains $B_2, \ldots, B_{k-1}$ and $\mathcal{Q}'$ contains $B_{k+1}, \ldots, B_m$. We will show by induction on k that $\alpha(\mathcal{Q}) = \alpha(\mathcal{P}_1) + \cdots + \alpha(\mathcal{P}_{k-1})$. In view of the remark at the beginning of the proof, the base case is $k = 3$: in this case, $\overline{OB_2}$ is a chord of $\mathcal{Q}$, and the result follows directly from the convex decomposition lemma. For the inductive step, suppose we have proved the claim for some $k' \geq 3$, and let $k = k' + 1$, so there are $k - 2$ points $B_2, \ldots, B_{k-1}$ on the same side of $\overleftrightarrow{B_1 B_k}$ as B_2. If we omit B_{k-1}, the inductive hypothesis shows that $\alpha(\mathcal{Q}) = \alpha(\mathcal{P}_1) + \cdots + \alpha(\mathcal{P}_{k-3}) + \alpha\left(\widetilde{\mathcal{P}}_{k-2}\right)$, where the $\mathcal{P}_i$'s coincide with the subpolygons in the statement of the lemma, but $\widetilde{\mathcal{P}}_{k-2}$ is a new one that contains $\overline{OB_{k-1}}$ as a chord. The convex decomposition lemma

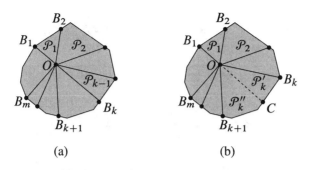

Fig. 11.5. The pizza lemma.

applied to the convex polygon $\widetilde{\mathscr{P}}_{k-2}$ then implies $\alpha\big(\widetilde{\mathscr{P}}_{k-2}\big) = \alpha(\mathscr{P}_{k-2}) + \alpha(\mathscr{P}_{k-1})$, and inserting this into the formula for $\alpha(\mathcal{Q})$ completes the induction. The same argument (with the labels adjusted accordingly) shows that $\alpha(Q') = \alpha(\mathscr{P}_k) + \cdots + \alpha(\mathscr{P}_m)$, and putting these two formulas together yields (11.1).

Now consider the case in which B_1 and O are not collinear with any of the other B_i's. The ray opposite $\overrightarrow{OB_1}$ intersects $\mathscr{P}$ at exactly one point C by Corollary 8.16 (Fig. 11.5(b)). Let k be the largest index such that B_k is on the same side of $\overleftrightarrow{B_1 C}$ as B_2; then an argument just like the one in the first paragraph of this proof shows that $B_1, \ldots, B_k$ lie on one side of that line and $B_{k+1}, \ldots, B_m$ lie on the other. The points $\{B_1, \ldots, B_k, C, B_{k+1}, \ldots, B_m\}$ satisfy the hypotheses of the preceding argument, so we can write

$$\alpha(\mathscr{P}) = \alpha(\mathscr{P}_1) + \cdots + \alpha(\mathscr{P}_{k-1}) + \alpha(\mathscr{P}_k') + \alpha(\mathscr{P}_k'') + \alpha(\mathscr{P}_{k+1}) + \cdots + \alpha(\mathscr{P}_m),$$

where the $\mathscr{P}_i$'s are the same as in the statement of the lemma, but $\mathscr{P}_k'$ and $\mathscr{P}_k''$ are two new polygons that have $\overline{OC}$ as a common edge. Note that the subpolygon $\mathscr{P}_k$ is convex: by definition, it is contained in one of the closed half-planes defined by each of the edge lines $\overleftrightarrow{OB_k}$ and $\overleftrightarrow{OB_{k+1}}$, and the convexity of $\mathscr{P}$ guarantees that it is contained in one half-plane defined by each of its other edge lines. Because $\{\mathscr{P}_k', \mathscr{P}_k''\}$ is exactly the admissible decomposition of $\mathscr{P}_k$ determined by the chord $\overline{OC}$, we have $\alpha(\mathscr{P}_k) = \alpha(\mathscr{P}_k') + \alpha(\mathscr{P}_k'')$, and (11.1) follows.   □

Lemma 11.3 (Parallelogram Decomposition Lemma). *Suppose $ABCD$ is a parallelogram and E, F, G, H are interior points on $\overline{AB}$, $\overline{BC}$, $\overline{CD}$, and $\overline{DA}$, respectively, such that $\overline{AH} \cong \overline{BF}$ and $\overline{AE} \cong \overline{DG}$ (Fig. 11.6). Then there is a point X where $\overline{HF}$ intersects $\overline{EG}$, and $AEXH$, $EBFX$, $HXGD$, and $XFCG$ are parallelograms such that*

$$\alpha(ABCD) = \alpha(AEXH) + \alpha(EBFX) + \alpha(HXGD) + \alpha(XFCG).$$

Proof. Theorem 10.25 shows that $\overline{AD} \cong \overline{BC}$ and $\overline{AB} \cong \overline{CD}$. It follows from the hypothesis and the segment subtraction theorem that $\overline{HD} \cong \overline{FC}$ and $\overline{EB} \cong \overline{GC}$ as well. Now, the polygon splitting theorem implies that $ABFH$ and $HFCD$ are both convex polygons, and since each one has a pair of congruent parallel sides, they are parallelograms. The same argument shows that $AEGD$ and $EBCG$ are also parallelograms.

Because $\overleftrightarrow{HF}$ meets $\overleftrightarrow{AD}$ and $\overleftrightarrow{EG} \parallel \overleftrightarrow{AD}$, it follows from Proclus's lemma that there is a point X where $\overleftrightarrow{HF}$ meets $\overleftrightarrow{EG}$. The fact that $A * E * B$ implies that A and B are on opposite sides of $\overleftrightarrow{EG}$; since $\overleftrightarrow{AD}$, $\overleftrightarrow{EG}$, and $\overleftrightarrow{BC}$ are all mutually parallel, Lemma 7.18

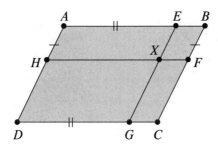

Fig. 11.6. Decomposition of a parallelogram.

shows that all points on $\overleftrightarrow{AD}$ are on the opposite side of $\overleftrightarrow{EG}$ from points on $\overleftrightarrow{BC}$. This implies that $H * X * F$. The polygon splitting theorem applied to $ABFH$ and $HFCD$ shows that $AEXH$, $EBFX$, $HXGD$, and $XFCG$ are all convex polygons, and since their opposite sides are all parallel, they are parallelograms. The pizza lemma then shows that

$$\alpha(ABCD) = \alpha(AEXH) + \alpha(EBFX) + \alpha(HXGD) + \alpha(XFCG). \qquad \square$$

Area Formulas

Now we are ready to begin deriving some familiar area formulas in Euclidean geometry. We begin with squares. Because any two squares with the same side lengths are congruent (by SASAS, for example), the area of a square of side length x depends only on the number x. Let us temporarily write this number as $s(x)$ and derive some important facts about it. The first fact, which follows immediately from the area postulate, is that $s(1) = 1$. The next fact, expressed in the following lemma, is that the area of a square increases as the size of the square increases.

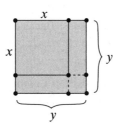

Fig. 11.7. $s(x)$ increases with x.

Lemma 11.4. *If $x < y$, then $s(x) < s(y)$.*

Proof. Start with a square of side length y (for example, we can start with a segment of length y and use Corollary 10.29 to construct a square with that segment as one of its sides). Choose points in the interiors of all four sides, as described in the parallelogram decomposition lemma, so as to cut off a smaller square of side length x (Fig. 11.7). Then that lemma implies that the area of the $y \times y$ square is equal to the area of the $x \times x$ square plus three terms that are all strictly positive. $\qquad \square$

Lemma 11.5. *Suppose $ABCD$ is a rectangle whose side lengths are $AB = nx$ and $BC = x$ for some positive real number x and positive integer n. Then $\alpha(ABCD) = n \cdot s(x)$.*

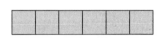

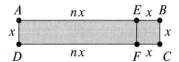

Fig. 11.8. The area of an $x \times nx$ rectangle.   **Fig. 11.9.** Proving the inductive step.

Proof. Let $x > 0$ be a fixed positive number. The intuitive idea behind the proof is shown in Fig. 11.8: an $x \times nx$ rectangle can be decomposed into n squares, each of which has sides of length x. Because this is a statement that something is true for each positive integer n, it is natural to prove it by induction on n.

The base case is $n = 1$. If $ABCD$ satisfies the hypotheses with $n = 1$, it follows that $ABCD$ is actually a square, and the result is true by definition of $s(x)$.

Now assume the theorem is true for some positive integer n, and let $ABCD$ be a rectangle whose side lengths are $AB = (n + 1)x$ and $BC = x$. Choose points E and F such that $A * E * B$ and $D * F * C$, with $AE = DF = nx$ (Fig. 11.9). Then $AEFD$ and $EBCF$ are convex polygons by the polygon splitting theorem, and they are rectangles by Theorem 10.27. Since $AEFD$ has side lengths nx and x, while $EBCF$ is a square with side length x, the inductive hypothesis and the convex decomposition lemma yield

$$\alpha(ABCD) = \alpha(AEFD) + \alpha(EBCF) = n \cdot s(x) + s(x) = (n + 1)s(x).$$

This completes the inductive step and thus the proof. $\qquad\square$

Lemma 11.6. *Let x be any positive real number. If n is a positive integer, then $s(nx) = n^2 s(x)$.*

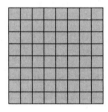

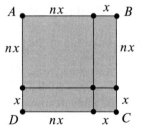

Fig. 11.10. The area of an $nx \times nx$ square.   **Fig. 11.11.** The inductive step of the proof.

Proof. Once again, the intuitive idea is to decompose an $nx \times nx$ square into n^2 smaller $x \times x$ squares (Fig. 11.10), and once again the official proof is by induction on n.

When $n = 1$, the theorem is true because $s(x) = s(x)$. Assume the theorem is true for some positive integer n, and let $ABCD$ be a square whose side lengths are $(n + 1)x$ (Fig. 11.11). We can choose points in all four sides as in Lemma 11.3, so as to cut off a smaller square of side length nx (Fig. 11.11). This decomposes $ABCD$ into an $nx \times nx$ square, an $x \times x$ square, and two $nx \times x$ rectangles. The area of the $nx \times nx$ square is $n^2 s(x)$ by the inductive hypothesis; the area of the $x \times x$ square is $s(x)$ by definition of s; and both rectangles have area $n \cdot s(x)$ by Lemma 11.5. Therefore, the parallelogram decomposition lemma gives

$$\alpha(ABCD) = n^2 s(x) + s(x) + 2n \cdot s(x)$$
$$= (n+1)^2 s(x).$$

This completes the inductive step and thus the proof.   □

The next proof is also rated PG.

Theorem 11.7 (Area of a Square). *The area of a square of side length x is x^2.*

Proof. The claim is that $s(x) = x^2$ for every positive real number x. We will prove this in several steps of increasing generality.

STEP 1: *x is a positive integer.* In this case, we write $x = n$ and use Lemma 11.6 to conclude that $s(n) = s(n \cdot 1) = n^2 s(1) = n^2$.

STEP 2: *x is a positive rational number.* Write $x = m/n$, where m and n are positive integers. Let $\mathscr{P}$ be any square of side length m; we will prove that $s(m/n) = (m/n)^2$ by computing the area of the auxiliary square $\mathscr{P}$ in two different ways (see Fig. 11.12).

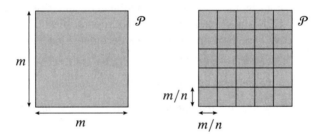

Fig. 11.12. Computing the area of an $m \times m$ square in two ways.

On the one hand, the result of Step 1 tells us that

$$\alpha(\mathscr{P}) = s(m) = m^2. \tag{11.2}$$

On the other hand, we can think of $\mathscr{P}$ as being decomposed into n^2 smaller squares of side lengths m/n. This amounts to writing $m = n(m/n)$ and using Lemma 11.6 to conclude that

$$\alpha(\mathscr{P}) = s\big(n(m/n)\big) = n^2 s(m/n). \tag{11.3}$$

Equating (11.2) and (11.3) and solving for $s(m/n)$, we obtain $s(m/n) = (m/n)^2$ as desired.

STEP 3: *x is any positive real number.* By trichotomy, there are three possibilities: $s(x) < x^2$, $s(x) > x^2$, or $s(x) = x^2$. First suppose $s(x) < x^2$. Then because $s(x)$ and x are both positive, it follows that $\sqrt{s(x)} < x$. Thus we can choose some rational number r such that $\sqrt{s(x)} < r < x$ and therefore $s(x) < r^2 < x^2$. On the one hand, Step 2 ensures us that $s(r) = r^2$, so by substitution we conclude that $s(x) < s(r)$. On the other hand, the fact that $r < x$ guarantees that $s(r) < s(x)$ by Lemma 11.4. This is a contradiction, which rules out the possibility that $s(x) < x^2$. The same argument, with the inequalities reversed, rules out $s(x) > x^2$, so the only remaining possibility is $s(x) = x^2$.   □

Theorem 11.8 (Area of a Rectangle). *The area of a rectangle is the product of the lengths of any two adjacent sides.*

Proof. Let $PQRS$ be a rectangle with side lengths x and y. We need to show that $\alpha(PQRS) = xy$. Since any two rectangles with the same side lengths are congruent by SASAS, it suffices to compute the area of *some* rectangle with the same side lengths.

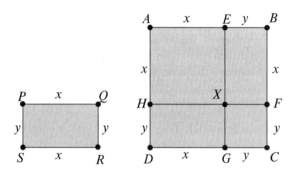

Fig. 11.13. Area of a rectangle.

Let $ABCD$ be a square of side length $x + y$. Choose points E, F, G, H as in the parallelogram decomposition lemma such that $AE = DG = AH = BF = x$, and let X be the point where $\overline{EG}$ and $\overline{HF}$ intersect (Fig. 11.13). This decomposes $ABCD$ into a square $AEXH$ with side length x, a square $XFCG$ with side length y, and two rectangles $EBFX$ and $HXGD$, both of which have side lengths x and y. On the one hand, $\alpha(ABCD)$ is equal to $(x + y)^2$ by Theorem 11.7. On the other hand, the parallelogram decomposition lemma shows that it is equal to the following sum:

$$\alpha(ABCD) = x^2 + y^2 + \alpha(EBFX) + \alpha(HXGD) = x^2 + y^2 + 2\alpha(HXGD).$$

Now set these two expressions for $\alpha(ABCD)$ equal to each other and solve for $\alpha(HXGD)$, giving

$$\alpha(HXGD) = \tfrac{1}{2}\big((x + y)^2 - x^2 - y^2\big) = xy. \qquad □$$

Next we consider triangles.

Lemma 11.9 (Area of a Right Triangle). *The area of a right triangle is one-half of the product of the lengths of its legs.*

Proof. For any positive real numbers x and y, since any two right triangles with leg lengths x and y are congruent by SAS, it suffices to compute the area of one such triangle. Let $ABCD$ be a rectangle with side lengths x and y (Fig. 11.14). Then the diagonal AC cuts $ABCD$ into two congruent right triangles, each with leg lengths x and y. Therefore, by the convex decomposition lemma, $\alpha(ABCD)$ is equal to twice the area of either triangle. Since $\alpha(ABCD) = xy$, it follows that each triangle has area $\tfrac{1}{2}xy$ as claimed.   □

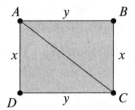

Fig. 11.14. The area of a right triangle.

Suppose $\triangle ABC$ is a triangle. Choose any side of $\triangle ABC$, and for the moment call that side the ***base of the triangle***. We define the ***height of the triangle*** corresponding to the chosen base to be the distance from the vertex opposite the base to the line containing the base. (Recall from Chapter 7 that a perpendicular segment from the opposite vertex to the line containing the base is called an ***altitude of the triangle***. Thus the height can also be characterized as the length of the altitude to the chosen base.)

Theorem 11.10 (Area of a Triangle). *The area of a triangle is equal to one-half the length of any base multiplied by the corresponding height.*

Proof. Exercise 11A.   □

The next corollary is Euclid's Proposition I.37. One way to remember what it says is to visualize a triangle with a fixed base, but with its top vertex allowed to slide along a line parallel to the base.

Corollary 11.11 (Triangle Sliding Theorem). *Suppose $\triangle ABC$ and $\triangle A'BC$ are triangles with a common side $\overline{BC}$, such that A and A' both lie on a line parallel to $\overleftrightarrow{BC}$ (Fig. 11.15). Then $\alpha(\triangle ABC) = \alpha(\triangle A'BC)$.*

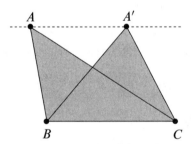

Fig. 11.15. The triangle sliding theorem.

Proof. Exercise 11B.   □

The previous corollary concerned triangles that share a side. The next one, which we will use in the next chapter when we study similar triangles, concerns instead triangles that share a vertex. It is Euclid's first proposition in Book VI.

Corollary 11.12 (Triangle Area Proportion Theorem). *Suppose $\triangle ABC$ and $\triangle AB'C'$ are triangles with a common vertex A, such that the points B, C, B', C' are collinear (Fig. 11.16). Then*

$$\frac{\alpha(\triangle ABC)}{\alpha(\triangle AB'C')} = \frac{BC}{B'C'}.$$

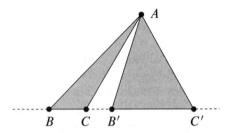

Fig. 11.16. The triangle area proportion theorem.

Proof. Exercise 11C. □

Finally, we consider trapezoids and parallelograms. Suppose $ABCD$ is a trapezoid. The two parallel sides of $ABCD$ are called the ***bases of the trapezoid***. Because parallel lines are equidistant, it makes sense to define the ***height of the trapezoid*** to be the distance between the lines containing the bases.

Since parallelograms are special cases of trapezoids, we can apply this definition to parallelograms as well. In this case, we can choose to call either pair of opposite sides the ***bases of the parallelogram***, and then the corresponding ***height*** is the distance between the lines containing those two parallel bases.

Theorem 11.13 (Area of a Trapezoid). *The area of a trapezoid is the average of the lengths of the bases multiplied by the height.*

Proof. Exercise 11D. □

Corollary 11.14 (Area of a Parallelogram). *The area of a parallelogram is the length of any base multiplied by the corresponding height.*

Proof. A parallelogram is a trapezoid in which both bases have the same length. □

Is the Area Postulate Independent of the Others?

We have observed that each of the postulates of Euclidean geometry that we introduced prior to this chapter is independent of the preceding ones, by showing that there is a model of the preceding postulates in which the new postulate is false (in addition to the Cartesian model, in which all of the postulates of Euclidean geometry are true). Can we do the same thing for the area postulate?

Somewhat surprisingly, it turns out that the Euclidean area postulate is *not* independent of the preceding ones: it follows logically from Postulates 1–9 and 10E that there exists a unique area function that assigns the area 1 to a unit square. We will not go into all the details of the X-rated proof, but versions of it can be found in [**MP91**, Section 10.2] and [**Har00**, Section 23]. Here is the outline of the argument:

STEP 1: We can *define* the area of a triangle to be $\frac{1}{2}bh$, where b is the length of any base and h is the corresponding height. It is possible to prove that this formula yields the same number regardless of which base is chosen.

STEP 2: It is possible to prove that every polygonal region has an admissible decomposition into triangular regions (called a *triangulation* of the region). We then wish to define the area of a polygonal region to be the sum of the areas of the triangular regions in any triangulation. In order to do so, it is necessary to prove that two different triangulations of the same polygonal region will necessarily yield the same area. Basically, this is done by showing that there is a third triangulation with the property that its triangles form admissible decompositions of each of the triangles in either of the original triangulations.

STEP 3: Once it is known that this function is well defined, it is straightforward to prove that it is in fact an area function that assigns the area 1 to a unit square and that it is the unique such function.

Exercises

11A. Prove Theorem 11.10 (the area of a triangle). [Be careful: you have to prove that the area formula holds no matter which base is chosen. There are several cases to consider, depending on whether the altitude meets the base at an interior point, at a vertex, or not at all.]

11B. Prove Corollary 11.11 (the triangle sliding theorem).

11C. Prove Corollary 11.12 (the triangle area proportion theorem).

11D. Prove Theorem 11.13 (the area of a trapezoid). [Hint: Use a diagonal to decompose the trapezoid into triangles.]

11E. Suppose $ABCD$ is a parallelogram and E, F, G, H are points satisfying the hypotheses of Lemma 11.3, and in addition suppose that the point X where $\overline{HF}$ meets $\overline{EG}$ lies on the diagonal $\overline{AC}$ (see Fig. 11.17). What is the relationship between $\alpha(EBFX)$ and $\alpha(GDHX)$? Prove your answer correct.

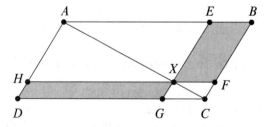

Fig. 11.17. The setup for Exercise 11E.

11F. Recall from Exercise 9F that a *kite* is a convex quadrilateral whose four sides consist of two pairs of congruent adjacent sides. Find a formula for the area of a kite in terms of the lengths of its diagonals, and prove your answer correct.

11G. Let $ABCD$ be a trapezoid with bases of lengths a and b ($a < b$) and with both angles adjacent to the base of length b measuring $45°$ (Fig. 11.18). Find a formula for $\alpha(ABCD)$ in terms of a and b, and prove your answer correct.

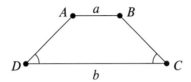

Fig. 11.18. The trapezoid of Exercise 11G.

11H. The two rectangles in Fig. 11.19 illustrate a famous "paradox" that demonstrates the risks of drawing conclusions from diagrams. The picture seems to show admissible decompositions of an 8×8 square and a 5×13 rectangle, with each figure decomposed into two triangles of area 12 and two trapezoids of area 20. Thus the area addition property implies that both figures have area $12 + 12 + 20 + 20 = 64$. But this contradicts Theorem 11.8, which implies that the 5×13 rectangle should have area 65. Find the flaw in this reasoning, and fix it: give a precise description of the decompositions shown in the two pictures and then use theorems in this chapter to compute the areas of the four subregions in each figure and verify that the area addition property holds in each case.

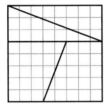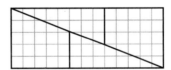

Fig. 11.19. An area paradox.

Chapter 12

Similarity

The idea of "scaling" geometric objects is ubiquitous in our everyday experience. When you draw a map or enlarge a photo or instruct your computer to use a different font size, you are creating a new geometric object that has the "same shape" as the old one but has all of its parts reduced or enlarged in size proportionally. In geometry, roughly speaking, two figures that have the same shape but not necessarily the same size are said to be *similar* to each other.

Definitions

Before we give a more precise definition of similarity, let us review some standard terminology regarding ratios and proportions, because these terms are frequently confused in everyday speech. If x and y are positive real numbers, the **ratio of x to y** is just the quotient x/y. A **proportion** is an equality between two ratios, such as $x_1/y_1 = x_2/y_2$. Two ordered pairs of positive numbers (x_1, y_1) and (x_2, y_2) are said to be **proportional** if their ratios are equal: $x_1/y_1 = x_2/y_2$. Simple algebra shows that this equation can be written in several equivalent ways:

$$\frac{x_1}{y_1} = \frac{x_2}{y_2} \quad \Leftrightarrow \quad \frac{y_1}{x_1} = \frac{y_2}{x_2} \quad \Leftrightarrow \quad \frac{x_1}{x_2} = \frac{y_1}{y_2} \quad \Leftrightarrow \quad \frac{x_2}{x_1} = \frac{y_2}{y_1} \quad \Leftrightarrow \quad x_1 y_2 = x_2 y_1.$$

We also say that three or more ordered pairs $(x_1, y_1), \ldots, (x_n, y_n)$ are proportional when their ratios are all equal: $x_1/y_1 = x_2/y_2 = \cdots = x_n/y_n$.

Here is our official definition of similarity. If $\mathcal{P}_1$ and $\mathcal{P}_2$ are polygons, we say they are **similar polygons** if there is a correspondence between their vertices such that consecutive vertices correspond to consecutive vertices, all pairs of corresponding interior angle measures are equal, and all pairs of corresponding side lengths are proportional. In particular, congruent polygons are always similar. The notation $\mathcal{P}_1 \sim \mathcal{P}_2$ means that $\mathcal{P}_1$ is similar to $\mathcal{P}_2$. When we name the polygons by listing their vertices, the order of the vertices indicates the correspondence, just as in the notation for congruence.

For example, if $\triangle ABC$ and $\triangle DEF$ are triangles, the notation $\triangle ABC \sim \triangle DEF$ means that $\triangle ABC$ is similar to $\triangle DEF$ under the correspondence $A \leftrightarrow D$, $B \leftrightarrow E$, and

$C \leftrightarrow F$ (Fig. 12.1), which is the case if and only if all of the following conditions hold:

$$\angle A \cong \angle D, \qquad \angle B \cong \angle E, \qquad \angle C \cong \angle F,$$

$$\frac{AB}{DE} = \frac{AC}{DF} = \frac{BC}{EF}.$$

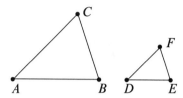

Fig. 12.1. Similar triangles.

Suppose $\mathcal{P}_1 \sim \mathcal{P}_2$, and let r denote the common ratio of lengths of sides of $\mathcal{P}_1$ to those of $\mathcal{P}_2$. Then it follows from the definition of similarity that each side of $\mathcal{P}_1$ has length equal to r times that of the corresponding side of $\mathcal{P}_2$. This common ratio r is called the **scale factor** of the similarity, or the **constant of proportionality**. The next theorem is essentially Euclid's Proposition VI.21. Our proof is similar in spirit to his but is translated into more algebraic language.

Theorem 12.1 (Transitive Property of Similarity). *Two polygons that are both similar to a third polygon are similar to each other.*

Proof. Suppose $\mathcal{P} = A_1 \ldots A_n$, $\mathcal{P}' = A'_1 \ldots A'_n$, and $\mathcal{P}'' = A''_1 \ldots A''_n$ are polygons such that $A_1 \ldots A_n \sim A'_1 \ldots A'_n$ and $A'_1 \ldots A'_n \sim A''_1 \ldots A''_n$. For each i, the interior angle measure at A_i is equal to that at A'_i, which in turn is equal to that at A''_i, so it follows from transitivity of equality that the interior angle measures at A_i and A''_i are equal. The hypothesis implies that there are positive numbers r and s such that the following equations hold for each i:

$$A_i A_{i+1} = r \cdot A'_i A'_{i+i},$$

$$A'_i A'_{i+1} = s \cdot A''_i A''_{i+i}.$$

By substitution, therefore, we conclude that

$$A_i A_{i+i} = rs \cdot A''_i A''_{i+1}.$$

This shows that all pairs of corresponding sides of $\mathcal{P}$ and $\mathcal{P}''$ are proportional, so the two triangles are similar. □

The Side-Splitter Theorem

Our first main goal is to develop some shortcuts for proving similarity of triangles, analogous to congruence theorems like SAS and AAS. The main tool for proving that such shortcuts work will be the following theorem, which shows that a line parallel to one side of a triangle cuts off proportional segments from the other two sides. This is the first half of Euclid's Proposition VI.2.

Theorem 12.2 (The Side-Splitter Theorem). *Suppose $\triangle ABC$ is a triangle and ℓ is a line parallel to $\overleftrightarrow{BC}$ that intersects $\overline{AB}$ at an interior point D (Fig. 12.2). Then ℓ also intersects $\overline{AC}$ at an interior point E, and the following proportions hold:*

$$\frac{AD}{AB} = \frac{AE}{AC} \quad and \quad \frac{AD}{DB} = \frac{AE}{EC}. \tag{12.1}$$

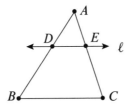

Fig. 12.2. The side-splitter theorem.

Proof. Because ℓ is parallel to $\overleftrightarrow{BC}$, it does not contain B or C; and because it intersects $\overline{AB}$ at an interior point, it does not contain A either. Therefore, Pasch's theorem guarantees that ℓ also intersects the interior of one other side of $\triangle ABC$. Since it cannot intersect $\overline{BC}$, it must intersect the interior of $\overline{AC}$. Let E denote the intersection point.

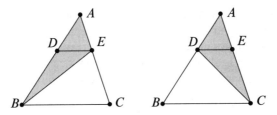

Fig. 12.3. Proof of the side-splitter theorem.

Draw $\overline{BE}$, and consider $\triangle AEB$ (Fig. 12.3). From the triangle area proportion theorem (Corollary 11.12), we conclude that

$$\frac{\alpha(\triangle ADE)}{\alpha(\triangle ABE)} = \frac{AD}{AB}. \tag{12.2}$$

Similarly, drawing $\overline{DC}$ and considering $\triangle ADC$, we obtain

$$\frac{\alpha(\triangle ADE)}{\alpha(\triangle ADC)} = \frac{AE}{AC}. \tag{12.3}$$

It follows from the convex decomposition theorem that

$$\alpha(\triangle ABE) = \alpha(\triangle ADE) + \alpha(\triangle DEB),$$
$$\alpha(\triangle ADC) = \alpha(\triangle ADE) + \alpha(\triangle DEC). \tag{12.4}$$

The triangle sliding theorem (Corollary 11.11) gives $\alpha(\triangle DEB) = \alpha(\triangle DEC)$. Substituting this into (12.4), we find

$$\alpha(\triangle ABE) = \alpha(\triangle ADC). \tag{12.5}$$

Combining (12.2), (12.3), and (12.5), we obtain

$$\frac{AD}{AB} = \frac{\alpha(\triangle ADE)}{\alpha(\triangle ABE)} = \frac{\alpha(\triangle ADE)}{\alpha(\triangle ADC)} = \frac{AE}{AC},$$

which proves the first equation in (12.1).

To prove the second equation, we invert both sides of the first equation in (12.1) and use the fact that D and E are interior points on their respective sides to obtain

$$\frac{AD + DB}{AD} = \frac{AE + EC}{AE}.$$

Subtracting 1 from both sides and simplifying, we obtain $DB/AD = EC/AE$, which is equivalent to the second equation in (12.1).   □

We have followed the lead of Euclid (as do most high-school texts) in using the theory of area to prove this theorem. In fact, it is possible to give a proof that does not use areas at all (and therefore does not require the area postulate)—see [**Ven05**, Chapter 7] for such a proof. That proof is considerably more involved than the one we have given here; because of that and because the area-based proof is perfectly rigorous and is much more likely to be found in high-school texts, we have chosen to stick with Euclid's approach. There is little to be gained by avoiding the use of area in the treatment of similar triangles.

Triangle Similarity Theorems

The next theorem is the most important criterion for similarity. The closest Euclid comes to this theorem is his Proposition VI.4. His statement is slightly weaker—he assumes all three pairs of corresponding angles are congruent, not just two—but, of course, once we have two congruences, the third follows from the angle-sum theorem. Our proof is somewhat different from Euclid's, but they are both based on the side-splitter theorem.

Theorem 12.3 (AA Similarity Theorem). *If there is a correspondence between the vertices of two triangles such that two pairs of corresponding angles are congruent, then the triangles are similar under that correspondence.*

Proof. Suppose $\triangle ABC$ and $\triangle DEF$ are triangles such that $\angle A \cong \angle D$ and $\angle B \cong \angle E$. By the angle-sum theorem for triangles, the measure of the third angle in each triangle is equal to $180°$ minus the sum of the measures of the other two, from which it follows that $\angle C \cong \angle F$. It remains only to prove that

$$\frac{AB}{DE} = \frac{AC}{DF} = \frac{BC}{EF}. \tag{12.6}$$

If any one of the ratios in (12.6) is equal to 1, then $\triangle ABC$ is congruent to $\triangle DEF$ by ASA, and the theorem is true because all three ratios are equal to 1. So let us suppose that all of these ratios are different from 1. We will prove the first equality in (12.6); the proof of the other equality is exactly analogous.

Since $AB \neq DE$, one of them is larger—say, $DE > AB$. Choose a point P in the interior of $\overline{DE}$ such that $\overline{DP} \cong \overline{AB}$, and let ℓ be the line through P and parallel to $\overleftrightarrow{EF}$ (Fig. 12.4). It follows from the side-splitter theorem that ℓ intersects $\overline{DF}$ at an interior point Q and that

$$\frac{DP}{DE} = \frac{DQ}{DF}. \tag{12.7}$$

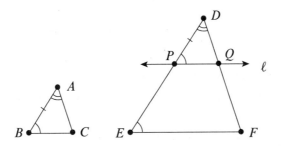

Fig. 12.4. Proof of the AA similarity theorem.

The Y-lemma shows that Q and F are on the same side of $\overleftrightarrow{DE}$, so $\angle DPQ$ and $\angle E$ are corresponding angles for the transversal $\overleftrightarrow{DE}$ to ℓ and $\overleftrightarrow{EF}$. By the converse to the corresponding angles theorem, $\angle DPQ \cong \angle E$, which by hypothesis is congruent in turn to $\angle B$. Since $\angle D \cong \angle A$ by hypothesis and $\overline{DP} \cong \overline{AB}$ by construction, we have $\triangle DPQ \cong \triangle ABC$ by ASA. Substituting $DP = AB$ and $DQ = AC$ into (12.7), we obtain the first equation in (12.6). $\qquad\square$

This is one of the most important theorems in Euclidean geometry. Nearly every geometry book has some version of it, although some high-school textbooks take it as an additional postulate instead of proving it.

The next theorem shows that similar triangles can be readily constructed in Euclidean geometry once a new size is chosen for one of the sides. It is an analogue for similar triangles of the triangle copying theorem, Theorem 5.10. (Euclid's Proposition VI.18 gives a similar result with a similar proof but stated for more general polygons, not just triangles.)

Theorem 12.4 (Similar Triangle Construction Theorem). *If $\triangle ABC$ is a triangle and $\overline{DE}$ is any segment, then on each side of $\overleftrightarrow{DE}$, there is a point F such that $\triangle ABC \sim \triangle DEF$.*

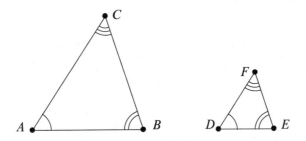

Fig. 12.5. The similar triangle construction theorem.

Proof. The AAA construction theorem (Theorem 10.17) shows that on either side of $\overleftrightarrow{DE}$, there is a point F such that $\triangle DEF$ has angle measures satisfying $m\angle D = m\angle A$, $m\angle E = m\angle B$, and $m\angle F = m\angle C$. By the AA similarity theorem, $\triangle ABC \sim \triangle DEF$. $\qquad\square$

It is important to observe that although Theorem 5.10 showed that it is always possible in neutral geometry to construct a triangle *congruent* to a given one, this construction of *similar* triangles only works in Euclidean geometry because it uses Euclid's Postulate 5 and the AA similarity theorem. In fact, we will see later that in hyperbolic geometry, it is impossible to construct noncongruent similar triangles!

Here are some other useful criteria for similarity. These are Euclid's Propositions VI.5 and VI.6, respectively.

Theorem 12.5 (SSS Similarity Theorem). *If $\triangle ABC$ and $\triangle DEF$ are triangles such that $AB/DE = AC/DF = BC/EF$, then $\triangle ABC \sim \triangle DEF$.*

Proof. Exercise 12A.   □

Theorem 12.6 (SAS Similarity Theorem). *If $\triangle ABC$ and $\triangle DEF$ are triangles such that $\angle A \cong \angle D$ and $AB/DE = AC/DF$, then $\triangle ABC \sim \triangle DEF$.*

Proof. Exercise 12B.   □

The next theorem describes a common situation in which similar triangles arise.

Theorem 12.7 (Two Transversals Theorem). *Suppose ℓ and ℓ' are parallel lines, and m and n are two distinct transversals to ℓ and ℓ' meeting at a point X that is not on either ℓ or ℓ'. Let M and N be the points where m and n, respectively, meet ℓ; and let M' and N' be the points where they meet ℓ' (Fig. 12.6). Then $\triangle XMN \sim \triangle XM'N'$.*

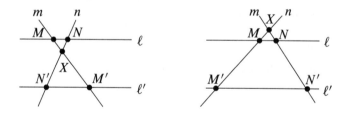

Fig. 12.6. Two possible configurations for the two transversals theorem.

Proof. Because X does not lie on either ℓ or ℓ', it follows that the three collinear points M, M', and X are all distinct. Thus one of them is between the other two.

CASE 1: $M * X * M'$. In this case, we will show that $N * X * N'$ as well. Suppose one of the other possible betweenness relations holds, for example $X * N * N'$. This implies that X and N' are on opposite sides of ℓ. However, the assumption $M * X * M'$ implies that X and M' are on the same side of ℓ (by the Y-lemma), and Lemma 7.18 implies that M' and N' are on the same side of ℓ. Thus X and N' are on the same side of ℓ, which is a contradiction. A similar contradiction follows from $X * N' * N$, so the only possibility is $N * X * N'$.

Because N and N' are on opposite sides of m, it follows that $\angle NMX$ and $\angle N'M'X$ are alternate interior angles for the transversal m to ℓ and ℓ'. Thus $\angle NMX \cong \angle N'M'X$ by the converse to the alternate interior angles theorem. The same argument shows that $\angle MNX \cong \angle M'N'X$. Therefore, $\triangle XMN \sim \triangle XM'N'$ by AA.

CASE 2: $X * M * M'$ or $X * M' * M$. The argument is the same in both cases, so we may as well assume the former. In this case, $\triangle XM'N'$ is a triangle and M is an interior point on the side $\overline{XM'}$. Because $\ell \parallel \ell'$, it follows from the side-splitter theorem that N is an interior point of $\overline{XN'}$ and $XM/XM' = XN/XN'$. Thus $\triangle XMN \sim \triangle XM'N'$ by the SAS similarity theorem. $\qquad\square$

The midsegment theorem (Theorem 10.30) showed that any segment that bisects two sides of a triangle is parallel to the third side. The next theorem is a generalization of that: it says that any segment that cuts two sides of a triangle in the same proportion is parallel to the third side. It is the second half of Euclid's Proposition VI.2.

Theorem 12.8 (Converse to the Side-Splitter Theorem). *Suppose $\triangle ABC$ is a triangle and D and E are interior points on $\overline{AB}$ and $\overline{AC}$, respectively, such that*

$$\frac{AD}{AB} = \frac{AE}{AC}.$$

Then $\overleftrightarrow{DE}$ is parallel to $\overleftrightarrow{BC}$.

Proof. Exercise 12C. $\qquad\square$

Proportion Theorems

Similar triangles can be used to prove many important theorems about proportionality. Here are two of them. The first one says that the bisector of an angle of a triangle divides the opposite side into segments that are proportional to the other two sides. Its proof is rated PG. This is the first half of Euclid's Proposition VI.3, with essentially the same proof. (The second half is the converse; see Exercise 12D.)

Theorem 12.9 (Angle Bisector Proportion Theorem). *Suppose $\triangle ABC$ is a triangle and D is the point where the bisector of $\angle BAC$ meets $\overline{BC}$ (Fig. 12.7). Then*

$$\frac{BD}{DC} = \frac{AB}{AC}. \tag{12.8}$$

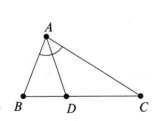

Fig. 12.7. The angle bisector proportion theorem.

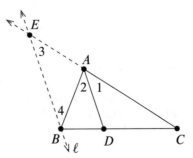

Fig. 12.8. Proof of the theorem.

Proof. Let ℓ be the line through B and parallel to $\overleftrightarrow{AD}$. By Proclus's lemma, since $\overleftrightarrow{AC}$ intersects $\overleftrightarrow{AD}$, it also intersects ℓ; let E denote the intersection point. The side-splitter theorem implies that A is an interior point of $\overline{EC}$ and that

$$\frac{AC}{EA} = \frac{DC}{BD}. \tag{12.9}$$

Let $\angle 1$, $\angle 2$, $\angle 3$, and $\angle 4$ be as shown in Fig. 12.8. Because $E * A * C$, it follows that C is on the opposite side of $\overleftrightarrow{AD}$ from E; and the Y-lemma applied to $\overleftrightarrow{CB}$ shows that B and D are on the same side of $\overleftrightarrow{EA}$. Thus $\angle 3$ and $\angle 1$ are corresponding angles for the transversal $\overleftrightarrow{AE}$ to the parallel lines $\overleftrightarrow{AD}$ and $\overleftrightarrow{EB}$, so $\angle 3 \cong \angle 1$.

On the other hand, betweenness vs. betweenness shows that $\overrightarrow{BA}$ is between $\overrightarrow{BE}$ and $\overrightarrow{BC}$, which implies that D and E are on opposite sides of $\overleftrightarrow{AB}$; thus $\angle 2$ and $\angle 4$ are alternate interior angles for the transversal $\overleftrightarrow{AB}$ to the same two parallel lines, which implies that $\angle 2 \cong \angle 4$. Combining the two angle congruences that we have just proved with the fact that $\angle 1 \cong \angle 2$ (by definition of angle bisector), we obtain $\angle 3 \cong \angle 1 \cong \angle 2 \cong \angle 4$. The converse to the isosceles triangle theorem then implies that $EA = AB$. Substituting this into (12.9), we obtain the following equation, which is equivalent to (12.8):

$$\frac{AC}{AB} = \frac{DC}{BD}. \qquad \square$$

Theorem 12.10 (Parallel Projection Theorem). *Suppose ℓ, m, n, t, and t' are distinct lines such that $\ell \parallel m \parallel n$; t intersects ℓ, m, and n at A, B, and C, respectively; and t' intersects the same three lines at A', B', and C', respectively. Then*

(a) $\dfrac{AB}{BC} = \dfrac{A'B'}{B'C'}.$

(b) *$A * B * C$ if and only if $A' * B' * C'$.*

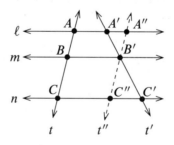

Fig. 12.9. Proof of the parallel projection theorem.

Proof. First we prove (a). If $B = B'$, then the theorem follows immediately from the two transversals theorem applied to t and t', so assume that B and B' are distinct points. Let t'' be the line through B' and parallel to t; by Proclus's lemma, there are points A'' and C'' where t'' intersects ℓ and n, respectively. (See Fig. 12.9. In the figure, B' is pictured between A' and C', but the proof works equally well if the points are ordered in other ways.) Then $AA''B'B$ and $BB'C''C$ are parallelograms by the parallelogram lemma, from which it follows that $AB = A''B'$ and $BC = B'C''$. Therefore, to prove the theorem, it suffices to show that

$$\frac{A''B'}{B'C''} = \frac{A'B'}{B'C'}. \tag{12.10}$$

If $C'' = C'$, then $t'' = t'$, and the result follows trivially. On the other hand, if $C'' \neq C'$, then we can apply the two transversals theorem to t' and t'' and obtain (12.10).

To prove (b), assume $A * B * C$. This implies that A and C are on opposite sides of m. On the other hand, it follows from Lemma 7.18 that A and A' are on the same side of m, as are C and C'. Thus A' and C' are on opposite sides of m, so $A' * B' * C'$. The reverse implication is proved in the same way. $\qquad\square$

This theorem is called the "parallel projection theorem" because it suggests parallel light rays projecting images of points of one line onto another line. In this context, part (a) of the theorem says that parallel projection preserves ratios of lengths, and part (b) says that it preserves betweenness.

Collinearity and Concurrence Theorems

The properties of similar triangles also yield some useful theorems about intersections between lines and triangles. Admittedly, the next two theorems are not particularly exciting in their own right; but they turn out to be essential tools for proving a number of interesting and surprising geometric results (e.g., Theorem 12.13 and Exercises 14Q and 14X).

The first such theorem was proved around 100 CE by an ancient Greek mathematician named Menelaus of Alexandria. Its proof is rated PG.

Theorem 12.11 (Menelaus's Theorem). *Let $\triangle ABC$ be a triangle. Suppose D, E, F are points different from A, B, C and lying on $\overleftrightarrow{AB}$, $\overleftrightarrow{BC}$, and $\overleftrightarrow{AC}$, respectively, such that either two of the points lie on $\triangle ABC$ or none of them do. Then $D, E,$ and F are collinear if and only if*

$$\left(\frac{AD}{DB}\right)\left(\frac{BE}{EC}\right)\left(\frac{CF}{FA}\right) = 1. \tag{12.11}$$

Remark. To make the statement of Menelaus's theorem easier to remember, it is useful to introduce the following definition: if $\triangle ABC$ is a triangle, a **transversal for** $\triangle ABC$ is a line that intersects each of the edge lines $\overleftrightarrow{AB}$, $\overleftrightarrow{BC}$, and $\overleftrightarrow{CA}$ but none of the vertices A, B, or C. Note that Corollary 5.3 to Pasch's theorem implies that a transversal must meet either two sides of the triangle or none (see Fig. 12.10). The "only if" part of Menelaus's theorem is equivalent to the statement that whenever a transversal meets $\overleftrightarrow{AB}$ at D, $\overleftrightarrow{BC}$ at E, and $\overleftrightarrow{CA}$ at F, then (12.11) is satisfied.

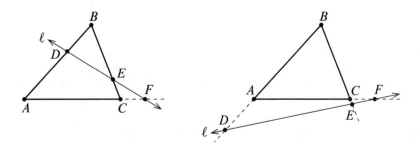

Fig. 12.10. Two types of transversals to a triangle.

Proof of Menelaus's Theorem.   Assume first that D, E, and F are collinear, and let ℓ be the line containing them. Let m be the line through B and parallel to $\overleftrightarrow{AC}$ (Fig. 12.11). (We have drawn the diagram with D and E lying on $\triangle ABC$ but not F; however, the proof works exactly the same in other cases as well.)

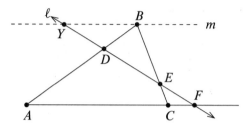

Fig. 12.11. Proof of Menelaus's theorem.

Because ℓ meets $\overleftrightarrow{AC}$ and is distinct from it, Proclus's lemma shows that ℓ also meets m. Let Y be the point of intersection. Now D cannot lie on m because B is the only point where $\overleftrightarrow{AB}$ meets m, and D cannot lie on $\overleftrightarrow{AC}$ because that would imply A, B, C were collinear. Thus the two transversals theorem can be applied to the transversals ℓ and $\overleftrightarrow{AB}$, showing that $\triangle DBY \sim \triangle DAF$, which implies

$$\frac{AD}{DB} = \frac{FA}{BY}. \tag{12.12}$$

A similar argument applied to the transversals ℓ and $\overleftrightarrow{BC}$ shows that $\triangle EBY \sim \triangle ECF$, and therefore

$$\frac{BE}{EC} = \frac{BY}{CF}. \tag{12.13}$$

When we multiply (12.12) and (12.13) together and cancel the BY factors, we obtain an equation equivalent to (12.11).

For the converse, let D, E, F be points satisfying the hypothesis and assume that (12.11) holds. At least one of the points D, E, F does not lie on $\triangle ABC$; assume without loss of generality that F is such a point. First, we will show that $\overleftrightarrow{DE}$ is not parallel to $\overleftrightarrow{AC}$. Suppose for the sake of contradiction that these two lines are parallel. Let m be the line through B and parallel to $\overleftrightarrow{DE}$ and $\overleftrightarrow{AC}$. The parallel projection theorem applied to the three parallel lines m, $\overleftrightarrow{DE}$, and $\overleftrightarrow{AC}$ implies

$$\frac{AD}{DB} = \frac{CE}{EB}. \tag{12.14}$$

When we substitute this into (12.11) and simplify, we obtain $CF/FA = 1$, which means that F is equidistant from A and C. Since F lies on $\overleftrightarrow{AC}$ by hypothesis, it follows from Lemma 3.26(b) that F is the midpoint of $\overline{AC}$, contradicting our assumption that $F \notin \triangle ABC$. Thus $\overleftrightarrow{DE}$ and $\overleftrightarrow{AC}$ are not parallel.

Let F' be the point where $\overleftrightarrow{DE}$ and $\overleftrightarrow{AC}$ intersect. Since D, E, and F' are collinear, the "only if" part of Menelaus's theorem (proved above) implies

$$\left(\frac{AD}{DB}\right)\left(\frac{BE}{EC}\right)\left(\frac{CF'}{F'A}\right) = 1.$$

Combining this with (12.11) yields

$$\frac{CF'}{F'A} = \frac{CF}{FA}. \tag{12.15}$$

Now, the fact that $F, F' \notin \overline{AC}$ means we have either $F * C * A$ or $F * A * C$, and either $F' * C * A$ or $F' * A * C$. Without loss of generality, we can assume $F * C * A$. Then $CF < FA$ (the whole segment is greater than the part), and (12.15) implies $CF' < F'A$. Since either $F' * C * A$ or $F' * A * C$ implies that A and C lie in the same ray starting at F', the inequality $CF' < F'A$ implies $F' * C * A$ by the ordering lemma for points. It follows from the betweenness theorem for points that

$$FC + CA = FA \qquad \text{and} \qquad F'C + CA = F'A. \tag{12.16}$$

Thus when we subtract 1 from both sides of (12.15) and use (12.16), we obtain

$$\frac{CF' - F'A}{F'A} = \frac{CF - FA}{FA},$$
$$\frac{-CA}{F'A} = \frac{-CA}{FA},$$

from which it follows that $F'A = FA$. Since both F and F' lie in the ray opposite $\overrightarrow{CA}$, the unique point theorem implies $F = F'$, and thus D, E, F are collinear. $\qquad \square$

Our next theorem is a kind of complement to Menelaus's theorem, in which the roles of points and lines are interchanged. It was discovered much later, though—the first proof was published in the eleventh century by the Arab mathematician Al-Mu'taman ibn Hud (who also happened to be the king of Saragossa, Spain!) [**Hog05**]. Western tradition has named it after the Italian mathematician Giovanni Ceva (pronounced "*chay-va*"), who independently discovered a proof in 1678. The following standard terminology honors Ceva's discovery: a segment that joins a vertex of a triangle to an interior point on the opposite side is called a ***cevian*** ("*chay*-vi-an") of the triangle. Two or more lines, rays, or segments are said to be ***concurrent*** if there is a point that lies on all of them.

Theorem 12.12 (Ceva's Theorem). *Suppose $\triangle ABC$ is a triangle and D, E, F are points in the interiors of $\overline{AB}$, $\overline{BC}$, and $\overline{CA}$, respectively (see Fig. 12.12). Then the three cevians $\overline{AE}$, $\overline{BF}$, and $\overline{CD}$ are concurrent if and only if*

$$\left(\frac{AD}{DB}\right)\left(\frac{BE}{EC}\right)\left(\frac{CF}{FA}\right) = 1. \tag{12.17}$$

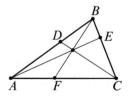

Fig. 12.12. Ceva's theorem.

Proof. Exercise 12H. $\qquad \square$

When the cevians are the medians of the triangle, we can say more.

Theorem 12.13 (Median Concurrence Theorem). *The medians of a triangle are concurrent, and the distance from the point of intersection to each vertex is twice the distance to the midpoint of the opposite side.*

Proof. Exercise 12I.   □

The point where the medians meet is called the ***centroid*** of the triangle. It is shown in physics and calculus courses that a physical triangle made of a uniform thin sheet of rigid material would balance if it were supported only at its centroid (see Fig. 12.13).

Fig. 12.13. The centroid of a triangle.

The Golden Ratio

Suppose $ABCD$ is a rectangle in which $AB > BC$ (see Fig. 12.14). We can choose points X and Y on the two long sides $\overline{AB}$ and $\overline{CD}$ in such a way that $AX = DY = AD$, and then $AXYD$ is a square and $BCYX$ is another rectangle. If it happens that $BCYX$ is similar to the original rectangle $ABCD$, then $ABCD$ is called a ***golden rectangle***.

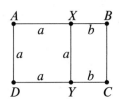

Fig. 12.14. A golden rectangle.

Theorem 12.14 (Golden Rectangle Theorem). *A rectangle is a golden rectangle if and only if the ratio of the longer side length to the shorter one is equal to the following number, called the **golden ratio**:*

$$\varphi = \frac{1 + \sqrt{5}}{2}. \tag{12.18}$$

Proof. Let $ABCD$ be a rectangle, and let a and b denote the lengths shown in Fig. 12.14. It follows from the definition that $ABCD$ is a golden rectangle if and only if

$$\frac{a}{b} = \frac{a+b}{a}. \tag{12.19}$$

Suppose first that $ABCD$ is a golden rectangle, so (12.19) is satisfied. This equation can also be written in the form $a/b = 1 + b/a$. Thus if we set $x = a/b$, (12.19) implies

$$x = 1 + \frac{1}{x}. \tag{12.20}$$

Multiplying both sides by x and moving all terms to the left-hand side yields

$$x^2 - x - 1 = 0. \tag{12.21}$$

This quadratic equation has two solutions, given by

$$x = \frac{1 \pm \sqrt{5}}{2}.$$

The only positive solution is $x = \varphi$, given by (12.18), so a/b must be equal to this number.

Conversely, if $ABCD$ is a rectangle in which the side lengths are a and b with $a = \varphi b$, then a straightforward computation shows that (12.19) is satisfied, which shows that $ABCD$ is a golden rectangle. $\qquad \square$

The golden ratio shows up in a surprising variety of contexts. Euclid was already aware of many of the special properties of segments whose lengths are related by this ratio, which he called ***extreme and mean ratio*** (see, for example, Propositions II.11, VI.30, XIII.1–6, XIII.8–9). It is an irrational number, $\varphi \approx 1.618034\ldots$, and the fact that it is a solution to equations (12.20) and (12.21) implies the following remarkable numerical relationships:

$$\varphi \approx 1.618034\ldots, \tag{12.22}$$

$$\frac{1}{\varphi} = \varphi - 1 \approx 0.618034\ldots, \tag{12.23}$$

$$\varphi^2 = \varphi + 1 \approx 2.618034\ldots. \tag{12.24}$$

It has close connections with such mathematical concepts as the Fibonacci numbers and logarithmic spirals, both of which arise frequently in mathematical descriptions of natural phenomena (if you don't know what these are, look them up). Ever since the Renaissance, the golden ratio has played a role in the philosophy of art as well, with many authors arguing that golden rectangles are the most aesthetically pleasing rectangles. The beautiful book [**Liv02**] by Mario Livio gives an engaging and thorough historical account of the golden ratio in mathematics and culture.

For now, we will content ourselves with presenting one important application of the golden ratio. (Others will appear in Chapter 16.) We have studied three special triangles so far: the 60-60-60, 30-60-90, and 45-45-90 triangles. The golden ratio allows us to analyze one more special triangle. A ***golden triangle*** is an isosceles triangle in which the ratio of the length of each leg to the length of the base is equal to the golden ratio (Fig. 12.15).

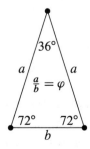

Fig. 12.15. A golden triangle.

Theorem 12.15 (36-72-72 Theorem). *A triangle is a golden triangle if and only if it has angle measures* 36°, 72°, *and* 72°.

Proof. See Exercise 12K.   □

Perimeters and Areas of Similar Figures

When two polygons are similar, it is reasonable to expect that other geometric quantities in addition to side lengths should be related in a simple way. One such quantity is the *perimeter of a polygon*, which is defined as the sum of the lengths of all its sides. We denote the perimeter of $\mathcal{P}$ by $\mathrm{perim}(\mathcal{P})$.

Theorem 12.16 (Perimeter Scaling Theorem). *If two polygons are similar, then the ratio of their perimeters is the same as the ratio of their corresponding side lengths.*

Proof. Suppose $\mathcal{P}$ and $\mathcal{P}'$ are similar polygons, and let r be the scale factor of the similarity, so each side of $\mathcal{P}$ is r times as long as the corresponding side of $\mathcal{P}'$. If the side lengths of $\mathcal{P}$ are $s_1, \ldots, s_n$ and those of $\mathcal{P}'$ are $s_1', \ldots, s_n'$, then simple algebra shows that

$$\mathrm{perim}(\mathcal{P}) = s_1 + \cdots + s_n = rs_1' + \cdots + rs_n' = r(s_1' + \cdots + s_n') = r \cdot \mathrm{perim}(\mathcal{P}'). \quad □$$

A geometric quantity associated with polygons is said to be *one-dimensional* if the corresponding quantities of similar polygons have the same ratio as the corresponding side lengths. Thus side lengths and perimeters are one-dimensional quantities. The next two theorems describe two more such quantities.

Theorem 12.17 (Height Scaling Theorem). *If two triangles are similar, their corresponding heights have the same ratio as their corresponding side lengths.*

Proof. Exercise 12L.   □

Theorem 12.18 (Diagonal Scaling Theorem). *If two convex quadrilaterals are similar, the lengths of their corresponding diagonals have the same ratio as their corresponding side lengths.*

Proof. Exercise 12M.   □

By contrast, areas scale differently. The next theorem is Euclid's Proposition VI.19.

Theorem 12.19 (Triangle Area Scaling Theorem). *If two triangles are similar, then the ratio of their areas is the square of the ratio of their corresponding side lengths; that is, if* $\triangle ABC \sim \triangle DEF$ *and* $AB = r \cdot DE$, *then*

$$\alpha(\triangle ABC) = r^2 \cdot \alpha(\triangle DEF). \tag{12.25}$$

Proof. Exercise 12N.   □

Because of the exponent 2 in (12.25), area is said to be a *two-dimensional quantity*.

Theorem 12.20 (Quadrilateral Area Scaling Theorem). *If two convex quadrilaterals are similar, then the ratio of their areas is the square of the ratio of their corresponding side lengths.*

Proof. Exercise 12O.   □

Exercises

12A. Prove Theorem 12.5 (the SSS similarity theorem). [Hint: Use the similar triangle construction theorem to construct a triangle $\triangle D'E'F'$ that is similar to $\triangle ABC$, but with $\overline{D'E'} \cong \overline{DE}$. Then use the hypothesis and a little algebra to show that $\triangle D'E'F' \cong \triangle DEF$.]

12B. Prove Theorem 12.6 (the SAS similarity theorem). [Hint: The proof is very similar to that of the SSS similarity theorem.]

12C. Prove Theorem 12.8 (the converse to the side-splitter theorem).

12D. CONVERSE TO THE ANGLE BISECTOR PROPORTION THEOREM: Suppose $\triangle ABC$ is a triangle and D is a point in the interior of $\overline{BC}$ such that $BD/DC = AB/AC$. Prove that $\overrightarrow{AD}$ is the bisector of $\angle BAC$.

12E. Prove that the three midsegments of a triangle yield an admissible decomposition of the triangle into four congruent triangles, each of which is similar to the original one and has one-quarter the area.

12F. Suppose $\triangle ABC$ and $\triangle A'B'C'$ are triangles such that $\overleftrightarrow{AB} \parallel \overleftrightarrow{A'B'}$, $\overleftrightarrow{BC} \parallel \overleftrightarrow{B'C'}$, and $\overleftrightarrow{AC} \parallel \overleftrightarrow{A'C'}$. Prove that $\triangle ABC \sim \triangle A'B'C'$.

12G. Given an integer $n \geq 3$, show that any two regular n-gons are similar to each other.

12H. Prove Theorem 12.12 (Ceva's theorem). [Hint: To prove the forward direction, apply Menelaus's theorem to $\triangle ABF$ and $\triangle CBF$ in Fig. 12.12. For the converse, show that there is a point $F' \in \text{Int } \overline{AC}$ such that the cevians $\overline{AE}$, $\overline{BF'}$, and $\overline{CD}$ are concurrent, and proceed as in the proof of the "if" part of Menelaus's theorem.]

12I. Prove Theorem 12.13 (the median concurrence theorem). [Hint: Use Ceva's theorem and the midsegment theorem.]

12J. Prove that the medians of a triangle give an admissible decomposition of the triangle into six triangles of equal area.

12K. Prove Theorem 12.15 (the 36-72-72 theorem). [Hint: If $\triangle ABC$ has $m\angle A = 36°$ and $m\angle B = m\angle C = 72°$, begin by letting D be the point where the bisector of $\angle B$ meets $\overline{AC}$ as in Fig. 12.16 and showing that $\triangle BCD \sim \triangle ABC$. Conversely, if $\triangle ABC$ is a golden triangle with base $\overline{BC}$, construct a 36-72-72 triangle with base congruent to $\overline{BC}$, and show that the two triangles are congruent.]

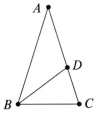

Fig. 12.16. Proof of the golden triangle theorem.

12L. Prove Theorem 12.17 (the height scaling theorem).

12M. Prove Theorem 12.18 (the diagonal scaling theorem).

12N. Prove Theorem 12.19 (the triangle area scaling theorem).

12O. Prove Theorem 12.20 (the quadrilateral area scaling theorem).

Chapter 13

Right Triangles

Right triangles are among the most important figures in Euclidean geometry. In previous chapters, we have already seen some important properties of right triangles:

- Every right triangle has two acute angles (Corollary 5.15).
- The hypotenuse is the longest side of a right triangle (Corollary 5.17).
- The altitude to the hypotenuse meets the interior of the hypotenuse (Corollary 7.5).
- The area is one-half the product of the lengths of the legs (Lemma 11.9).

In this chapter, we will develop many more such properties.

The Pythagorean Theorem

Our first fact about right triangles is arguably the most important theorem in all of Euclidean geometry and indeed one of the most important in all of mathematics.

Theorem 13.1 (The Pythagorean Theorem). *Suppose $\triangle ABC$ is a right triangle with right angle at C, and let a, b, and c denote the lengths of the sides opposite A, B, and C, respectively (Fig. 13.1). Then $a^2 + b^2 = c^2$.*

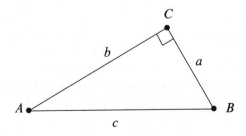

Fig. 13.1. The Pythagorean theorem: $a^2 + b^2 = c^2$.

This result appears to have been known, at least as an empirical observation, in ancient Mesopotamia, India, China, and Egypt. For at least the last two millennia, tradition has ascribed the first proof to Pythagoras of Samos, a Greek mathematician who lived around 500 BCE, although no proof by Pythagoras has survived to modern times. The earliest complete proofs we have are from Euclid's *Elements*; in fact, Euclid gives two different proofs (Propositions I.47 and VI.31). Since Euclid's time, literally hundreds of different proofs of the Pythagorean theorem have been discovered. (An early twentieth-century book by a mathematician named Elisha Loomis [**Loo68**] contains 370 proofs. It is easy to find many proofs on the Internet, often with clever animations or interactive diagrams.) Because of the central importance of this theorem, we will present three proofs, with two more outlined in Exercises 13A and 13B. (For a theorem of this significance, it never hurts to see another proof!)

We begin with a proof that closely parallels Euclid's proof from Book I. Here is how Euclid stated the theorem (in Heath's translation):

Euclid's Proposition I.47. *In right-angled triangles the square on the side subtending the right angle is equal to the squares on the sides containing the right angle.*

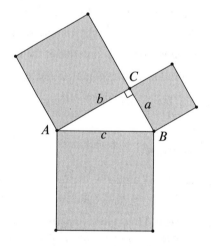

Fig. 13.2. Euclid's version of the Pythagorean theorem.

To Euclid, this theorem was literally about squares: his conclusion means that the area of the square constructed on the hypotenuse is equal to the area of the union of the two squares constructed on the legs (Fig. 13.2). (Recall that what Euclid means by saying that two polygonal regions are "equal" is what we would express by saying they have equal area; since he does not use numbers to measure area, his proofs of "equality of figures" are all based on finding admissible decompositions into congruent pieces.) The following proof is almost identical to his, except that we express it using the modern terminology of area in place of Euclid's notion of "equality of figures."

Euclid's first proof of the Pythagorean theorem. Let $\triangle ABC$ be a right triangle with its right angle at C. Let $ABDE$, $ACFG$, and $BCJK$ be squares that have $\overline{AB}$, $\overline{AC}$, and $\overline{BC}$, respectively, as edges, with each square constructed on the opposite side of its edge from the remaining vertex of $\triangle ABC$ (Fig. 13.3). The theorem will be proved once we show that

$$\alpha(ACFG) + \alpha(BCJK) = \alpha(ABDE). \tag{13.1}$$

Let X be the foot of the perpendicular from C to $\overleftrightarrow{AB}$. Since $\overleftrightarrow{CX}$ is perpendicular to $\overleftrightarrow{AB}$, it is also perpendicular to $\overleftrightarrow{ED}$ by Theorem 10.5, so $\overleftrightarrow{CX}$ intersects $\overleftrightarrow{ED}$ at a point Y. Because $\overleftrightarrow{CX}$ and $\overleftrightarrow{AE}$ have $\overleftrightarrow{AB}$ as a common perpendicular, they are parallel. It follows from Corollary 7.5 that X is an interior point of $\overline{AB}$, and then it follows from the parallel projection theorem (Theorem 12.10) that Y is an interior point of $\overline{ED}$.

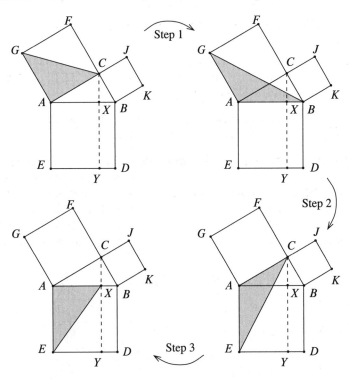

Fig. 13.3. Euclid's first proof of the Pythagorean theorem.

Note that $\alpha(ABDE) = \alpha(AXYE) + \alpha(BXYD)$ by the convex decomposition lemma. So if we can show that $\alpha(ACFG) = \alpha(AXYE)$ and $\alpha(BCJK) = \alpha(BXYD)$, then (13.1) follows by substitution and we are done. The proofs of both equalities are essentially the same, so we will show that $\alpha(ACFG) = \alpha(AXYE)$.

Because a diagonal of a rectangle divides it into two congruent triangles, each of which has area equal to half that of the rectangle, to prove $\alpha(ACFG) = \alpha(AXYE)$ it suffices to prove that $\alpha(\triangle ACG) = \alpha(\triangle AEX)$. We will prove this in several steps (illustrated in Fig. 13.3).

STEP 1: $\alpha(\triangle ACG) = \alpha(\triangle ABG)$. Because $\overleftrightarrow{CF}$ is parallel to $\overleftrightarrow{AG}$ and F, C, and B are collinear by the partial converse to the linear pair theorem (Theorem 4.16), this follows from the triangle sliding theorem (Corollary 11.11).

STEP 2: $\alpha(\triangle ABG) = \alpha(\triangle AEC)$. We will prove this by showing that these two triangles are congruent. (You can visualize this congruence by imagining $\triangle ABG$ rotating clockwise around A until it coincides with $\triangle AEC$.) By construction, $\overline{AB} \cong \overline{AE}$ and $\overline{AG} \cong \overline{AC}$. Once we verify that $\overrightarrow{AE} * \overrightarrow{AB} * \overrightarrow{AC}$ and $\overrightarrow{AB} * \overrightarrow{AC} * \overrightarrow{AG}$, it will follow that $m\angle CAE = m\angle BAC + m\angle BAE = m\angle BAC + 90°$ and $m\angle BAG = m\angle BAC + m\angle CAG = m\angle BAC + 90°$. Thus $\triangle ABG \cong \triangle AEC$ by SAS.

To show that $\overrightarrow{AE} * \overrightarrow{AB} * \overrightarrow{AC}$, note that X and C are on the same side of $\overleftrightarrow{AE}$ because $\overleftrightarrow{XC} \parallel \overleftrightarrow{AE}$. Thus $\overrightarrow{AE}$, $\overrightarrow{AC}$, and $\overrightarrow{AB} = \overrightarrow{AX}$ all lie in HR $\left(\overleftrightarrow{AE}, C\right)$. Because E and C are on opposite sides of $\overleftrightarrow{AB}$ by construction, the adjacency lemma implies that $\overrightarrow{AE} * \overrightarrow{AB} * \overrightarrow{AC}$. The same argument shows that $\overrightarrow{AB} * \overrightarrow{AC} * \overrightarrow{AG}$.

STEP 3: $\alpha(\triangle AEC) \cong \alpha(\triangle AEX)$. This follows from the triangle sliding theorem, because $\overleftrightarrow{CX} \parallel \overleftrightarrow{AE}$.

This completes the proof that $\alpha(\triangle ACG) = \alpha(\triangle AXE)$, from which we conclude that $\alpha(ACFG) = \alpha(AXYE)$. The same argument shows that $\alpha(BCJK) = \alpha(BXYD)$ and thus completes the proof of the Pythagorean theorem.   □

The previous proof is elaborate and imaginative but perhaps not the most intuitively obvious. The next proof is conceptually one of the simplest.

Proof of the Pythagorean theorem by decomposition. Let $\triangle ABC$ be a right triangle with its right angle at C, and let a, b, c be the lengths of the sides opposite A, B, and C, respectively. Let $PQRS$ be a square of side length $a + b$. The idea of this proof is to compute the area of $PQRS$ in two different ways, as illustrated in Fig. 13.4. In each of these figures, the white regions have combined area $2ab$, so the remaining shaded areas must also be equal. We just need to make this pictorial argument into a rigorous proof.

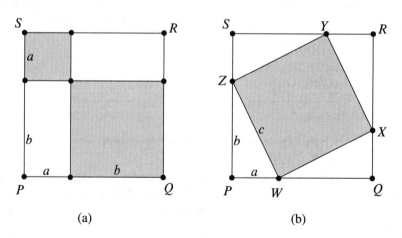

(a)   (b)

Fig. 13.4. Proof of the Pythagorean theorem by decomposition.

On the one hand, the usual formula for the area of a square shows that

$$\alpha(PQRS) = (a+b)^2 = a^2 + b^2 + 2ab. \tag{13.2}$$

(This can be thought of as representing the areas of the two squares and two rectangles in Fig. 13.4(a).)

On the other hand, we can choose points W, X, Y, and Z in the interiors of $\overline{PQ}$, $\overline{QR}$, $\overline{RS}$, and $\overline{SP}$, respectively, such that $PW = QX = RY = SZ = a$ (Fig. 13.4(b)).

Repeated application of the convex decomposition lemma shows that

$$\begin{aligned}
\alpha(PQRS) &= \alpha(WQRSZ) + \alpha(\triangle ZPW) \\
&= \alpha(WXRSZ) + \alpha(\triangle ZPW) + \alpha(\triangle WQX) \\
&= \alpha(WXYSZ) + \alpha(\triangle ZPW) + \alpha(\triangle WQX) + \alpha(\triangle XRY) \\
&= \alpha(WXYZ) + \alpha(\triangle ZPW) + \alpha(\triangle WQX) + \alpha(\triangle XRY) + \alpha(\triangle YSZ).
\end{aligned}$$

It follows from SAS that each of the triangles in the sum above is congruent to $\triangle ABC$. This implies, first, that each such triangle has area equal to $\frac{1}{2}ab$ and, second, that each hypotenuse has the same length as the hypotenuse of $\triangle ABC$, namely c. Once we show that $WXYZ$ is a square, we can conclude that

$$\alpha(PQRS) = c^2 + 4\left(\tfrac{1}{2}ab\right) = c^2 + 2ab. \tag{13.3}$$

Equating (13.2) and (13.3) and subtracting $2ab$ from both sides, we obtain $a^2 + b^2 = c^2$.

It remains only to prove that $WXYZ$ is a square. We showed above that its side lengths are all equal to c, so we only need to show that its angles are right angles. First we will show that $\angle QWX$, $\angle XWZ$, and $\angle ZWP$ satisfy the hypotheses of the linear triple theorem. Clearly $\overrightarrow{WP}$ and $\overrightarrow{WQ}$ are opposite rays. Because $\angle QWX$ is acute and $\angle QWZ$ is obtuse (being supplementary to the acute angle $\angle ZWP$), it follows from the ordering lemma for rays that $\overrightarrow{WQ} * \overrightarrow{WX} * \overrightarrow{WZ}$. The same argument shows that $\overrightarrow{WX} * \overrightarrow{WZ} * \overrightarrow{WP}$, and therefore the linear triple theorem applies: $m\angle QWX + m\angle XWZ + m\angle ZWP = 180°$. Because $\angle QWX$ and $\angle ZWP$ are congruent to the two acute angles of the right triangle $\triangle ABC$, they are complementary, so it follows that $\angle XWZ$ is a right angle. The same argument applies to each of the other three angles of $WXYZ$. $\qquad\square$

Later in this chapter, we will give an even shorter proof based on similar triangles.

It is important to know that the converse to the Pythagorean theorem is also true. It is Euclid's Proposition I.48, the last proposition in Book I.

Theorem 13.2 (Converse to the Pythagorean Theorem). *Suppose $\triangle ABC$ is a triangle with side lengths a, b, and c. If $a^2 + b^2 = c^2$, then $\triangle ABC$ is a right triangle, and its hypotenuse is the side of length c.*

Proof. Exercise 13C. $\qquad\square$

Applications of the Pythagorean Theorem

The Pythagorean theorem abounds with applications. We present just a few of them here.

Theorem 13.3 (Side Lengths of 30-60-90 Triangles). *In a triangle with angle measures $30°$, $60°$, and $90°$, the longer leg is $\sqrt{3}$ times as long as the shorter leg, and the hypotenuse is twice as long as the shorter leg.*

Proof. Exercise 13D. $\qquad\square$

Theorem 13.4 (Side Lengths of 45-45-90 Triangles). *In a triangle with angle measures $45°$, $45°$, and $90°$, the legs are congruent, and the hypotenuse is $\sqrt{2}$ times as long as either leg.*

Proof. Exercise 13E.   □

Theorem 13.5 (Diagonal of a Square). *In a square, each diagonal is $\sqrt{2}$ times as long as each side.*

Proof. The diagonal is the hypotenuse of a right triangle whose legs are two sides of the square.   □

The next application, at face value, does not appear to have anything to do with the Pythagorean theorem, or even with right triangles. It gives an answer to the following question: which triples of positive numbers can be the side lengths of a triangle? Obviously they have to satisfy the constraints imposed by the triangle inequality: each number must be strictly less than the sum of the other two. The next theorem, whose proof is rated PG, says that this is the only limitation. It is Euclid's Proposition I.22, but with a decidedly different proof.

Theorem 13.6 (SSS Existence Theorem). *Suppose a, b, and c are positive real numbers such that each one is strictly less than the sum of the other two. Then there exists a triangle with side lengths a, b, and c.*

Proof. To figure out how to proceed, we will reason heuristically for a moment: let us pretend that we have already found such a triangle and see what the Pythagorean theorem tells us. Suppose $\triangle ABC$ is a triangle in which the sides opposite A, B, and C have lengths a, b, and c, respectively. Relabeling if necessary, we may assume that $a \leq b \leq c$. Let F be the foot of the perpendicular from C to $\overline{AB}$ (Fig. 13.5); since we have arranged for $\overline{AB}$

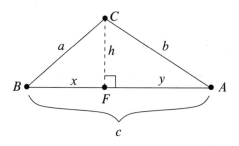

Fig. 13.5. The SSS existence theorem.

to be the longest side, it follows from Corollary 7.4 that F lies in the interior of $\overline{AB}$. Let $h = CF$, $x = BF$, and $y = FA$. Then the betweenness theorem for points implies that

$$c = x + y, \tag{13.4}$$

and the Pythagorean theorem implies

$$b^2 = y^2 + h^2, \tag{13.5}$$

$$a^2 = x^2 + h^2. \tag{13.6}$$

Subtracting (13.6) from (13.5) and substituting (13.4) for $x + y$, we obtain

$$b^2 - a^2 = y^2 - x^2 = (y + x)(y - x) = c(y - x). \tag{13.7}$$

Now (13.7) and (13.4) are a pair of linear equations for the unknowns x and y, which can easily be solved to yield

$$x = \frac{c^2 - b^2 + a^2}{2c}, \qquad y = \frac{c^2 - a^2 + b^2}{2c}. \tag{13.8}$$

It then follows from (13.6) that

$$h = \sqrt{a^2 - x^2}. \tag{13.9}$$

With these computations as motivation, let us proceed with the formal proof. As above, we can assume that the three numbers are labeled so that $a \le b \le c$. Then the hypothesis implies, in particular, that $c < a + b$.

We begin by defining real numbers x and y by (13.8). The first thing we need to check is that x and y are positive. The assumption that c is the largest of the three numbers implies that $c^2 - b^2 \ge 0$, and thus $c^2 - b^2 + a^2 \ge a^2 > 0$, so x is positive. A similar analysis shows that y is positive.

Next, we need to check that $x < a$. We compute

$$a - x = a - \frac{c^2 - b^2 + a^2}{2c} = \frac{2ac - c^2 + b^2 - a^2}{2c} = \frac{b^2 - (c - a)^2}{2c}. \tag{13.10}$$

The fact that $a + b > c$ implies that $b > c - a$. Because b and $c - a$ are nonnegative, this in turn implies $b^2 > (c - a)^2$, so (13.10) shows that $a - x$ is positive. Thus $a > x$, which implies $a^2 > x^2$, so we can define a positive real number h by (13.9). A direct (albeit tedious) computation then shows that (13.4), (13.5), and (13.6) are all satisfied by these choices of x, y, and h.

Let $\overline{AB}$ be any segment of length c. Because $0 < x < a \le c$, there is a point F in the interior of $\overline{AB}$ such that $BF = x$. Let m be the line perpendicular to $\overleftrightarrow{AB}$ at F, and let C be a point on m such that $CF = h$. Then the betweenness theorem for points together with (13.4) shows that $FA = c - x = y$, and the Pythagorean theorem together with (13.5) and (13.6) yields

$$BC = \sqrt{x^2 + h^2} = a, \qquad AC = \sqrt{y^2 + h^2} = b.$$

Thus $\triangle ABC$ has the desired side lengths. $\qquad\square$

The preceding proof relies on the Pythagorean theorem, which is true only in Euclidean geometry. In fact, the SSS existence theorem is actually true in neutral geometry, but its proof is considerably harder. One source for the proof is [**Mar96**, Theorem 20.15].

In practice, the next corollary, which can be viewed as a complement to the AAA construction theorem (Theorem 10.17), is the most useful application of the SSS existence theorem.

Corollary 13.7 (SSS Construction Theorem). *Suppose a, b, and c are positive real numbers such that each one is strictly less than the sum of the other two, and $\overline{AB}$ is a segment of length c. Then on either side of $\overleftrightarrow{AB}$, there exists a unique point C such that $\triangle ABC$ has side lengths a, b, and c opposite vertices A, B, and C, respectively.*

Proof. The preceding theorem implies that there is a triangle with side lengths a, b, and c, so on each side of $\overleftrightarrow{AB}$ we can make a copy of that triangle with $\overline{AB}$ as one of its sides. Uniqueness follows from the unique triangle theorem (Theorem 5.11). $\qquad\square$

Similarity Relations in Right Triangles

The theory of similar triangles yields some important relations involving the altitude to the hypotenuse of a right triangle. The first one is Euclid's Proposition VI.8, with the same proof.

Theorem 13.8 (Right Triangle Similarity Theorem). *The altitude to the hypotenuse of a right triangle cuts it into two triangles that are similar to each other and to the original triangle.*

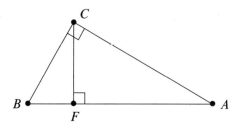

Fig. 13.6. The right triangle similarity theorem.

Proof. Let $\triangle ABC$ be a right triangle with right angle at C, and let F be the foot of the altitude from C to the hypotenuse (Fig. 13.6). By Corollary 7.5, F is an interior point of $\overline{AB}$. We will show that $\triangle CBF \sim \triangle ACF \sim \triangle ABC$.

Because $\triangle CBF$ and $\triangle ABC$ share $\angle B$ and have right angles at F and C, respectively, it follows from the AA similarity theorem that $\triangle CBF \sim \triangle ABC$. Similarly, $\triangle ACF$ and $\triangle ABC$ share $\angle A$ and have right angles at F and C, so $\triangle ACF \sim \triangle ABC$. Since similarity is transitive, we also have $\triangle CBF \sim \triangle ACF$. $\qquad\square$

The most important application of the preceding theorem is to derive relationships among the lengths of the various line segments in Fig. 13.6. To state these relationships concisely, let us introduce some standard terminology. If $\triangle ABC$ is a right triangle with right angle at C and F is the foot of the altitude from C, then $\overline{BF}$ is called the ***projection of $\overline{BC}$ onto the hypotenuse***, and $\overline{AF}$ is called the ***projection of $\overline{AC}$ onto the hypotenuse***. (If you visualize the triangle sitting upright on its hypotenuse, you can think of these projections as representing the shadows of the legs cast onto the hypotenuse by an overhead light source.)

If x and y are positive real numbers, the number $m = \sqrt{xy}$ is called the ***geometric mean of x and y***. It is the unique positive number such that

$$\frac{x}{m} = \frac{m}{y}.$$

The geometric mean is also sometimes called the ***mean proportional***. For brevity, we will often say "one segment is the geometric mean of two others" to mean that the *length* of the first segment is the geometric mean of the *lengths* of the other two.

Theorem 13.9 (Right Triangle Proportion Theorem). *In every right triangle, the following proportions hold:*

(a) *The altitude to the hypotenuse is the geometric mean of the projections of the two legs onto the hypotenuse.*

(b) *Each leg is the geometric mean of its projection onto the hypotenuse and the whole hypotenuse.*

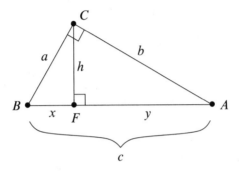

Fig. 13.7. Proportions in a right triangle.

Proof. Let $\triangle ABC$ be a right triangle with right angle at C, let F be the foot of the altitude from C to the hypotenuse, and label the lengths of the segments as shown in Fig. 13.7. We have to prove that the following three proportions hold:

$$\frac{x}{h} = \frac{h}{y}, \qquad \frac{x}{a} = \frac{a}{c}, \qquad \text{and} \qquad \frac{y}{b} = \frac{b}{c}. \tag{13.11}$$

These follow, respectively, from the similarities $\triangle CBF \sim \triangle ACF$, $\triangle CBF \sim \triangle ABC$, and $\triangle ACF \sim \triangle ABC$. $\qquad\square$

Using the proportionality relations of the preceding theorem, we can present a very short proof of the Pythagorean theorem. This is essentially the same as the second proof offered by Euclid (Proposition VI.31) but translated into modern algebraic language.

Proof of the Pythagorean theorem by similar triangles. Let $\triangle ABC$ be a right triangle with its right angle at C, let a, b, and c denote the lengths of the sides as in the statement of the theorem, and let x, y, and h be defined as in the proof of Theorem 13.9. Cross-multiplying the last two equations in (13.11) and adding them together, we obtain

$$a^2 = cx,$$
$$b^2 = cy,$$
$$a^2 + b^2 = c(x + y).$$

Since $x + y = c$, this proves the theorem. $\qquad\square$

Trigonometry

Trigonometric ratios like sine and cosine are usually introduced in the context of the Cartesian plane. However, they can also be defined purely in terms of axiomatic Euclidean geometry, where they have some important geometric applications. In this section, we introduce the main definitions and the most important results.

Suppose θ is any positive real number less than 90. By the AAA construction theorem, there exists a right triangle in which one angle has measure θ. If $\triangle ABC$ and $\triangle A'B'C'$ are any two such triangles with $m\angle C = m\angle C' = 90°$ and $m\angle A = m\angle A' = \theta$, then $\triangle ABC \sim \triangle A'B'C'$ by the AA similarity theorem (Fig. 13.8). Thus the following ratios are equal:

$$\frac{BC}{AB} = \frac{B'C'}{A'B'}, \qquad \frac{AC}{AB} = \frac{A'C'}{A'B'}.$$

Therefore, these ratios (length of opposite leg to length of hypotenuse and length of adjacent leg to length of hypotenuse) are independent of the choice of right triangle with an angle of measure θ and depend only on the number θ.

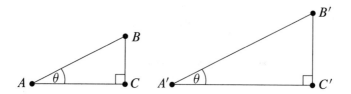

Fig. 13.8. Defining trigonometric ratios.

We can thus make the following definitions: for any real number θ such that $0° < \theta < 90°$, we define the **sine of θ**, denoted by $\sin\theta$, to be the ratio of the length of the leg opposite the θ angle to the length of the hypotenuse in any right triangle with an angle whose measure is θ; similarly, we define the **cosine of θ**, denoted by $\cos\theta$, to be the ratio of the length of the leg adjacent to the θ angle to the length of the hypotenuse in any such right triangle. Succinctly, we can write

$$\sin\theta = \frac{\text{opposite}}{\text{hypotenuse}}, \qquad \cos\theta = \frac{\text{adjacent}}{\text{hypotenuse}}.$$

For angle measures that are not between 0° and 90°, we have to make special definitions. First of all, for obtuse angle measures, we define

$$\sin\theta = \sin(180° - \theta) \quad \text{and} \quad \cos\theta = -\cos(180° - \theta) \qquad \text{if } 90° < \theta < 180°.$$

Then for the special measures 0°, 90°, and 180°, we define

$$\sin 0° = 0, \qquad \cos 0° = 1;$$
$$\sin 90° = 1, \qquad \cos 90° = 0; \qquad\qquad (13.12)$$
$$\sin 180° = 0, \qquad \cos 180° = -1.$$

These definitions will be justified by the simple statements of the theorems we will prove below.

We can also define the sine and cosine of an *angle* to be the sine and cosine of its measure: if $\angle A$ is any angle, the **sine of $\angle A$**, denoted by $\sin\angle A$, is defined to be the sine of $m\angle A$, with a similar definition for the *cosine of* $\angle A$. Thus, for example, if $\triangle ABC$ is a right triangle with its right angle at C, we can write

$$\sin\angle A = \frac{BC}{AB}, \qquad \cos\angle A = \frac{AC}{AB}.$$

There are also other standard trigonometric ratios—the tangent, cotangent, secant, and cosecant—that can be defined similarly; but for our purposes the sine and cosine will be sufficient.

Some simple properties of the sine and cosine follow easily from the definitions.

Theorem 13.10. *Suppose* $\theta \in [0, 180]$. *Then*

$$0 < \sin\theta < 1 \quad \text{if and only if} \quad 0° < \theta < 90° \text{ or } 90° < \theta < 180°; \tag{13.13}$$

$$0 < \cos\theta < 1 \quad \text{if and only if} \quad 0° < \theta < 90°; \tag{13.14}$$

$$-1 < \cos\theta < 0 \quad \text{if and only if} \quad 90° < \theta < 180°. \tag{13.15}$$

Proof. These facts follow immediately from the definitions and the fact that the hypotenuse of a right triangle is strictly longer than either leg. $\square$

Theorem 13.11. *The cosine function is injective.*

Proof. Suppose α and β are numbers in $[0, 180]$ such that $\cos\alpha = \cos\beta$. Let x denote the common value of these two cosines. Suppose first that $0 < x < 1$. It follows from Theorem 13.10 that α and β are both strictly between $0°$ and $90°$. Construct right triangles $\triangle ABC$ and $\triangle A'B'C'$ with $AB = A'B' = 1$, $m\angle A = \alpha$, $m\angle B = \beta$, and $m\angle C = m\angle C' = 90°$ (Fig. 13.9). (Such a triangle is easily constructed by first constructing an angle with the desired measure, marking a point on one ray at a distance 1 from the vertex, and dropping a perpendicular to the other ray.) Our assumption implies that $AC = A'C' = x$. Therefore, $\triangle ABC \cong \triangle A'B'C'$ by HL, and so $\alpha = \beta$.

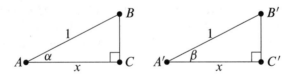

Fig. 13.9. Proving that the cosine function is injective.

Next, suppose $-1 < x < 0$. Then α and β are both strictly between $90°$ and $180°$, and we have

$$\cos(180° - \alpha) = -\cos\alpha = -\cos\beta = \cos(180° - \beta).$$

It follows from the argument in the previous paragraph that $180° - \alpha = 180° - \beta$, and thus $\alpha = \beta$.

Finally, if x is one of the special values 1, 0, or -1, then it follows from the definitions and Theorem 13.10 that both α and β must be equal to $0°$, $90°$, or $180°$, respectively, and thus they are equal to each other. $\square$

The most important fact about sines and cosines is the following identity. We use the traditional notation for powers of the sine and cosine functions: for any positive integer n, $\sin^n \theta$ and $\cos^n \theta$ mean $(\sin\theta)^n$ and $(\cos\theta)^n$, respectively.

Theorem 13.12 (The Pythagorean Identity). *If θ is any real number in the interval* $[0, 180]$, *then* $\sin^2\theta + \cos^2\theta = 1$.

Proof. First suppose $0° < \theta < 90°$, and let $\triangle ABC$ be a right triangle with right angle at C and with $m\angle A = \theta$. (Such a triangle exists by the AAA construction theorem.) If we let a, b, c denote the lengths of the sides opposite A, B, and C, respectively, then $a^2 + b^2 = c^2$ by the Pythagorean theorem, and $\sin\theta = a/c$ and $\cos\theta = b/c$ by definition. Using a little algebra, we compute

$$\sin^2\theta + \cos^2\theta = \left(\frac{a}{c}\right)^2 + \left(\frac{b}{c}\right)^2 = \frac{a^2 + b^2}{c^2} = 1.$$

Next, if $90° < \theta < 180°$, the preceding computation applies to $180° - \theta$, so the definitions of sine and cosine for obtuse angles yield

$$\sin^2\theta + \cos^2\theta = \big(\sin(180° - \theta)\big)^2 + \big(-\cos(180° - \theta)\big)^2 = 1.$$

Finally, when θ is one of the special angle measures $0°$, $90°$, or $180°$, the formula follows directly from the definitions (13.12). $\qquad\qquad\qquad\qquad\qquad\qquad\qquad\qquad\qquad\square$

The next theorem is a generalization of the Pythagorean theorem to nonright triangles. It will be used in the next section to prove that the postulates for Euclidean geometry are categorical.

Theorem 13.13 (Law of Cosines). *Let $\triangle ABC$ be any triangle, and let a, b, and c denote the lengths of the sides opposite A, B, and C, respectively. Then*

$$a^2 + b^2 = c^2 + 2ab\cos\angle C. \qquad (13.16)$$

Proof. Write $\theta = m\angle C$. Note that $\triangle ABC$ has at least two acute angles, so at least one of the angles $\angle A$ and $\angle B$ is acute. Since formula (13.16) is unchanged when a and b are interchanged, we can interchange the names of the vertices A and B if necessary to ensure that $\angle B$ is acute. Let F be the foot of the perpendicular from A to $\overleftrightarrow{BC}$, and define real numbers h and x by $h = AF$ and $x = CF$ (Fig. 13.10). We consider three cases, depending on the size of θ.

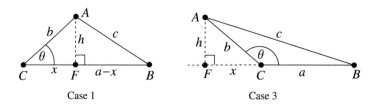

Fig. 13.10. Proof of the law of cosines.

CASE 1: $0° < \theta < 90°$. Then $F \in \text{Int}\,\overline{BC}$ by Theorem 7.3, and therefore $BF = a - x$. By definition, $\cos\angle C = x/b$; and by the Pythagorean theorem,

$$c^2 = h^2 + (a - x)^2,$$
$$b^2 = h^2 + x^2.$$

Subtracting the second equation from the first and simplifying, we obtain

$$c^2 - b^2 = a^2 - 2ax.$$

Substituting $x = b \cos \angle C$ and rearranging, we obtain (13.16).

CASE 2: $\theta = 90°$. In this case, $\cos \angle C = 0$ by definition, and (13.16) just reduces to the Pythagorean theorem.

CASE 3: $90° < \theta < 180°$. In this case, F lies in the interior of the ray opposite $\overrightarrow{CB}$ by Theorem 7.2. This implies $F * C * B$ and therefore $FB = a + x$. Now we have $\cos \angle C = -\cos(180° - \theta) = -x/b$, and the Pythagorean theorem gives

$$c^2 = h^2 + (a + x)^2,$$
$$b^2 = h^2 + x^2.$$

Again subtracting the second equation from the first and simplifying, we obtain

$$c^2 - b^2 = a^2 + 2ax = a^2 - 2ab \cos \angle C$$

as before. $\square$

Theorem 13.14 (Law of Sines). *Let $\triangle ABC$ be any triangle, and let a, b, and c denote the lengths of the sides opposite A, B, and C, respectively. Then*

$$\frac{\sin \angle A}{a} = \frac{\sin \angle B}{b} = \frac{\sin \angle C}{c}. \tag{13.17}$$

Proof. Exercise 13H. $\square$

The next theorem should be familiar from your prior study of trigonometry. It is actually true for more general angles than we prove here, but for simplicity we prove it only for angle measures less than $90°$. This theorem (or actually Corollary 13.17, which follows from it) will play an important role in Chapter 16. Its proof is rated PG.

Theorem 13.15 (Sum Formulas). *Suppose α and β are real numbers such that α, β, and $\alpha + \beta$ are all strictly between $0°$ and $90°$. Then*

$$\sin(\alpha + \beta) = \sin \alpha \cos \beta + \cos \alpha \sin \beta, \tag{13.18}$$
$$\cos(\alpha + \beta) = \cos \alpha \cos \beta - \sin \alpha \sin \beta. \tag{13.19}$$

Proof. Let $\overrightarrow{r}$ be a ray starting at a point O, and let $\overrightarrow{s}$ and $\overrightarrow{t}$ be rays starting at O, both on the same side of $\overrightarrow{r}$, such that $m\angle rs = \alpha$ and $m\angle rt = \alpha + \beta$. It then follows from the ordering lemma for rays that $\overrightarrow{r} * \overrightarrow{s} * \overrightarrow{t}$, and thus $m\angle st = m\angle rt - m\angle rs = \beta$. (See Fig. 13.11.)

Let A be the point on $\overrightarrow{t}$ such that $OA = 1$; let F and G be the feet of the perpendiculars from A to $\overleftrightarrow{s}$ and $\overleftrightarrow{r}$, respectively; and let H and K be the feet of the perpendiculars from F to $\overleftrightarrow{r}$ and $\overleftrightarrow{AG}$. Because $\triangle AOF$ and $\triangle AOG$ are right triangles, the angle-sum theorem for triangles gives $m\angle OAF = 90° - \beta$ and $m\angle OAG = 90° - \alpha - \beta < m\angle OAF$. Since $\overrightarrow{AG}$ and $\overrightarrow{AF}$ lie on the same side of $\overleftrightarrow{OA}$ (why?), it follows from the ordering lemma for rays that $\overrightarrow{AO} * \overrightarrow{AG} * \overrightarrow{AF}$, and therefore

$$m\angle GAF = m\angle OAF - m\angle OAG = \alpha.$$

We need to verify two facts that are "obvious" from the diagram: $O * G * H$ and $A * K * G$. For the first, $\overrightarrow{AO} * \overrightarrow{AG} * \overrightarrow{AF}$ implies that O and F are on opposite sides of

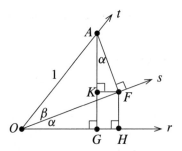

Fig. 13.11. Proof of the sum formulas.

$\overleftrightarrow{AG}$. Because $\overleftrightarrow{AG} \parallel \overleftrightarrow{FH}$, it follows that F and H are on the same side of $\overleftrightarrow{AG}$, and thus O and H are on opposite sides, so $O * G * H$. For the second, note that K lies in the interior of $\overrightarrow{AG}$ by Theorem 7.2. On the other hand, F lies in the interior of $\angle rt$ by the interior lemma, and thus it lies on the same side of $\overleftrightarrow{r}$ as A. Because $\overleftrightarrow{FK} \parallel \overleftrightarrow{r}$, it follows that K also lies on the same side of $\overleftrightarrow{r}$ as A, and thus in $\mathrm{Int}\,\overrightarrow{GA}$. These two facts imply that $K \in \mathrm{Int}\,\overrightarrow{AG} \cap \mathrm{Int}\,\overrightarrow{GA} = \mathrm{Int}\,\overline{AG}$, and thus $A * K * G$. Note also that $KFHG$ is a rectangle, and therefore

$$GH = KF, \qquad KG = FH.$$

Now we can complete the proof. It follows from the definitions of the sine and cosine that

$$\sin\alpha = \frac{FH}{OF} = \frac{KF}{AF}, \qquad \sin\beta = AF,$$

$$\cos\alpha = \frac{OH}{OF} = \frac{AK}{AF}, \qquad \cos\beta = OF.$$

Therefore,

$$\sin(\alpha + \beta) = AG = AK + KG = AK + FH$$
$$= \left(\frac{AK}{AF}\right) AF + \left(\frac{FH}{OF}\right) OF$$
$$= \cos\alpha \sin\beta + \sin\alpha \cos\beta.$$

Similarly,

$$\cos(\alpha + \beta) = OG = OH - GH = OH - KF$$
$$= \left(\frac{OH}{OF}\right) OF - \left(\frac{KF}{AF}\right) AF$$
$$= \cos\alpha \cos\beta - \sin\alpha \sin\beta. \qquad \square$$

The next two corollaries follow easily from the sum formulas.

Corollary 13.16 (Double Angle Formulas). *Suppose α is a real number strictly between $0°$ and $45°$. Then*

$$\sin 2\alpha = 2\sin\alpha \cos\alpha,$$

$$\cos 2\alpha = \cos^2\alpha - \sin^2\alpha.$$

Proof. Exercise 13I. □

Corollary 13.17 (Triple Angle Formulas). *Suppose α is a real number strictly between $0°$ and $30°$. Then*

$$\sin 3\alpha = 3\sin\alpha - 4\sin^3\alpha,$$
$$\cos 3\alpha = 4\cos^3\alpha - 3\cos\alpha.$$

Proof. Exercise 13J. □

Because any two triangles with the same side lengths are congruent, it should come as no surprise that all of the important geometric properties of a triangle can be computed purely in terms of its side lengths. The law of cosines shows how the cosine of each angle can be expressed in terms of side lengths. The next formula, which was written down by the Greek mathematician Heron of Alexandria in the first century CE, gives such an expression for the area.

Theorem 13.18 (Heron's Formula). *Let $\triangle ABC$ be a triangle, and let a,b,c denote the lengths of the sides opposite A, B, and C, respectively. Then*

$$\alpha(\triangle ABC) = \sqrt{s(s-a)(s-b)(s-c)},$$

*where $s = (a+b+c)/2$ (called the **semiperimeter of** $\triangle ABC$).*

Proof. Exercise 13K. □

Categoricity of the Euclidean Postulates

Recall from Chapter 6 that an axiomatic system is said to be ***categorical*** if given any two models of it, there is an isomorphism between them. For models of Euclidean geometry, an isomorphism means a one-to-one correspondence between the points of one model and the points of the other, which takes lines to lines, distances to distances, and angle measures to angle measures. We now have all the tools we need to prove that the postulates of Euclidean geometry are categorical. The complete (R-rated) proof is not terribly hard, but it is a little long, so we will merely sketch the main ideas here.

Theorem 13.19. *The postulates for Euclidean geometry are categorical.*

Sketch of proof. Suppose $\mathcal{M}$ and $\mathcal{M}'$ are two models of Euclidean geometry. We begin by defining a map F from the points of $\mathcal{M}$ to the points of $\mathcal{M}'$.

Choose an arbitrary point O in $\mathcal{M}$, and choose a line ℓ containing O. Let m be the line perpendicular to ℓ at O, and choose points $X \in \ell$ and $Y \in m$ such that $OX = OY = 1$. Now make similar choices O', ℓ', m', X', and Y' in $\mathcal{M}'$ (see Fig. 13.12).

First, if P is any point of $\mathcal{M}$ that lies on ℓ, we define $P' = F(P)$ as follows: if $P = O$, then $P' = O'$; if $P \in \text{Int}\,\overrightarrow{OX}$, then P' is the unique point in $\text{Int}\,\overrightarrow{O'X'}$ such that $O'P' = OP$; and if P is in the interior of the ray opposite $\overrightarrow{OX}$, then P' is the unique point in the interior of the ray opposite $\overrightarrow{O'X'}$ such that $O'P' = OP$.

Now if Q is a point of $\mathcal{M}$ that does not lie on ℓ, we define $Q' = F(Q)$ as follows. Let P be the foot of the perpendicular from Q to ℓ, let $P' = F(P)$ as above, and let n' be the line in $\mathcal{M}'$ perpendicular to ℓ' at P'. If Q is on the same side of ℓ as Y, then let Q' be

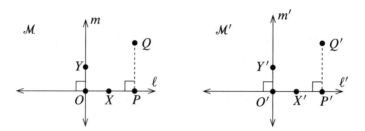

Fig. 13.12. Proof that any two models of Euclidean geometry are isomorphic.

the point of n' on the same side of ℓ' as Y', such that $P'Q' = PQ$; while if Q is on the opposite side of ℓ from Y, let Q' be the unique point of n' on the opposite side of ℓ' from Y', such that $P'Q' = PQ$.

To prove that this map defines an isomorphism, we need to check several things. For brevity, for any point Q in the model $\mathcal{M}$, we will let Q' denote the corresponding point $F(Q)$ in $\mathcal{M}'$.

STEP 1: *F preserves distances.* More specifically, we have to show that if Q_1 and Q_2 are any points of $\mathcal{M}$, then $Q'_1 Q'_2 = Q_1 Q_2$. This is done in several stages: first, when Q_1 and Q_2 both lie on ℓ (using the definition of F); second, when they both lie on a line perpendicular to ℓ (again using the definition of F); third, when they both lie on a line parallel to ℓ (using the fact that opposite sides of a rectangle are congruent); and finally, when none of the above are true (by drawing a right triangle in $\mathcal{M}$ with Q_1 and Q_2 as vertices and legs parallel to ℓ and m, and an analogous triangle in $\mathcal{M}'$, and using the fact that the Pythagorean theorem holds in both $\mathcal{M}$ and $\mathcal{M}'$).

STEP 2: *F preserves betweenness of points.* This means that if $Q_1 * Q_2 * Q_3$, then $Q'_1 * Q'_2 * Q'_3$. This follows easily from the fact that F preserves distances, by virtue of the betweenness theorem for points and its converse.

STEP 3: *F takes lines to lines, rays to rays, and segments to segments.* In other words, if k is any line, segment, or ray in $\mathcal{M}$, then the set of all points of the form Q' for $Q \in k$ is a line, segment, or ray, respectively, in $\mathcal{M}'$. This is an easy consequence of Step 2, because lines, segments, and rays can be characterized in terms of betweenness.

STEP 4: *F maps points of $\mathcal{M}$ bijectively to points of $\mathcal{M}'$.* if Q_1 and Q_2 are distinct points of $\mathcal{M}$, then $Q'_1 Q'_2 = Q_1 Q_2 \neq 0$, so Q'_1 and Q'_2 are also distinct; thus F is injective. To show that F is surjective, let Q' be any point of $\mathcal{M}'$, and let P' be the foot of the perpendicular from Q' to ℓ'; then it is an easy matter to find points P and Q in $\mathcal{M}$ such that $F(P) = P'$ and $F(Q) = Q'$.

STEP 5: *F preserves angle measures.* The fact that F preserves zero angles and straight angles is an easy consequence of the fact that it preserves betweenness. Because F preserves distances, the law of cosines together with the fact that the cosine function is injective shows that it preserves measures of proper angles. $\square$

The significance of this theorem is that we can now be sure that every model of Euclidean geometry is isomorphic to the Cartesian plane. Thus the Cartesian plane is, in a sense, the *only* model of Euclidean geometry—in any other model, the points, lines, distances, angles, and areas will have exactly the same properties, with only the names of the objects changed.

Exercises

13A. Use the idea suggested by Fig. 13.13 to give a proof of the Pythagorean theorem. [Hint: Because the only figure given to you by the hypothesis is an arbitrary right triangle, first you have to explain how a figure like the one in the diagram can be constructed and justify any claims you make about relationships that the diagram suggests.] (This proof was discovered by James A. Garfield shortly before he became the twentieth president of the United States.)

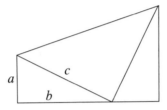

Fig. 13.13. President Garfield's proof of the Pythagorean theorem.

13B. Use the idea suggested by Fig. 13.14 to give a proof of the Pythagorean theorem. [Hint: As in the preceding exercise, you first have to construct the figure and justify your claims about it. In addition, in this case, proving that the area of the large square is the sum of the smaller areas is a bit trickier than usual.]

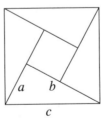

Fig. 13.14. The setup for Exercise 13B.

13C. Prove Theorem 13.2 (the converse to the Pythagorean theorem). [Hint: Construct a right triangle whose legs have lengths a and b, and show that it is congruent to $\triangle ABC$.]

13D. Prove Theorem 13.3 (side lengths of 30-60-90 triangles).

13E. Prove Theorem 13.4 (side lengths of 45-45-90 triangles).

13F. If $\mathcal{R}$ is a rectangle and $\mathcal{S}$ is a square, prove that $\alpha(\mathcal{R}) = \alpha(\mathcal{S})$ if and only if the side length of $\mathcal{S}$ is the geometric mean of the side lengths of $\mathcal{R}$.

13G. Prove that the cosine function is bijective from $[0, 180]$ to $[-1, 1]$.

13H. Prove Theorem 13.14 (the law of sines). [Hint: It suffices to prove one of the stated equations. You will need to consider separately the cases in which both angles that appear in the chosen equation are acute, one is right, and one is obtuse.]

13I. Prove Corollary 13.16 (the double angle formulas).

13J. Prove Corollary 13.17 (the triple angle formulas).

13K. Prove Theorem 13.18 (Heron's formula). [Hint: One proof uses the formulas for x, y, and h that were derived in the proof of Theorem 13.6; another one uses the law of cosines and the Pythagorean identity.]

13L. In Fig. 13.15, we are given only that $ABCD$ is a trapezoid with bases $\overline{AB}$ and $\overline{CD}$ and with side lengths as shown. Find the area of $ABCD$, and prove your answer correct. [Hint: First prove that $\angle B$ and $\angle D$ are obtuse, by using Theorem 7.2 and the Pythagorean theorem to show that any other possibilities lead to contradictions.]

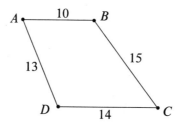

Fig. 13.15. Exercise 13L.

Chapter 14

Circles

The next objects we will study are *circles*. Euclid placed circles on a high pedestal in his development of geometry: one of his five axioms asserted, in essence, the possibility of constructing a circle that has any given point as center and goes through any other given point, and his first three propositions were constructions making essential use of circles.

In our approach to geometry, circles play a less pivotal role, but they are important nonetheless. For us, circles are sets of points, and their existence follows immediately from the definitions (see Theorem 3.29). But we will have a number of significant theorems describing their properties.

Definitions

We begin by recalling some definitions. In Chapter 3, we defined a ***circle*** as a set consisting of all points whose distance from a given point is equal to a certain given positive number. The given distance is called the ***radius of the circle***, and the given point is called its ***center***. If O is any point and r is any positive number, the circle with center O and radius r is denoted by $\mathcal{C}(O,r)$:

$$\mathcal{C}(O,r) = \{P : OP = r\}.$$

When we do not need to refer specifically to the center and radius, we will name circles by script letters near the beginning of the alphabet, such as $\mathcal{C}$, $\mathcal{D}$, $\mathcal{E}$ (most often $\mathcal{C}$).

Here is some important terminology regarding circles. Suppose $\mathcal{C} = \mathcal{C}(O,r)$ is the circle with center O and radius r.

- Any segment that has O as one endpoint and a point on $\mathcal{C}$ as the other endpoint is called a ***radius of*** $\mathcal{C}$ (plural: ***radii***).
- A ***chord of*** $\mathcal{C}$ is any segment whose endpoints both lie on $\mathcal{C}$.
- A ***diameter of*** $\mathcal{C}$ is a chord that contains the center of $\mathcal{C}$.
- Two circles are said to be ***concentric*** if they have the same center.

- A point P is an ***interior point of*** $\mathcal{C}$ if $OP < r$. The set of all interior points of $\mathcal{C}$ is called the ***interior of*** $\mathcal{C}$ and is denoted by $\operatorname{Int}\mathcal{C}$.

- A point P is an ***exterior point of*** $\mathcal{C}$ if $OP > r$. The set of all exterior points of $\mathcal{C}$ is called the ***exterior of*** $\mathcal{C}$ and is denoted by $\operatorname{Ext}\mathcal{C}$.

Note that the term *radius* is used in two different senses: it can mean a positive real number, the common distance from the center to each point on the circle (usually called "the radius" of the circle); or it can mean a line segment from the center to a point on the circle (usually called "a radius" of the circle). It should always be clear from the context which is meant. It follows from the definitions that every radius of $\mathcal{C}(O,r)$ has length r.

Similarly, the word *diameter* is used with two different meanings: it can mean either a chord containing the center, as defined above (usually called "a diameter" of $\mathcal{C}$); and it can also mean the length of a chord containing the center, which is a real number (usually called "the diameter" of $\mathcal{C}$). The next lemma shows that all diameters have the same length, so the phrase "the diameter of $\mathcal{C}$" is justified.

Lemma 14.1. *If $\mathcal{C}$ is a circle of radius r, then every diameter of $\mathcal{C}$ has length $2r$, and the center of $\mathcal{C}$ is the midpoint of each diameter.*

Proof. Write $\mathcal{C} = \mathcal{C}(O,r)$. Suppose $\overline{AB}$ is a diameter of $\mathcal{C}$, which means that A and B lie on $\mathcal{C}$ and $O \in \overline{AB}$. Clearly O cannot be equal to A or B, because points on $\mathcal{C}$ have distance r from O. It follows that $A * O * B$, so the betweenness theorem for points implies that $AB = AO + OB = r + r = 2r$. Since O is an interior point of $\overline{AB}$ that is equidistant from A and B, it is the midpoint. $\qquad\square$

The following lemma expresses a simple but important fact about circles.

Lemma 14.2. *Two concentric circles that have a point in common are equal.*

Proof. Suppose $\mathcal{C}(O,r)$ and $\mathcal{C}(O',r')$ are concentric circles and P is a point on both. Then $O = O'$ because they are concentric, and $r = OP = O'P = r'$, which shows that they have the same radius. By definition of circles, they are the same circle. $\qquad\square$

Circles and Lines

In Euclid's *Elements*, we frequently encounter situations in which circles and lines are supposed to intersect and in which it is obvious from a diagram that they do intersect; but Euclid's postulates are often not sufficient to guarantee the existence of an intersection point.

In this section, we explore the question of when a line and a circle can be shown to intersect. The first theorem shows that the possibilities for intersections are limited. It is Euclid's Proposition III.10.

Theorem 14.3. *No circle contains three distinct collinear points.*

Proof. Let $\mathcal{C} = \mathcal{C}(O,r)$ be a circle, and assume for the sake of contradiction that there are three distinct points $A,B,C \in \mathcal{C}$ that are all collinear. One of them is between the other two, so without loss of generality say that $A * B * C$ (Fig. 14.1). By definition of a circle,

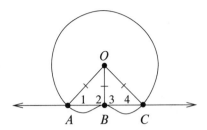

Fig. 14.1. A circle cannot contain three collinear points.

$OA = OB = OC = r$, which means that $\triangle AOB$, $\triangle AOC$, and $\triangle BOC$ are all isosceles triangles. From the isosceles triangle theorem we conclude that

$$\angle 2 \cong \angle 1, \qquad \angle 1 \cong \angle 4, \qquad \text{and} \qquad \angle 4 \cong \angle 3,$$

where the angles are labeled as in Fig. 14.1. It follows by transitivity that $\angle 2 \cong \angle 3$. Because A, B, and C are collinear, $\angle 2$ and $\angle 3$ are congruent angles forming a linear pair, so they are both right angles. But then $\angle 1$ and $\angle 4$ are also right angles, so $\triangle AOC$ contains two right angles, which is impossible. $\square$

As a consequence of the preceding theorem, every line that intersects a circle contains either one or two points of the circle. There is a name for each type of line. Suppose $\mathcal{C}$ is a circle and ℓ is a line. We say ℓ is ***tangent to*** $\mathcal{C}$ if $\ell \cap \mathcal{C}$ contains exactly one point (Fig. 14.2) and ℓ is a ***secant line for*** $\mathcal{C}$ if $\ell \cap \mathcal{C}$ contains exactly two points (Fig. 14.3). If ℓ is tangent to $\mathcal{C}$ and P is the unique point in $\ell \cap \mathcal{C}$, then P is called the ***point of tangency***, and we say ℓ is ***tangent to $\mathcal{C}$ at P***.

Fig. 14.2. A tangent line.

Fig. 14.3. A secant line.

First we study secant lines. Note that secant lines and chords are closely related: from the definitions, it follows that every secant line contains a unique chord (the segment whose endpoints are the two points where the line intersects the circle), and every chord is contained in a unique secant line (the line containing its endpoints).

Theorem 14.4 (Properties of Secant Lines). *Suppose $\mathcal{C}$ is a circle and ℓ is a secant line that intersects $\mathcal{C}$ at A and B. Then every interior point of the chord $\overline{AB}$ is in the interior of $\mathcal{C}$, and every point of ℓ that is not in $\overline{AB}$ is in the exterior of $\mathcal{C}$.*

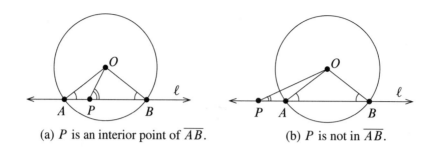

(a) P is an interior point of $\overline{AB}$.   (b) P is not in $\overline{AB}$.

Fig. 14.4. Proof of Theorem 14.4.

Proof. We will prove the first statement and leave the second to Exercise 14A. Suppose P is an interior point of $\overline{AB}$, so $A * P * B$ (Fig. 14.4(a)). Let O be the center of $\mathcal{C}$, and let r be its radius. First consider the case in which $O \in \overleftrightarrow{AB}$, so $AB = 2r$. The point O is certainly an interior point of $\mathcal{C}$. If P is an interior point of $\overline{AB}$ other than O, then either $P \in \operatorname{Int} \overline{OA}$ or $P \in \operatorname{Int} \overline{OB}$. In the former case, $OP < OA = r$ (the whole segment is greater than the part), so $P \in \operatorname{Int} \mathcal{C}$; the proof in the latter case is the same.

Now consider the case in which $O \notin \overleftrightarrow{AB}$, so A, B, and O are noncollinear. Because $OA = OB$, it follows from the isosceles triangle theorem that $\angle OAB \cong \angle OBA$. On the other hand, $\angle OPB$ is an exterior angle for $\triangle OAP$, from which it follows that $m\angle OPB > m\angle OAP = m\angle OBA$. Thus the scalene inequality applied to $\triangle OPB$ implies that $OB > OP$. This means that $OP < r$, so P is an interior point of $\mathcal{C}$.   □

The following theorem is essentially equivalent to Euclid's Propositions III.1 and III.3.

Theorem 14.5 (Properties of Chords). *Suppose $\mathcal{C}$ is a circle and $\overline{AB}$ is a chord of $\mathcal{C}$.*

(a) *The perpendicular bisector of $\overline{AB}$ passes through the center of $\mathcal{C}$.*

(b) *If $\overline{AB}$ is not a diameter, a radius of $\mathcal{C}$ is perpendicular to $\overline{AB}$ if and only if it bisects $\overline{AB}$.*

Proof. Exercise 14B.   □

The next theorem rectifies another one of the omissions in Euclid's logic, by giving conditions under which a circle and a line can be guaranteed to intersect. Euclid tacitly used the result of this theorem, for example, in his proof of Proposition I.12, his version of the theorem on dropping a perpendicular (see Exercise 14C).

Theorem 14.6 (Line-Circle Theorem). *Suppose $\mathcal{C}$ is a circle and ℓ is a line that contains a point in the interior of $\mathcal{C}$. Then ℓ is a secant line for $\mathcal{C}$, and thus there are exactly two points where ℓ intersects $\mathcal{C}$.*

Proof. Let O be the center of $\mathcal{C}$ and let r be its radius. If ℓ contains O, then by the segment construction theorem, there are two distinct points on ℓ whose distance from O is r (one on each ray starting at O), and these points lie on $\mathcal{C}$, so ℓ is a secant line. For the remainder of the proof, assume that $O \notin \ell$.

Let P be a point on ℓ that lies in the interior of $\mathcal{C}$. If F is the foot of the perpendicular from O to ℓ, then Theorem 7.13 together with our hypothesis implies that $OF \leq OP < r$.

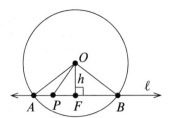

Fig. 14.5. Proof of the line-circle theorem.

Let $h = OF$ and $x = \sqrt{r^2 - h^2}$; the fact that $h < r$ implies that x is a positive real number, and elementary algebra shows that $x^2 + h^2 = r^2$. As before, there are two distinct points $A, B \in \ell$ such that $AF = FB = x$ (Fig. 14.5). Because $\triangle OFA$ is a right triangle with hypotenuse OA, the Pythagorean theorem shows that $OA = \sqrt{x^2 + h^2} = r$, so $A \in \mathcal{C}$. A similar computation shows that $B \in \mathcal{C}$. Since A and B are distinct points, ℓ is a secant line as claimed. $\square$

Now we turn our attention to tangent lines. Note that the definition of a tangent line to a circle stipulates only that it is a line that intersects the circle in exactly one point. But like many other concepts in Euclidean geometry, the intuitive idea of "tangency" comes freighted with other associations about how the line and the circle are situated with respect to each other—for example, we think of the "direction" of a tangent line as being determined by the circle in a certain way, and we think of a tangent line as touching the circle but not going inside it. The next two theorems justify some of those associations. The first is equivalent to Euclid's Propositions III.16 and III.18.

Theorem 14.7 (Tangent Line Theorem). *Suppose $\mathcal{C}$ is a circle and ℓ is a line that intersects $\mathcal{C}$ at a point P. Then ℓ is tangent to $\mathcal{C}$ if and only if ℓ is perpendicular to the radius through P.*

Proof. Let O be the center of $\mathcal{C}$ and let r be its radius. Suppose first that $\ell \perp \overline{OP}$; we will show that ℓ is tangent to $\mathcal{C}$ at P. Because P is the foot of the perpendicular from O to ℓ, it follows from Theorem 7.13 that any other point $Q \in \ell$ distinct from P satisfies $OQ > OP$ (Fig. 14.6). This means that $OQ \neq r$, so $Q \notin \mathcal{C}$; thus P is the only point of $\ell \cap \mathcal{C}$, which means that ℓ is tangent to $\mathcal{C}$.

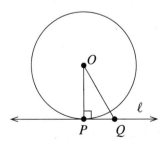

Fig. 14.6. Proof of the tangent line theorem.

Conversely, suppose ℓ is tangent to $\mathcal{C}$ at P. If Q is any point of ℓ other than P, then Q cannot lie on $\mathcal{C}$ because then ℓ would be a secant line by definition; and Q cannot lie in the interior of $\mathcal{C}$ because then ℓ would be a secant line by Theorem 14.6. Thus Q is exterior to $\mathcal{C}$, which means $OQ > r$. Since $OP = r$, it follows that P is the closest point to O on ℓ, and thus $\overline{OP} \perp \ell$ by Theorem 7.13. $\square$

Corollary 14.8. *If $\mathcal{C}$ is a circle and $A \in \mathcal{C}$, there is a unique line tangent to $\mathcal{C}$ at A.*

Proof. Given $\mathcal{C}$ and A as in the hypothesis, the preceding theorem shows that a line is tangent to $\mathcal{C}$ at A if and only if it is perpendicular to $\overline{OA}$ at A. Since there is a unique such perpendicular, there is a unique tangent line at A. $\square$

The next theorem is an analogue for tangent lines of Theorem 14.4.

Theorem 14.9 (Properties of Tangent Lines). *If $\mathcal{C}$ is a circle and ℓ is a line that is tangent to $\mathcal{C}$ at P, then every point of ℓ except P lies in the exterior of $\mathcal{C}$, and every point of $\mathcal{C}$ except P lies on the same side of ℓ as the center of $\mathcal{C}$.*

Proof. Suppose ℓ is tangent to $\mathcal{C}$ at P, and let O be the center of $\mathcal{C}$. The tangent line theorem implies that P is the foot of the perpendicular from O to ℓ. If Q is any point of ℓ other than P, then $OQ > OP$ by Theorem 7.13. This means Q is in the exterior of $\mathcal{C}$, which proves the first claim.

To prove the second, let Q be a point of $\mathcal{C}$ different from P. We need to show that $\overline{OQ}$ does not meet ℓ. Suppose for the sake of contradiction that there is a point $X \in \overline{OQ} \cap \ell$. Then X cannot be equal to Q because then ℓ would intersect $\mathcal{C}$ at the two distinct points P and Q. Thus either $X = O$ or $O * X * Q$. In either case, it follows that $OX < OQ = r$, so $X \in \mathrm{Int}\,\mathcal{C}$, contradicting the result of the preceding paragraph. $\square$

Intersections between Circles

Next we address the question of determining when two circles intersect. The next theorem fills another glaring gap in Euclid's logic. In his first proof (Proposition I.1), he constructs two circles whose radii are equal to the distance between their centers and concludes that there is a point where they intersect. The next theorem justifies this conclusion (see Exercise 14D).

Theorem 14.10 (Two Circles Theorem). *Suppose $\mathcal{C}$ and $\mathcal{D}$ are two circles. If either of the following conditions is satisfied, then there exist exactly two points where the circles intersect, one on each side of the line containing their centers.*

(a) *The following inequalities all hold:*

$$d < r + s, \qquad s < r + d, \qquad r < d + s,$$

where r and s are the radii of $\mathcal{C}$ and $\mathcal{D}$, respectively, and d is the distance between their centers.

(b) *$\mathcal{D}$ contains a point in the interior of $\mathcal{C}$ and a point in the exterior of $\mathcal{C}$.*

Proof. Write $\mathcal{C} = \mathcal{C}(O, r)$ and $\mathcal{D} = \mathcal{C}(Q, s)$, and let $d = OQ$ denote the distance between the two centers. First assume that the hypotheses of (a) are satisfied. If d were equal to zero, then the last two inequalities would imply $s < r$ and $r < s$, which is a contradiction.

Thus d is positive, and both r and s are positive by definition of circles. A point P is in the intersection of $\mathcal{C}$ and $\mathcal{D}$ if and only if $OP = r$ and $QP = s$. Because the three inequalities hold, the SSS construction theorem (Corollary 13.7) guarantees that there are exactly two such points, one on each side of $\overleftrightarrow{OQ}$. This proves that condition (a) is sufficient.

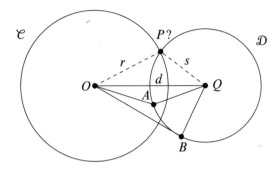

Fig. 14.7. Proof of the two circles theorem.

Now suppose that condition (b) is satisfied. We will prove that all three inequalities of (a) hold, from which the conclusion follows. The hypothesis is that there is a point $A \in \mathcal{D}$ that is in the interior of $\mathcal{C}$ and another point $B \in \mathcal{D}$ that is in the exterior of $\mathcal{C}$ (Fig. 14.7). These facts can be summarized as follows:

$$OA < r, \qquad OB > r, \qquad QA = QB = s.$$

The general triangle inequality applied to the points O, A, and Q implies

$$d = OQ \leq OA + AQ < r + s,$$
$$s = AQ \leq AO + OQ < r + d,$$

so the first two inequalities in (a) hold. (We could have written down a third triangle inequality with OA on the left-hand side, but it is not helpful in proving (a), as you can verify for yourself.) On the other hand, we can apply the general triangle inequality to O, B, and Q to obtain

$$r < OB \leq OQ + QB = d + s.$$

Thus the third inequality in (a) is satisfied, so condition (b) is also sufficient. $\qquad\square$

There is another way that two circles can intersect: we say that two circles are ***tangent to each other*** if they intersect at exactly one point (see Fig. 14.8).

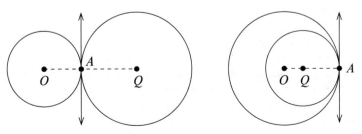

Fig. 14.8. Tangent circles.

Theorem 14.11 (Tangent Circles Theorem). *Suppose $\mathcal{C}(O,r)$ and $\mathcal{C}(Q,s)$ are tangent to each other at A. Then O, Q, and A are distinct and collinear, and $\mathcal{C}(O,r)$ and $\mathcal{C}(Q,s)$ have a common tangent line at A.*

Proof. Exercise 14E.   □

Arcs and Inscribed Angles

Roughly speaking, an *arc* is a part of a circle cut off by two points on the circle. More precisely, we make the following definitions (see Fig. 14.9). Suppose $\mathcal{C} = \mathcal{C}(O,r)$ is a

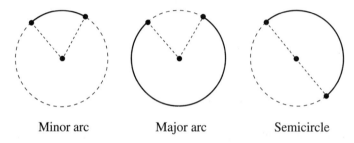

Minor arc   Major arc   Semicircle

Fig. 14.9. Arcs.

circle and A and B are two distinct points on $\mathcal{C}$. If $\angle AOB$ is a proper angle, we define the *minor arc bounded by A and B* to be the set consisting of A, B, and all points on $\mathcal{C}$ that are in the interior of $\angle AOB$; and we define the *major arc bounded by A and B* to be the set consisting of A, B, and all points on $\mathcal{C}$ that are in the exterior of $\angle AOB$. If $\angle AOB$ is a straight angle (or, equivalently, if $\overline{AB}$ is a diameter), each set consisting of A, B, and all points of $\mathcal{C}$ on one side of $\overleftrightarrow{AB}$ is called a *semicircle*. In general, an *arc* is a subset of a circle that is either a major arc, a minor arc, or a semicircle. The points A and B defining the arc are called its *endpoints*, and all other points of the arc are called its *interior points*. (We will prove shortly that endpoints and interior points are well defined, independently of the notation used to specify the arc.) It is immediate from the definitions that any two distinct points on a circle are the endpoints of exactly two different arcs that intersect only at those two points, and the entire circle is the union of those two arcs. These two arcs are called *conjugate arcs*; thus for every arc, there is a unique arc of the same circle that is conjugate to it.

Given a circle $\mathcal{C}$, an arc of $\mathcal{C}$ can be unambiguously designated by specifying its two endpoints and one point in its interior. The notation $\overset{\frown}{ACB}$ means the arc with endpoints A and B and with C as an interior point. More specifically, if $\angle AOB$ is proper and $C \in \text{Int}\angle AOB$, then $\overset{\frown}{ACB}$ means the minor arc with endpoints A and B; if $\angle AOB$ is proper and $C \in \text{Ext}\angle AOB$, then $\overset{\frown}{ACB}$ means the major arc with endpoints A and B; and if $\angle AOB$ is a straight angle, then $\overset{\frown}{ACB}$ means the semicircle defined by the intersection of $\mathcal{C}$ with $\text{CHP}\left(\overleftrightarrow{AB},C\right)$. If it is clear from the context or unimportant which of the two arcs bounded by A and B is meant, we will sometimes use the simpler notation $\overset{\frown}{AB}$ to denote one of the arcs with endpoints A and B. (Some writers always use a three-letter notation like $\overset{\frown}{ACB}$ to denote a major arc and a two-letter notation like $\overset{\frown}{AB}$ to denote a minor one; but since it is important for us to be able to refer to a generic arc $\overset{\frown}{AB}$ without knowing in advance whether it is major or minor, we do not follow that convention.)

Notice that the notation $\overset{\frown}{ACB}$ does not mention which circle is to be used to define the arc. Usually, the circle in question will be clear from the context. In any case, if an arc is identified by giving three points, it is not necessary to specify the circle because, as Corollary 14.27 will show later in this chapter, given any three noncollinear points, there is a *unique* circle that contains them all.

There is also a useful characterization of arcs in terms of half-planes. The key to proving it is the following lemma.

Lemma 14.12. *Suppose $\mathcal{C}$ is a circle and $\overset{\frown}{AB}$ is a minor arc of $\mathcal{C}$. If X is any point on $\mathcal{C}$, then X lies on $\overset{\frown}{AB}$ if and only if the radius $\overline{OX}$ meets the chord $\overline{AB}$.*

Proof. Let $\overset{\frown}{AB}$ be a minor arc of $\mathcal{C}$, and let F be the foot of the perpendicular from O to $\overleftrightarrow{AB}$. The isosceles triangle altitude theorem implies that F is the midpoint of $\overline{AB}$ (see Fig. 14.10).

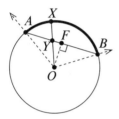
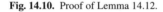
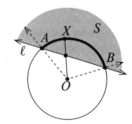

Fig. 14.10. Proof of Lemma 14.12. **Fig. 14.11.** Proof of Theorem 14.13.

Let X be an arbitrary point on $\mathcal{C}$, and assume first that $X \in \overset{\frown}{AB}$. If $X = A$ or $X = B$, then certainly $\overline{OX}$ meets $\overline{AB}$, so assume X is an interior point of the arc. This means that $\overrightarrow{OA} * \overrightarrow{OX} * \overrightarrow{OB}$, and thus by the crossbar theorem $\overrightarrow{OX}$ meets $\overline{AB}$ at a point $Y \in \mathrm{Int}\,\overline{AB}$. Then $Y \in \mathrm{Int}\,\mathcal{C}$ by Theorem 14.4, so $OY < OX$. This implies that $O * Y * X$ by the ordering lemma for points and therefore that $Y \in \overline{OX}$.

Conversely, assume that $\overline{OX}$ meets $\overline{AB}$ at a point Y. Then betweenness vs. betweenness implies that $\overrightarrow{OA} * \overrightarrow{OY} * \overrightarrow{OB}$. Because X lies on $\overrightarrow{OY}$, it is in the interior of $\angle AOB$ and therefore on the minor arc $\overset{\frown}{AB}$. $\qquad \square$

The proof of the next theorem is rated PG.

Theorem 14.13. *Suppose $\mathcal{C}$ is a circle with center O, and A, B are two distinct points on $\mathcal{C}$. Then the two arcs bounded by A and B are the intersections with $\mathcal{C}$ of the closed half-planes determined by $\overleftrightarrow{AB}$.*

Proof. Let $\ell = \overleftrightarrow{AB}$. If $O \in \ell$, then the intersections of $\mathcal{C}$ with the two closed half-planes determined by ℓ are the semicircles determined by A and B by definition. So let us assume that $O \notin \ell$. Let S denote the closed half-plane determined by ℓ that does not contain O (Fig. 14.11). We will prove first that $S \cap \mathcal{C}$ is equal to the minor arc bounded by A and B, which we denote by $\overset{\frown}{AB}$.

First let X be a point of $\overset{\frown}{AB}$. Then Lemma 14.12 shows that $\overline{OX}$ intersects $\overline{AB}$. If the intersection point is X itself, then X is equal to A or B (since these are the only points of $\overline{AB}$ on the circle), so it is certainly the case that $X \in S$. If the intersection point is not X, then it must be an interior point of $\overline{OX}$, which shows that X and O are on opposite sides of ℓ, and so $X \in S$. In either case, $X \in S \cap \mathcal{C}$.

Conversely, let X be a point of $S \cap \mathcal{C}$. If $X \in \ell$, then $X = A$ or $X = B$, which implies $X \in \widehat{AB}$. If not, then X is on the opposite side of ℓ from O, which means that the radius $\overline{OX}$ meets ℓ at a point Y. The fact that $Y \in \overline{OX}$ means that $OY \leq OX$, so Y is either on or inside $\mathcal{C}$. Theorem 14.4 shows that $Y \in \overline{AB}$, and then Lemma 14.12 shows that X lies on the minor arc $\widehat{AB}$. This completes the proof that $S \cap \mathcal{C}$ is equal to the minor arc $\widehat{AB}$.

Now we have to show that $\mathrm{CHP}(\ell, O) \cap \mathcal{C}$ is equal to the arc conjugate to $\widehat{AB}$. If $X \in \mathrm{CHP}(\ell, O) \cap \mathcal{C}$, then either it is equal to A or B, in which case it is on the conjugate arc by definition, or it is a point of $\mathcal{C}$ on the same side of ℓ as O; by the previous argument, this implies that $X \notin \widehat{AB}$, so it must be on the conjugate arc. Conversely, if X is on the conjugate arc, then either it is equal to A or B, both of which are in $\mathrm{CHP}(\ell, O) \cap \mathcal{C}$, or it is a point of $\mathcal{C}$ that is not in $\widehat{AB}$, which by the previous argument implies it is not in the closed half-plane S, so it is in $\mathrm{CHP}(\ell, O) \cap \mathcal{C}$.   □

Now we can show that endpoints (and thus also interior points) of arcs are well defined. Note that we cannot characterize the endpoints of an arc as its only extreme points, because an arc has no passing points. This proof is rated PG.

Theorem 14.14 (Consistency of Endpoints of Arcs). *Suppose A, B, C, D are points on a circle $\mathcal{C}$, with $A \neq B$ and $C \neq D$. If $\widehat{AB} = \widehat{CD}$, then $\{A, B\} = \{C, D\}$.*

Proof. Let X be any point of $\widehat{AB}$ other than A and B, and let S denote the closed half-plane $\mathrm{CHP}\left(\overleftrightarrow{AB}, X\right)$. Then the previous theorem shows that $S \cap \mathcal{C} = \widehat{AB} = \widehat{CD}$. In particular, this means that both C and D lie in S. Suppose for the sake of contradiction that one of them, say C, is distinct from A and B. Note that A and B are the only points of $\overleftrightarrow{AB}$ that lie on $\mathcal{C}$, so C must lie in the *open* half-plane $\mathrm{OHP}\left(\overleftrightarrow{AB}, X\right)$. Because S is convex, the chord $\overline{CD}$ is contained in S, and in fact all of the chord except perhaps D is contained in $\mathrm{OHP}\left(\overleftrightarrow{AB}, X\right)$ (Fig. 14.12). Let M be the midpoint of $\overline{CD}$; thus M lies in the interior of $\mathcal{C}$ and on the same side of $\overleftrightarrow{AB}$ as C and X. Because $\overleftrightarrow{AM}$ contains the point $M \in \mathrm{Int}\,\mathcal{C}$, it is a secant line and thus has another intersection point with $\mathcal{C}$; call that point Y. The Y-lemma shows that Y lies on the same side of $\overleftrightarrow{AB}$ as M (and thus also as X and C), so $Y \in \widehat{AB}$. Because M is an interior point of $\overline{AY}$ by Theorem 14.4, it follows that Y and A are on opposite sides of $\overleftrightarrow{CD}$. But this contradicts Theorem 14.13, which implies that $\widehat{AB} = \widehat{CD}$ is contained in one half-plane determined by $\overleftrightarrow{CD}$.   □

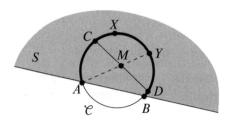

Fig. 14.12. Proving consistency of arc endpoints.

Central Angles and Inscribed Angles

Every arc is naturally associated with an angle whose vertex is at the center of the circle, and we can use this angle to measure the "size" of the arc in degrees, as follows. If $\overarc{AB}$ is any arc on $\mathcal{C}(O,r)$, the angle $\angle AOB$ is called the **central angle associated with** $\overarc{AB}$. We define the **arc measure** (or just **measure**) **of** $\overarc{AB}$, denoted by $m\overarc{AB}$, as follows:

- $m\overarc{AB} = m\angle AOB$ (the standard measure of $\angle AOB$) if $\overarc{AB}$ is a minor arc;
- $m\overarc{AB} = rm\angle AOB$ (the reflex measure of $\angle AOB$) if $\overarc{AB}$ is a major arc;
- $m\overarc{AB} = 180°$ if $\overarc{AB}$ is a semicircle.

Lemma 14.15. *Any two conjugate arcs have measures adding up to 360°.*

Proof. Two conjugate arcs are either a minor arc and its conjugate major arc, in which case their measures add up to 360° by definition, or two semicircles, in which case both measures are 180°. □

There is another kind of angle that appears in the study of circles and is closely related to arcs. If $\mathcal{C}$ is a circle, an angle is said to be **inscribed in** $\mathcal{C}$ if its vertex lies on $\mathcal{C}$ and the interiors of both of its rays intersect $\mathcal{C}$ (Fig. 14.13).

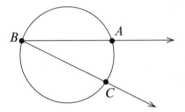

Fig. 14.13. An inscribed angle.

Suppose $\angle ABC$ is an angle inscribed in $\mathcal{C}$, and suppose further that A and C are the two points where the sides of the angle meet $\mathcal{C}$. The points A and C determine two arcs of $\mathcal{C}$: the arc $\overarc{ABC}$, containing B, and its conjugate arc, which we denote simply by $\overarc{AC}$. We call $\overarc{ABC}$ the **inscribed arc** for $\angle ABC$, and we call the conjugate arc $\overarc{AC}$ its **intercepted arc**. In this situation, we also say that $\angle ABC$ is **inscribed in** $\overarc{ABC}$. In particular, an **angle inscribed in a semicircle** is an inscribed angle whose vertex lies on the circle and whose sides intersect the circle at the endpoints of a diameter.

The oldest known relationship between inscribed angles and arcs is very old indeed: the first proof of the following theorem is generally attributed to the Greek mathematician Thales (pronounced "*Thay*-leez") of Miletus, who lived in Asia Minor almost 300 years before Euclid. This result appears in Euclid as part of Proposition III.31, with essentially the same proof as the one given here.

Theorem 14.16 (Thales' Theorem). *Any angle inscribed in a semicircle is a right angle.*

Proof. Suppose $\angle ABC$ is inscribed in a semicircle, and let $\mathcal{C} = \mathcal{C}(O,r)$ be the circle containing the semicircle (Fig. 14.14). Suppose further that A and C have been chosen to lie on $\mathcal{C}$. Then $\overline{AC}$ is a diameter of $\mathcal{C}$, and $OA = OB = OC = r$. Because $\triangle AOB$ and $\triangle BOC$ are both isosceles, we have $\angle A \cong \angle 1$ and $\angle C \cong \angle 2$, where $\angle 1 = \angle ABO$ and

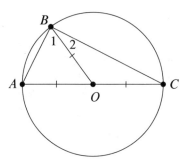

Fig. 14.14. The proof of Thales' theorem.

$\angle 2 = \angle OBC$ as in the diagram. Betweenness vs. betweenness implies that $m\angle ABC = m\angle 1 + m\angle 2$. Therefore, by the angle-sum theorem for triangles,

$$180° = m\angle ABC + m\angle A + m\angle C$$
$$= m\angle ABC + m\angle 1 + m\angle 2$$
$$= m\angle ABC + m\angle ABC.$$

It follows that $m\angle ABC = 90°$ as claimed. $\square$

Thales' theorem has the following important converse.

Theorem 14.17 (Converse to Thales' Theorem). *The hypotenuse of a right triangle is a diameter of a circle that contains all three vertices.*

Proof. Exercise 14I. $\square$

One important application of Thales' theorem is to prove the existence of tangent lines containing a point outside a circle. (This can be viewed as a complement to Corollary 14.8, which proved the existence of a tangent line containing a point *on* the circle.) This result appears in Euclid as Proposition III.17, but with a different proof.

Theorem 14.18 (Existence of Tangent Lines through an Exterior Point). *Let $\mathcal{C}$ be a circle, and let A be a point in the exterior of $\mathcal{C}$. Then there are exactly two distinct tangent lines to $\mathcal{C}$ containing A. The two points of tangency X and Y are equidistant from A, and the center of $\mathcal{C}$ lies on the bisector of $\angle XAY$.*

Proof. Let O be the center of $\mathcal{C}$. To see how to proceed, suppose X is a point on $\mathcal{C}$ such that $\overleftrightarrow{AX}$ is tangent to $\mathcal{C}$ at X. Then $\angle AXO$ is a right angle by the tangent line theorem, and therefore AO is a diameter of a circle that contains A, O, and X by the converse to Thales' theorem. Thus we can look for tangent lines by constructing a circle whose diameter is AO and finding the points where that circle intersects $\mathcal{C}$. (See Fig. 14.15.)

Let $\mathcal{D}$ be the circle whose center is the midpoint of $\overline{AO}$ and whose radius is $\frac{1}{2}AO$; thus $\overline{AO}$ is a diameter of $\mathcal{D}$. Then $\mathcal{D}$ contains the point O in the interior of $\mathcal{C}$ and the point A in the exterior of $\mathcal{C}$, so the two circles theorem shows that $\mathcal{C}$ and $\mathcal{D}$ have exactly two points of intersection—call them X and Y. Because $\angle AXO$ is inscribed in a semicircle of $\mathcal{D}$, it is a right angle by Thales' theorem. Thus $\overleftrightarrow{AX}$ is tangent to $\mathcal{C}$ by the tangent line theorem. The same argument shows that $\overleftrightarrow{AY}$ is tangent to $\mathcal{C}$. Because $OX = OY$, the

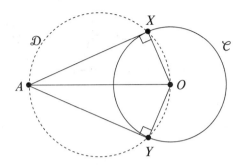

Fig. 14.15. Proof of Theorem 14.18.

triangles $\triangle AXO$ and $\triangle AYO$ are congruent by HL. Therefore, $AX = AY$, which means that the points of tangency are equidistant from A. Moreover, $\overline{OX}$ consists of interior points of $\mathcal{C}$ except for X, and $\overleftrightarrow{AY}$ consists of exterior points except for Y, so $\overline{OX}$ does not intersect $\overleftrightarrow{AY}$ and thus O is on the same side of $\overleftrightarrow{AY}$ as X; similarly O is on the same side of $\overleftrightarrow{AX}$ as Y. This means $O \in \operatorname{Int} \angle XAY$, and since $\angle OAX \cong \angle OAY$, it follows that O lies on the bisector of $\angle XAY$.

To prove that these are the only tangent lines, note that the argument in the first paragraph shows that if ℓ is any line through A and tangent to $\mathcal{C}$, then the point of tangency has to lie on both $\mathcal{C}$ and $\mathcal{D}$, so it must be equal to either X or Y. □

The next theorem is a significant generalization of Thales' theorem. Its proof is rated PG.

Theorem 14.19 (Inscribed Angle Theorem). *The measure of a proper angle inscribed in a circle is one-half the measure of its intercepted arc.*

Proof. Suppose $\angle ABC$ is a proper angle inscribed in the circle $\mathcal{C} = \mathcal{C}(O, r)$, with A and C chosen to lie on the circle. Let D be the other endpoint of the diameter containing B, and let ℓ be the line tangent to $\mathcal{C}$ at B (see the upper left diagram in Fig. 14.16). Note that the points A, C, and D all lie on the same side of ℓ by Theorem 14.9. It follows that $\overrightarrow{BA}$, $\overrightarrow{BC}$, and $\overrightarrow{BD}$ all lie in one half-rotation, so if they are all distinct, one of them must be between the other two. The proof of the theorem is divided into five cases, depending on the arrangement of these rays. All five cases are illustrated in Fig. 14.16.

CASE 1: $\overrightarrow{BD}$ *is equal to* $\overrightarrow{BA}$ *or* $\overrightarrow{BC}$. Without loss of generality, suppose $\overrightarrow{BD} = \overrightarrow{BC}$. Then $B * O * C$, and $\overline{OB}$ and $\overline{OA}$ are congruent because they are both radii of $\mathcal{C}$. It follows that $\angle ABC$ and $\angle OAB$ are congruent by the isosceles triangle theorem. Since $\angle AOC$ is an exterior angle for $\triangle OAB$, we have

$$m\angle AOC = m\angle ABC + m\angle OAB = 2m\angle ABC.$$

Because B lies in the exterior of $\angle AOC$, it follows that the arc $\overarc{AC}$ intercepted by $\angle ABC$ is a minor arc, so $m\overarc{AC} = m\angle AOC = 2m\angle ABC$.

CASE 2: *Either* $\overrightarrow{BA} * \overrightarrow{BC} * \overrightarrow{BD}$ *or* $\overrightarrow{BC} * \overrightarrow{BA} * \overrightarrow{BD}$. The argument is the same in both cases, so we may as well assume the former. By the argument in Case 1 applied to

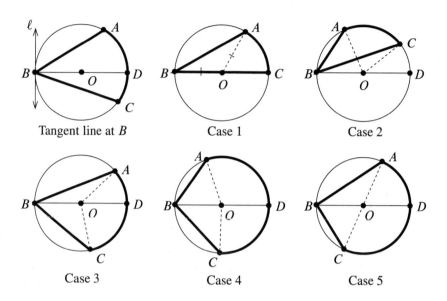

Fig. 14.16. Proof of the inscribed angle theorem.

the inscribed angles $\angle ABD$ and $\angle CBD$, we have

$$m\angle AOD = 2m\angle ABD,$$
$$m\angle COD = 2m\angle CBD. \tag{14.1}$$

Because $\overrightarrow{BA} * \overrightarrow{BC} * \overrightarrow{BD}$, we have $m\angle ABD > m\angle CBD$ (the whole angle is greater than the part), and therefore it follows from (14.1) that $m\angle AOD > m\angle COD$ as well. By the ordering lemma for rays, therefore, we also have $\overrightarrow{OA} * \overrightarrow{OC} * \overrightarrow{OD}$. Subtracting the second equation in (14.1) from the first and using the betweenness theorem for rays, we find $m\angle AOC = 2m\angle ABC$. Since B is exterior to $\angle AOC$, it follows that the intercepted arc $\overset{\frown}{AC}$ is a minor arc, so again we conclude that $m\overset{\frown}{AC} = 2m\angle ABC$.

The remaining possibility is that $\overrightarrow{BD}$ is between $\overrightarrow{BA}$ and $\overrightarrow{BC}$. We need to consider this possibility in three separate cases, depending on the relationship between $\angle AOC$ and D.

CASE 3: $\overrightarrow{BA} * \overrightarrow{BD} * \overrightarrow{BC}$, $\angle AOC$ *is a proper angle, and* $D \in \text{Int}\angle AOC$. Again by the argument in Case 1, the two equations in (14.1) hold. In this case, the hypotheses imply $m\angle ABC = m\angle ABD + m\angle CBD$ and $m\angle AOC = m\angle AOD + m\angle COD$, so (14.1) implies $m\angle AOC = 2m\angle ABC$. Since D is in the interior of $\angle AOC$, the intercepted arc $\overset{\frown}{ADC}$ is again a minor arc, so $m\overset{\frown}{ADC} = m\angle AOC = 2m\angle ABC$.

CASE 4: $\overrightarrow{BA} * \overrightarrow{BD} * \overrightarrow{BC}$, $\angle AOC$ *is a proper angle, and* $D \in \text{Ext}\angle AOC$. Because A and C are on opposite sides of $\overleftrightarrow{BD}$ (by Theorem 4.12), A cannot be in the interior of $\angle COD$, nor can C be in the interior of $\angle AOD$. Thus the hypotheses of the 360 theorem are satisfied by the rays $\overrightarrow{OA}$, $\overrightarrow{OD}$, and $\overrightarrow{OC}$, from which we conclude that

$$360° - m\angle AOC = m\angle AOD + m\angle COD.$$

On the other hand, (14.1) holds by the same argument as before. Because the arc $\overset{\frown}{ADC}$ intercepted by $\angle ABC$ is a major arc,

$$m\overset{\frown}{ADC} = 360° - m\angle AOC$$
$$= m\angle AOD + m\angle COD$$
$$= 2m\angle ABD + 2m\angle CBD = 2m\angle ABC.$$

CASE 5: $\overrightarrow{BA} * \overrightarrow{BD} * \overrightarrow{BC}$, *and* $\angle AOC$ *is a straight angle.* In this case, $\overline{AC}$ is a diameter and thus $\overset{\frown}{ABC}$ and $\overset{\frown}{ADC}$ are semicircles. Thales' theorem then implies that $m\angle ABC = 90°$, which is half the measure of the intercepted arc $\overset{\frown}{ADC}$. □

The inscribed angle theorem has several important corollaries.

Corollary 14.20 (Arc Addition Theorem). *Suppose A, B, and C are three distinct points on a circle* $\mathcal{C}$, *and* $\overset{\frown}{AB}$ *and* $\overset{\frown}{BC}$ *are arcs that intersect only at B. Then* $m\overset{\frown}{ABC} = m\overset{\frown}{AB} + m\overset{\frown}{BC}$.

Proof. Exercise 14L. □

Corollary 14.21 (Intersecting Chords Theorem). *Suppose* $\overline{AB}$ *and* $\overline{CD}$ *are two distinct chords of a circle* $\mathcal{C}$ *that intersect at a point* $P \in \mathrm{Int}\,\mathcal{C}$ *(Fig. 14.17). Then*

$$(PA)(PB) = (PC)(PD). \tag{14.2}$$

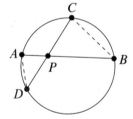

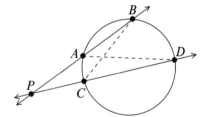

Fig. 14.17. The intersecting chords theorem.   **Fig. 14.18.** The intersecting secants theorem.

Proof. The Y-lemma shows that B and D are on the same side of $\overleftrightarrow{AC}$ as P, and thus they both lie in the same arc bounded by A and C by Lemma 14.12. It follows that $\overset{\frown}{ABC} = \overset{\frown}{ADC}$, and the arc $\overset{\frown}{AC}$ conjugate to $\overset{\frown}{ABC}$ is intercepted by both $\angle ABC$ and $\angle ADC$. Thus the inscribed angle theorem yields

$$m\angle ADC = \tfrac{1}{2} m\overset{\frown}{AC} = m\angle ABC.$$

Because $\angle APD$ and $\angle CPB$ are vertical angles, it follows from the AA similarity theorem that $\triangle PAD \sim \triangle PCB$. Thus $PA/PC = PD/PB$, which is equivalent to (14.2). □

Corollary 14.22 (Intersecting Secants Theorem). *Suppose two distinct secant lines of a circle* $\mathcal{C}$ *intersect at a point* P *exterior to* $\mathcal{C}$. *Let* A, B *be the points where one of the secants meets* $\mathcal{C}$, *and let* C, D *be the points where the other one does (Fig. 14.18). Then*

$$(PA)(PB) = (PC)(PD). \tag{14.3}$$

Proof. By interchanging the names of A and B if necessary (which doesn't affect (14.3)), we can arrange that $P * A * B$. Similarly, we may assume $P * C * D$. Then B and D are on the opposite side of $\overleftrightarrow{AC}$ from P, and so on the same side as each other. Reasoning as in the previous proof, we conclude that the same arc $\overparen{AC}$ is intercepted by both $\angle ABC$ and $\angle ADC$. Thus these angles are congruent, so $\triangle PAD \sim \triangle PCB$ by AA similarity, and the result follows. □

Inscribed and Circumscribed Polygons

Next we explore the ways in which polygons and circles interact. Suppose $\mathcal{P}$ is a polygon and $\mathcal{C}$ is a circle. We say that $\mathcal{P}$ *is inscribed in* $\mathcal{C}$, or equivalently that $\mathcal{C}$ *is circumscribed about* $\mathcal{P}$, if all of the vertices of $\mathcal{P}$ lie on $\mathcal{C}$ (Fig. 14.19).

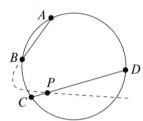

Fig. 14.19. A polygon inscribed in a circle.   **Fig. 14.20.** Semiparallel chords.

Not every polygon can be inscribed in a circle. For example, as the next theorem shows, a necessary condition is that the polygon be convex.

Theorem 14.23. *Every polygon inscribed in a circle is convex.*

Proof. Suppose $\mathcal{C}$ is a circle and $\mathcal{P}$ is a polygon inscribed in $\mathcal{C}$. To show that $\mathcal{P}$ is convex, it suffices to show that any two nonadjacent edges are semiparallel. Since nonadjacent edges are nonintersecting chords of $\mathcal{C}$, it suffices to show that nonintersecting chords are semiparallel. Suppose, therefore, that $\overline{AB}$ and $\overline{CD}$ are nonintersecting chords, and assume for the sake of contradiction that they are not semiparallel. Without loss of generality, we may assume that $\overleftrightarrow{AB}$ intersects $\overline{CD}$ at a point P (Fig. 14.20). Now P cannot be in $\overline{AB}$ because we are assuming the chords do not intersect. By Theorem 14.4 applied to the secant line $\overleftrightarrow{AB}$, this means that P is an exterior point of $\mathcal{C}$. On the other hand, because $P \in \overline{CD}$, the same theorem applied to the secant line $\overleftrightarrow{CD}$ implies that P is an interior point of $\mathcal{C}$ (if $C * P * D$) or on $\mathcal{C}$ (if $P = C$ or $P = D$). In either case, this is a contradiction, so $\overline{AB}$ and $\overline{CD}$ are semiparallel. □

A polygon that can be inscribed in a circle is called a *cyclic polygon*. If $\mathcal{P}$ is a cyclic polygon, then a circle that is circumscribed about $\mathcal{P}$ (or, equivalently, in which $\mathcal{P}$ is inscribed) is called a *circumcircle for* $\mathcal{P}$, and the center of the circumcircle is called a *circumcenter for* $\mathcal{P}$. The next theorem shows, among other things, that every cyclic polygon has a *unique* circumcircle and circumcenter and describes how to find them. Recall that two or more lines, rays, or segments are said to be *concurrent* if there is a point that lies on all of them.

Theorem 14.24 (Circumcircle Theorem). *A polygon $\mathcal{P}$ is cyclic if and only if the perpendicular bisectors of all of its edges are concurrent. If this is the case, the point O where these perpendicular bisectors intersect is the unique circumcenter for $\mathcal{P}$, and the circle with center O and radius equal to the distance from O to any vertex is the unique circumcircle.*

Proof. Suppose first that $\mathcal{P}$ is cyclic, and let $\mathcal{C}$ be a circumcircle for $\mathcal{P}$ (Fig. 14.21). Then each edge of $\mathcal{P}$ is a chord for $\mathcal{C}$. By Theorem 14.5, the perpendicular bisector of each edge passes through the center of $\mathcal{C}$, so it follows that the perpendicular bisectors are all concurrent.

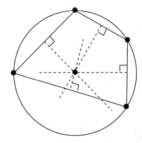

Fig. 14.21. The circumcircle theorem.

Conversely, suppose that the perpendicular bisectors of all of the edges of $\mathcal{P}$ are concurrent, and let O be a point where they all meet. By the perpendicular bisector theorem (Theorem 7.8), O is equidistant from both endpoints of each edge. If we list the vertices in sequence as $A_1, \ldots, A_n$, this implies $OA_1 = OA_2$, $OA_2 = OA_3$, $\ldots$, and $OA_{n-1} = OA_n$. If we let $r = OA_1$, it then follows that all of the vertices lie on the circle $\mathcal{C}(O,r)$, so $\mathcal{P}$ is cyclic.

To prove that the circumcircle is unique, suppose $\mathcal{C}(O,r)$ and $\mathcal{C}(O',r')$ are both circumcircles for $\mathcal{P}$. Then the perpendicular bisectors of all the edges of $\mathcal{P}$ pass through both O and O'. If $O' \neq O$, this means that all of the perpendicular bisectors are the same line, which means that all of the edges of $\mathcal{P}$ are either collinear or parallel. Since adjacent edges intersect and are noncollinear, this is impossible. Thus $O = O'$. Because the two circumcircles are concentric and both pass through the vertices, they are the same circle by Lemma 14.2. $\qquad\square$

Using this theorem, it is easy to come up with examples of convex quadrilaterals that are not cyclic (see Exercise 14U). For triangles, however, the situation is much simpler. This theorem is Euclid's Proposition IV.5, and our proof is essentially the same as his.

Theorem 14.25 (Cyclic Triangle Theorem). *Every triangle is cyclic.*

Proof. Let $\triangle ABC$ be any triangle (Fig. 14.22). Let ℓ be the perpendicular bisector of $\overline{AB}$, and let m be the perpendicular bisector of $\overline{BC}$. If ℓ were parallel to m, then $\overleftrightarrow{AB}$ would be parallel to $\overleftrightarrow{BC}$ by Corollary 10.6, which is impossible by definition of a triangle. Thus ℓ and m must intersect.

Let O be a point where ℓ and m intersect. Because O is on the perpendicular bisector of $\overline{AB}$, it is equidistant from A and B by the perpendicular bisector theorem; and because it

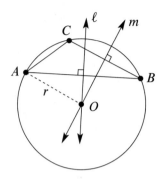

Fig. 14.22. Proof that every triangle is cyclic.

is also on the perpendicular bisector of $\overline{BC}$, it is equidistant from B and C. In other words, $OA = OB = OC$. If we set $r = OA$, it follows that $\triangle ABC$ is inscribed in $\mathcal{C}(O,r)$, so it is cyclic.   □

Corollary 14.26 (Perpendicular Bisector Concurrence Theorem). *In any triangle, the perpendicular bisectors of all three sides are concurrent.*

Proof. The circumcenter is a point where the perpendicular bisectors meet.   □

Corollary 14.27. *Given three noncollinear points, there is a unique circle that contains all of them.*

Proof. Three noncollinear points determine a triangle, and the circle in question is the triangle's circumcircle.   □

As we mentioned above, not every quadrilateral is cyclic, but there is a simple necessary and sufficient criterion. The "only if" part of the next theorem is Euclid's Proposition III.22. The proof of the theorem is rated PG.

Theorem 14.28 (Cyclic Quadrilateral Theorem). *A quadrilateral $ABCD$ is cyclic if and only if it is convex and both pairs of opposite angles are supplementary:* $m\angle A + m\angle C = 180°$ *and* $m\angle B + m\angle D = 180°$.

Proof. Suppose first that $ABCD$ is a cyclic quadrilateral. Theorem 14.23 shows that it is convex. Let $\mathcal{C}$ be its circumcircle (Fig. 14.23); then $\angle A$ is inscribed in $\mathcal{C}$, and its intercepted arc is $\overset{\frown}{BCD}$. Similarly, $\angle C$ is inscribed in $\mathcal{C}$ with intercepted arc $\overset{\frown}{DAB}$. Since the measures of conjugate arcs add up to 360° by Lemma 14.15, the inscribed angle theorem yields

$$m\angle A + m\angle C = \tfrac{1}{2} m\overset{\frown}{BCD} + \tfrac{1}{2} m\overset{\frown}{DAB} = \tfrac{1}{2}\left(m\overset{\frown}{BCD} + m\overset{\frown}{DAB}\right) = 180°.$$

Exactly the same argument shows that $\angle B$ and $\angle D$ are supplementary. This proves that the opposite angle condition is necessary.

Conversely, suppose $ABCD$ is a quadrilateral that is not cyclic. We need to prove that either it is nonconvex or it has a pair of nonsupplementary opposite angles. If it is nonconvex, we are done, so assume henceforth that $ABCD$ is convex. Let $\mathcal{C}$ be the circumcircle for $\triangle ABC$. Because $ABCD$ is not cyclic, D does not lie on $\mathcal{C}$ (Fig. 14.24).

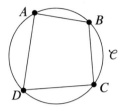

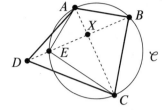

Fig. 14.23. A cyclic quadrilateral.   **Fig. 14.24.** Proving sufficiency.

The diagonals $\overline{AC}$ and $\overline{BD}$ have a point X in common by Theorem 9.4. Because X is an interior point of the chord $\overline{AC}$, it is in the interior of $\mathcal{C}$ by Theorem 14.4. Because $\overleftrightarrow{BD}$ contains X, the line-circle theorem shows that it is a secant line, and therefore it meets $\mathcal{C}$ at a second point E. Because all points of $\mathcal{C}$ are on one side of the tangent to $\mathcal{C}$ at B (Theorem 14.9), E must lie on $\overrightarrow{BD}$ and not its opposite ray. Thus either $B * E * D$ or $B * D * E$. The argument is nearly the same in both cases, so we will just give the details for the former case. The quadrilateral $ABCE$ is cyclic, so the result of the first paragraph shows that its opposite angles are supplementary. In particular, this means

$$m\angle BCE + m\angle BAE = 180°. \tag{14.4}$$

From betweenness vs. betweenness, we conclude that $\overrightarrow{AB} * \overrightarrow{AE} * \overrightarrow{AD}$, which implies that $m\angle BAD > m\angle BAE$. Similar reasoning shows that $m\angle BCD > m\angle BCE$. Adding these inequalities and using (14.4), we conclude that $m\angle BAD + m\angle BCD > 180°$, so $\angle BAD$ and $\angle BCD$ are not supplementary. □

Circumscribed Polygons

Now we introduce a concept that is, in a certain sense, complementary to inscribed polygons. If $\overline{AB}$ is a segment and $\mathcal{C}$ is a circle, we say that $\overline{AB}$ *is tangent to* $\mathcal{C}$ if the line $\overleftrightarrow{AB}$ is tangent to $\mathcal{C}$ and the point of tangency lies on $\overline{AB}$. If $\mathcal{P}$ is a polygon and $\mathcal{C}$ is a circle, we say that $\mathcal{P}$ *is circumscribed about* $\mathcal{C}$, or equivalently that $\mathcal{C}$ *is inscribed in* $\mathcal{P}$, if every edge of $\mathcal{P}$ is tangent to $\mathcal{C}$ (Fig. 14.25). If $\mathcal{C}$ is inscribed in $\mathcal{P}$, we say that $\mathcal{C}$ is an *incircle for* $\mathcal{P}$ and its center is an *incenter for* $\mathcal{P}$. A polygon that has an incircle is said to be a *tangential polygon*.

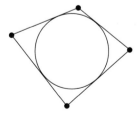

Fig. 14.25. A tangential polygon.

Lemma 14.29. *If $\mathcal{C}$ is an incircle for a polygon $\mathcal{P}$, then the point of tangency for each edge is an interior point of the edge.*

Proof. Let $\mathcal{C}$ be an incircle for $\mathcal{P}$, and assume for contradiction that some edge is tangent to $\mathcal{C}$ at a vertex V. Then the other edge that shares the vertex V is also tangent to $\mathcal{C}$ at V, and by Corollary 14.8, this means that the two edges are collinear. This contradicts the definition of a polygon.   □

Just like cyclic polygons, tangential polygons are always convex. This fact is considerably harder to prove than its cyclic counterpart. We begin with a technical lemma, which will also be useful in the next chapter. This proof is rated PG.

Lemma 14.30. *Let $\mathcal{P}$ be a polygon circumscribed about a circle $\mathcal{C}$. Suppose A is any vertex of $\mathcal{P}$, and E and F are the points of tangency of the two edges containing A. Then there are no points of $\mathcal{P}$ in the interior of $\triangle AEF$.*

Proof. Suppose for the sake of contradiction that there is a point of $\mathcal{P}$ in $\operatorname{Int}\triangle AEF$. Label the vertices of $\mathcal{P}$ as $A_1,\ldots,A_n$ with the labels chosen so that $A = A_1$. Let i be the smallest index such that $\overline{A_iA_{i+1}}$ contains a point in $\operatorname{Int}\triangle AEF$, and let X be such a point (Fig. 14.26). We wish to show that A_i is an exterior point of $\triangle AEF$. Note that E and F are interior points of their respective edges by Lemma 14.29, so A is the only vertex of $\mathcal{P}$ on $\triangle AEF$. Now A_i cannot be equal to A, because the two edges containing A intersect the triangle $\triangle AEF$ itself but not its interior; thus A_i does not lie on $\triangle AEF$. It is also the case that $A_i \notin \operatorname{Int}\triangle AEF$, because A_i is also on the edge $\overline{A_{i-1}A_i}$, which does not contain any interior points of $\triangle AEF$ by our choice of i. Therefore A_i is an exterior point of $\triangle AEF$ and X is an interior point, which implies that $\overline{A_iX}$ contains a point on $\triangle AEF$ by Lemma 8.11. Since $\overline{A_iX} \subseteq \overline{A_iA_{i+1}}$, we also conclude that $\overline{A_iA_{i+1}}$ intersects $\triangle AEF$. Now, the edge $\overline{A_iA_{i+1}}$ is contained in a tangent line, so it contains no interior points of $\mathcal{C}$ and therefore no interior points of the chord $\overline{EF}$ by Theorem 14.5. On the other hand, $\overline{A_iA_{i+1}}$ cannot intersect either $\overline{AE}$ or $\overline{AF}$, because distinct edges of $\mathcal{P}$ can intersect only at a common vertex. This is a contradiction, and thus the lemma is proved.   □

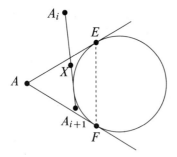

Fig. 14.26. Proof of Lemma 14.30.

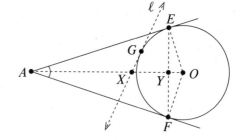

Fig. 14.27. Proof of Theorem 14.31.

The next proof is rated R.

Theorem 14.31. *Every tangential polygon is convex.*

Proof. Suppose $\mathcal{P}$ is a tangential polygon, and let $\mathcal{C} = \mathcal{C}(O,r)$ be its inscribed circle. Let ℓ be an edge line of $\mathcal{P}$; we will show that all vertices of $\mathcal{P}$ are contained in $\operatorname{CHP}(\ell,O)$. Assume for the sake of contradiction that this is not the case: then there is a vertex A that is on the opposite side of ℓ from O. This implies that there is a point X where $\overrightarrow{AO}$ meets ℓ (Fig. 14.27). Let G be the point of tangency of ℓ with $\mathcal{C}$, so G is a point of $\mathcal{P}$. We will show that G lies in the interior of $\triangle AEF$, which contradicts the result of Lemma 14.30.

Because G lies on $\mathcal{C}$ and is not equal to E or F, Theorem 14.9 guarantees that G lies on the same side of $\overleftrightarrow{AE}$ as F and on the same side of $\overleftrightarrow{AF}$ as E. To prove that $G \in \operatorname{Int} \triangle AEF$, it remains only to show that G lies on the same side of $\overleftrightarrow{EF}$ as A. Since O lies in the interior of $\angle EAF$, the crossbar theorem shows that there is a point Y where the ray $\overrightarrow{AO}$ meets $\overline{EF}$. We need to show that Y is actually an interior point of the segment $\overline{AO}$. It is easy to check that $\triangle AEO \cong \triangle AFO$ by SSS, which implies $\angle EAO \cong \angle FAO$ and therefore $\triangle EAY \cong \triangle FAY$ by SAS. Consequently, $\angle EYA$ and $\angle FYA$ are congruent to each other and thus are both right angles. Since $\overline{EY}$ is the altitude to the hypotenuse of the right triangle $\triangle AEO$, it follows that $Y \in \operatorname{Int} \overline{AO}$ as claimed.

Because we are assuming that A and O lie on opposite sides of ℓ, we have $A * X * O$; and since Y is an interior point of $\mathcal{C}$ and X is exterior, we have $OX < OY$ and therefore $X * Y * O$ by the ordering lemma for points. Together, these two relations imply $A * X * Y * O$ by Theorem 3.16, which in turn implies $A * X * Y$. Thus A and X are on the same side of $\overleftrightarrow{EF}$. If G and X were on opposite sides, then $\overline{GX}$ would intersect $\overleftrightarrow{EF}$ at a point Z. Now, G and X are both interior points of $\angle EAF$, which is a convex set, so this means that Z would also be in $\operatorname{Int} \angle EAF$. The portion of $\overleftrightarrow{EF}$ in $\operatorname{Int} \angle EAF$ is exactly $\operatorname{Int} \overline{EF}$, so Z lies on $\overline{EF}$. This is a contradiction, because Z lies on the tangent line ℓ, which contains no interior points of $\mathcal{C}$.

We have shown that for each edge line of $\mathcal{P}$, all of the vertices of $\mathcal{P}$ lie in one of the closed half-planes determined by that edge line. Since closed half-planes are convex sets, it follows that all of $\mathcal{P}$ lies in the same closed half-plane, and thus $\mathcal{P}$ is a convex polygon. $\qquad\square$

Corollary 14.32. *If $\mathcal{P}$ is a tangential polygon, then the interior of its incircle is contained in the interior of $\mathcal{P}$.*

Proof. Suppose $\mathcal{P}$ is a tangential polygon and $\mathcal{C}$ is its incircle. Let X be any point in $\operatorname{Int} C$. Because $\mathcal{P}$ is convex, to prove that $X \in \operatorname{Int} \mathcal{P}$, we need only show that X lies on the $\mathcal{P}$-side of each of the edge lines of $\mathcal{P}$. Let ℓ be one of the edge lines, so ℓ is a tangent line to $\mathcal{C}$ (Fig. 14.28). It follows from Theorem 14.9 that every point of $\mathcal{C}$ lies on the $\mathcal{P}$-side

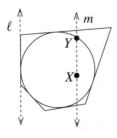

Fig. 14.28. Proof of Corollary 14.32.

of ℓ. If we let m be the line through X and parallel to ℓ, then m is a secant line by the line-circle theorem. Let Y be one of the points where m meets $\mathcal{C}$. Then Y lies on the $\mathcal{P}$-side of ℓ, and $\overline{XY}$ does not meet ℓ, so X lies on the $\mathcal{P}$-side as well. $\qquad\square$

We are now ready for the main theorem about inscribed circles. Because most of the groundwork has already been laid, we leave the proof as an exercise.

Theorem 14.33 (Incircle Theorem). *A polygon $\mathcal{P}$ is tangential if and only if it is convex and the bisectors of all of its angles are concurrent. If this is the case, the point O where these bisectors intersect is the unique incenter for $\mathcal{P}$, and the circle with center O and radius equal to the distance from O to any edge line is the unique incircle.*

Proof. Exercise 14O.   □

The next theorem is Euclid's Proposition IV.4, with the same proof.

Theorem 14.34 (Tangential Triangle Theorem). *Every triangle is tangential.*

Proof. Let $\triangle ABC$ be a triangle. Let $\vec{r}$ be the bisector of $\angle A$ and let $\vec{s}$ be the bisector of $\angle B$ (Fig. 14.29). By virtue of the incircle theorem, we just need to show that these rays intersect and that their intersection point also lies on the bisector of $\angle C$.

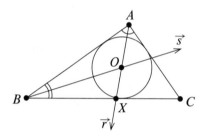

Fig. 14.29. Proof that every triangle is tangential.

To see that they intersect, first note that the crossbar theorem implies that $\vec{r}$ intersects $\overline{BC}$ at an interior point X. Then the crossbar theorem applied to $\triangle ABX$ shows that $\vec{s}$ intersects $\overline{AX}$ at a point O. Thus O lies on the bisector of $\angle A$, so it is equidistant from $\overleftrightarrow{AB}$ and $\overleftrightarrow{AC}$ by the angle bisector theorem; and it also lies on the bisector of $\angle B$, so it is equidistant from $\overleftrightarrow{AB}$ and $\overleftrightarrow{BC}$. By transitivity, it is equidistant from $\overleftrightarrow{AC}$ and $\overleftrightarrow{BC}$. If we can show that O is in the interior of $\angle C$, it will then follow from the converse to the angle bisector theorem that it lies on the bisector of $\angle C$ as well, and then the incircle theorem will show that $\triangle ABC$ is tangential with O as its incenter.

The fact that $\vec{r}$ is in the interior of $\angle A$ means, in particular, that O lies on the same side of $\overleftrightarrow{AC}$ as B; and the fact that $\vec{s}$ is in the interior of $\angle B$ means that O lies on the same side of $\overleftrightarrow{BC}$ as A. Thus O is in the interior of $\angle C$ as required.   □

Corollary 14.35 (Angle Bisector Concurrence Theorem). *In any triangle, the bisectors of the three angles are concurrent.*

Proof. The incenter is a point where all three bisectors meet.   □

We have now seen three different concurrence theorems for triangles, and associated with each concurrence theorem is a "center" where three lines meet: the medians meet at the *centroid* (Theorem 12.13), the perpendicular bisectors of the sides meet at the *circumcenter* (Corollary 14.26), and the angle bisectors meet at the *incenter* (Corollary 14.35). There is one more important center associated with triangles, described in the next theorem.

Theorem 14.36 (Altitude Concurrence Theorem). *In any triangle, the lines containing the three altitudes are concurrent.*

Proof. Exercise 14P. □

The point where the lines containing the altitudes meet is called the **orthocenter** of the triangle. In general, the four centers of a triangle (centroid, incenter, circumcenter, orthocenter) are all distinct, except in the case of equilateral triangles (see Exercise 14R).

Next we explore the question of which quadrilaterals are tangential. We begin with the following simple special case.

Lemma 14.37. *Every kite is a tangential quadrilateral.*

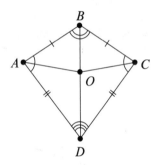

Fig. 14.30. A kite is a tangential quadrilateral.

Proof. Suppose $ABCD$ is a kite with $AB = BC$ and $CD = DA$ (Fig. 14.30). Then the diagonal BD cuts $ABCD$ into two congruent triangles $\triangle ABD$ and $\triangle CBD$ by SSS, and thus $\angle ABD \cong \angle CBD$ and $\angle ADB \cong \angle CDB$. Let O be the point where the bisector of $\angle A$ meets BD. By SAS, $\triangle ABO \cong \triangle CBO$ and $\triangle ADO \cong \triangle CDO$, so it follows that $\overrightarrow{CO}$ bisects $\angle C$. Therefore the bisectors of all four angles of $ABCD$ meet at O, so $ABCD$ is tangential by the incircle theorem. □

For general quadrilaterals, there is a simple necessary and sufficient criterion, analogous to the one for cyclic quadrilaterals. This proof is also rated R.

Theorem 14.38 (Tangential Quadrilateral Theorem). *A quadrilateral $ABCD$ is tangential if and only if it is convex and both pairs of opposite side lengths have the same sum:* $AB + CD = AD + BC.$

Proof. First suppose $ABCD$ is a tangential quadrilateral, and let $\mathcal{C}$ be its incircle. Theorem 14.31 shows that $ABCD$ is convex. Let E, F, G, and H denote the points of tangency of the sides $\overline{AB}$, $\overline{BC}$, $\overline{CD}$, and $\overline{DA}$, respectively (Fig. 14.31). By Theorem 14.18, A is equidistant from E and H, and similarly each of the other vertices is equidistant from the

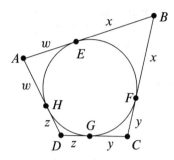

Fig. 14.31. Proving necessity of the condition in Theorem 14.38.

points of tangency on the two edges that meet at that vertex. Let w, x, y, and z denote the lengths of the segments shown in the diagram. Then $AB + CD$ and $AD + BC$ are both equal to $w + x + y + z$, so they are equal to each other.

To prove the converse, suppose $ABCD$ is a convex quadrilateral in which $AB + CD = AD + BC$. On the one hand, if $AB = BC$, then the hypothesis implies that $CD = AD$ as well, so $ABCD$ is a kite and thus is tangential by Lemma 14.37. On the other hand, suppose $AB \neq BC$. Renaming the vertices if necessary, we may assume that $AB > BC$. The hypothesis implies $AD - CD = AB - BC > 0$, so it follows that $AD > CD$ as well.

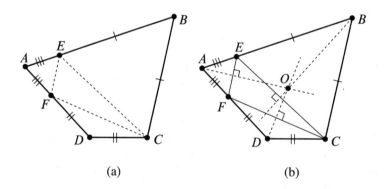

(a)   (b)

Fig. 14.32. Proving sufficiency of the condition in Theorem 14.38.

Let E be the point in the interior of $\overline{AB}$ such that $BE = BC$, and let F be the point in the interior of $\overline{AD}$ such that $DF = DC$ (see Fig. 14.32(a)). The fact that $AB - BC = AD - CD$ together with the betweenness theorem for points implies that $AF = AE$ also. Thus $\triangle BCE$, $\triangle DCF$, and $\triangle AFE$ are all isosceles triangles. It follows from Theorem 7.6 that in each of these isosceles triangles, the altitude to the base is contained in both the perpendicular bisector of the base and the bisector of the angle opposite the base. Thus the bisectors of $\angle A$, $\angle B$, and $\angle D$ are contained in the perpendicular bisectors of the three sides of $\triangle EFC$ (see Fig. 14.32(b)). Because the perpendicular bisectors of the sides of a triangle are all concurrent (by Corollary 14.26), these three lines all intersect at a point O.

We need to check that O lies on each of the angle bisectors themselves and not their opposite rays. This is where the assumption of convexity comes in. Because $ABCD$ is convex, both rays $\overrightarrow{AD}$ and $\overrightarrow{BC}$ lie on the same side of $\overleftrightarrow{AB}$. The bisectors of $\angle A$ and $\angle B$ make acute angles with $\overrightarrow{AB}$ and $\overrightarrow{BA}$, respectively (because their measures are one-half of the measures of proper angles), so Euclid's fifth postulate shows that the lines containing those rays intersect on the same side of $\overleftrightarrow{AB}$ as the points C and D, and thus the intersection point O lies on the bisectors of $\angle A$ and $\angle B$. A similar argument starting with $\overrightarrow{DA}$ shows that O lies on the bisector of $\angle D$. It remains only to show that O also lies on the bisector of $\angle C$.

The angle bisector theorem shows that O is equidistant from $\overleftrightarrow{CD}$ and $\overleftrightarrow{DA}$, from $\overleftrightarrow{DA}$ and $\overleftrightarrow{AB}$, and from $\overleftrightarrow{AB}$ and $\overleftrightarrow{BC}$; thus by transitivity it is equidistant from $\overleftrightarrow{CD}$ and $\overleftrightarrow{BC}$. Because $O \in \operatorname{Int} \angle B$, it is on the same side of $\overleftrightarrow{BC}$ as A, which is also the same side as D because $\overline{AD}$ and $\overline{BC}$ are semiparallel. A similar argument shows that O is on the same side of $\overleftrightarrow{CD}$ as B. Thus $O \in \operatorname{Int} \angle C$, so it lies on the bisector of $\angle C$ by the partial converse to the angle bisector theorem. It follows from the incircle theorem that $ABCD$ is tangential. $\qquad\square$

Regular Polygons and Circles

There is a close relationship between regular polygons and circles. The next theorem shows why.

Theorem 14.39. *Every regular polygon is both cyclic and tangential, and its incenter is equal to its circumcenter.*

Proof. Exercise 14T. $\qquad\square$

If $\mathscr{P}$ is a regular polygon, the point O that is both the incenter and circumcenter is called the **center of $\mathscr{P}$**. Any angle of the form $\angle AOB$, where A and B are consecutive vertices of $\mathscr{P}$, is called a **central angle of $\mathscr{P}$**; it is a central angle of both the inscribed and circumscribed circles.

Theorem 14.40 (Central Angles of a Regular Polygon). *Every central angle of a regular n-gon has measure $360°/n$.*

Proof. Let $\mathscr{P}$ be a regular n-gon, let O be its center, and let $\mathscr{C}$ be its circumcircle. Let A, B, C be any three consecutive vertices of $\mathscr{P}$, and draw $\overline{OA}$, $\overline{OB}$, and $\overline{OC}$. Label $\angle 1$ through $\angle 4$ as shown in Fig. 14.33.

The isosceles triangles $\triangle AOB$ and $\triangle BOC$ are congruent to each other by SSS, and thus the base angles $\angle 1$, $\angle 2$, $\angle 3$, and $\angle 4$ are all congruent. Because O is in the interior of $\mathscr{P}$ (since O is also the circumcenter of $\mathscr{P}$), it follows that $\overrightarrow{BO}$ is the bisector of $\angle ABC$, and thus

$$m\angle 2 = \frac{1}{2}m\angle ABC = \frac{1}{2}\left(\frac{n-2}{n}360°\right).$$

Therefore, by the angle-sum theorem for $\triangle AOB$, the central angle $\angle AOB$ satisfies

$$m\angle AOB = 180° - m\angle 1 - m\angle 2 = 180° - \frac{1}{2}\left(\frac{n-2}{n}360°\right) - \frac{1}{2}\left(\frac{n-2}{n}360°\right) = \frac{360°}{n}.$$

The same argument applies to each of the central angles. $\qquad\square$

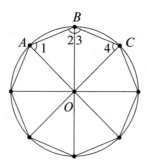

Fig. 14.33. A regular polygon inscribed in a circle.

So far, all of our theorems about inscribed and circumscribed polygons have started with a given polygon and asked whether there was a circle inscribed in it or circumscribed about it. Now we reverse the process: we start with a given circle and look for regular polygons that can be inscribed in it or circumscribed about it. The next lemma simplifies the process of proving that an inscribed polygon is regular.

Lemma 14.41. *Every equilateral polygon inscribed in a circle is regular.*

Proof. Suppose $\mathcal{P}$ is an equilateral polygon inscribed in a circle $\mathcal{C}$ (see Fig. 14.33 again, but now we are assuming only that $\mathcal{P}$ is equilateral, not necessarily regular). Then $\mathcal{P}$ is convex by Theorem 14.23, so we need only prove it is equiangular. Let O be the center of $\mathcal{C}$, and let A, B, C be any three consecutive vertices of $\mathcal{P}$. Label $\angle 1$ through $\angle 4$ as in the preceding proof. Then as before, $\triangle AOB \cong \triangle BOC$ by SSS, so $\angle 1 \cong \angle 2 \cong \angle 3 \cong \angle 4$. Let β denote the common measure of these angles.

The vertices A and C must be on opposite sides of $\overleftrightarrow{OB}$, because otherwise they would have to coincide with each other by the unique triangle theorem. Thus $\angle 2$ and $\angle 3$ are adjacent angles. Because they are base angles of isosceles triangles, they are both acute, so it follows from the adjacency lemma that $m\angle ABC = m\angle 2 + m\angle 3 = 2\beta$.

Now, the same argument applies to any three consecutive vertices of $\mathcal{P}$; since all of the vertices are equidistant from O and all of the sides of $\mathcal{P}$ are congruent, all of the triangles formed by O and two consecutive vertices are congruent to each other. Thus all of the base angles of these triangles have measure β, and so each angle of $\mathcal{P}$ has measure 2β. It follows that $\mathcal{P}$ is equiangular and thus regular. □

Theorem 14.42. *Suppose $\mathcal{C}$ is a circle, n is an integer greater than or equal to 3, and P is any point on $\mathcal{C}$. Then there is a regular n-gon inscribed in $\mathcal{C}$ that has P as one of its vertices.*

Proof. We consider two cases, depending on whether n is even or odd. First suppose n is even, and write $n = 2k$ with k an integer greater than 1. Let O be the center of $\mathcal{C}$, let Q be the point on $\mathcal{C}$ such that $Q * O * P$, and denote the two sides of $\overleftrightarrow{OP}$ by S_1 and S_2. Let $\theta = 360/n$. For each $i = 1, \ldots, k-1$, let $\overrightarrow{r_i}$ be the ray starting at O on side S_1 that makes an angle with $\overrightarrow{OP}$ of measure $i\theta$, and let $\overrightarrow{s_i}$ be the ray starting at O on side S_2 that makes the same angle measure with $\overrightarrow{OP}$. Let A_i be the point where $\overrightarrow{r_i}$ meets $\mathcal{C}$, and let B_i be the point where $\overrightarrow{s_i}$ meets $\mathcal{C}$ (Fig. 14.34(a)). Then $\mathcal{P} = PA_1A_2 \ldots A_{k-1}QB_{k-1} \ldots B_2B_1$ is an n-gon inscribed in $\mathcal{C}$. We will show that it is a regular n-gon.

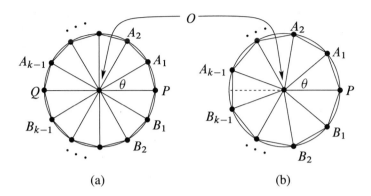

Fig. 14.34. Proving the existence of inscribed regular n-gons.

Note that $\angle POA_1$ and $\angle POB_1$ have measure θ by construction. The ordering lemma for rays shows that $\overrightarrow{OP} * \overrightarrow{OA_i} * \overrightarrow{OA_j}$ whenever $1 \le i < j$, and similarly $\overrightarrow{OP} * \overrightarrow{OB_i} * \overrightarrow{OB_j}$. Therefore, the betweenness theorem for rays allows us to conclude that $\angle A_i OA_{i+1}$ and $\angle B_i OB_{i+1}$ both have measure θ for $i = 1, \ldots, k-2$. Finally, $\angle POA_{k-1}$ and $\angle A_{k-1}OQ$ form a linear pair, so $m\angle A_{k-1}OQ = 180° - m\angle POA_{k-1} = 180° - (k-1)\theta = \theta$, and similarly $m\angle B_{k-1}OQ = \theta$. Therefore, each of the n triangles formed by O and two consecutive vertices of $\mathcal{P}$ is an isosceles triangle with angle θ at O, so all n triangles are congruent to each other. This implies immediately that $\mathcal{P}$ is equilateral.

In case n is odd, the argument must be modified in the following ways. First, write $n = 2k - 1$ for some integer $k > 1$, and write $\theta = 360/n$. Then define $A_1, \ldots, A_{k-1}$ and $B_1, \ldots, B_{k-1}$ exactly as before, and let $\mathcal{P} = PA_1A_2 \ldots A_{k-1}B_{k-1} \ldots B_2B_1$ (Fig. 14.34(b)). The same argument shows that all of the triangles formed by O and two consecutive vertices of $\mathcal{P}$ are congruent, except that we need a special argument for the triangle $A_{k-1}OB_{k-1}$. In this case, it is easy to check that the three rays $\overrightarrow{OP}$, $\overrightarrow{OA_{k-1}}$, and $\overrightarrow{OB_{k-1}}$ satisfy the hypotheses of the 360 theorem, so

$$m\angle A_{k-1}OB_{k-1} = 360° - m\angle POA_{k-1} - m\angle POB_{k-1} = 360° - 2(k-1)\theta = \theta.$$

The rest of the proof proceeds as before. In either case, we have proved that $\mathcal{P}$ is equilateral, and it follows from Lemma 14.41 that is is regular. $\square$

Using the preceding theorem, we can also show that *circumscribed* regular polygons exist.

Theorem 14.43. *Suppose $\mathcal{C}$ is a circle, n is an integer greater than or equal to 3, and P is any point on $\mathcal{C}$. Then there is a regular n-gon circumscribed about $\mathcal{C}$ that has one of its vertices on $\overrightarrow{OP}$.*

Proof. Let $\mathcal{P} = A_1 \ldots A_n$ be a regular n-gon inscribed in $\mathcal{C}$, with $P = A_1$. Let r denote the radius of $\mathcal{C}$ and let $\theta = 360/n$. The fact that $n \ge 3$ implies $\theta \le 120°$, so $\cos(\theta/2) > 0$. Let $s = r/\cos(\theta/2)$. For each i, let B_i be the point on $\overrightarrow{OA_i}$ whose distance from O is equal to s (Fig. 14.35), and let $\mathcal{Q} = B_1 \ldots B_n$. Each of the triangles $\triangle B_i OB_{i+1}$ is similar to $\triangle A_i OA_{i+1}$ by the SAS similarity theorem, and thus all of these triangles are congruent to each other. This shows that $\mathcal{Q}$ is equilateral. Because it is inscribed in the circle $\mathcal{C}(O, s)$, it follows from Lemma 14.41 that it is regular.

Next we show that $\mathcal{Q}$ is circumscribed about $\mathcal{C}$. Let B_i and B_{i+1} be any two consec-

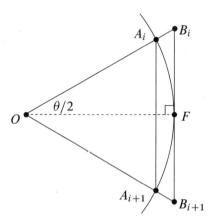

Fig. 14.35. Proving the existence of circumscribed regular polygons.

utive vertices of $\mathcal{Q}$ (where for convenience we use B_{n+1} as an alternative notation for B_1), and let F be the foot of the altitude from O to $\overline{B_i B_{i+1}}$. By the isosceles triangle altitude theorem, $\overrightarrow{OF}$ bisects $\angle B_i O B_{i+1}$, and so $\angle F O B_i$ has measure $\theta/2$. Therefore, because $OB_i = s = r/\cos(\theta/2)$, we have $OF = s\cos(\theta/2) = r$. This means that F lies on $\mathcal{C}$, and since $\overline{B_i B_{i+1}}$ is perpendicular to the radius $\overline{OF}$, it is tangent to $\mathcal{C}$. The same argument works for each side of $\mathcal{Q}$ and shows that $\mathcal{Q}$ is circumscribed about $\mathcal{C}$. $\qquad\square$

Theorem 14.44. *Let $\mathcal{C}$ be a circle. Given an integer $n \geq 3$, any two regular n-gons inscribed in $\mathcal{C}$ are congruent to each other, as are any two regular n-gons circumscribed about $\mathcal{C}$.*

Proof. Exercise 14V. $\qquad\square$

Exercises

14A. Complete the proof of Theorem 14.4 by showing that points of a secant line that are not on the corresponding chord are exterior points of the circle. [Hint: See Fig. 14.4(b).]

14B. Prove Theorem 14.5 (properties of chords).

14C. Suppose that ℓ is a line, C and D are points on opposite sides of ℓ, and $r = CD$. Prove that $\mathcal{C}(C, r)$ intersects ℓ in exactly two points. (This is what's needed in Euclid's proof of his Proposition I.12.)

14D. Suppose that A and B are distinct points and $r = AB$. Prove that $\mathcal{C}(A, r)$ and $\mathcal{C}(B, r)$ intersect. (This is what's needed in Euclid's proof of his Proposition I.1.)

14E. Prove Theorem 14.11 (the tangent circles theorem). [Hint: First prove that O, Q, and A are collinear by assuming they are noncollinear and proving that the two circles have two points of intersection.]

14F. Suppose $\mathcal{C} = \mathcal{C}(O, r)$ is a circle and P is a point in the interior of $\mathcal{C}$ but different from O. Show that the point where $\overrightarrow{OP}$ intersects $\mathcal{C}$ is the unique closest point to P on $\mathcal{C}$.

14G. Suppose $\mathcal{C} = \mathcal{C}(O, r)$ is a circle and P is a point in the exterior of $\mathcal{C}$. Show that the point where $\overline{OP}$ intersects $\mathcal{C}$ is the unique closest point to P on $\mathcal{C}$.

14H. Suppose $\overline{AB}$ and $\overline{CD}$ are two chords in a circle $\mathcal{C}(O, r)$. Show that $AB = CD$ if and only if O is equidistant from $\overleftrightarrow{AB}$ and $\overleftrightarrow{CD}$, and $AB < CD$ if and only if $d\left(O, \overleftrightarrow{AB}\right) > d\left(O, \overleftrightarrow{CD}\right)$.

14I. Prove Theorem 14.17 (the converse to Thales' theorem). [Hint: If $\triangle ABC$ is a right triangle with right angle at C, begin by letting $\overrightarrow{r}$ be a ray that makes an angle with $\overrightarrow{CA}$ that is congruent to $\angle A$.]

14J. In Fig. 14.36, the small circle is centered at O, the large one is centered at P, and the two circles intersect only at A. The segments $\overline{AB}$ and $\overline{OP}$ have lengths as shown. Find the length of $\overline{CD}$, and prove your answer correct.

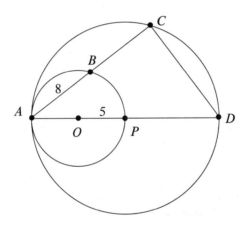

Fig. 14.36. The setup for Exercise 14J.

14K. Prove that the interior of a circle is a convex set.

14L. Prove Corollary 14.20 (the arc addition theorem). [Hint: There is a reason this is a corollary of the inscribed angle theorem.]

14M. Suppose $\mathcal{C}$ is a circle, P is an exterior point of $\mathcal{C}$, ℓ is a line through P and tangent to $\mathcal{C}$ at B, and m is a line through P that meets $\mathcal{C}$ at two distinct points C and D (Fig. 14.37). Find a relation among the lengths PB, PC, and PD, and prove your answer correct.

14N. Suppose $\overline{AB}$ and $\overline{CD}$ are two distinct chords of a circle $\mathcal{C}$ that intersect at a point $P \in \text{Int}\,\mathcal{C}$ (Fig. 14.38). Let $\overset{\frown}{AC}$ denote the arc conjugate to $\overset{\frown}{ABC}$, and let $\overset{\frown}{BD}$ denote the arc conjugate to $\overset{\frown}{BCD}$. Find a relation among $m\angle APC$, $m\overset{\frown}{AC}$, and $m\overset{\frown}{BD}$, and prove your answer correct.

14O. Prove Theorem 14.33 (the incircle theorem). [Hint: Follow the same outline as the proof of the circumcircle theorem, but use the angle bisector theorem in place of the perpendicular bisector theorem and the tangent line theorem in place of Theorem 14.5.]

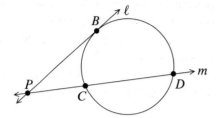

Fig. 14.37. The setup for Exercise 14M.

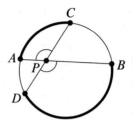

Fig. 14.38. The setup for Exercise 14N.

14P. Prove Theorem 14.36 (the altitude concurrence theorem). [Hint: Start by copying the triangle three times as in Fig. 14.39.]

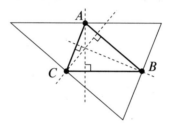

Fig. 14.39. Proof of the altitude concurrence theorem.

14Q. Suppose $\triangle ABC$ is a triangle and $\mathcal{C}$ is its incircle. Show that the three cevians joining the vertices to the points where $\mathcal{C}$ is tangent to the edges are concurrent. (The point where they meet is called the ***Gergonne point*** of the triangle, after the French mathematician Joseph Gergonne.)

14R. Suppose $\triangle ABC$ is a triangle. Prove that the centroid, incenter, circumcenter, and orthocenter are the same point if and only if $\triangle ABC$ is equilateral. What can you say if $\triangle ABC$ is isosceles but not equilateral?

14S. EULER LINE THEOREM: Prove that the orthocenter, centroid, and circumcenter of any triangle are collinear. [Hint: Let O be the circumcenter and let G be the centroid. It suffices to assume that $O \neq G$ (why?). Let H be the point such that $O * G * H$ and $GH = 2OG$. Use the median concurrence theorem and similar triangles to show that H lies on each altitude and thus is equal to the orthocenter. See Fig. 14.40.]

14T. Prove Theorem 14.39 (every regular polygon is both cyclic and tangential). [Hint: Begin by letting O be the center of the circle passing through any three consecutive vertices of the polygon.]

14U. Let $ABCD$ be a parallelogram.
   (a) Prove that $ABCD$ is cyclic if and only if it is a rectangle, in which case its circumcenter is the point where its diagonals intersect.
   (b) Prove that $ABCD$ is tangential if and only if it is a rhombus, in which case its incenter is the point where its diagonals intersect.

14V. Prove Theorem 14.44 (regular n-gons inscribed in or circumscribed about a fixed circle are congruent).

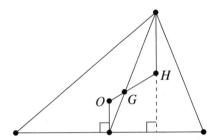

Fig. 14.40. The Euler line theorem.

14W. THE NINE-POINT CIRCLE THEOREM: Let $\triangle ABC$ be any triangle. Prove that the following nine points all lie on a single circle: the midpoints of the triangle's three sides, the feet of its three altitudes, and the midpoints of the segments connecting the orthocenter to the three vertices (see Fig. 14.41). [Hint: Let M and N be the

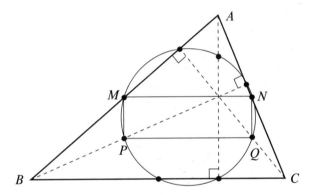

Fig. 14.41. The nine-point circle theorem.

midpoints of $\overline{AB}$ and $\overline{AC}$, respectively, and let P and Q be the midpoints of the segments connecting the orthocenter to B and C, respectively. Use the midsegment theorem four times to prove that $MNQP$ is a rectangle and thus has a circumscribed circle. Now find two other rectangles, and show that all three rectangles have the same circumscribed circle. Use this to conclude that all six midpoints lie on one circle. Finally, use the converse to Thales' theorem to conclude that the feet of the three altitudes also lie on the same circle.]

14X. PASCAL'S MYSTIC HEXAGON THEOREM: *Suppose $ABCDEF$ is a cyclic hexagon such that the lines $\overleftrightarrow{AB}$, $\overleftrightarrow{BC}$, and $\overleftrightarrow{CD}$ meet $\overleftrightarrow{DE}$, $\overleftrightarrow{EF}$, and $\overleftrightarrow{FA}$, respectively. Then the three points of intersection are collinear.* (See Fig. 14.42.) Prove this theorem under the additional assumption that no two sides of the hexagon are parallel. [Hint: The additional assumption guarantees that the lines $\overleftrightarrow{AB}$, $\overleftrightarrow{CD}$, and $\overleftrightarrow{EF}$ intersect to form a triangle $\triangle XYZ$ as shown in Fig. 14.43. Apply Menelaus's theorem to the three transversals to this triangle, and apply the intersecting secants theorem to the secant lines through X, Y, and Z. Finally, use Menelaus's theorem again to conclude that the intersection points P, Q, and R are collinear.]

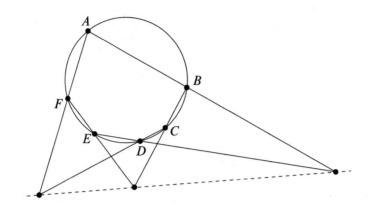

Fig. 14.42. Pascal's mystic hexagon.

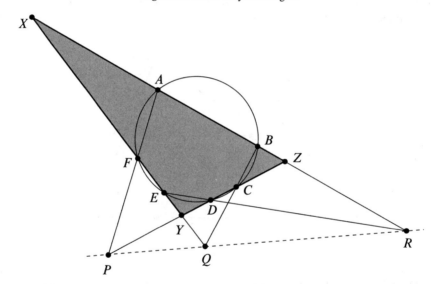

Fig. 14.43. Proof of Pascal's theorem.

Chapter 15

Circumference and Circular Area

You undoubtedly have a sense of what it means to measure or compute the *circumference* of a circle: intuitively, it is the length of a string that would be required to wrap once around the circle, or the distance the circle would roll on a flat surface between successive times when a given point on the circle touches the ground. Similarly, it is intuitively clear that a circular region (that is, a circle together with its interior) should, like polygonal regions, have a well-defined *area*, expressing how much of the plane is included inside the circle.

In the context of our axiomatic system, however, our theories of length and area are not yet equipped to make sense of the circumference or area of a circle. The length of a *segment* is defined to be the distance between its endpoints, but that definition does not give us any help in deciding how to assign a length to a circle. Also, the area postulate guarantees that every *polygonal region* has a well-defined area, but that does not permit us to assign areas to sets with other sorts of boundaries such as circles. We have to introduce new definitions for these concepts.

The problem of finding relationships among the radius, circumference, and area of a circle is among the oldest in geometry. Euclid, to be sure, included very few propositions about circumferences and areas of circles (Propositions III.30 and XII.2, for example). But the subject was brought to dramatic fruition by the work of the brilliant Greek mathematician Archimedes of Syracuse (ca. 287 BCE–212 BCE). Archimedes used a technique called the *method of exhaustion* to derive the relationship between areas of circles and their circumferences (see [**Arc02**] or [**Arc97**] for Heath's translation of Archimedes' work). In this chapter, we present modern definitions of circumferences and circular areas that are directly inspired by the work of Archimedes, and our proofs of the main theorems are based directly on his.

Circumference

Archimedes approximated circumferences of circles by inscribing regular polygons in them and increasing the number of sides to get better and better approximations. This idea is at the root of our modern definition of circumference and arc length, but we will extend it by considering *all* inscribed polygons rather than just regular ones.

First we recall some fundamental properties of the real numbers. (See Appendix H.) If S is a set of real numbers, an ***upper bound for*** S is a real number b such that $b \geq x$ for every $x \in S$. (An upper bound might be an element of S or it might not be.) A number b is called a ***least upper bound for*** S if b is an upper bound and if in addition no number less than b is also an upper bound. For example, if S is the closed interval $[0, 1]$, then 2 is an upper bound, while 1 is the least upper bound. If T is the set of all negative real numbers, then every nonnegative number is an upper bound, and 0 is the least upper bound. On the other hand, some sets of real numbers, such as the set of integers, have no upper bounds, and therefore no least upper bounds.

One of the most important properties of the real numbers is the ***least upper bound property***: *if S is any nonempty set of real numbers and S has an upper bound, then S has a unique least upper bound.* The least upper bound of a set S is also called its ***supremum*** and is often denoted by $\sup S$. We will use the least upper bound property to define circumferences and areas of circles.

Let $\mathcal{C}$ be a circle. Following Archimedes, we would like to consider inscribed polygons as reasonable approximations to $\mathcal{C}$. Intuitively, it appears that the perimeter of any inscribed polygon should be strictly less than the circumference of $\mathcal{C}$, and as we choose polygons with more and smaller sides, we expect to be able to make their perimeters as close as we like to the circumference of $\mathcal{C}$. Physically, this corresponds to finding an approximate measure for the circumference by measuring distances between closely spaced points on the circle and adding them up.

All of this is just vague intuition so far, because we do not have a mathematical definition of circumference. But it suggests what the official definition should be. Suppose $\mathcal{C}$ is a circle. We wish to define the circumference of $\mathcal{C}$ to be the least upper bound of the set of perimeters of all polygons inscribed in $\mathcal{C}$. In order for this definition to make sense, we need to show that this set is nonempty and has an upper bound. It is certainly nonempty, because if we choose any three distinct points $A, B, C \in \mathcal{C}$, then $\triangle ABC$ is inscribed in $\mathcal{C}$.

Before we prove that it has an upper bound, we need to prove a couple of lemmas that will be useful several times throughout the chapter.

Lemma 15.1. *Suppose $\mathcal{C}$ is a circle. Given any polygon inscribed in $\mathcal{C}$, there is another inscribed polygon with larger perimeter and larger area; and given any polygon circumscribed about $\mathcal{C}$, there is another circumscribed polygon with smaller perimeter and smaller area.*

Proof. First suppose $\mathcal{P} = A_1 A_2 \ldots A_n$ is inscribed in $\mathcal{C}$. The line $\overleftrightarrow{A_1 A_2}$ cuts $\mathcal{C}$ into two arcs by Theorem 14.13, one that is on the $\mathcal{P}$-side of the line and one that is not. Let C be any point in the interior of the arc on the non-$\mathcal{P}$-side, and let $\mathcal{P}' = A_1 C A_2 \ldots A_n$ (Fig. 15.1). By our choice of C, neither $\overline{A_1 C}$ nor $\overline{C A_2}$ can intersect any edge of $\mathcal{P}$ except at A_1 and A_2, so $\mathcal{P}'$ is a polygon. Because it is inscribed in $\mathcal{C}$, it is convex.

To see that $\text{perim}(\mathcal{P}') > \text{perim}(\mathcal{P})$, note that both polygons share the same edges, except that $\mathcal{P}'$ has the two edges $\overline{A_1C}$ and $\overline{CA_2}$ where $\mathcal{P}$ has only $\overline{A_1A_2}$. Since $A_1A_2 < A_1C + CA_2$ by the triangle inequality, the result follows. To see that $\alpha(\mathcal{P}') > \alpha(\mathcal{P})$, note that $\overline{A_1A_2}$ is a chord of $\mathcal{P}'$, so the convex decomposition lemma gives

$$\alpha(\mathcal{P}') = \alpha(\mathcal{P}) + \alpha(\triangle A_1A_2C) > \alpha(\mathcal{P}). \tag{15.1}$$

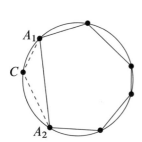

Fig. 15.1. A larger inscribed polygon.

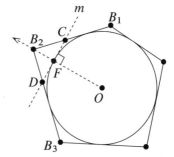

Fig. 15.2. A smaller circumscribed polygon.

Next consider a circumscribed polygon $\mathcal{Q} = B_1B_2\ldots B_n$, and let O be the center of $\mathcal{C}$. The ray $\overrightarrow{OB_2}$ meets $\mathcal{C}$ at a point F. Since $B_2 \in \text{Ext}\,\mathcal{C}$, it follows that $O * F * B_2$. Let m be the line perpendicular to $\overleftrightarrow{OB_2}$ at F. Then m is tangent to $\mathcal{C}$ by the tangent line theorem. Because $\angle B_1B_2O$ is acute (since its measure is half that of the proper angle $\angle B_1B_2B_3$ by Theorem 14.18), Euclid's fifth postulate guarantees that m meets $\overrightarrow{B_2B_1}$ at a point C on the same side of $\overleftrightarrow{B_2O}$ as B_1. Since B_2 is on the opposite side of m from points of $\mathcal{C}$ by Theorem 14.9, it follows that C is in the interior of $\overline{B_1B_2}$. Similarly, there is a point D on the other side of $\overleftrightarrow{B_2O}$ where m meets $\text{Int}\,\overline{B_2B_3}$. Define $\mathcal{Q}' = B_1CDB_3\ldots B_n$. It follows from Lemma 14.30 that $\overline{CD}$ does not meet any of the edges of $\mathcal{Q}$ except at its endpoints, so $\mathcal{Q}'$ is a polygon. The new edge $\overline{CD}$ is tangent to $\mathcal{C}$ by construction, as are all of the original edges of $\mathcal{Q}$. Because the point of tangency of $\overline{B_1B_2}$ lies on the opposite side of m from B_2, it follows that $\overline{B_1C}$ is tangent to $\mathcal{C}$, and the same argument shows that $\overline{DB_3}$ is also tangent. Thus $\mathcal{Q}'$ is a polygon circumscribed about $\mathcal{C}$.

Next we verify that the perimeter of $\mathcal{Q}'$ is smaller than that of $\mathcal{Q}$. We have

$$\begin{aligned}\text{perim}(\mathcal{Q}) &= B_1B_2 + B_2B_3 + \cdots + B_nB_1, \\ \text{perim}(\mathcal{Q}') &= B_1C + CD + DB_3 + \cdots + B_nB_1.\end{aligned} \tag{15.2}$$

By the triangle inequality, $CD < CB_2 + B_2D$. Because C and D are interior points of $\overline{B_1B_2}$ and $\overline{B_2B_3}$, respectively, the triangle inequality and the betweenness theorem for points imply

$$B_1C + CD + DB_3 < B_1C + CB_2 + B_2D + DB_3 = B_1B_2 + B_2B_3.$$

Substituting this into (15.2), we see that $\text{perim}(\mathcal{Q}') < \text{perim}(\mathcal{Q})$.

Finally, because $\overline{CD}$ is a chord of the convex polygon $\mathcal{Q}$, the convex decomposition lemma gives

$$\alpha(\mathcal{Q}) = \alpha(\mathcal{Q}') + \alpha(\triangle B_2CD) > \alpha(\mathcal{Q}'). \qquad \square$$

The next lemma is a slight refinement of the preceding one for inscribed polygons.

Lemma 15.2. *Suppose $\mathcal{C}$ is a circle with center O, and $\mathcal{P}$ is a polygon inscribed in $\mathcal{C}$ such that $O \notin \operatorname{Int} \mathcal{P}$. Then there is another polygon $\mathcal{P}'$ inscribed in $\mathcal{C}$ that has larger area and perimeter than $\mathcal{P}$ and that satisfies $O \in \operatorname{Int} \mathcal{P}'$.*

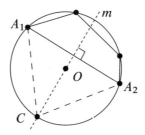

Fig. 15.3. Proof of Lemma 15.2.

Proof. The assumption that $O \notin \operatorname{Int} \mathcal{P}$ means that there is at least one edge of $\mathcal{P}$ such that O is either on the edge or on the side opposite the $\mathcal{P}$-side of the edge (Fig. 15.3). Label the vertices of $\mathcal{P}$ as $A_1 \dots A_n$ in such a way that $\overline{A_1 A_2}$ is the edge in question. Let m be the perpendicular bisector of $\overline{A_1 A_2}$; then m contains O by Theorem 14.5. The line m intersects $\mathcal{C}$ exactly twice, once on the $\mathcal{P}$-side of $\overleftrightarrow{A_1 A_2}$ and once on the opposite side. Let C be the point of intersection that is *not* on the $\mathcal{P}$-side, and let $\mathcal{P}' = A_1 C A_2 \dots A_n$. It follows just as in the preceding lemma that $\mathcal{P}'$ is an inscribed polygon with larger area and perimeter than $\mathcal{P}$.

We need to show that $O \in \operatorname{Int} \mathcal{P}'$. Note that $\overline{A_1 A_2}$ is a chord of $\mathcal{P}'$. If $O \in \overline{A_1 A_2}$, then it is in the interior of $\mathcal{P}'$ by Lemma 8.8. On the other hand, suppose $O \notin \overline{A_1 A_2}$. Because O lies on the bisector of $\angle A_1 C A_2$ (by the isosceles triangle altitude theorem), it is in the interior of that angle, so it is on the same side of $\overleftrightarrow{A_1 C}$ as A_2 and on the same side of $\overleftrightarrow{A_2 C}$ as A_1; in addition, it is on the same side of $\overleftrightarrow{A_1 A_2}$ as C by construction. Thus it is in the interior of $\triangle A_1 A_2 C$. By the convex decomposition lemma, $\operatorname{Reg} \mathcal{P}' = \operatorname{Reg} \mathcal{P} \cup \operatorname{Reg} \triangle A_1 A_2 C$, so $O \in \operatorname{Reg} \mathcal{P}'$. Since O does not lie on $\mathcal{P}'$ itself, it follows that $O \in \operatorname{Int} \mathcal{P}'$. $\square$

Now we can prove that the set of perimeters of inscribed polygons has an upper bound. It turns out that the perimeter of any *circumscribed* polygon will do. The next proof is rated PG.

Theorem 15.3. *The perimeter of every polygon circumscribed about a circle is strictly larger than the perimeter of every polygon inscribed in the same circle.*

Proof. Let $\mathcal{C} = \mathcal{C}(O, r)$ be a circle. Suppose $\mathcal{P} = A_1 \dots A_n$ is an inscribed polygon in $\mathcal{C}$ and $\mathcal{Q} = B_1 \dots B_m$ is a circumscribed one. It follows from Corollary 14.32 that O is an interior point of $\mathcal{Q}$. It might or might not be an interior point of $\mathcal{P}$, but for starters, let us assume that it is. (We'll show how to remove this assumption later.) Thus each ray starting at O intersects $\mathcal{P}$ and $\mathcal{Q}$ exactly once each by Corollary 8.16.

For each vertex A_i of $\mathcal{P}$, let C_i denote the point where $\overrightarrow{OA_i}$ meets $\mathcal{Q}$ (see Fig. 15.4). Suppose A_i and A_{i+1} are consecutive vertices of $\mathcal{P}$ (with A_{n+1} interpreted to mean A_1),

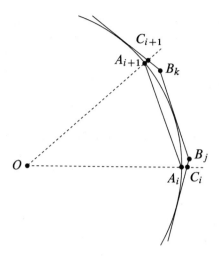

Fig. 15.4. Comparing perimeters of inscribed and circumscribed polygons.

and let $B_j, \ldots, B_k$ denote the vertices of $\mathcal{Q}$ that lie in the interior of $\angle C_i O C_{i+1}$, arranged in consecutive order as they appear in $\mathcal{Q}$, with B_j an endpoint of the edge containing C_i and B_k an endpoint of the edge containing C_{i+1}. Let b_i denote the following sum of lengths:

$$b_i = C_i B_j + B_j B_{j+1} + \cdots + B_{k-1} B_k + B_k C_{i+1}.$$

A straightforward argument using the betweenness theorem for points shows that the perimeter of $\mathcal{Q}$ is equal to $b_1 + b_2 + \cdots + b_n$. Because $\operatorname{perim}(\mathcal{P}) = A_1 A_2 + \cdots + A_n A_{n+1}$, the theorem will be proved if we can show that $A_i A_{i+1} < b_i$ for each i. We consider several cases, illustrated in Fig. 15.5.

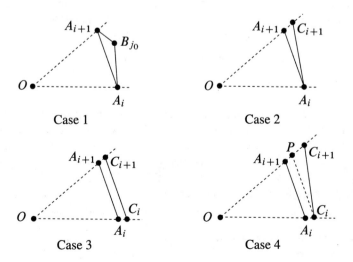

Fig. 15.5. The different cases in the proof of Theorem 15.3.

CASE 1: $C_i = A_i$ *and* $C_{i+1} = A_{i+1}$. In this case, there must be at least one vertex $B_{j_0} \in \text{Int}\angle C_i O C_{i+1}$ that does not lie on $\overline{A_i A_{i+1}}$, for otherwise the chord $\overline{A_i A_{i+1}}$ would be contained in $\mathcal{Q}$, which is impossible because a circumscribed polygon contains no interior points of the circle. Using first the ordinary triangle inequality and then the general triangle inequality, we obtain

$$A_i A_{i+1} = C_i C_{i+1} < C_i B_{j_0} + B_{j_0} C_{i+1} \le b_i.$$

In all of the remaining cases, we can assume that either $C_i \ne A_i$ or $C_{i+1} \ne A_{i+1}$; without loss of generality, let us say it is the latter. In these cases, the general triangle inequality yields $C_i C_{i+1} \le b_i$, so by transitivity it suffices to show that $A_i A_{i+1} < C_i C_{i+1}$. Note that $OA_i = OA_{i+1} = r$ because both A_i and A_{i+1} lie on $\mathcal{C}$, and that $OC_{i+1} > OA_{i+1}$ and $OC_i \ge OA_i$ because no point of $\mathcal{Q}$ is in the interior of $\mathcal{C}$.

CASE 2: $C_i = A_i$ *but* $C_{i+1} \ne A_{i+1}$. Because $\angle OA_{i+1}A_i$ is one of the base angles of the isosceles triangle $A_i O A_{i+1}$, it is acute, and therefore $\angle A_i A_{i+1} C_{i+1}$ is obtuse, so it is the largest angle in $\triangle A_i A_{i+1} C_{i+1}$. By the scalene inequality,

$$A_i A_{i+1} < A_i C_{i+1} = C_i C_{i+1} \le b_i.$$

CASE 3: $C_i \ne A_i$, $C_{i+1} \ne A_{i+1}$, *and* $OC_i = OC_{i+1}$. In this case, $\triangle A_i O A_{i+1} \sim \triangle C_i O C_{i+1}$ by the SAS similarity theorem. Thus $A_i A_{i+1}/C_i C_{i+1} = OA_{i+1}/OC_{i+1} < 1$, so

$$A_i A_{i+1} < C_i C_{i+1} \le b_i.$$

CASE 4: $C_i \ne A_i$, $C_{i+1} \ne A_{i+1}$, *and* $OC_i \ne OC_{i+1}$. Without loss of generality, we can assume that $OC_{i+1} > OC_i$. Let P be the point on $\overline{OC_{i+1}}$ such that $OP = OC_i$. Then the argument of Case 2 shows that $A_i A_{i+1} < C_i P$, and the argument of Case 3 shows that $C_i P < C_i C_{i+1}$. Thus

$$A_i A_{i+1} < C_i P < C_i C_{i+1} \le b_i.$$

This completes the proof under the assumption that $O \in \text{Int}\,\mathcal{P}$.

Finally, we have to consider the possibility that $O \notin \text{Int}\,\mathcal{P}$. Lemma 15.2 shows that there is another inscribed polygon $\mathcal{P}'$ that has larger perimeter than $\mathcal{P}$ and such that $O \in \text{Int}\,\mathcal{P}'$; then the preceding argument applied to $\mathcal{P}'$ shows that

$$\text{perim}(\mathcal{P}) < \text{perim}(\mathcal{P}') < \text{perim}(\mathcal{Q}),$$

so the theorem is true in this case as well. $\square$

Theorem 15.3 shows that the perimeter of any circumscribed polygon is an upper bound for the set of perimeters of inscribed polygons. Now we can give the official definition of the circumference of a circle. Let $\mathcal{C}$ be a circle, and let $\mathcal{S}(\mathcal{C})$ be the set of perimeters of all polygons inscribed in $\mathcal{C}$. The ***circumference of*** $\mathcal{C}$, denoted by $\text{circum}(\mathcal{C})$, is defined as the least upper bound of the set $\mathcal{S}(\mathcal{C})$. Because this set is nonempty and has an upper bound, $\text{circum}(\mathcal{C})$ is a well-defined real number.

The following corollary is a useful tool for estimating circumferences.

Corollary 15.4. *Let* $\mathcal{C}$ *be a circle. If* $\mathcal{P}$ *is any polygon inscribed in* $\mathcal{C}$ *and* $\mathcal{Q}$ *is any polygon circumscribed about* $\mathcal{C}$, *then*

$$\text{perim}(\mathcal{P}) < \text{circum}(\mathcal{C}) < \text{perim}(\mathcal{Q}).$$

Proof. Since circum($\mathcal{C}$) is defined as the least upper bound of the set $S(\mathcal{C})$ and perim($\mathcal{P}$) is an element of this set, perim($\mathcal{P}$) $\leq$ circum($\mathcal{C}$). Similarly, since perim($\mathcal{Q}$) is an upper bound for $S(\mathcal{C})$ by Theorem 15.3 and circum($\mathcal{C}$) is the *least* upper bound, we have circum($\mathcal{C}$) $\leq$ perim($\mathcal{Q}$).

To show that both of the inequalities are strict, we must show that neither perim($\mathcal{P}$) nor perim($\mathcal{Q}$) can be equal to the least upper bound of $S(\mathcal{C})$. Lemma 15.1 shows that there is another inscribed polygon $\mathcal{P}'$ whose perimeter is larger than that of $\mathcal{P}$; thus perim($\mathcal{P}$) is not an upper bound for $S(\mathcal{C})$, so it cannot be equal to circum($\mathcal{C}$). Similarly, the same lemma shows that there is another circumscribed polygon $\mathcal{Q}'$ with perim($\mathcal{Q}'$) < perim($\mathcal{Q}$). Theorem 15.3 shows that perim($\mathcal{Q}'$) is also an upper bound for $S(\mathcal{C})$, so perim($\mathcal{Q}$) cannot be equal to the least upper bound circum($\mathcal{C}$). $\square$

Approximations by Regular Polygons

Although the definition of circumference is phrased in terms of arbitrary inscribed polygons, for many computational and theoretical purposes it is useful to restrict our attention to regular polygons as Archimedes did. Theorem 14.42 shows that they are abundant.

Recall that Corollary 14.32 showed that the interior of the inscribed circle is entirely contained in the interior of a circumscribed polygon. This is certainly not the case for inscribed polygons, as simple examples show. But for *regular* inscribed polygons, the center of the circle, at least, is an interior point, because it is the same as the center of the circumscribed circle by Theorem 14.39.

We will show below that perimeters of inscribed regular polygons are sufficient to determine the circumference of a circle. The key is the following theorem, which shows that perimeters of inscribed and circumscribed regular polygons get closer and closer to each other as the number of sides increases. This proof is rated R.

Theorem 15.5. *Suppose* $\mathcal{C}$ *is a circle of radius* r, $\mathcal{P}$ *is a regular* n-*gon inscribed in* $\mathcal{C}$, *and* $\mathcal{Q}$ *is a regular* n-*gon circumscribed about* $\mathcal{C}$, *with* $n \geq 12$. *Then*

$$\text{perim}(\mathcal{Q}) - \text{perim}(\mathcal{P}) < \frac{800r}{n^2}.$$

Proof. Let O be the center of $\mathcal{C}$. By replacing $\mathcal{Q}$ with another circumscribed regular n-gon congruent to the first, we may assume that each ray starting at O and passing through a vertex of $\mathcal{P}$ also contains a vertex of $\mathcal{Q}$. Let A and A' be two consecutive vertices of $\mathcal{P}$, and let B and B' be the vertices of $\mathcal{Q}$ on the same rays. Let F be the foot of the perpendicular from O to $\overline{BB'}$, and let G be the point where $\overline{OF}$ meets $\overline{AA'}$ (Fig. 15.6). Note that $\triangle OAG$ and $\triangle OBF$ are similar right triangles (by AA) and $\overline{OA}$ and $\overline{OF}$ have length r because they are both radii of $\mathcal{C}$. If we set $\alpha = 360°/(2n)$, then $m\angle AOG = \alpha$. Define the following lengths:

$$a = AG, \qquad b = BF, \qquad c = FG.$$

Then

$$\text{perim}(\mathcal{P}) = 2na, \qquad \text{perim}(\mathcal{Q}) = 2nb. \tag{15.3}$$

We need to derive an upper bound for $2nb - 2na$.

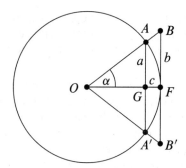

Fig. 15.6. Comparing perimeters of inscribed and circumscribed regular polygons.

First, by the triangle inequality applied to $\triangle OAG$, we have $r < (r - c) + a$, from which it follows that

$$c < a. \tag{15.4}$$

Next, because $OG/OF < 1$, similarity shows that $AG/BF < 1$, or

$$a < b. \tag{15.5}$$

The Pythagorean theorem applied to $\triangle OAG$ yields $(r - c)^2 + a^2 = r^2$, which simplifies to $2rc = a^2 + c^2$. With (15.4) and (15.5), this implies $2rc = a^2 + c^2 < 2a^2 < 2b^2$, or

$$c < \frac{b^2}{r}. \tag{15.6}$$

Finally, the fact that $\triangle OAG \sim \triangle OBF$ implies

$$\frac{a}{r-c} = \frac{b}{r}.$$

Cross-multiplying, we obtain $ar = br - bc$, which simplifies to $b - a = bc/r$. Using (15.6), we conclude

$$b - a < \frac{b^3}{r^2}. \tag{15.7}$$

Now we need to estimate b in terms of n and r. Given an integer $n \geq 12$, let k be the largest integer less than $n/6$; thus

$$\frac{n}{6} - 1 \leq k < \frac{n}{6}. \tag{15.8}$$

The fact that $n \geq 12$ implies $n - n/2 \geq 6$, which is equivalent to $n - 6 \geq n/2$. Therefore, the first inequality in (15.8) implies

$$k \geq \frac{n-6}{6} \geq \frac{n}{12}. \tag{15.9}$$

On the other hand, the second inequality in (15.8) implies

$$k\alpha = k \cdot \frac{360°}{2n} < \frac{n}{6} \cdot \frac{360°}{2n} = 30°. \tag{15.10}$$

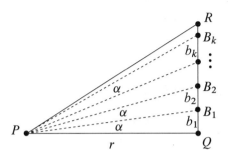

Fig. 15.7. Estimating b.

Now, let $\triangle PQR$ be a right triangle with right angle at Q, $m\angle P = 30°$, $PQ = r$, and $QR = r/\sqrt{3}$. (Such a triangle is easy to construct using the result of Theorem 13.3.) Let $\vec{r_1}, \vec{r_2}, \ldots, \vec{r_k}$ be the rays starting at P and making angles with $\overrightarrow{PQ}$ whose measures are $\alpha, 2\alpha, \ldots, k\alpha$, and let $B_1, B_2, \ldots, B_k$ be the points where these rays meet $\overline{QR}$. By virtue of (15.10) and the betweenness vs. betweenness theorem, we have $Q * B_1 * B_2 * \cdots * B_k * R$. Define numbers $b_1, \ldots, b_k$ by $b_1 = QB_1$, $b_2 = B_1 B_2$, $\ldots$, $b_k = B_{k-1} B_k$, so $b_1 + \cdots + b_k = QB_k < QR = r/\sqrt{3}$ (Fig. 15.7). Note that $\triangle PQB_1$ is congruent to $\triangle OFB$ in Fig. 15.6 by AAS, so $b_1 = b$. On the other hand, Theorem 7.13 shows that $PQ < PB_1 < PB_2 < \cdots < PB_k$, and then it follows from the angle bisector proportion theorem that $b_1 < b_2 < \cdots < b_k$. Therefore, using (15.9), we have

$$\frac{r}{\sqrt{3}} > b_1 + \cdots + b_k > kb \geq \frac{nb}{12}.$$

Solving this inequality for b yields

$$b < \frac{12r}{\sqrt{3}\,n}. \tag{15.11}$$

Combining this with (15.7) and using the rough estimates $12^3 < 1800$ and $\sqrt{3} > 1.5$, we finally obtain

$$2n(b-a) < \frac{2nb^3}{r^2} < \frac{2n \cdot 12^3 r^3}{3\sqrt{3}\,n^3 r^2} < \frac{800r}{n^2}, \tag{15.12}$$

which was to be proved. $\qquad\qquad\square$

The estimate given by this theorem is not very useful for practical computations, because it is necessary to use a very large value of n in order to counteract the effect of the 800 in the numerator. But by judiciously choosing the number of sides of the polygon, it is possible to derive much more accurate estimates of the circumference. For example, if n is a multiple of 6, then the estimate (15.9) can be replaced by $k = n/6$, which has the effect of replacing 12^3 in (15.12) by 6^3, thus yielding a final error bound of $100r/n^2$ instead of $800r/n^2$.

The main significance of Theorem 15.5 is theoretical: it guarantees we can make the difference between the perimeter of $\mathcal{P}$ and the circumference of $\mathcal{C}$ as small as we wish by taking n large enough. One consequence of this is that we can express the circumference as a limit, as the following corollary shows.

Corollary 15.6. *Let $\mathcal{C}$ be a circle, and for each n let $\mathcal{P}_n$ be a regular n-gon inscribed in $\mathcal{C}$. Then*

$$\text{circum}(\mathcal{C}) = \lim_{n \to \infty} \text{perim}(\mathcal{P}_n).$$

Proof. For each n, let $\mathcal{Q}_n$ be a regular n-gon circumscribed about $\mathcal{C}$. Then Corollary 15.9 shows that

$$\text{perim}(\mathcal{P}_n) < \text{circum}(\mathcal{C}) < \text{perim}(\mathcal{Q}_n).$$

Given $\varepsilon > 0$, the preceding theorem shows that for all n larger than $\sqrt{800r/\varepsilon}$ (and at least 12), we have

$$
\begin{aligned}
\left| \text{circum}(\mathcal{C}) - \text{perim}(\mathcal{P}_n) \right| &= \text{circum}(\mathcal{C}) - \text{perim}(\mathcal{P}_n) \\
&< \text{perim}(\mathcal{Q}_n) - \text{perim}(\mathcal{P}_n) \\
&< \frac{800r}{n^2} < \varepsilon.
\end{aligned}
$$

This proves the required limit statement. □

The Definition of Pi

The next theorem shows how to relate circumferences of different circles to their radii.

Theorem 15.7 (Circumference Scaling Theorem). *For any two circles, the ratio of their circumferences is the same as the ratio of their radii.*

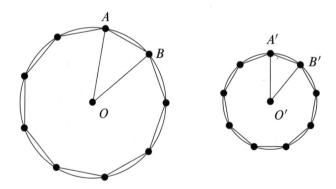

Fig. 15.8. Proof of the circumference scaling theorem.

Proof. Let $\mathcal{C} = \mathcal{C}(O, r)$ and $\mathcal{C}' = \mathcal{C}(O', r')$ be two circles. Suppose $\mathcal{P}_n$ is a regular n-gon inscribed in $\mathcal{C}$ and $\mathcal{P}'_n$ is a regular n-gon inscribed in $\mathcal{C}'$ (Fig. 15.8). Let A, B denote any two consecutive vertices of $\mathcal{P}_n$, and let A', B' denote any two consecutive vertices of $\mathcal{P}'_n$. The central angles $\angle AOB$ and $A'O'B'$ both measure $360°/n$ by Theorem 14.40. Since $OA/O'A' = OB/O'B' = r/r'$, it follows from the SAS similarity theorem that $\triangle AOB \sim \triangle A'O'B'$, and thus $AB/A'B' = r/r'$. Since this is true for all the edges, it follows that $\mathcal{P}_n$ and $\mathcal{P}'_n$ are similar with scale factor r/r', and therefore by Theorem 12.16, their perimeters have the same ratio. Thus by Corollary 15.6, we have

$$\begin{aligned}
\text{circum}(\mathcal{C}) &= \lim_{n\to\infty} \text{perim}(\mathcal{P}_n) \\
&= \lim_{n\to\infty} (r/r')\,\text{perim}(\mathcal{P}'_n) \\
&= (r/r') \lim_{n\to\infty} \text{perim}(\mathcal{P}'_n) \\
&= (r/r')\,\text{circum}(\mathcal{C}'). \qquad \square
\end{aligned}$$

Suppose $\mathcal{C}(O,r)$ and $\mathcal{C}(O',r')$ are any two circles, and denote their circumferences by C and C', respectively. The preceding theorem shows that $C'/C = r'/r$, which is equivalent to $C/r = C'/r'$. Since this is true for any two circles, it follows that the ratio of circumference to radius is the same number for all circles. Since the diameter of every circle is twice its radius by Lemma 14.1, it also follows that the ratio of circumference to diameter is the same for all circles: $C/d = C/(2r) = C'/(2r') = C'/d'$. This common ratio of circumference to diameter is a positive real number called *pi* and denoted by π.

Corollary 15.8 (Circumference of a Circle). *The circumference of a circle of radius r is* $2\pi r$.

Proof. Let $\mathcal{C}$ be a circle of radius r, and let d denote its diameter and C its circumference. Then the definition of π gives $\pi = C/d$, so $C = \pi d = 2\pi r$. $\qquad \square$

In particular, if $\mathcal{C}$ is a circle of radius 1 (called a *unit circle*), the preceding corollary shows that its circumference is exactly 2π.

Approximations of π

The preceding discussion established a theoretical definition of π, but it did not give us any idea of its value. It was proved by Johann Lambert in the eighteenth century that π is an irrational number, so to evaluate it numerically, the best we can hope for is an approximation. The key to estimating it is the following corollary, which follows immediately from Corollary 15.4.

Corollary 15.9. *Let $\mathcal{C}$ be a unit circle. If $\mathcal{P}$ is any polygon inscribed in $\mathcal{C}$ and $\mathcal{Q}$ is any polygon circumscribed about $\mathcal{C}$, then*

$$\tfrac{1}{2}\text{perim}(\mathcal{P}) \le \pi \le \tfrac{1}{2}\text{perim}(\mathcal{Q}). \qquad \square$$

We can use this idea to give a very rough approximation to π.

Theorem 15.10. $3 < \pi < 4$.

Proof. Let $\mathcal{C}$ be a unit circle, and let $\mathcal{P}$ be an inscribed regular hexagon and $\mathcal{Q}$ a circumscribed square (see Fig. 15.9). Then an easy argument shows that the perimeter of $\mathcal{P}$ is exactly 6 and that of $\mathcal{Q}$ is 8. The result follows from Corollary 15.9. $\qquad \square$

Of course, by examining polygons with more sides, we can do much better than this crude estimate. By using inscribed and circumscribed regular 96-gons together with some amazingly ingenious rational approximations to square roots, Archimedes [**Arc97, Arc02**] was able to prove that the ratio of circumference to diameter lies strictly between $3\frac{10}{71}$

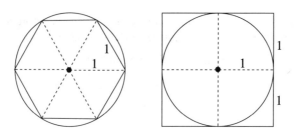

Fig. 15.9. A rough estimate of π.

and $3\frac{1}{7}$, which yields $3.1408 < \pi < 3.1429$. (It is important to note that Archimedes did not consider π to be an exact number, but he considered circumferences and diameters to be "magnitudes" in the same sense that Euclid used the word, and it made sense to compare ratios of two magnitudes to ratios of other magnitudes. This is the sense in which Archimedes' results can be considered as approximations to π.)

Since that time, many innovative geometric and analytic methods for estimating π have been devised, and although it is hard to imagine a practical use for any accuracy beyond one or two dozen decimal digits ($\pi \approx 3.14159265358979323846264\overline{3}\ldots$), setting new accuracy records in the calculation has become a popular way to test the limits of computing power. As of mid-2012, π has been calculated to an accuracy of 10 trillion (10^{13}) decimal digits.

The first use of the symbol π in this context was in the eighteenth century, and it caught on quickly after that. It is unfortunate, though, that the eighteenth-century authors chose to give a name to the ratio of circumference to diameter, because in many ways the more fundamental ratio is that of circumference to radius, or 2π. In many, many mathematical formulas, from geometry to trigonometry to complex analysis to probability to Fourier analysis, π occurs almost exclusively in the combination 2π. Several mathematicians have argued recently for the introduction of a new fundamental circle constant, τ (the Greek letter tau, representing "one turn"), which is equal to 2π and represents the circumference of a unit circle (see [**Har12**] for an eloquent defense of the proposal). The idea makes good logical and pedagogical sense, but centuries-old mathematical customs are hard to change.

Area of a Circular Region

A *circular region* is the union of a circle and its interior. If $\mathcal{C}$ is a circle, the region consisting of $\mathcal{C}$ and its interior is called the *circular region bounded by $\mathcal{C}$* and is denoted by $\operatorname{Reg}\mathcal{C}$. In order to speak of the area of a circular region, we need a new definition, analogous to the definition of circumference. If $\mathcal{C}$ is any circle, we wish to define the area of $\operatorname{Reg}\mathcal{C}$ to be the least upper bound of the set of areas of inscribed polygons. To verify that this is well defined, we need to show that the set of such areas is nonempty and has an upper bound. It is certainly nonempty, because there are inscribed polygons. The next theorem shows that the area of any circumscribed polygon will serve as an upper bound.

Theorem 15.11. *Let $\mathcal{C}$ be a circle, let $\mathcal{P}$ be a polygon inscribed in $\mathcal{C}$, and let $\mathcal{Q}$ be a polygon circumscribed about $\mathcal{C}$. Then $\alpha(\mathcal{Q}) > \alpha(\mathcal{P})$.*

Proof. Write $\mathcal{P} = A_1 \ldots A_n$ and $\mathcal{Q} = B_1 \ldots B_m$ as in the proof of Theorem 15.3. As we did in that proof, we begin by assuming that O lies in the interior of $\mathcal{P}$. For each $i = 1, \ldots, n$, let C_i be the point where $\overrightarrow{OA_i}$ meets $\mathcal{Q}$. Let $\mathcal{P}_i = \triangle A_i O A_{i+1}$, and let $\mathcal{Q}_i$ be the polygon formed by $\overline{OC_i}$, $\overline{OC_{i+1}}$, and the portion of $\mathcal{Q}$ in the interior of $\angle C_i O C_{i+1}$. Then the pizza lemma (Lemma 11.2) shows that

$$\alpha(\mathcal{P}) = \alpha(\mathcal{P}_1) + \cdots + \alpha(\mathcal{P}_n),$$
$$\alpha(\mathcal{Q}) = \alpha(\mathcal{Q}_1) + \cdots + \alpha(\mathcal{Q}_n). \tag{15.13}$$

For each i, the segment $\overline{A_i A_{i+1}}$ is a chord of the convex polygon $\mathcal{Q}_i$, and one of the polygons in the admissible decomposition of $\mathcal{Q}_i$ that it defines is exactly $\mathcal{P}_i$. If we let $\mathcal{Q}_i'$ denote the other, then the convex decomposition lemma gives

$$\alpha(\mathcal{Q}_i) = \alpha(\mathcal{P}_i) + \alpha(\mathcal{Q}_i') > \alpha(\mathcal{P}_i).$$

Inserting this into (15.13) shows that $\alpha(\mathcal{Q}) > \alpha(\mathcal{P})$ as claimed.

If $O \notin \operatorname{Int} \mathcal{P}$, Lemma 15.2 shows that we can find an inscribed polygon $\mathcal{P}'$ that has O in its interior and satisfies $\alpha(\mathcal{P}') > \alpha(\mathcal{P})$. The argument in the preceding paragraph then applies to $\mathcal{P}'$, and we conclude that $\alpha(\mathcal{Q}) > \alpha(\mathcal{P}') > \alpha(\mathcal{P})$. □

Because of the preceding theorem, for any circle $\mathcal{C}$, we can define the *area of* **Reg** $\mathcal{C}$ to be the least upper bound of the set of areas of inscribed polygons. This approach, first introduced by the ancient Greeks and developed to a high art by Archimedes, is known as the *method of exhaustion*, because as we consider inscribed polygons with more and smaller sides, their areas eventually use up, or "exhaust," all of the available space inside the circle.

As we do for polygons, for simplicity we often refer to the "area of $\mathcal{C}$," denoted by $\alpha(\mathcal{C})$, with the understanding that this means the area of the circular region bounded by $\mathcal{C}$.

The next three results are the analogues for areas of Theorem 15.5, Corollary 15.6, and Theorem 15.7.

Theorem 15.12. *Suppose $\mathcal{C}$ is a circle of radius r, $\mathcal{P}$ is a regular n-gon inscribed in $\mathcal{C}$, and $\mathcal{Q}$ is a regular n-gon circumscribed about $\mathcal{C}$, with $n \geq 12$. Then*

$$\alpha(\mathcal{Q}) - \alpha(\mathcal{P}) < \frac{800r^2}{n^2}.$$

Proof. We use the same notation as in the proof of Theorem 15.5. From the pizza lemma, it follows that the areas of $\mathcal{P}$ and $\mathcal{Q}$ can be computed by summing the areas of n isosceles triangles, each of which shares two consecutive vertices with $\mathcal{P}$ or $\mathcal{Q}$ and has its other vertex at O. Referring to Fig. 15.6, we see that each of the triangles in $\mathcal{Q}$ has a base of length $2b$ and height r, while each one in $\mathcal{P}$ has a base of length $2a$ and height $r - c$. Thus

$$\alpha(\mathcal{Q}) = n \cdot \tfrac{1}{2}(2b)r = nbr,$$
$$\alpha(\mathcal{P}) = n \cdot \tfrac{1}{2}(2a)(r - c) = nar - nac,$$

from which it follows that

$$\alpha(\mathcal{Q}) - \alpha(\mathcal{P}) = nr(b - a) + nac. \tag{15.14}$$

From (15.7), we conclude

$$nr(b - a) < \frac{nrb^3}{r^2} = \frac{nb^3}{r},$$

and from (15.5) and (15.6),

$$nac < nbc < \frac{nb^3}{r}.$$

Therefore, (15.14) and (15.11) yield

$$\alpha(\mathcal{Q}) - \alpha(\mathcal{P}) < \frac{2nb^3}{r} < \frac{2n \cdot 12^3 r^3}{3\sqrt{3} n^3 r} < \frac{800 r^2}{n^2}. \qquad \square$$

Corollary 15.13. *Let $\mathcal{C}$ be a circle, and for each $n \geq 3$, let $\mathcal{P}_n$ be a regular n-gon inscribed in $\mathcal{C}$. Then*

$$\alpha(\mathcal{C}) = \lim_{n \to \infty} \alpha(\mathcal{P}_n).$$

Proof. This follows exactly as in the proof of Corollary 15.6. $\square$

Now that we have defined the area of a circle, we need to figure out how to compute it. The key insight is the following beautiful theorem first proved by Archimedes. The geometric idea of the proof is illustrated in Fig. 15.10: when n is even, the area of an inscribed regular n-gon is equal to the area of a certain parallelogram whose base is half the perimeter of the polygon and whose height is slightly less than the radius of the circle. (When n is odd, the parallelogram is replaced by a trapezoid whose two bases have lengths that add up to the perimeter.) As the number of sides increases, the base approaches half the circumference of the circle, and the height approaches the radius. The proof of the theorem is a more algebraic version of the same idea.

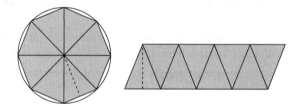

Fig. 15.10. The idea behind Archimedes' theorem.

Theorem 15.14 (Archimedes' Theorem). *For any circle $\mathcal{C}$ of radius r,*

$$\alpha(\mathcal{C}) = \tfrac{1}{2} r \cdot \text{circum}(\mathcal{C}). \qquad (15.15)$$

Proof. Let $\mathcal{C}$ be a circle of radius r, and let $\mathcal{P}_n$ be an inscribed regular n-gon. As in the proof of Theorem 15.12, the area of $\mathcal{P}_n$ is the sum of the areas of n isosceles triangles, each of which has base of length $2a$ and height $r - c$, with a and c as in the proof of Theorem 15.5. Therefore $\alpha(\mathcal{P}_n) = \frac{1}{2}(2na)(r - c)$. Note that $2na$ is also the perimeter of $\mathcal{P}_n$, so we can write

$$\alpha(\mathcal{P}_n) = \tfrac{1}{2}(r - c) \cdot \text{perim}(\mathcal{P}_n).$$

From (15.6) and (15.11), we have the estimate

$$c < \frac{b^2}{r} < \frac{12^2 r}{3\sqrt{3} n^2}.$$

Because r is fixed, we can make this expression as small as desired by taking n large enough. Therefore, when we take the limit as $n \to \infty$, we find that $c \to 0$, and therefore

$\frac{1}{2}(r-c) \to \frac{1}{2}r$. From Corollaries 15.6 and 15.13, we have $\mathrm{perim}(\mathcal{P}_n) \to \mathrm{circum}(\mathcal{C})$ and $\alpha(\mathcal{P}_n) \to \alpha(\mathcal{C})$ as $n \to \infty$. Putting this all together, we obtain

$$\alpha(\mathcal{C}) = \lim_{n\to\infty} \alpha(\mathcal{P}_n) = \lim_{n\to\infty} \left(\tfrac{1}{2}(r-c) \cdot \mathrm{perim}(\mathcal{P}_n) \right)$$

$$= \left(\lim_{n\to\infty} \tfrac{1}{2}(r-c) \right) \cdot \left(\lim_{n\to\infty} \mathrm{perim}(\mathcal{P}_n) \right)$$

$$= \tfrac{1}{2} r \cdot \mathrm{circum}(\mathcal{C}). \qquad \square$$

Corollary 15.15 (Area of a Circle). *The area of a circle of radius r is πr^2.*

Proof. Just substitute $\mathrm{circum}(\mathcal{C}) = 2\pi r$ into Archimedes' formula (15.15). $\qquad \square$

Generalizations

The approach we used for defining the area of a circle as the least upper bound of areas of inscribed polygons can easily be adapted to other closed curves such as ellipses or to figures bounded by combinations of curves and line segments such as the region between a parabola and a line. Archimedes was able to compute many such areas by adapting the method of exhaustion. However, for each different type of figure, new tricks and insights are needed for simplifying the formulas and computing the appropriate limits, so this approach is only applicable to rather special regions. Exercise 15C outlines the results for sectors of circles (regions bounded by two radii and an arc). With the development of integral calculus in the seventeenth century, the method of exhaustion was systematized and developed into a method for computing areas of very general figures.

Generalizing the circumference of a circle turns out to be a little trickier. For closed curves that bound convex regions, such as ellipses, circumference can be defined in a manner very similar to our definition of circumference of a circle, as the least upper bound of the perimeters of inscribed polygons. However, for curves that are not convex, this method does not work because perimeters of inscribed polygons can actually be longer than the curves they are meant to approximate. And for nonclosed curves such as arcs of circles or portions of parabolas, inscribed polygons do not even make sense.

It is possible to develop a theory of length for more or less arbitrary curves in the plane by approximating them with polygonal paths and taking limits in a suitable sense, but the details are tricky. This is exactly how lengths of curves are defined in calculus, and the definition leads to the familiar formula for arc length that is studied in multivariable calculus courses.

Exercises

15A. Prove that a circular region is a convex set. [Hint: See Exercise 14K.]

15B. Although we have not given an official definition of the length of an arc, it is possible to do so. Suppose we know only that for every arc $\overset{\frown}{AB}$ there is a positive number $L\left(\overset{\frown}{AB}\right)$ called its ***arc length***, satisfying the following properties:
   (i) Arcs with the same radius and same measure have the same length.
   (ii) If $\overset{\frown}{ABC}$ is the union of two arcs $\overset{\frown}{AB}$ and $\overset{\frown}{BC}$ that intersect only at B, then
   $$L\left(\overset{\frown}{ABC}\right) = L\left(\overset{\frown}{AB}\right) + L\left(\overset{\frown}{BC}\right).$$

(iii) The lengths of any two conjugate arcs add up to the circumference of the circle. Let $\mathcal{C}$ be a circle of radius r. Use the following outline to prove that the length of every arc of $\mathcal{C}$ is $2\pi r/360$ times the measure of the arc.

  (a) Given any real number x such that $0 < x < 360$, let $a(x)$ denote the length of any arc of $\mathcal{C}$ whose measure is x. Show that it suffices to show that $a(x) = 2\pi rx/360$ whenever $0 < x \le 180$.

  (b) Show that $a(180) = \pi r$.

  (c) Show that if x is a real number and n is a positive integer such that both x and nx are strictly between 0 and 180, then $a(nx) = n \cdot a(x)$. [Hint: Induction.]

  (d) Show that if n is any positive integer, then $a(180/n) = \pi r/n$.

  (e) Show that if m and n are positive integers with $m \le n$, then $a(180m/n) = m\pi r/n$.

  (f) Show that $a(x) = \pi rx/360$ if x is any real number between 0 and 180. [Hint: Look at the proof of Theorem 11.7.]

15C. Suppose $\mathcal{C} = \mathcal{C}(O, r)$ is a circle and $\widehat{AB}$ is an arc on $\mathcal{C}$. The *sector determined by $\mathbf{AB}$*, denoted by $\text{Sec}(AB)$, is the union of all radii from the center to points on $\widehat{AB}$:

$$\text{Sec}(AB) = \left\{ P : P \text{ lies on a radius } \overline{OX} \text{ for some point } X \in \widehat{AB} \right\}.$$

Although we have not given an official definition of the area of a sector, it is possible to do so. Suppose we know only that for every sector $\text{Sec}(AB)$ there is a positive number $\alpha(\text{Sec}(AB))$ called its *area*, satisfying the following properties:

  (i) Sectors determined by arcs with the same radius and same measure have the same area.

  (ii) If $\widehat{ABC}$ is the union of two arcs $\widehat{AB}$ and $\widehat{BC}$ that intersect only at B, then
$$\alpha(\text{Sec}(ABC)) = \alpha(\text{Sec}(AB)) + \alpha(\text{Sec}(BC)).$$

  (iii) The areas of the sectors determined by any two conjugate arcs add up to the area of the circle.

Prove that the area of every sector of a circle of radius r is $\pi r^2/360$ times the measure of the arc that determines it. [Hint: See Exercise 15B.]

Chapter 16

Compass and Straightedge Constructions

A prominent theme in the history of Euclidean geometry is the construction of geometric figures using and straightedge. As we remarked in Chapter 1, Euclid based his first three postulates on the types of figures (circles and lines) that can be constructed using a collapsible compass and an unmarked straightedge. A substantial number of propositions in the *Elements* are solutions to construction problems. As a result, a favorite preoccupation of geometers since Euclid has been searching for ways to construct new figures.

Constructions played a vital role in Greek mathematics because for the ancient Greeks, the only way to assert that a mathematical object (such as a particular equilateral triangle, square, or angle bisector) exists was to give an algorithm for constructing it. In our modern axiomatic approach to geometry, we have not placed a similar emphasis on constructions because our modern understanding of the axiomatic method gives us a different way of justifying the existence of mathematical objects: the existence of certain objects is explicitly stated in the postulates, and the existence of others can then be derived from those.

Compass and straightedge constructions should not be confused with the types of theorems that we have called "construction theorems," such as the angle construction theorem (Theorem 4.5) and the SSS construction theorem (Corollary 13.7). Those theorems merely assert the *existence* of certain geometric objects, but they are mute on the question of how those objects might be constructed by hand. By contrast, every compass and straightedge construction theorem describes an algorithm for producing a particular geometric figure while following the strict limitations imposed by Euclid's first three postulates.

Many clever and useful constructions are described in the *Elements*, including ones involving triangles, quadrilaterals, parallel lines, areas, circles, and inscribed and circumscribed polygons. Many of these constructions are suitable for high-school students and offer opportunities for deepening one's understanding of fundamental geometric relationships. These techniques of construction are also useful in practical settings such as drafting or navigation or whenever computer graphics are unavailable or impractical.

In this chapter, we can introduce only a selected sampling of constructions, to illustrate how they can be solved within the context of our modern axiomatic system. Many of the constructions are easier to do than to read, so after the first few we leave many of them to the reader as exercises.

At the end of the chapter, we address one of the most important issues in the history of geometric constructions: the fact that some construction problems are impossible to solve using only compass and straightedge.

Basic Constructions

We begin with a careful description of what is meant by compass and straightedge constructions in Euclidean geometry. The following operations, based on Euclid's first three postulates, are allowed:

- Given two distinct points A, B, construct the line segment $\overline{AB}$.
- Given a line segment in the plane, extend it in either direction to form a line segment longer than any predetermined length.
- Given two distinct points in the plane, construct the circle centered at the first point and passing through the second point.

In addition, although they are not mentioned in Euclid's postulates, the following operations are allowable in compass and straightedge constructions:

- Given two nonparallel lines, locate the point where they intersect.
- Given a circle and a secant line, locate the two points where they intersect.
- Given two nontangential circles, locate the points (if any) where they intersect.

All other constructions (such as cutting off a portion of a longer line segment congruent to a shorter one or bisecting an angle) must be built up out of these basic operations. Throughout this chapter, we assume all the postulates of Euclidean geometry.

As we mentioned in the introduction, Euclid begins Book I with the solution to a construction problem (Proposition I.1). We begin in the same place.

Construction Problem 16.1 (Equilateral Triangle on a Given Segment). *Given a segment $\overline{AB}$ and a side of $\overleftrightarrow{AB}$, construct a point C on the chosen side such that $\triangle ABC$ is equilateral.*

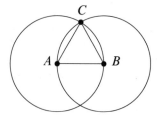

Fig. 16.1. Constructing an equilateral triangle.

Solution. Let $\overline{AB}$ be the given segment. Draw the circle with center A and passing through B and the circle with center B and passing through A (Fig. 16.1). Because the radii of both circles and the distance AB are all equal to each other, the three inequalities of Theorem 14.10(a) are satisfied, and thus the two circles have two intersection points, one on either side of $\overleftrightarrow{AB}$. Let C be the intersection point on the chosen side. Then the fact that B and C lie on a circle centered at A implies $AB = AC$, and the fact that A and C lie on a circle centered at B implies $AB = CB$. Thus $\triangle ABC$ is equilateral. $\qquad\square$

As you can see, this is the same as Euclid's construction, but we have used Theorem 14.10 (the two circles theorem) to justify the existence of the intersection point C. You will find that most of the constructions we discuss can be done along the same lines as in the *Elements*, but they frequently require some extra justifications to be made completely rigorous.

The next two constructions are Euclid's Propositions I.2 and I.3.

Construction Problem 16.2 (Copying a Line Segment to a Given Endpoint). *Given a line segment $\overline{AB}$ and a point C, construct a point X such that $\overline{CX} \cong \overline{AB}$.*

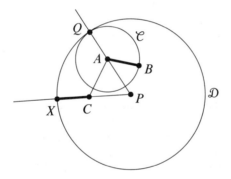

Fig. 16.2. Copying a line segment to a given endpoint.

Solution. Let $\overline{AB}$ be the given segment and let C be the given point (Fig. 16.2). If $C = A$, we can just take $X = B$ and there is nothing more to do, so assume $C \neq A$. Using the solution to Construction Problem 16.1, construct a point P (on either side of $\overleftrightarrow{AC}$) such that $\triangle APC$ is equilateral. Draw the circle $\mathcal{C}$ with center A and passing through B. Because $\overleftrightarrow{AP}$ contains the center of $\mathcal{C}$, it is a secant line. Extend $\overline{PA}$ past A, and let Q be the point of intersection between the extended segment and $\mathcal{C}$, chosen so that $P * A * Q$. Now draw the circle $\mathcal{D}$ with center P and passing through Q. Draw $\overline{PC}$, extend it past C, and let X be the point where the extended segment intersects $\mathcal{D}$.

We will show that $\overline{CX} \cong \overline{AB}$. Notice that $\overline{PQ} \cong \overline{PX}$ because they are both radii of $\mathcal{D}$ and $\overline{PA} \cong \overline{PC}$ because they are sides of an equilateral triangle. Since $P * A * Q$ and $P * C * X$ by construction, it follows from the segment subtraction theorem that $\overline{CX} \cong \overline{AQ}$. On the other hand, since $\overline{AQ}$ and $\overline{AB}$ are both radii of $\mathcal{C}$, they are also congruent. By transitivity, $\overline{CX} \cong \overline{AQ} \cong \overline{AB}$ as claimed. $\qquad\square$

Construction Problem 16.3 (Cutting Off a Segment). *Given two segments $\overline{AB}$ and $\overline{CD}$ such that $CD > AB$, construct a point E in the interior of $\overline{CD}$ such that $\overline{CE} \cong \overline{AB}$.*

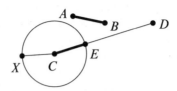

Fig. 16.3. Cutting off a segment.

Solution. Given $\overline{AB}$ and $\overline{CD}$ satisfying the hypothesis, by the preceding construction we can find a point X such that $\overline{CX} \cong \overline{AB}$ (see Fig. 16.3). Draw the circle $\mathcal{C}$ with center C and passing through X, and let E be the point where $\mathcal{C}$ meets the interior of $\overline{CD}$. (To see that there is such a point, note that $\overrightarrow{CD}$ contains a point on $\mathcal{C}$ by the segment construction theorem and this point lies in the interior of $\overline{CD}$ because CD is larger than the radius of $\mathcal{C}$.) Then $\overline{AB} \cong \overline{CX}$ by construction, and $\overline{CX} \cong \overline{CE}$ because they are both radii of $\mathcal{C}$. Thus $\overline{AB} \cong \overline{CE}$ by transitivity.   □

This justifies what we regularly do with a physical (noncollapsing) compass: opening the compass to the length of a given line segment, lifting the compass off the paper, and using it to mark off a segment of the given length on a longer line segment somewhere else. The preceding construction shows how to accomplish the same thing with a "collapsing compass" of the type Euclid's axioms envision. Henceforth, we will simply refer to this operation as "cutting off a segment on $\overline{CD}$ congruent to $\overline{AB}$," with the understanding that it is justified by Construction Problem 16.3.

When we use a physical compass to find an intersection point between a circle and a segment or another circle, it is usually unnecessary to draw the entire circle; all that is needed is a small arc of the circle near the desired intersection point. In order to make our drawings less cluttered from now on, we will often draw only enough of the circles to locate the desired points. Similarly, if there are line segments that are needed only to prove the correctness of a construction but not to complete it, we will indicate them with dashed lines.

The next construction is Euclid's Proposition I.9.

Construction Problem 16.4 (Bisecting an Angle). *Given a proper angle, construct its bisector.*

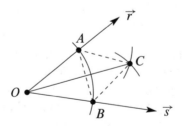

Fig. 16.4. Bisecting an angle.

Solution.   Suppose $\angle rs$ is the given proper angle and O is its vertex. Choose any point A in the interior of $\overrightarrow{r}$, and draw the circle $\mathcal{C}$ with center O and passing through A

(Fig. 16.4). Let B be the point where $\mathcal{C}$ meets $\vec{s}$. Now construct a point C on the opposite side of $\overleftrightarrow{AB}$ from O such that $\triangle ABC$ is equilateral. Draw $\overline{OC}$.

We need to show that $\overrightarrow{OC}$ bisects $\angle rs$. First, we have to verify that C lies in the interior of the angle. Note that $\angle OAB$ is acute (being a base angle of the isosceles triangle $\triangle OAB$) and $m\angle CAB = 60°$ because $\triangle ABC$ is equilateral. Because $\angle OAB$ and $\angle CAB$ are adjacent by construction, it follows from the adjacency lemma that $\overrightarrow{AO} * \overrightarrow{AB} * \overrightarrow{AC}$ and thus that B and C are on the same side of $\overleftrightarrow{OA}$. A similar argument shows that A and C are on the same side of $\overleftrightarrow{OB}$, and thus $C \in \operatorname{Int} \angle AOB$ as required.

Finally, we note that $\triangle OAC$ and $\triangle OBC$ are congruent by SSS, so $m\angle BOC = m\angle AOC$, and thus $\overrightarrow{OC}$ is the bisector of $\angle AOB$ as claimed. $\square$

The statement of the preceding construction referred to the bisector of an angle as the object to be constructed. Since a bisector is technically a ray, which is infinitely long, we cannot of course draw the entire angle bisector. Whenever we talk about "constructing a ray," what is really required is to construct enough points to completely determine the ray: the endpoint and at least one interior point of the ray. Similarly, to "construct a line" means to construct two points on the line. Once this is done, the rules of compass and straightedge constructions allow us to draw as much of the ray or line as needed.

We leave the next few elementary constructions as exercises. For some of them, we have provided diagrams that suggest an approach. For many of them, the proof of the corresponding existence theorem can easily be adapted to provide a construction algorithm. If you get stuck, you can always look at the way Euclid performed the same constructions.

When you write up a solution to a construction problem, make sure you include the following: (a) a careful description of each step of the construction; (b) a justification for the existence of each intersection point unless the justification will be obvious to your reader (for example, no justification is needed for locating the intersection point between two lines that can easily be shown to be nonparallel); (c) a proof that the given construction does indeed satisfy the required conditions.

The next four constructions are Euclid's Propositions I.10, I.11, I.12, and I.22, respectively. The first one is most commonly used to locate the midpoint of a segment.

Construction Problem 16.5 (Perpendicular Bisector). *Given a segment, construct its perpendicular bisector.*

Solution. Exercise 16A. (See Fig. 16.5.) $\square$

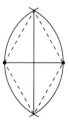

Fig. 16.5. Constructing a perpendicular bisector.

Construction Problem 16.6 (Perpendicular through a Point on a Line). *Given a line ℓ and a point $A \in \ell$, construct the line through A and perpendicular to ℓ.*

Solution. Exercise 16B.   □

Construction Problem 16.7 (Perpendicular through a Point Not on a Line). *Given a line ℓ and a point $A \notin \ell$, construct the line through A and perpendicular to ℓ.*

Solution. Exercise 16C.   □

Construction Problem 16.8 (Triangle with Given Side Lengths). *Given three segments such that the length of the longest is less than the sum of the lengths of the other two, construct a triangle whose sides are congruent to the three given segments.*

Solution. Exercise 16D.   □

The next construction does not appear explicitly in the *Elements*, but it is frequently useful.

Construction Problem 16.9 (Copying a Triangle to a Given Segment). *Given a triangle $\triangle ABC$, a segment $\overline{DE}$ congruent to $\overline{AB}$, and a side of $\overleftrightarrow{DE}$, construct a point F on the given side such that $\triangle DEF \cong \triangle ABC$.*

Solution. Exercise 16E.   □

Here is Euclid's Proposition I.23.

Construction Problem 16.10 (Copying an Angle to a Given Ray). *Given a proper angle $\angle ab$, a ray $\overrightarrow{c}$, and a side of $\overleftrightarrow{c}$, construct the ray $\overrightarrow{d}$ with the same endpoint as $\overrightarrow{c}$ and lying on the given side of $\overleftrightarrow{c}$ such that $\angle cd \cong \angle ab$.*

Solution. Exercise 16F.   □

Construction Problem 16.11 (Copying a Convex Quadrilateral to a Given Segment). *Given a convex quadrilateral $ABCD$, a segment $\overline{EF}$ congruent to $\overline{AB}$, and a side of $\overrightarrow{EF}$, construct points G and H on the given side such that $EFGH \cong ABCD$.*

Solution. Exercise 16G.   □

Construction Problem 16.12 (Rectangle with Given Side Lengths). *Given any two segments $\overline{AB}$ and $\overline{EF}$ and a side of $\overleftrightarrow{AB}$, construct points C and D on the given side such that $ABCD$ is a rectangle with $\overline{BC} \cong \overline{EF}$.*

Solution. Exercise 16H.   □

Next come Euclid's Propositions I.46 and I.31.

Construction Problem 16.13 (Square on a Given Segment). *Given a segment $\overline{AB}$ and a side of $\overleftrightarrow{AB}$, construct points C and D on the chosen side such that $ABCD$ is a square.*

Solution. Exercise 16I.   □

Construction Problem 16.14 (Parallel through a Point Not on a Line). *Given a line ℓ and a point $A \notin \ell$, construct the line through A and parallel to ℓ.*

Solution. Exercise 16J.   □

Ratio Constructions

A number of propositions in *Elements*, mostly in Book VI, show how to construct segments whose lengths have certain desired ratios. Here we present four of those constructions. The first two are Euclid's Propositions VI.9 and VI.10.

Construction Problem 16.15 (Cutting a Segment into n Equal Parts). *Given a segment* $\overline{AB}$ *and an integer* $n \geq 2$, *construct points* $C_1, \ldots, C_{n-1} \in \operatorname{Int} \overline{AB}$ *such that* $A * C_1 * \cdots * C_{n-1} * B$ *and* $AC_1 = C_1 C_2 = \cdots = C_{n-1} B$.

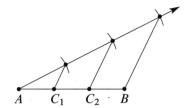

Fig. 16.6. Cutting a segment into three equal parts.

Solution. Exercise 16K. (See Fig. 16.6.) □

Construction Problem 16.16 (Cutting a Segment in a Rational Ratio). *Given a segment* $\overline{AB}$ *and a rational number* x *strictly between* 0 *and* 1, *construct a point* $D \in \operatorname{Int} \overline{AB}$ *such that* $AD = x \cdot AB$.

Solution. Because x is rational, we can write $x = m/n$ for some integers m and n with $0 < m < n$. Use the previous construction to mark points $C_1, \ldots, C_{n-1}$ that divide $\overline{AB}$ into n equal parts, and then let D be the mth point starting from A. Because the distance between successive points is $(1/n)AB$, it follows that $AD = (m/n)AB = x \cdot AB$. □

The next construction appears twice in the *Elements*, once as part of Proposition II.14 and once on its own as Proposition VI.13.

Construction Problem 16.17 (Geometric Mean of Two Segments). *Given two segments* $\overline{AB}$ *and* $\overline{CD}$, *construct a third segment that is their geometric mean.*

Solution. Exercise 16L. (See Fig. 16.7.) □

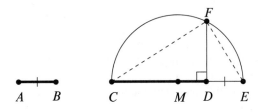

Fig. 16.7. Constructing the geometric mean of $\overline{AB}$ and $\overline{CD}$.

The following construction also appears twice in the *Elements*, as Propositions II.11 and VI.30.

Construction Problem 16.18 (The Golden Ratio). *Given a line segment $\overline{AB}$, construct a point $E \in \text{Int }\overline{AB}$ such that AB/AE is equal to the golden ratio.*

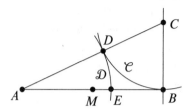

Fig. 16.8. Constructing the golden ratio.

Solution. Let $\overline{AB}$ be the given segment. Locate the midpoint M of $\overline{AB}$ by constructing its perpendicular bisector. Draw the line perpendicular to $\overleftrightarrow{AB}$ at B, and mark point C on that line such that $\overline{BC} \cong \overline{BM}$ (see Fig. 16.8). Draw $\overline{CA}$. Because $CB < CA$, we can draw the circle $\mathcal{C}$ with center C and passing through B, and let D be the point where $\mathcal{C}$ meets $\text{Int}\overline{CA}$. Then draw the circle $\mathcal{D}$ with center A and passing through D, and let E be the point where $\mathcal{D}$ meets $\overrightarrow{AB}$.

We will show that AB/AE is equal to the golden ratio φ, defined by (12.18). Note that $BC = BM = \frac{1}{2}AB$ by construction, and therefore $AC = \frac{1}{2}\sqrt{5}\,AB$ by the Pythagorean theorem. Note also that $CD = CB = \frac{1}{2}AB$ and $AD = AE$ by construction. It follows from equation (12.23) that

$$AE = AD = AC - CD = \tfrac{1}{2}\sqrt{5}\,AB - \tfrac{1}{2}AB = \frac{\sqrt{5}-1}{2}AB = (\varphi - 1)AB = \frac{AB}{\varphi}.$$

This implies $AB/AE = \varphi$. Because $\varphi > 1$, it follows that $AE < AB$, and therefore E is an interior point of $\overline{AB}$.   □

Area Constructions

The *Elements* included a large number of constructions involving areas. Most of these describe how to construct a polygon of a certain type with the same area as another given polygon. To illustrate how they work, we present just a few of them here. The first is Euclid's Proposition I.42.

Construction Problem 16.19 (Parallelogram with the Same Area as a Triangle). *Suppose $\triangle ABC$ is a triangle and $\angle rs$ is a proper angle. Construct a parallelogram with the same area as $\triangle ABC$ and with one of its angles congruent to $\angle rs$.*

Solution. Given $\triangle ABC$, construct the line ℓ through A and parallel to $\overleftrightarrow{BC}$ (Fig. 16.9). Construct the ray $\overrightarrow{r'}$ starting at B, lying on the same side of $\overleftrightarrow{BC}$ as A, and making an angle with $\overleftrightarrow{BC}$ congruent to $\angle rs$. Let D be the point where $\overrightarrow{r'}$ meets ℓ (extended if necessary). Locate the midpoint M of $\overline{BC}$, and draw the line m through M and parallel to $\overrightarrow{r'}$. Locate the point E where m meets ℓ.

Because $\overline{DE} \parallel \overline{BM}$ and $\overline{ME} \parallel \overline{BD}$, it follows from the parallelogram lemma that $DEMB$ is a parallelogram. Its area is equal to $(BM)h$, where h is the distance between the lines ℓ and $\overleftrightarrow{BC}$, while $\alpha(\triangle ABC) = \frac{1}{2}(BC)h$, which in turn is equal to $(BM)h$ because M is the midpoint of $\overline{BC}$. The angle $\angle DBM$ is congruent to $\angle rs$ by construction.   □

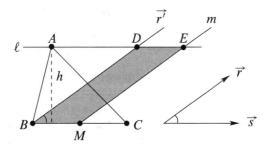

Fig. 16.9. Constructing a parallelogram with the same area as a triangle.

The next construction is a special case of Euclid's Proposition I.44, but with a slightly simpler proof.

Construction Problem 16.20 (Rectangle with a Given Area and Edge). *Given a rectangle $ABCD$, a segment $\overline{EF}$, and a side of $\overleftrightarrow{EF}$, construct a new rectangle with the same area as $ABCD$ with $\overline{EF}$ as one of its edges and with its opposite edge on the given side of $\overleftrightarrow{EF}$.*

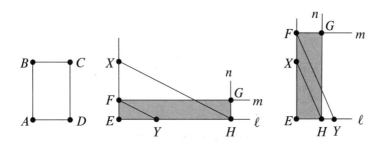

Fig. 16.10. Two possible configurations in Construction Problem 16.20.

Solution. Let $ABCD$ be the given rectangle and let $\overline{EF}$ be the given segment. If $\overline{EF} \cong \overline{AB}$, then we can just construct a rectangle congruent to $ABCD$ with $\overline{EF}$ as one of its sides (Construction Problem 16.12), so assume henceforth that $\overline{EF} \not\cong \overline{AB}$. Construct the line ℓ perpendicular to $\overleftrightarrow{EF}$ through E. Mark the point $X \in \overrightarrow{EF}$ (extended past F if necessary) such that $\overline{EX} \cong \overline{AB}$, and mark the point $Y \in \ell$ on the chosen side of $\overleftrightarrow{EF}$ such that $\overline{EY} \cong \overline{AD}$ (see Fig. 16.10). Draw $\overline{FY}$. Then draw the line through X and parallel to $\overleftrightarrow{FY}$, and let H be the point where this line meets ℓ (extended if necessary). (These two lines meet by Proclus's lemma.) Now draw the line m through F and perpendicular to $\overleftrightarrow{EF}$ and the line n through H and perpendicular to ℓ, and mark the point G where m and n meet. (Because $m \parallel \ell$ by the common perpendicular theorem, this point also exists by Proclus's lemma.)

The common perpendicular theorem shows that $\overline{FG} \parallel \overline{EH}$ and $\overline{GH} \parallel \overline{EF}$. Thus $EFGH$ is a parallelogram by the parallelogram lemma, and because $\angle E$ is a right angle, $EFGH$ is a rectangle. We will show that $\alpha(EFGH) = \alpha(ABCD)$. Note that $\angle EFY \cong \angle EXH$ by the corresponding angles theorem, so $\triangle EFY \sim \triangle EXH$ by AA. It follows that $EF/EX = EY/EH$. This implies $(EF)(EH) = (EX)(EY)$. Since $EX = AB$ and $EY = AD$ by construction, we conclude that

$$\alpha(EFGH) = (EF)(EH) = (EX)(EY) = (AB)(AD) = \alpha(ABCD). \qquad \square$$

Recall that Euclid did not use numbers to measure areas; instead, areas were considered as "magnitudes" (like line segments and angles), which could only be added, subtracted, and compared to other magnitudes of the same type. Thus the closest Euclid could come to "calculating the area" of a given geometric figure was to construct a square that had the same area, a process that is traditionally called "squaring" the figure. The next two propositions are special cases of Euclid's Proposition II.14.

Construction Problem 16.21 (Squaring a Rectangle). *Given a rectangle, construct a square with the same area as the rectangle.*

Solution. Exercise 16M. $\qquad \square$

Construction Problem 16.22 (Squaring a Convex Polygon). *Given a convex polygon, construct a square with the same area as the polygon.*

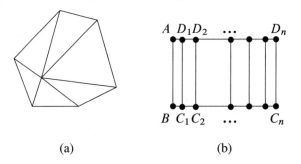

(a)   (b)

Fig. 16.11. Squaring a convex polygon.

Proof. Let $\mathcal{P}$ be a convex n-gon. Choose a random point in Int $\mathcal{P}$ and connect it to each of the vertices of $\mathcal{P}$, giving an admissible decomposition of $\mathcal{P}$ into n triangles $\mathcal{T}_1, \ldots, \mathcal{T}_n$ by the pizza lemma (see Fig. 16.11(a)). For each triangle $\mathcal{T}_i$, construct a rectangle $\mathcal{R}_i$ with the same area (Construction Problem 16.19). Next, construct a segment $\overline{AB}$, and use Construction Problem 16.20 to construct a rectangle ABC_1D_1 with the same area as $\mathcal{R}_1$ and with $\overline{AB}$ as one of its edges. Then construct a second rectangle $C_1D_1D_2C_2$ with the same area as $\mathcal{R}_2$, with $\overline{C_1D_1}$ as one of its edges, and with C_2 and D_2 on the opposite side of $\overleftrightarrow{C_1D_1}$ from $\overline{AB}$. Continuing in this way for $\mathcal{R}_3, \ldots, \mathcal{R}_n$, we obtain a large rectangle ABC_nC_n with the same area as $\mathcal{P}$, as shown in Fig. 16.11(b). Finally, use Construction Problem 16.21 to construct a square with the same area as the large rectangle. $\qquad \square$

(The full strength of Euclid's Proposition II.14 applies to arbitrary polygonal regions, not just convex ones; the only missing ingredient for proving the general case is the fact that arbitrary polygonal regions admit admissible decompositions into triangular regions, which we have not proved.)

Our last area construction does not appear explicitly in the *Elements*, but Euclid could easily have done it using the theorems he had already proved.

Construction Problem 16.23 (Doubling a Square). *Given a square, construct a new square whose area is twice that of the original one.*

Solution. Exercise 16N.   □

Circle Constructions

Many constructions in the *Elements* involve circles. Here we present just a few of them. These are Euclid's Propositions III.1, IV.4, and IV.5.

Construction Problem 16.24 (Center of a Circle). *Given a circle, construct its center.*

Solution. Exercise 16O.   □

Construction Problem 16.25 (Inscribed Circle). *Given a triangle, construct its inscribed circle.*

Solution. Exercise 16P.   □

Construction Problem 16.26 (Circumscribed Circle). *Given a triangle, construct its circumscribed circle.*

Solution. Exercise 16Q.   □

Constructing Regular Polygons

Our last six construction problems all focus on regular polygons inscribed in circles. Theorem 14.42 showed it is always possible to inscribe a regular polygon with any number of sides in a circle. However, as we will see, constructing them with compass and straightedge can be a much more difficult problem.

Our first regular polygon construction is Euclid's Proposition IV.6.

Construction Problem 16.27 (Square Inscribed in a Circle). *Given a circle and a point A on the circle, construct a square inscribed in the circle that has one vertex at A.*

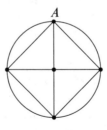

Fig. 16.12. A square inscribed in a circle.

Solution. Exercise 16R. (See Fig. 16.12.)   □

Next we consider pentagons. The construction of a regular pentagon inscribed in a circle appears in the *Elements* as Proposition IV.11 and is one of the gems of Euclidean geometry. To motivate the construction, consider that the central angles of a regular pentagon all measure 72°, which is twice the measure of the smaller angle of a golden triangle.

Thus it should not be surprising that the following algorithm begins by constructing two golden triangles.

Construction Problem 16.28 (Regular Pentagon Inscribed in a Circle). *Given a circle and a point A on the circle, construct a regular pentagon inscribed in the circle that has one vertex at A.*

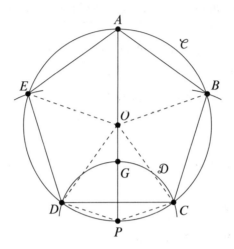

Fig. 16.13. Inscribing a regular pentagon.

Solution. Let $\mathcal{C}$ be the given circle and A the given point (Fig. 16.13). Locate the center O of $\mathcal{C}$. Draw $\overline{AO}$, extend it past O, and mark the point P where the extended segment intersects $\mathcal{C}$. Locate the point G in the interior of $\overline{PO}$ such that PO/PG is equal to the golden ratio (Construction Problem 16.18). Draw the circle $\mathcal{D}$ with center P and passing through G, and let C and D be the two points where $\mathcal{D}$ intersects $\mathcal{C}$. (These points exist because $\mathcal{C}$ contains the points $P \in \text{Int}\,\mathcal{D}$ and $A \in \text{Ext}\,\mathcal{D}$.) Then draw the circle with center C and passing through D, and let B be the other point where this circle meets $\mathcal{C}$; and similarly, draw the circle with center D and passing through C, and let E be the other point where this circle meets $\mathcal{C}$. Draw the pentagon $ABCDE$, which is inscribed in $\mathcal{C}$ by construction.

We will show that $ABCDE$ is equilateral. Note that $\overline{PC} \cong \overline{PG}$ because both are radii of the same circle, and similarly $\overline{OP} \cong \overline{OC}$. Since $PO/PC = PO/PG$, which is the golden ratio, it follows that $\triangle POC$ is a golden triangle, and thus $m\angle POC = 36°$. Similarly, $m\angle POD = 36°$. Since D and C lie on opposite sides of $\overleftrightarrow{OP}$ by the two circles theorem, $\angle POC$ and $\angle POD$ are adjacent angles, and thus $m\angle COD = 72°$.

Now, $\triangle BOC$ and $\triangle DOE$ are both congruent to $\triangle COD$ by SSS, and therefore $m\angle BOC = m\angle DOE = 72°$. The linear triple theorem applied to $\angle AOB$, $\angle BOC$, and $\angle COP$ implies that $m\angle AOB = 72°$, and similarly $m\angle AOE = 72°$; thus $\triangle AOB$ and $\triangle AOE$ are both congruent to the three triangles $\triangle BOC$, $\triangle COD$, and $\triangle DOE$ by SAS. It follows that $AB = BC = CD = DE = EA$, so $ABCDE$ is equilateral, and thus it is regular by Lemma 14.41. $\square$

Next we have Euclid's Proposition IV.15.

Construction Problem 16.29 (Regular Hexagon Inscribed in a Circle). *Given a circle and a point A on the circle, construct a regular hexagon inscribed in the circle that has one vertex at A.*

Fig. 16.14. Inscribing a regular hexagon.   **Fig. 16.15.** Inscribing an equilateral triangle.

Solution. Exercise 16T. (See Fig. 16.14.)   □

The next two constructions do not appear explicitly in the *Elements*, but they can be derived easily from other constructions that do appear.

Construction Problem 16.30 (Equilateral Triangle Inscribed in a Circle). *Given a circle and a point A on the circle, construct an equilateral triangle inscribed in the circle that has one vertex at A.*

Solution. Exercise 16U. (See Fig. 16.15.)   □

Construction Problem 16.31 (Regular Octagon Inscribed in a Circle). *Given a circle and a point A on the circle, construct a regular octagon inscribed in the circle that has one vertex at A.*

Solution. Let $\mathcal{C}$ be the circle and let O be its center. First inscribe a square with one vertex at A, and then bisect each of its central angles and add a new vertex where the bisector meets $\mathcal{C}$ (see Fig. 16.16). Each of the eight resulting central angles has measure $45°$, so all eight triangles formed by O and two consecutive vertices are congruent to each other. Thus the resulting octagon is equilateral and hence regular.   □

Finally, we have Euclid's Proposition IV.16.

Construction Problem 16.32 (Regular 15-gon Inscribed in a Circle). *Given a circle and a point A on the circle, construct a regular fifteen-sided polygon inscribed in the circle that has one vertex at A.*

Solution. Exercise 16V. (See Fig. 16.17.)   □

Impossible Constructions

Even though Euclid was able to produce constructions for a great many geometric figures, there were a number of seemingly simple construction problems that he was unable to solve. Four of the most famous are the following:

- TRISECTING AN ANGLE: Given an arbitrary proper angle $\angle ABC$, construct an angle whose measure is $\frac{1}{3}m\angle ABC$.

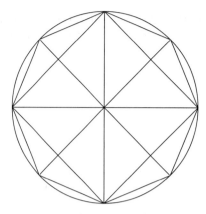 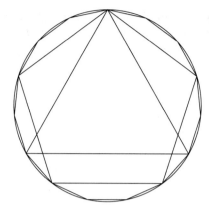

Fig. 16.16. Inscribing a regular octagon.   **Fig. 16.17.** Inscribing a regular 15-gon.

- DOUBLING A CUBE: Given an arbitrary segment $\overline{AB}$, construct a segment $\overline{CD}$ such that the cube with side length CD has twice the volume of that with side length AB.

- SQUARING A CIRCLE: Given an arbitrary circle, construct a square with the same area as the circle.

- CONSTRUCTING A REGULAR HEPTAGON: Given an arbitrary circle $\mathcal{C}$, construct a regular seven-sided polygon inscribed in $\mathcal{C}$.

During the two millennia after Euclid, many mathematicians strove to find algorithms to solve these construction problems, but no one succeeded. Finally, in the nineteenth century, mathematicians were able to prove rigorously that none of these construction problems have solutions in general. (Of course, for some particular starting configurations, it might be possible to perform the construction—for example, a 90° angle can be trisected by first constructing an equilateral triangle and then bisecting one of its angles. But the nineteenth-century mathematicians showed that there is no general construction algorithm that will precisely trisect every angle.) In this section, we examine the arguments leading to these impossibility proofs.

Constructible Points, Lines, Circles, and Numbers

Every construction starts with some given data. For the purpose of proving that some constructions are impossible, let us assume that to begin with, we are given two points O and I such that the distance between them is equal to 1. We say that a line, circle, or point is ***constructible*** if it can be produced by a finite (perhaps empty) sequence of compass and straightedge construction steps of the six types described at the beginning of this chapter, starting only with the points O and I. (Whenever it is necessary to choose a "random" point satisfying certain properties, it is always possible to choose a specific constructible point satisfying the desired properties. For example, to choose a random point in the interior of a segment, you can construct its midpoint. We leave it to you to work out the details in each specific case.)

We say that a real number r is a ***constructible number*** if there exist a pair of constructible points A and B such that $AB = |r|$. (Thus $-r$ is constructible if and only if r is constructible.) Let K denote the set of all constructible numbers. The key to understanding which constructions are impossible is to study the algebraic properties of K. For this purpose, we need to introduce a few basic concepts from abstract algebra. For much more detail about the algebraic ideas touched upon here, see any good abstract algebra text, such as [**Her99**].

A subset $F \subseteq \mathbb{R}$ is called a *field* if it contains 0 and 1 and is closed under addition, subtraction, multiplication, and division by nonzero elements of F. (Algebraists study a more general notion of fields that do not need to be subsets of $\mathbb{R}$, but this limited definition will be ample for our purposes.) If $F_1, F_2 \subseteq \mathbb{R}$ are fields such that $F_1 \subseteq F_2$, then we say that F_1 is a *subfield* of F_2 and that F_2 is an *extension* of F_1. For example, two obvious subfields of $\mathbb{R}$ are $\mathbb{R}$ itself and the field $\mathbb{Q}$ of rational numbers. Obviously $\mathbb{R}$ is the largest subfield of $\mathbb{R}$; the next lemma shows that $\mathbb{Q}$ is the smallest.

Lemma 16.33. *Every subfield of $\mathbb{R}$ contains $\mathbb{Q}$.*

Proof. Suppose F is a subfield of $\mathbb{R}$. Then by definition, F contains 0 and 1; and if it contains a positive integer n, then it contains $n + 1$ because it is closed under addition. Thus by induction it contains all positive integers. Since it is also closed under subtraction, it also contains all negative integers, and since it is closed under division by nonzero elements, it contains all rational numbers. $\qquad\qquad\square$

Theorem 16.34. *The set K of constructible numbers is a subfield of $\mathbb{R}$.*

Proof. It follows from the definitions that 0 is constructible because $OO = 0$, and 1 is constructible because $OI = 1$. Suppose x and y are constructible numbers. This means that there are constructible points A, B, C, D such that $AB = |x|$ and $CD = |y|$ (Fig. 16.18). If $x = 0$, then it is immediate that $x + y$, $x - y$, and xy are constructible, as is x/y provided $y \neq 0$; and a similar remark holds if $y = 0$. So henceforth we assume that both x and y are nonzero.

To show that $x + y$ is constructible, we need to show how to construct two points whose distance is $|x + y|$. Assume without loss of generality that $|y| \leq |x|$. Draw $\overline{AB}$ and extend it past B to form a longer segment such that the extended part is longer than $\overline{CD}$. Then by drawing a circle with center B and length CD and marking the two points where this circle meets the line $\overleftrightarrow{AB}$, we can construct segments of lengths $|x| + |y|$ and $|x| - |y|$. Since $|x + y|$ is equal to $|x| + |y|$ when x and y have the same sign and to $|x| - |y|$ when they don't, this shows that $x + y$ is constructible. Since $-x$ and $-y$ are also constructible, so are $x - y = x + (-y)$ and $y - x = y + (-x)$.

Next, to show that xy is constructible, we need to show how to construct two points whose distance is $|xy|$. First construct a rectangle whose side lengths are $|x|$ and $|y|$, and then construct a new rectangle with the same area but with one side of length 1 (Construction Problem 16.20). The other side has length $|xy|$ as desired. (See Fig. 16.19(a).)

To complete the proof that K is a field, we need to show that x/y is constructible. Construct a rectangle of side lengths $|x|$ and 1, and then construct a new rectangle with the same area (namely $|x|$) but with one side of length $|y|$. The other side of this rectangle has length $|x/y|$ as desired (Fig. 16.19(b)). $\qquad\qquad\square$

Thanks to Theorem 16.34 and Lemma 16.33, every rational number is constructible. The next theorem shows that K contains more than just the rational numbers.

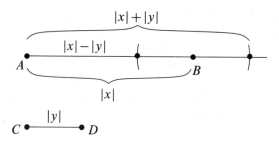

Fig. 16.18. The set of constructible numbers is closed under addition and subtraction.

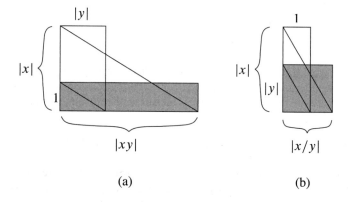

(a)   (b)

Fig. 16.19. The set of constructible numbers is closed under multiplication and division.

Theorem 16.35. *If x is a positive constructible number, then so is $\sqrt{x}$.*

Proof. Suppose x is constructible and positive. Construct a rectangle whose side lengths are x and 1, and then construct a square with the same area (Construction Problem 16.21). The sides of this square have length $\sqrt{x}$.   □

Thus the field K of constructible numbers contains square roots of positive rational numbers, and square roots of square roots, etc. It also contains numbers like $\sqrt{1+\sqrt{2}}$ and much more complicated ones. In order to analyze what this means, we need the following definition. Suppose $F \subseteq \mathbb{R}$ is a field and e is any positive element of F whose square root is not in F. We define $F\left(\sqrt{e}\right)$ to be the following subset of $\mathbb{R}$:

$$F\left(\sqrt{e}\right) = \{a + b\sqrt{e} : a, b \in F\}.$$

Any subset of this form is called a ***quadratic extension of F***. For example, because $\sqrt{2}$ is irrational, the set $\mathbb{Q}\left(\sqrt{2}\right)$ is a quadratic extension of $\mathbb{Q}$.

The next lemma gives a useful property of elements of a quadratic extension.

Lemma 16.36. *Suppose $F \subseteq \mathbb{R}$ is a field and e is a positive element of F whose square root is not in F. Then a number $a + b\sqrt{e} \in F\left(\sqrt{e}\right)$ is zero if and only if $a = b = 0$.*

Proof. Obviously, if $a = b = 0$, then $a + b\sqrt{e} = 0$. Suppose, conversely, that $a + b\sqrt{e} = 0$. If $b \neq 0$, then $\sqrt{e} = -a/b \in F$, contradicting the hypothesis. Thus $b = 0$, which implies also that $a = 0$. $\qquad\square$

Lemma 16.37. *If $F \subseteq \mathbb{R}$ is a field and e is a positive element of F whose square root is not in F, then $F\left(\sqrt{e}\right)$ is an extension field of F containing $\sqrt{e}$.*

Proof. Clearly, $F\left(\sqrt{e}\right)$ contains $0 + 1\sqrt{e} = \sqrt{e}$, and it contains F because each $x \in F$ can be written in the form $x + 0\sqrt{e}$. To see that $F\left(\sqrt{e}\right)$ is a field, we just need to verify that it is closed under the basic algebraic operations. Closure under addition and subtraction is easy to verify, and closure under multiplication follows from the following computation:

$$\left(a + b\sqrt{e}\right)\left(c + d\sqrt{e}\right) = (ac + bde) + (ad + bc)\sqrt{e}.$$

To check closure under division, suppose $a + b\sqrt{e}$ and $c + d\sqrt{e}$ are elements of $F\left(\sqrt{e}\right)$ with $c + d\sqrt{e} \neq 0$. Then $c - d\sqrt{e}$ is also nonzero by Lemma 16.36, so

$$\frac{a + b\sqrt{e}}{c + d\sqrt{e}} = \left(\frac{a + b\sqrt{e}}{c + d\sqrt{e}}\right)\left(\frac{c - d\sqrt{e}}{c - d\sqrt{e}}\right) = \left(\frac{ac - bde}{c^2 - d^2 e}\right) + \left(\frac{bc - ad}{c^2 - d^2 e}\right)\sqrt{e},$$

which is also an element of $F\left(\sqrt{e}\right)$. $\qquad\square$

Theorems 16.34 and 16.35 together show that every quadratic extension of $\mathbb{Q}$ is contained in K. But more is true: let us say that a subfield $F \subseteq \mathbb{R}$ is an ***iterated quadratic extension of*** $\mathbb{Q}$ if there is a finite sequence of subfields $F_0 \subseteq F_1 \subseteq \cdots \subseteq F_n \subseteq \mathbb{R}$ such that $F_0 = \mathbb{Q}$, $F_n = F$, and each F_i is a quadratic extension of F_{i-1}. It follows immediately from Theorems 16.34 and 16.35 that every number in an iterated quadratic extension of $\mathbb{Q}$ is constructible.

An important step on the way to the proofs of impossibility of the ancient construction problems was the realization that the converse is true. The following theorem was essentially proved by René Descartes in 1637 [**Des54**]. Its proof is rated PG.

Theorem 16.38 (Characterization of Constructible Numbers). *A real number is constructible if and only if it is contained in some iterated quadratic extension of $\mathbb{Q}$.*

Proof. As we noted above, Theorems 16.34 and 16.35 imply that every number in an iterated quadratic extension of $\mathbb{Q}$ is constructible. Thus we need only prove the converse. Suppose r is a constructible number. This means that there is a finite sequence of construction steps starting with two points O and I whose distance is 1 and ending with points P and Q whose distance is $|r|$.

The key is to work in the Cartesian model. Using the method described in the proof of Theorem 13.19, we can define an isomorphism between the Euclidean plane and the Cartesian plane that sends O to $(0,0)$ and I to $(1,0)$. Each construction step then yields a certain line, circle, or point in the Cartesian plane. Our plan is to keep careful track of the fields containing the coordinates of all points that we produce along the way.

Suppose F is any subfield of $\mathbb{R}$. Let us say that a point (x, y) in the Cartesian plane is an ***F-point*** if both x and y are elements of F. Then a line is called an ***F-line*** if it passes through two F-points, and a circle is called an ***F-circle*** if its center is an F-point and it passes through at least one F-point.

It will be helpful to see what can be said about the equations of F-lines and F-circles. If ℓ is a nonvertical F-line passing through F-points (x_1, y_1) and (x_2, y_2), then it is described by an equation of the form $y = mx + b$ with m and b given by (2.1) (see p. 34). Because F is closed under the basic arithmetic operations, it follows that both m and b are in F. Similarly, a vertical F-line is given by an equation $x = c$ with $c \in F$. If $\mathcal{C}$ is an F-circle centered at an F-point (x_1, y_1) and passing through another F-point (x_2, y_2), then it is described by an equation of the form

$$(x - x_1)^2 + (y - y_1)^2 = (x_2 - x_1)^2 + (y_2 - y_1)^2,$$

which can be simplified to

$$x^2 + y^2 + Ax + By = C$$

for some numbers A, B, C, all of which are elements of F.

The heart of the proof is the establishment of the following facts:

(a) If ℓ_1 and ℓ_2 are nonparallel F-lines, then they intersect at an F-point.

(b) If $\mathcal{C}$ is an F-circle and ℓ is an F-line that is a secant line for $\mathcal{C}$, then the coordinates of their intersection points lie either in F or in some quadratic extension of F.

(c) If $\mathcal{C}_1$ and $\mathcal{C}_2$ are distinct F-circles that intersect at two points, then the coordinates of their intersection points lie either in F or in some quadratic extension of F.

To prove (a), suppose ℓ_1 and ℓ_2 are nonparallel F-lines. If they are both nonvertical, they are given by formulas $y = m_1 x + b_1$ and $y = m_2 x + b_2$, with $m_1, m_2, b_1, b_2 \in F$. The assumption that they are nonparallel means $m_1 \neq m_2$ by Lemma 2.22. Then a little algebra shows that the coordinates of their intersection point (x, y) are given by

$$x = -\frac{b_2 - b_1}{m_2 - m_1}, \qquad y = \frac{b_1 m_2 - b_2 m_1}{m_2 - m_1},$$

both of which are in F. The case in which one of the lines is vertical is similar but easier.

To prove (b), suppose ℓ is a nonvertical F-line given by $y = mx + b$ and $\mathcal{C}$ is an F-circle given by $x^2 + y^2 + Ax + By = C$. Substituting $mx + b$ for y yields a quadratic equation for x whose coefficients all lie in F. The assumption that ℓ is a secant line means that this equation has solutions, and the quadratic formula shows that they lie either in F (in case the expression under the square root sign in the quadratic formula has a square root in F) or in a quadratic extension of F (if not). Inserting these solutions for x into the equation $y = mx + b$ yields solutions for y in the same subfield. Again, the case in which ℓ is vertical is easy.

Finally, to prove (c), suppose $\mathcal{C}_1$ and $\mathcal{C}_2$ are F-circles described by equations

$$x^2 + y^2 + A_1 x + B_1 y = C_1,$$
$$x^2 + y^2 + A_2 x + B_2 y = C_2,$$

with $A_1, B_1, C_1, A_2, B_2, C_2 \in F$. Subtracting these equations shows that any intersection point (x, y) has to satisfy

$$(A_1 - A_2)x + (B_1 - B_2)y = (C_1 - C_2). \tag{16.1}$$

If we assume $\mathcal{C}_1$ and $\mathcal{C}_2$ have two intersection points, then this linear equation must have a solution, so $A_1 - A_2$ and $B_1 - B_2$ cannot both be zero (unless $C_1 = C_2$ as well, which would contradict our assumption that the circles are distinct). Thus (16.1) is the equation for an F-line (vertical if $B_1 = B_2$, nonvertical otherwise), and we are back in the situation of part (b).

Recall that we are assuming r is a constructible number, which means that $|r|$ is equal to the distance between two constructible points P and Q. Thus there is a finite sequence of construction steps starting with O and I and leading to P and Q. We will show by induction on the number of steps that P and Q have coordinates lying in some iterated quadratic extension of $\mathbb{Q}$. Because O and I are $\mathbb{Q}$-points and at the start of the process there are no lines or circles to intersect, the first construction step can only yield a $\mathbb{Q}$-line or a $\mathbb{Q}$-circle. The algorithm continues to produce $\mathbb{Q}$-lines and $\mathbb{Q}$-circles until the first time we construct an intersection point, which must be either a $\mathbb{Q}$-point or an F_1-point for some quadratic extension $F_1 \supseteq \mathbb{Q}$. Suppose by induction that after the kth step of the construction there is an iterated quadratic extension $F \supseteq \mathbb{Q}$ such that everything we have constructed so far is an F-line, F-circle, or F-point. If the next step is to construct a line or circle using already-identified points, then it will be an F-line or F-circle by definition. If the next step is to identify an intersection point, then it will be either an F-point or an F'-point for some quadratic extension $F' \supseteq F$, either of which is an iterated quadratic extension of $\mathbb{Q}$. This completes the induction.

Let G denote an iterated quadratic extension of $\mathbb{Q}$ that contains the coordinates of all points produced in the construction algorithm. The formula (6.2) for distance in the Cartesian plane shows that $|r|$ is equal to the square root of an element of G, and therefore it lies either in G or in some quadratic extension of G, and thus in some iterated quadratic extension of $\mathbb{Q}$. $\qquad\square$

The Trisection Problem

With these algebraic tools in hand, it is now possible to attack the classical construction problems. We begin with the trisection problem. Let us say that an angle is a ***constructible angle*** if both of the lines containing its rays are constructible lines.

Theorem 16.39. *Let* $\theta \in [0, 180]$ *be arbitrary. Then there is a constructible angle with measure* $\theta°$ *if and only if* $\cos\theta$ *is a constructible number.*

Proof. If θ is equal to 0, 90, or 180, then it is obvious that there is a constructible $\theta°$ angle, and $\cos\theta$ is equal to 1, 0, or -1, each of which is a constructible number; so the theorem is true in those cases. Moreover, if $90 < \theta < 180$, there is a a constructible angle with measure θ if and only if there is one with measure $180° - \theta$ (starting with either angle, just extend one of its rays to construct a linear pair), and $\cos\theta$ is constructible if and only if $\cos(180° - \theta) = -\cos\theta$ is constructible. Thus it suffices to restrict our attention to the case $0 < \theta < 90$, and we assume this is the case henceforth.

Suppose first that $\cos\theta$ is constructible. The assumption that $0 < \theta < 90$ implies $0 < \cos\theta < 1$. Let $\overline{AB}$ be a segment whose length is $\cos\theta$. Construct the line ℓ perpendicular to $\overleftrightarrow{AB}$ through B and the circle $\mathcal{C}$ with center A and radius 1. Because $\cos\theta < 1$, B lies inside $\mathcal{C}$, so ℓ is a secant line for $\mathcal{C}$. Let C be one of the points where ℓ and $\mathcal{C}$ intersect (Fig. 16.20). Then $\triangle ABC$ is a right triangle in which $\cos\angle CAB = AB/AC = \cos\theta$. Since the cosine function is injective, it follows that $m\angle CAB = \theta$.

Conversely, suppose there is a constructible $\theta°$ angle $\angle rs$, and let A denote its vertex. Mark off the point $C \in \operatorname{Int} \overrightarrow{r}$ such that $AC = 1$, and drop a perpendicular from C to $\overleftrightarrow{s}$. Let

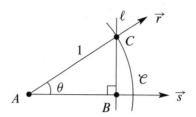

Fig. 16.20. A constructible angle.

B be the foot of the perpendicular (which lies in Int $\vec{s}$ by Theorem 7.2). Then $AB = \cos\theta$, so $\cos\theta$ is a constructible number. □

Let us pause to look back at what we have done so far. Theorem 16.38 shows that a number is constructible if and only if it is contained in some iterated quadratic extension of $\mathbb{Q}$; and Theorem 16.39 shows that an angle is constructible if and only if its cosine is a constructible number. Our next task is to find an angle whose cosine does not lie in any such extension of $\mathbb{Q}$. As we will see below, a 20° angle will serve this purpose.

The preceding theorem shows that there is a constructible 20° angle if and only if $\cos 20°$ is a constructible number. To see if this is the case, let us write $r = \cos 20°$. The triple angle formula for the cosine (Corollary 13.17) reads

$$\cos 3\theta = 4\cos^3\theta - 3\cos\theta.$$

Applying this with $\theta = 20°$ and using the fact that $\cos 60° = \frac{1}{2}$, we obtain

$$\frac{1}{2} = 4r^3 - 3r. \tag{16.2}$$

To make this a little easier to analyze, let us set $x = 2r$ and note that x is constructible if and only if r is. Making this substitution in (16.2) and multiplying through by 2, we get

$$x^3 - 3x - 1 = 0. \tag{16.3}$$

In order to analyze this equation, we will use some standard definitions and facts about polynomials, most of which should already be familiar to you. A **real polynomial in one variable** is a function $f : \mathbb{R} \to \mathbb{R}$ that can be written in the form

$$f(x) = a_n x^n + a_{n-1} x^{n-1} + \cdots + a_1 x + a_0, \tag{16.4}$$

for some nonnegative integer n and some real numbers $a_0, \ldots, a_n$. Suppose f is such a polynomial. We need the following standard definitions:

- The highest power of x that appears in the formula for f (i.e., the number n) is called the **degree of f**.
- The numbers $a_0, \ldots, a_n$ multiplying the powers of x are called the **coefficients of f**.
- Polynomials of degrees 1, 2, and 3 are called **linear**, **quadratic**, and **cubic polynomials**, respectively.
- A number x such that $f(x) = 0$ is called a **root** of the polynomial.
- A polynomial whose highest-degree term has a coefficient of 1 is called **monic**.

Here are the facts we need. The proofs of these can be found in most advanced calculus textbooks.

- Two polynomials are equal (i.e., determine the same function) if and only if all of their corresponding coefficients are equal.
- A polynomial of degree n has at most n real roots.
- Every polynomial of odd degree has at least one real root.
- If f is a monic polynomial of degree n that has n distinct real roots $r_1, \ldots, r_n$, then f can be factored as follows:

$$f(x) = (x - r_1)(x - r_2) \cdots (x - r_n).$$

- If a and b are real numbers such that $f(a) < 0$ and $f(b) > 0$, then f has at least one real root strictly between a and b.

The reason we need these facts is to prove the following lemma.

Lemma 16.40. *Equation* (16.3) *has no constructible solutions.*

Proof. First we will show that (16.3) has no rational solutions. Suppose for the sake of contradiction that $x = p/q$ is such a solution, and assume it has been written in lowest terms, meaning that p and q have no common prime factors. Substituting this into (16.3) and multiplying through by q^2, we get

$$\frac{p^3}{q} = 3pq + q^2.$$

Since the right-hand side is an integer, this shows that p^3 is divisible by q. Thus any prime that divides q must also divide p, which contradicts our assumption that p and q have no common factors unless $q = \pm 1$. It follows that $x = p/q$ is an integer. On the other hand, writing the equation in the form $x(x^2 - 3) = 1$ shows that x itself must be ± 1 (since these are the only integer factors of 1). However, neither $x = 1$ nor $x = -1$ solves the equation, so this is a contradiction.

Let f be the cubic polynomial $f(x) = x^3 - 3x - 1$. A direct computation shows that $f(-2) = -3$, $f(-1) = 1$, $f(1) = -3$, and $f(2) = 1$ (see Fig. 16.21), so in fact f has three distinct roots, lying in the intervals $(-2, -1)$, $(-1, 1)$, and $(1, 2)$. If we denote the three roots by r, s, t, then we can factor f as follows:

$$f(x) = (x - r)(x - s)(x - t) = x^3 - (r + s + t)x^2 + (rs + rt + st)x - rst. \tag{16.5}$$

Because these polynomial functions are equal, all of their corresponding coefficients are equal. The x^2-coefficient of $f(x)$ is zero, which implies that

$$r + s + t = 0. \tag{16.6}$$

Now suppose that one of the roots is constructible, say $x = r$. This means that r lies in some iterated quadratic extension of $\mathbb{Q}$. Let $F_m \supseteq F_{m-1} \supseteq \cdots \supseteq F_1 \supseteq F_0 = \mathbb{Q}$ be a sequence of quadratic extension fields such that $r \in F_m$; we can assume that these fields have been chosen such that m is as small as possible. Because f has no rational roots, we know that $m \geq 1$. We can write r in the form $r = a + b\sqrt{e}$, with a, b, e elements of F_{m-1} such that $\sqrt{e} \notin F_{m-1}$. Substituting this value for r and expanding, we obtain

$$0 = f\left(a + b\sqrt{e}\right) = (a^3 + 3ab^2 e - 3a - 1) + (3a^2 b + b^3 e - 3b)\sqrt{e}.$$

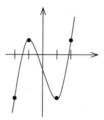

Fig. 16.21. The graph of $f(x) = x^3 - 3x - 1$.

By Lemma 16.36, this implies that $(a^3 + 3ab^2e - 3a - 1) = (3a^2b + b^3e - 3b) = 0$. This in turn implies that $a - b\sqrt{e}$ is also a root, because

$$f\left(a - b\sqrt{e}\right) = (a^3 + 3ab^2e - 3a - 1) - (3a^2b + b^3e - 3b)\sqrt{e}.$$

If we write $r = a + b\sqrt{e}$, $s = a - b\sqrt{e}$, and let t denote the third root of f, then (16.6) implies

$$t = -r - s = -2a.$$

However, $-2a \in F_{m-1}$, so we have shown that f has a root in F_{m-1}. This contradicts our assumption that m was the smallest number of quadratic extensions needed to find a root of f. □

The next theorem was first proved by the French mathematician Pierre Wantzel in 1837.

Theorem 16.41. *There is no constructible 20° angle.*

Proof. Assume for the sake of contradiction that there is a constructible 20° angle. Theorem 16.39 shows that $r = \cos 20°$ is a constructible number, and thus $x = 2r$ is also constructible. But x is a solution to (16.3), so this is a contradiction. □

Corollary 16.42. *There is no algorithm for trisecting an arbitrary angle with compass and straightedge.*

Proof. If there were such an algorithm, we could start with points O and I whose distance is 1, construct an equilateral triangle with $\overline{OI}$ as one of its sides, and then trisect one of its 60° angles, yielding a constructible 20° angle, which is impossible. □

Despite the fact that it has been known for 175 years that trisection with straightedge and compass is impossible, there is always a steady stream of well-meaning amateur mathematicians who claim that they have discovered a trisection algorithm, probably because the problem is so easy to state while the impossibility proof requires some rather sophisticated mathematical background to understand. Most such attempts either produce close approximations but not exact trisections or use the compass and straightedge in some illegitimate way; but you can be sure that all of them are wrong if they claim to have produced an algorithm that trisects an arbitrary angle while following the classical rules for using straightedge and compass. Underwood Dudley has written an amusing book [**Dud94**] exploring some of the many failed attempts and the people who continue to attempt the impossible.

Doubling a Cube and Squaring a Circle

With the same tools, we can address two more of the ancient construction problems.

Theorem 16.43. *There is no algorithm for doubling an arbitrary cube with compass and straightedge.*

Proof. If there were such an algorithm, then we could start with a segment of length 1 and construct a segment of length $\sqrt[3]{2}$, implying that the number $x = \sqrt[3]{2}$ is a constructible number. Let $F_m \supseteq \cdots \supseteq F_0 = \mathbb{Q}$ be a sequence of quadratic extension fields such that $x \in F_m$, again with m chosen as small as possible. As before, we can write $x = a + b\sqrt{e}$ with $a, b, e \in F_{m-1}$ and $\sqrt{e} \notin F_{m-1}$. Then $b \neq 0$, because $b = 0$ would mean $x \in F_{m-1}$, contradicting our assumption that m is the smallest possible such integer. We compute

$$0 = x^3 - 2 = \left(a + b\sqrt{e}\right)^3 - 2 = (a^3 + 3ab^2e - 2) + (3a^2b + b^3e)\sqrt{e}.$$

As before, Lemma 16.33 implies that $(a^3 + 3ab^2e - 2) = (3a^2b + b^3e) = 0$, from which it follows that $a - b\sqrt{e}$ is another solution to the same equation. Since real numbers have unique cube roots, this is a contradiction. $\qquad\square$

To attack the problem of squaring the circle, we need to introduce one more bit of algebraic terminology. A real number x is said to be ***algebraic*** if there is some polynomial f with rational coefficients such that $f(x) = 0$; it is said to be ***transcendental*** otherwise. For example, if x is any number in a quadratic extension field $\mathbb{Q}\left(\sqrt{e}\right)$, then $x = a + b\sqrt{e}$ with a, b, and e rational, and a simple computation shows that $x^2 - 2ax + a^2 - b^2 = 0$, so any such number is algebraic. Somewhat more generally, we have the following lemma. It is not terribly hard to prove, but the R-rated proof would require a detour into algebra that would take us a little too far afield, so we will just state it. You can find a proof in [**Her99**, Chapter 5].

Lemma 16.44. *Every constructible number is algebraic.*

In 1882, it was proved by the German mathematician Ferdinand von Lindemann that π is transcendental. The X-rated proof, which is quite hard, can be found in [**HW08**]. As a consequence, we have the following theorem.

Theorem 16.45. *There is no algorithm for squaring an arbitrary circle with compass and straightedge.*

Proof. If there were such an algorithm, we could begin by constructing a circle of radius 1, which has area π, and then construct a square with the same area. The side of this square would have length $\sqrt{\pi}$, and since the set of constructible numbers is closed under multiplication, we could then construct a segment of length π. This would imply that π is constructible and hence algebraic, a contradiction. $\qquad\square$

Constructible Polygons

Finally, we address the problem of constructing regular polygons. The next lemma shows that it is intimately connected with the problem of constructing specific angles.

Lemma 16.46. *Let n be an integer greater than or equal to 3. Then there is a constructible regular n-gon if and only if there is a constructible angle whose measure is $360°/n$.*

Proof. First suppose there is a constructible regular n-gon $\mathcal{P}$. By Theorem 14.39 it has a circumcircle $\mathcal{C}$. This circle is the same as the circumcircle of any three consecutive vertices, so we can draw the circle, locate its center O, and draw the segments connecting O to two consecutive vertices of $\mathcal{P}$. The central angle formed by these two segments has measure $360°/n$ by Theorem 14.40.

Conversely, suppose there is a constructible angle with measure $360°/n$. Let $\angle POQ$ be such an angle, and draw any circle $\mathcal{C}$ with center O. Then by repeatedly copying this angle on both sides of $\overleftrightarrow{OP}$ and marking the points where the various rays meet $\mathcal{C}$, we can recreate the configuration shown in Fig. 14.34 (see p. 273). The proof of Theorem 14.42 shows that the resulting polygon is a regular n-gon. $\square$

Using this lemma, we can immediately identify at least one nonconstructible polygon.

Theorem 16.47. *No regular nine-sided polygon can be constructed using straightedge and compass.*

Proof. By Lemma 16.46, if it were possible to construct a regular nine-sided polygon, then it would be possible to construct an angle whose measure is $360°/9 = 40°$. But then we could bisect this angle and construct a $20°$ angle, which we know to be impossible. $\square$

Euclid showed how to construct regular polygons with 3, 4, 5, 6, and 15 sides. Moreover, the technique we used for constructing a regular octagon (bisecting each central angle) can be applied to any regular polygon to construct one with twice the number of sides. Thus Euclid's methods can also be used to construct regular polygons with 8, 10, 12, 16, 20, 24, or 30 sides or in fact any regular polygon with 2^k times as many sides as an already-constructed one. All of this was known in ancient times.

Long before Wantzel's solution to the trisection problem, Carl Friedrich Gauss made a surprising discovery. In 1796, when he was nineteen years old, Gauss recorded in his diary the first major contribution to the theory of constructible polygons in more than 2000 years: he showed that a regular 17-gon (a ***heptadecagon***) can be constructed with straightedge and compass. What he actually proved was that the cosine of $360°/17$ can be written in the following form:

$$\frac{1}{16}\left(-1+\sqrt{17}+\sqrt{34-2\sqrt{17}}+2\sqrt{17+3\sqrt{17}-\sqrt{34-2\sqrt{17}}-2\sqrt{34+2\sqrt{17}}}\right).$$

This is a horrible-looking expression, but the only thing you need to notice about it is that it is built up from integers using only addition, subtraction, multiplication, division, and square roots; thus it is a constructible number. (Gauss did not actually write down an algorithm for constructing a regular heptadecagon; this was first done by Johannes Erchinger a few years later. In principle, you could construct such an algorithm by using this formula and following the algorithms used to solve the construction problems in this chapter, although the resulting algorithm might not be very efficient.)

In addition, Gauss noticed that the techniques that led to this formula for $\cos(360°/17)$ also led to analogous formulas for $\cos(360°/p)$ whenever p is a prime number of the form $p = 2^{(2^n)} + 1$ for some nonnegative integer n. Such prime numbers are called ***Fermat primes*** after the French mathematician Pierre de Fermat (1601–1665), who first studied them. The only known Fermat primes are 3, 5, 17, 257, and 65537, which are of the form $2^{(2^n)} + 1$ for $n = 0, 1, 2, 3, 4$. It is known (as of 2012) that $2^{(2^n)} + 1$ is not prime for

$5 \le n \le 32$, but it is not known if there are any other Fermat primes or even whether there are finitely many or infinitely many of them.

The same technique Euclid used to construct a regular 15-gon can easily be adapted to construct a regular polygon with pq sides given a regular p-gon and a regular q-gon, provided that p and q have no common prime factors. Using this, Gauss showed that a regular n-gon can be constructed whenever n is either of the form 2^k for some integer $k \ge 2$ or of the form $2^k p_1 \cdots p_m$ for $k \ge 0$ and some distinct Fermat primes $p_1, \ldots, p_m$.

In the same 1837 paper in which he solved the trisection problem, Pierre Wantzel proved that the converse is also true, leading to the following theorem, whose proof is decidedly X-rated. (See [**Har00**, Theorem 29.4] for a proof.)

Theorem 16.48 (Characterization of Constructible Regular Polygons). *A regular n-gon can be constructed with straightedge and compass if and only if n is either a power of 2 or a power of 2 times a product of distinct Fermat primes.*

Corollary 16.49. *There is no constructible regular heptagon.* □

Exercises

16A. Solve Construction Problem 16.5 (constructing a perpendicular bisector).

16B. Solve Construction Problem 16.6 (constructing a perpendicular through a point on a line).

16C. Solve Construction Problem 16.7 (constructing a perpendicular through a point not on a line).

16D. Solve Construction Problem 16.8 (constructing a triangle with given side lengths).

16E. Solve Construction Problem 16.9 (copying a triangle to a given segment).

16F. Solve Construction Problem 16.10 (copying an angle to a given ray). [Hint: Use Construction Problem 16.9.]

16G. Solve Construction Problem 16.11 (copying a convex quadrilateral to a given segment).

16H. Solve Construction Problem 16.12 (constructing a rectangle with given side lengths).

16I. Solve Construction Problem 16.13 (constructing a square).

16J. Solve Construction Problem 16.14 (constructing a parallel).

16K. Solve Construction Problem 16.15 (cutting a segment into n equal parts).

16L. Solve Construction Problem 16.17 (constructing the geometric mean of two segments).

16M. Solve Construction Problem 16.21 (squaring a rectangle). [Hint: Each side of the square should be the geometric mean of the sides of the rectangle.]

16N. Solve Construction Problem 16.23 (doubling a square).

16O. Solve Construction Problem 16.24 (locating the center of a circle).

16P. Solve Construction Problem 16.25 (inscribing a circle in a triangle).

16Q. Solve Construction Problem 16.26 (circumscribing a circle around a triangle).

16R. Solve Construction Problem 16.27 (inscribing a square in a circle).

16S. Using an actual compass and straightedge, draw a large circle, choose a point on it, and carry out the construction of a regular pentagon inscribed in the circle described in the solution to Construction Problem 16.28.

16T. Solve Construction Problem 16.29 (inscribing a regular hexagon in a circle).

16U. Solve Construction Problem 16.30 (inscribing an equilateral triangle in a circle).

16V. Solve Construction Problem 16.32 (inscribing a regular 15-gon in a circle).

16W. Sometimes beginning students attempt to use the following construction to trisect an arbitrary angle. Let $\angle rs$ be a proper angle, and let O be its vertex. Choose a point $A \in \operatorname{Int} \vec{r}$ arbitrarily, and use a compass to mark $B \in \operatorname{Int} \vec{s}$ such that $OA = OB$. Locate points C and D in $\operatorname{Int} \overline{AB}$ such that $A * C * D * B$ and $AC = CD = DB$ (Construction Problem 16.15), and draw $\overline{OC}$ and $\overline{OD}$ (Fig. 16.22). Prove that no matter what angle $\angle rs$ we started with, it is never the case that $\angle AOC$, $\angle COD$, and $\angle DOB$ are all congruent. [Hint: Use Theorem 12.9.]

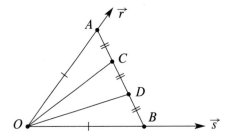

Fig. 16.22. How not to trisect an angle.

Chapter 17

The Parallel Postulate Revisited

As we have noted before, a major preoccupation of geometers from the time of Euclid until the nineteenth century was trying to show that Euclid's fifth postulate could be proved from his other postulates. Many "proofs" were offered, only to be found later to be flawed. We now know that no such proof exists, because the Euclidean parallel postulate is independent of the other postulates of Euclidean geometry, as the models of Chapter 6 demonstrate.

Nonetheless, in the process of searching for a proof, mathematicians were able to show that a great number of theorems can be proved without the fifth postulate. Even more importantly, they discovered a number of alternative postulates that could serve just as well as the fifth postulate. In this chapter, we will explore some of those other postulates.

To that end, let us turn our clock back to the moment before we introduced the Euclidean parallel postulate: just before the beginning of Chapter 10. All of the results of the present chapter are results in neutral geometry. Throughout the chapter, we assume the postulates of neutral geometry (Postulates 1–9), but no parallel postulate, Euclidean or otherwise. Thus we have at our disposal only the things that we proved in Chapters 3 through 9, when we were not relying on the parallel postulate. Some notable results that we cannot use are the converse to the alternate interior angles theorem and its corollaries, the angle-sum theorems, theorems about similarity, and the Pythagorean theorem.

Postulates Equivalent to the Euclidean Parallel Postulate

The postulate we chose to use in our axiomatic treatment of Euclidean geometry is Postulate 10E, which we are calling the Euclidean parallel postulate. In the remainder of this chapter, we will refer to this postulate as the EPP. Our main task in this chapter is to explore some of the postulates that have been discovered to be equivalent to the EPP, meaning that if one of them is added to the postulates of neutral geometry as an additional postulate, then the others follow as theorems. In this chapter, we will describe a number of such postulates, which we call *Euclidean postulates*.

We begin with the oldest one of all.

Euclidean Postulate 1 (Euclid's Fifth Postulate). *If ℓ and ℓ' are two lines cut by a transversal t in such a way that the measures of two consecutive interior angles add up to less than 180°, then ℓ and ℓ' intersect on the same side of t as those two angles.*

Our first theorem shows that the EPP and Euclid's fifth postulate are equivalent.

Theorem 17.1. *In the context of neutral geometry, Euclid's fifth postulate is equivalent to the Euclidean parallel postulate.*

Proof. We have already shown in Chapter 10 that the nine postulates of neutral geometry plus the EPP are sufficient to prove Euclid's fifth postulate. (Although we cannot simply quote the theorems of Chapters 10–15 the way we could in Euclidean geometry, we can refer to the proofs of those theorems to show that certain statements follow if the Euclidean parallel postulate is assumed.) Thus we need only prove the converse. Assume therefore that Euclid's fifth postulate is true, along with all the postulates of neutral geometry. We need to show that the EPP holds, or, more precisely, that given any line ℓ and any point A not on ℓ, there is one and only one line m that contains A and is parallel to ℓ.

Let ℓ be a line and let A be a point not on ℓ. Theorem 7.25 (which is valid in neutral geometry) shows that there is at least one line m through A and parallel to ℓ, so we have to show that there cannot be more than one such line. Theorem 7.25 also shows that we can choose our parallel line m so that ℓ and m admit a common perpendicular that contains A. Let t be such a common perpendicular, and let F be the point where t meets ℓ (the foot of the perpendicular from A to ℓ).

Suppose m' is another line through A and not equal to m. We will show that m' must intersect ℓ. If $m' = t$, then m' certainly intersects ℓ, so we can assume $m' \neq t$. Because m is the unique perpendicular to t at A, m' is not perpendicular to t. The two rays in m' starting at A make angles with $\overrightarrow{AF}$ that form a linear pair, and neither is a right angle, so one of them must be acute. Choose a point S in the interior of the ray that makes an acute angle with $\overrightarrow{AF}$, and choose $G \in \ell$ on the same side of t as S (Fig. 17.1).

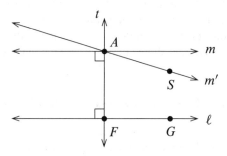

Fig. 17.1. Proof that Euclid's fifth postulate follows from the EPP.

Now, $\angle SAF$ and $\angle AFG$ form a pair of consecutive interior angles for the transversal $\overleftrightarrow{AF}$ to ℓ and m'. Because $\angle AFG$ is a right angle and $\angle SAF$ is acute, the sum of their angle measures is less than 180°. Therefore, Euclid's fifth postulate implies that the lines m' and ℓ meet, so m' is not parallel to ℓ.   □

As explained in Chapter 1, a common mistake in early attempts to prove that Euclid's fifth postulate follows from the other four was assuming that parallel lines are equidistant. The next theorem shows that this assumption is actually equivalent to the EPP.

Euclidean Postulate 2 (The Equidistance Postulate). *If two lines are parallel, then each one is equidistant from the other.*

Theorem 17.2. *In the context of neutral geometry, the equidistance postulate is equivalent to the Euclidean parallel postulate.*

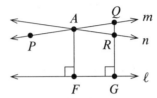

Fig. 17.2. Proof that the equidistance postulate implies the EPP.

Proof. Theorem 10.8 showed that the equidistance postulate follows from the EPP. To prove the converse, assume that the equidistance postulate holds. We will prove that the EPP holds.

Suppose ℓ is a line and A is a point not on ℓ, and assume for the sake of contradiction that m and n are two distinct lines through A and parallel to ℓ (see Fig. 17.2). Let F be the foot of the perpendicular from A to ℓ. If P and Q are any points on m such that $P * A * Q$, then P and Q are on opposite sides of n (because m and n are distinct), so one of them, say Q, must also be on the opposite side of n from F. Let G be the foot of the perpendicular from Q to ℓ. Because $\ell \parallel n$, both G and F are on the same side of n, so G and Q are on opposite sides of n. That means $\overline{QG}$ must intersect n at a point R. Because $m \parallel \ell$, the equidistance postulate implies $QG = AF$, and because $n \parallel \ell$, the equidistance postulate implies $RG = AF$. Thus by transitivity, $QG = RG$. This means that Q and R are both points in the interior of $\overrightarrow{GQ}$ at the same distance from G, so the unique point theorem (Corollary 3.36) implies that $Q = R$. But since m and n are distinct lines, they intersect only at A, so this is a contradiction. $\square$

Here are four other postulates that can readily be shown to be equivalent to the EPP. We leave the proofs as exercises.

Euclidean Postulate 3 (Playfair's Postulate). *Two straight lines cannot be drawn through the same point, parallel to the same straight line, without coinciding with one another.*

Euclidean Postulate 4 (The Alternate Interior Angles Postulate). *If two parallel lines are cut by a transversal, then both pairs of alternate interior angles are congruent.*

Euclidean Postulate 5 (Proclus's Postulate). *If ℓ and ℓ' are parallel lines and t is a line that is distinct from ℓ but intersects ℓ, then t also intersects ℓ'.*

Euclidean Postulate 6 (The Transitivity Postulate). *If ℓ, m, and n are distinct lines such that $\ell \parallel m$ and $m \parallel n$, then $\ell \parallel n$.*

Theorem 17.3. *In the context of neutral geometry, each of the four preceding postulates is equivalent to the Euclidean parallel postulate.*

Proof. See Exercises 17A–17D.   □

Next we present a postulate on the existence of similar triangles that was proposed by the English mathematician John Wallis in 1693. As we will see in the next chapter, similarity is a characteristically Euclidean phenomenon.

Euclidean Postulate 7 (Wallis's Postulate). *Given any triangle $\triangle ABC$ and any positive real number r, there exists a triangle $\triangle DEF$ similar to $\triangle ABC$ with scale factor $r = DE/AB = DF/AC = EF/BC$.*

The next proof is rated PG.

Theorem 17.4. *Wallis's postulate is equivalent to the Euclidean parallel postulate.*

Proof. The similar triangle construction theorem (Theorem 12.4) shows that the EPP implies Wallis's postulate: given $\triangle ABC$ and r, let $\overline{DE}$ be any segment whose length is r times the length of $\overline{AB}$, and then Theorem 12.4 guarantees the existence of a point F such that $\triangle DEF \sim \triangle ABC$ with scale factor r.

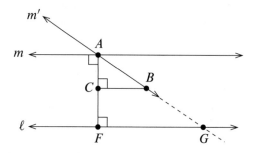

Fig. 17.3. Proof that Wallis's postulate implies the EPP.

Conversely, assume that Wallis's postulate holds. We will show directly that the EPP holds. To that end, suppose ℓ is a line and A is a point not on ℓ. Let F be the foot of the perpendicular from A to ℓ. We know there is one line m through A and parallel to ℓ, namely the line perpendicular to $\overleftrightarrow{AF}$ at A. Suppose m' is a different line through A (Fig. 17.3); we need to show that m' is not parallel to ℓ.

If $m' = \overleftrightarrow{AF}$, then it intersects ℓ at F. Otherwise, choose a point B on m' that lies on the same side of m as F, and let C be the foot of the perpendicular from B to $\overleftrightarrow{AF}$. Because $\overleftrightarrow{BC}$ and m have $\overleftrightarrow{AF}$ as a common perpendicular, they are parallel, which guarantees that C is on the same side of m as B and therefore also as F. If we let $r = AF/AC$, then Wallis's postulate guarantees that there is a triangle $\triangle XYZ$ similar to $\triangle ACB$ with scale factor r. By Theorem 5.10, there is a point G on the same side of $\overleftrightarrow{AF}$ as B such that $\triangle AFG \cong \triangle XYZ$, and therefore $\triangle AFG \sim \triangle ACB$. In particular, this means that $\angle AFG$ is a right angle, so by uniqueness of perpendiculars, $\overleftrightarrow{FG} = \ell$. Because $\angle FAG \cong \angle CAB$, the unique ray theorem implies that $\overrightarrow{AG} = \overrightarrow{AB}$. Thus we have shown that m' and ℓ both contain G, so they are not parallel.   □

Angle Sums and Defects

One of the most important consequences of the Euclidean parallel postulate is the angle-sum theorem for triangles—as we have seen in the last several chapters, it leads to a host of consequences in Euclidean geometry and seems to be a quintessentially Euclidean theorem. As the next theorem will show, it turns out that it is true *only* in Euclidean geometry, because it is equivalent to the EPP.

Euclidean Postulate 8 (The Angle-Sum Postulate). *The angle sum of every triangle is equal to* 180°.

In a moment, we will prove that the angle-sum postulate is equivalent to the EPP. Before we do so, we need the following lemma. The proof of this lemma and that of the theorem that follows it are both rated R.

Lemma 17.5 (Small Angle Lemma). *In neutral geometry, suppose* $\overrightarrow{FG}$ *is a ray and A is a point not on* $\overleftrightarrow{FG}$. *Given any positive number* ε, *there exists a point X in the interior of* $\overrightarrow{FG}$ *such that* $m\angle AXF < \varepsilon$.

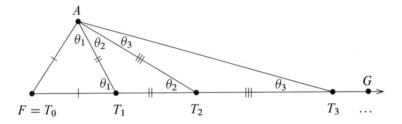

Fig. 17.4. Proof of the small angle lemma.

Proof. Construct a sequence of points $T_1, T_2, T_3, \ldots$ on $\overrightarrow{FG}$ as follows. First let T_1 be the point on $\overrightarrow{FG}$ such that $\overline{FT_1} \cong \overline{AF}$. Then let T_2 be the point such that $F * T_1 * T_2$ and $\overline{T_1 T_2} \cong \overline{AT_1}$. In general, having defined $T_1, \ldots, T_n$, let T_{n+1} be the point such that $T_{n-1} * T_n * T_{n+1}$ and $\overline{AT_n} \cong \overline{T_n T_{n+1}}$. For notational convenience, let $T_0 = F$.

For each $n \geq 1$, we have $\overline{AT_{n-1}} \cong \overline{T_{n-1} T_n}$ and thus $\angle T_{n-1} A T_n \cong \angle T_{n-1} T_n A$ by the isosceles triangle theorem; let θ_n denote the common measure of these two angles. We will show by induction that the following equation holds for each n:

$$m\angle FAT_n = \theta_1 + \cdots + \theta_n. \tag{17.1}$$

For the base case $n = 1$, this is true by definition of θ_1. Assuming that (17.1) holds for some $n \geq 1$, it follows from betweenness vs. betweenness, the betweenness theorem for rays, and the induction hypothesis that

$$m\angle FAT_{n+1} = m\angle FAT_n + m\angle T_n A T_{n+1} = (\theta_1 + \cdots + \theta_n) + \theta_{n+1}.$$

This proves that (17.1) holds for each positive integer n.

For each n, the points F, A, and T_n are noncollinear, so

$$\theta_1 + \cdots + \theta_n = m\angle FAT_n < 180°.$$

If we choose n large enough that $n\varepsilon > 180$, then at least one of the n angle measures $\theta_1, \ldots, \theta_n$ must be strictly less than ε (because otherwise their sum would be at least $n\varepsilon$, which is not less than 180°). Thus we can choose i such that $\theta_i < \varepsilon$ and set $X = T_i$, so that $m\angle AXF = \theta_i < \varepsilon$ as desired.   □

Theorem 17.6. *The angle-sum postulate is equivalent to the Euclidean parallel postulate.*

Proof. Theorem 10.11 showed that the angle-sum postulate follows from the EPP. To prove that the angle-sum postulate implies the EPP, we will actually prove the contrapositive: assuming the EPP is false, we will prove that there is a triangle whose angle sum is not equal to 180°.

The assumption that the EPP is false means that there exist at least one line ℓ and one point A through which there are two or more lines parallel to ℓ. Let F be the foot of the perpendicular from A to ℓ. At least one of the lines through A and parallel to ℓ, say m, is not perpendicular to $\overleftrightarrow{AF}$. Arguing just as in the proof of Theorem 17.1, we see that one of the rays in m must make an acute angle with $\overleftrightarrow{AF}$. Choose a point S in the interior of that ray and a point $G \in \ell$ on the same side of $\overleftrightarrow{AF}$ as S (Fig. 17.5). Define real numbers α and β by

$$\alpha = m\angle FAS, \qquad \beta = 90° - \alpha.$$

Then our assumptions guarantee that both α and β are strictly between 0° and 90°, and they satisfy $\alpha + \beta = 90°$. We will show that there must exist a right triangle whose two acute angles have measures smaller than α and β, respectively, which means that it violates the angle-sum postulate.

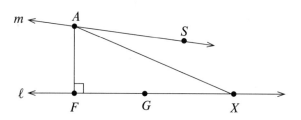

Fig. 17.5. Constructing a triangle with angle sum less than 180°.

By the small angle lemma, there is a point $X \in \text{Int } \overrightarrow{FG}$ such that $m\angle AXF < \beta$. Since $X \in \ell$, which is parallel to $\overleftrightarrow{AS}$, it follows that X is on the same side of $\overleftrightarrow{AS}$ as F; and by the Y-lemma, X is on the same side of $\overleftrightarrow{AF}$ as G and thus also as S. Thus by the interior lemma, $\overrightarrow{AS} * \overrightarrow{AX} * \overrightarrow{AF}$, from which it follows that $m\angle XAF < m\angle SAF = \alpha$. It follows that $\triangle AFX$ has angle sum

$$\sigma(\triangle AFX) = m\angle XAF + m\angle AXF + m\angle AFX < \alpha + \beta + 90° = 180°.$$

Thus the angle-sum postulate is false.   □

Notice that the triangle we constructed in the previous proof has angle sum strictly *less* than 180°. Our next theorem will show that this is the only way the angle-sum postulate can fail: it is impossible in neutral geometry for angle sums of triangles to be *greater* than 180°. This theorem was first stated and proved by Giovanni Saccheri in his 1733 book *Euclid Freed of Every Flaw* [**Sac86**], in which he attempted to prove Euclid's fifth postulate by contradiction. We will give a proof that was published by Adrien-Marie Legendre in 1794. It is rated R.

Theorem 17.7 (Saccheri–Legendre). *In neutral geometry, the angle sum of every triangle is less than or equal to* 180°.

Proof. Let $\triangle ABC$ be an arbitrary triangle, and label its angle measures as follows:

$$\alpha = m\angle A, \qquad \beta = m\angle B, \qquad \gamma = m\angle C.$$

Assume for the sake of contradiction that $\alpha + \beta + \gamma > 180°$.

Let P be a point such that $B * C * P$, and let $\overrightarrow{CQ}$ be the ray on the same side of $\overleftrightarrow{BC}$ as A such that $m\angle QCP = \beta$. Let D be the point on $\overrightarrow{CQ}$ such that $CD = BA$, and let θ denote the measure of $\angle ACD$ (Fig. 17.6).

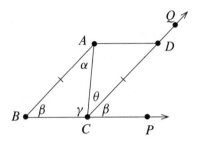

Fig. 17.6. Constructing a quadrilateral from an arbitrary triangle.

Our choices of D and P guarantee that $\angle B$ and $\angle DCP$ are corresponding angles for the transversal $\overleftrightarrow{BC}$ to $\overrightarrow{BA}$ and $\overrightarrow{CD}$, so $\overrightarrow{BA} \parallel \overrightarrow{CD}$ by the corresponding angles theorem. This implies that $ABCD$ is a trapezoid and thus a convex quadrilateral, and so $\overrightarrow{CB} * \overrightarrow{CA} * \overrightarrow{CD}$. On the other hand, $\angle ACP$ is an exterior angle for $\triangle ABC$, so $m\angle ACP > \beta$ and therefore $\overrightarrow{CA} * \overrightarrow{CD} * \overrightarrow{CP}$ by the ordering lemma for rays. Thus $\angle BCA$, $\angle ACD$, and $\angle DCP$ satisfy the hypotheses of the linear triple theorem, which implies that

$$\gamma + \theta + \beta = 180°. \tag{17.2}$$

Subtracting this from our assumed inequality $\alpha + \beta + \gamma > 180°$ yields $\alpha > \theta$, and then the hinge theorem implies that $BC > AD$. The idea of the proof is that if we make a large number of copies of $ABCD$, this small inequality will add up to a large inequality that violates the general triangle inequality.

Label the side lengths of $ABCD$ as

$$a = BC, \qquad c = AB = CD, \qquad e = AD,$$

so that $a - e > 0$. Let n be a positive integer large enough that $n(a - e) > 2c$.

Now let $\overline{A_0 B_0}$ be any segment of length c, and choose points A_1 and B_1 on one side of $\overrightarrow{A_0 B_0}$ so that $A_0 B_0 B_1 A_1 \cong ABCD$. (This is possible by Theorem 9.14.) On the opposite side of $\overleftrightarrow{A_1 B_1}$ from A_0 and B_0, choose points A_2 and B_2 such that $A_1 B_1 B_2 A_2 \cong ABCD$. Continue constructing n trapezoids in this way (Fig. 17.7).

For each $i = 1, \ldots, n-1$, the two angles at B_i have angle measures $\gamma + \theta$ and β, so they are supplementary by (17.2). Therefore, by the converse to the linear pair theorem, B_{i-1}, B_i, and B_{i+1} are collinear. Since this is true for each i, all of the points $B_0, \ldots, B_n$ are collinear. (We are not claiming that the A_i's are collinear, though.) It follows by

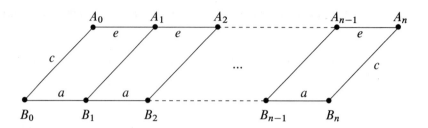

Fig. 17.7. Proof of the Saccheri–Legendre theorem.

induction from the betweenness theorem for points that $B_0 B_n = B_0 B_1 + \cdots + B_{n-1} B_n = na$, and then the general triangle inequality gives

$$na = B_0 B_n \le B_0 A_0 + A_0 A_1 + A_1 A_2 + \cdots + A_{n-1} A_n + A_n B_n = c + ne + c.$$

Rearranging this inequality, we conclude ultimately that $n(a-e) \le 2c$, which contradicts our choice of n such that $n(a-e) > 2c$. This contradiction shows that our assumption of an angle sum greater than $180°$ is impossible. □

In order to delve deeper into the relationships between angle sums and parallel postulates, let us study this phenomenon more closely. The Saccheri–Legendre theorem shows that in neutral geometry, the angle sum of a triangle is never more than $180°$, but it might be less. We will define a quantity called the *defect* of a triangle, which measures how much less it is for any given triangle. Specifically, if $\triangle ABC$ is a triangle, we define its **defect**, denoted by $\delta(\triangle ABC)$, to be the difference

$$\delta(\triangle ABC) = 180° - \sigma(\triangle ABC) = 180° - (m\angle A + m\angle B + m\angle C).$$

The Saccheri–Legendre theorem shows that the defect of every triangle is nonnegative; it is equal to zero if and only if the triangle satisfies the angle-sum formula of Euclidean geometry. The angle-sum postulate can be rephrased as the assertion that every triangle has zero defect.

More generally, we can define an analogous quantity for convex polygons with any number of sides. If $\mathcal{P}$ is a convex n-gon, we define its **defect**, denoted by $\delta(\mathcal{P})$, to be the difference

$$\delta(\mathcal{P}) = (n-2) \times 180° - \sigma(\mathcal{P}).$$

In particular, for a convex quadrilateral $DEFG$, this means

$$\delta(DEFG) = 360° - \sigma(DEFG).$$

A convex polygon has zero defect if and only if it satisfies the angle-sum formula of Euclidean geometry.

The next theorem expresses the most important property of defects. We will use it in the next section to prove a surprising equivalence with the EPP. Its proof is rated R.

Theorem 17.8 (Defect Addition). *In neutral geometry, suppose $\mathcal{P}$ is a convex polygon and $\overline{BC}$ is a chord of $\mathcal{P}$. Let $\mathcal{P}_1$ and $\mathcal{P}_2$ be the two convex polygons formed by $\overline{BC}$ as in the polygon splitting theorem (Theorem 8.9). Then $\delta(\mathcal{P}) = \delta(\mathcal{P}_1) + \delta(\mathcal{P}_2)$.*

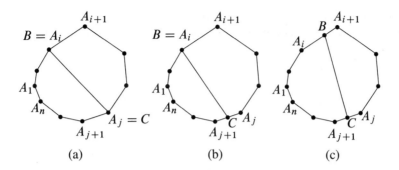

Fig. 17.8. The three cases in the proof of the defect addition theorem.

Proof. Depending on the positions of B and C, there are three slightly different arguments, corresponding to the three configurations of Fig. 17.8.

First consider the case in which both B and C are vertices (Fig. 17.8(a)). As in the proof of Theorem 8.9, we can label the vertices as $A_1, \ldots, A_n$ such that $B = A_i$ and $C = A_j$ and write

$$\mathcal{P}_1 = A_1 \ldots A_{i-1} B C A_{j+1} \ldots A_n \qquad \text{and} \qquad \mathcal{P}_2 = B A_{i+1} \ldots A_{j-1} C.$$

The assumption that $\mathcal{P}$ is convex means that $\overrightarrow{BC}$ is in the interior of $\angle A_{i-1} B A_{i+1}$ and $\overrightarrow{CB}$ is in the interior of $\angle A_{j-1} C A_{j+1}$. Therefore the betweenness theorem for rays implies that

$$m\angle A_{i-1} B C + m\angle C B A_{i+1} = m\angle A_{i-1} B A_{i+1}, \tag{17.3}$$

$$m\angle A_{j-1} C B + m\angle B C A_{j+1} = m\angle A_{j-1} C A_{j+1}. \tag{17.4}$$

The formula for angle sums gives

$$\sigma(\mathcal{P}_1) = m\angle A_1 + \cdots + m\angle A_{i-1} + m\angle A_{i-1} B C + m\angle B C A_{j+1} + \cdots + m\angle A_n \tag{17.5}$$

and

$$\sigma(\mathcal{P}_2) = m\angle C B A_{i+1} + m\angle A_{i+1} + \cdots + m\angle A_{j-1} + m\angle A_{j-1} C B. \tag{17.6}$$

Adding these two equations together and using (17.3) and (17.4), we get

$$\sigma(\mathcal{P}_1) + \sigma(\mathcal{P}_2) = m\angle A_1 + \cdots + m\angle A_n = \sigma(\mathcal{P}). \tag{17.7}$$

Now, the vertices of $\mathcal{P}_1$ are $A_1, \ldots, A_i$ (i vertices) and $A_j, \ldots, A_n$ ($n - j + 1$ vertices), for a total of $n + i - j + 1$. Thus its defect is

$$\delta(\mathcal{P}_1) = \big((n + i - j + 1) - 2\big) \times 180° - \sigma(\mathcal{P}_1).$$

Similarly, $\mathcal{P}_2$ has vertices $A_i, \ldots, A_j$ ($j - i + 1$ vertices), so its defect is

$$\delta(\mathcal{P}_2) = \big((j - i + 1) - 2\big) \times 180° - \sigma(\mathcal{P}_1).$$

Adding these last two equations together and using (17.7), we get

$$\delta(\mathcal{P}_1) + \delta(\mathcal{P}_2) = (n - 2) \times 180° - \sigma(\mathcal{P}) = \delta(\mathcal{P}).$$

Next consider the case in which one of the points (say B) is a vertex but C is not (Fig. 17.8(b)). As in the proof of Theorem 8.9, choose vertex labels so that $B = A_i$ and C is in the interior of $\overline{A_j A_{j+1}}$, so

$$\mathcal{P}_1 = A_1 \ldots A_{i-1} B C A_{j+1} \ldots A_n \quad \text{and} \quad \mathcal{P}_2 = B A_{i+1} \ldots A_{j-1} A_j C.$$

In this case, three things have to change in the preceding argument: first, in place of (17.4), the linear pair theorem gives

$$m\angle A_j C B + m\angle B C A_{j+1} = 180°.$$

Second, in $\mathcal{P}_2$, the single vertex $A_j = C$ is now replaced by two vertices A_j and C, so the term $m A_{j-1} C B$ in (17.6) has to be replaced by two terms $m\angle A_j + m\angle A_j C B$ on the right-hand side of (17.5). Finally, the number of vertices of $\mathcal{P}_2$ is now $j - i + 2$. When we insert these changes into the calculations above, we obtain two additional 180° terms that cancel each other out, and the result remains true as before.

Finally, when neither point is a vertex (Fig. 17.8(c)), we can label the vertices so that $B \in \text{Int } \overline{A_i A_{i+1}}$ and $C \in \text{Int } \overline{A_j A_{j+1}}$ and write

$$\mathcal{P}_1 = A_1 \ldots A_i B C A_{j+1} \ldots A_n \quad \text{and} \quad \mathcal{P}_2 = B A_{i+1} \ldots A_j C.$$

In addition to the changes in the preceding paragraph, (17.3) now has to be replaced by

$$m\angle A_i B C + m\angle C B A_{i+1} = 180°,$$

the term $m\angle A_{i-1} B C$ has to be replaced by $m\angle A_i + m\angle A_i B C$ in (17.6), and the number of vertices of $\mathcal{P}_1$ becomes $n + i - j + 2$. In this case there are two more 180° terms that also cancel, yielding the desired result.   □

Corollary 17.9 (Saccheri–Legendre Theorem for Convex Polygons). *In neutral geometry, if $\mathcal{P}$ is a convex n-gon, then the angle sum of $\mathcal{P}$ satisfies*

$$\sigma(\mathcal{P}) \le (n - 2) \times 180°.$$

Therefore, the defect of a convex polygon is always nonnegative.

Proof. Exercise 17F.   □

The preceding corollary is true for nonconvex polygons as well, but it is a little more work to prove and we do not need it.

Clairaut's Postulate

A basic fact of Euclidean geometry is that rectangles exist in abundance—see, for example, Theorem 10.28, which shows that a rectangle can be constructed with any given segment as one of its sides. The proof of that theorem relied on the EPP in several ways, as you can check. Note that rectangles all have angle sums of 360°. Now that we know that the angle-sum theorem is closely related to the validity of the EPP, we might expect that it would be difficult to construct rectangles without the EPP. In fact, the relationship is much deeper than you might have suspected.

In the mid-eighteenth century, a French mathematician named Alexis Clairaut proposed the following surprising substitute for Euclid's fifth postulate.

Euclidean Postulate 9 (Clairaut's Postulate). *There exists a rectangle.*

The proof of the next theorem is rated R.

Theorem 17.10. *Clairaut's postulate is equivalent to the Euclidean parallel postulate.*

Proof. Theorem 10.28 showed that the EPP implies Clairaut's postulate. Conversely, we will prove in several steps that Clairaut's postulate implies the angle-sum postulate. Since we have already shown that the angle-sum postulate implies the EPP, that suffices.

Assume there exists a rectangle. Note that the common perpendicular theorem implies that rectangles are parallelograms in neutral geometry. However, the proof that opposite sides of parallelograms are congruent uses the converse to the alternate interior angles theorem and thus does not work in neutral geometry, so we cannot conclude that rectangles have congruent opposite sides.

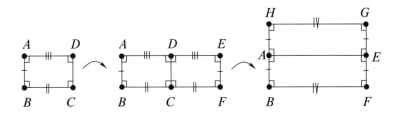

Fig. 17.9. Constructing a larger rectangle.

STEP 1: *If m is any positive real number, there exists a rectangle whose side lengths are all greater than m.* The key observation is that given one rectangle, we can construct another rectangle whose shortest side is at least twice as long as the shortest side of the original one. Let $ABCD$ be a rectangle, and without loss of generality suppose $\overline{CD}$ is its shortest side. Using Theorem 9.14, we can find points E and F on the opposite side of $\overleftrightarrow{CD}$ from A and B such that $CDEF \cong CDAB$ (Fig. 17.9).

Because $\angle ADC$ and $\angle CDE$ are right angles that share $\overrightarrow{DC}$ and lie on opposite sides of it, the partial converse to the linear pair theorem shows that $\overrightarrow{DA}$ and $\overrightarrow{DE}$ are opposite rays, so $A * D * E$. A similar argument shows that $B * C * F$. Thus $ABFE$ is a rectangle. Its side lengths satisfy

$$EF = AB \geq CD,$$
$$AE = AD + DE = 2AD \geq 2CD,$$
$$BF = BC + CF = 2BC \geq 2CD.$$

Now do the same construction again using $\overline{AE}$: let G and H be points on the opposite side of $\overleftrightarrow{AE}$ from B and F such that $AEGH \cong AEFB$. The same argument as in the preceding paragraph shows that $BFGH$ is a rectangle, and all four of its side lengths are greater than or equal to $2CD$.

To complete Step 1, we just start with any rectangle and apply the construction above repeatedly, at least doubling the minimum side length each time. After finitely many steps, we obtain a rectangle whose side lengths are all strictly greater than m.

STEP 2: *Every right triangle has zero defect.* Let $\triangle ABC$ be a right triangle with right angle at C. Let m be the length of the longest leg of $\triangle ABC$, and let $PQRS$ be a rectangle whose side lengths are all greater than m. On the ray $\overrightarrow{CB}$, let D be the point such that $\overline{CD} \cong \overline{PQ}$. Because $CD > CB$, the ordering lemma for points implies that $C * B * D$. Let E and F be points on the same side of $\overleftrightarrow{CD}$ as A such that $CDEF \cong PQRS$ (Fig. 17.10). Thus $CDEF$ is a rectangle with $CF > CA$ and $CD > CB$. Because both $\angle BCA$ and $\angle BCF$ are right angles, the unique ray theorem shows that $\overrightarrow{CA} = \overrightarrow{CF}$, so the ordering lemma for points implies $C * A * F$. Thus the hypotheses of the defect addition theorem are satisfied, and we conclude that

$$\delta(CDEF) = \delta(\triangle ABC) + \delta(ABDEF).$$

Since $\delta(CDEF) = 0$ and all defects are nonnegative, it follows that both terms on the right-hand side must be zero. In particular, $\triangle ABC$ has zero defect as claimed.

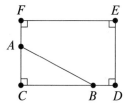

Fig. 17.10. Every right triangle has zero defect.

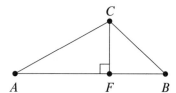

Fig. 17.11. Every triangle has zero defect.

STEP 3: *Every triangle has zero defect.* Let $\triangle ABC$ be a triangle, and assume that the vertices have been labeled so that $\overline{AB}$ is the longest side. Then the altitude from C meets $\overline{AB}$ at an interior point F and cuts $\triangle ABC$ into two right triangles, $\triangle AFC$ and $\triangle CFB$. Step 2 and the defect addition theorem imply that $\delta(\triangle ABC) = \delta(\triangle AFC) + \delta(\triangle CFB) = 0 + 0$. $\square$

The remarkable thing about Clairaut's postulate is that it seems to have finally achieved at least part of what geometers had been seeking for so many centuries: although Clairaut did not succeed in proving that Euclid's fifth postulate followed from the other four (as we now know he could not possibly have done), he did show that the fifth postulate can be replaced by an assumption that seems just as straightforward and self-evident as the other four postulates and does not mention lines being parallel or being extended arbitrarily far. We might speculate that if Euclid had realized he could use Clairaut's postulate in place of his own fifth, nobody would have expended much effort trying to improve on Euclid's postulates. (After all, is there really a big difference between assuming the existence of a rectangle and assuming the existence of a circle with any center and any radius?) Had this been the case, the entire course of mathematics history might have been radically different. One is even tempted to wonder whether non-Euclidean geometry—and with it our modern understanding of the axiomatic method—would ever have been discovered without the motivation provided by the awkward fifth postulate.

However, even though Clairaut's postulate provided an entirely reasonable resolution to the nonintuitiveness of Euclid's fifth postulate, it still did not resolve the question of whether the fifth postulate was independent of the others. The answer to that question would not come until a century after Clairaut's death.

Our last Euclidean postulate is a minor variation on Clairaut's postulate, recast in terms of triangles.

Euclidean Postulate 10 (The Weak Angle-Sum Postulate). *There exists a triangle with zero defect.*

Theorem 17.11. *The weak angle-sum postulate is equivalent to the Euclidean parallel postulate.*

Proof. The EPP implies the weak angle-sum postulate, because it implies that *every* triangle has zero defect (and there exists a triangle because there are three noncollinear points). Conversely, assume that $\triangle ABC$ is a triangle with zero defect. We may assume that the vertices have been labeled so that $\overline{AB}$ is the longest side. Let F be the foot of the altitude to $\overline{AB}$ as in the proof of Theorem 17.10 (see Fig. 17.11). The defect addition theorem guarantees that $\delta(\triangle AFC) + \delta(\triangle CFB) = \delta(\triangle ABC) = 0$. Since both terms on the left-hand side are nonnegative, they must both be zero. In particular $\triangle AFC$ is a right triangle with zero defect, so $m\angle FCA + m\angle CAF = 90°$.

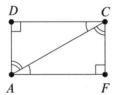

Fig. 17.12. Proof that the weak angle-sum postulate implies the EPP.

Now let D be a point on the opposite side of $\overleftrightarrow{AC}$ from F such that $\triangle CDA \cong \triangle AFC$ (Fig. 17.12). The alternate interior angles theorem shows that $\overleftrightarrow{DC} \parallel \overleftrightarrow{AF}$ and $\overleftrightarrow{DA} \parallel \overleftrightarrow{CF}$, so $AFCD$ is a parallelogram and thus convex. Because C lies in the interior of $\angle DAF$, it follows that

$$m\angle DAF = m\angle DAC + m\angle CAF = m\angle FCA + m\angle CAF = 90°.$$

Similarly, $\angle DCF$ is a right angle, so $AFCD$ is a rectangle. Thus Clairaut's postulate is satisfied, so the EPP follows from Theorem 17.10. □

It's All or Nothing

The results we have developed in this chapter now enable us to answer a question that might have occurred to you when we first started talking about parallel postulates in neutral geometry: is it possible for a model of neutral geometry to satisfy *neither* the Euclidean nor the hyperbolic parallel postulate? Of course, for any given line ℓ and point A not on ℓ, either the parallel to ℓ through A is unique or it is not. But can it happen that for some lines and some points the parallels are unique, while for others they are not? The next theorem shows that it is not possible. Its proof is rated PG.

Theorem 17.12 (The All-or-Nothing Theorem). *In neutral geometry, if there exist one line ℓ_0 and one point $A_0 \notin \ell_0$ such that there are two or more distinct lines parallel to ℓ_0 through A_0, then for every line ℓ and every point $A \notin \ell$, there are two or more distinct lines parallel to ℓ through A.*

Proof. Suppose there exist such a line ℓ_0 and point A_0. This implies that the Euclidean parallel postulate is false. Therefore, by Theorem 17.10, no rectangle exists.

Now suppose ℓ is an arbitrary line and A is a point not on ℓ. We will show that there are at least two distinct lines m and m' through A and parallel to ℓ. Let F be the foot of the perpendicular from A to ℓ, and let m be the line perpendicular to $\overleftrightarrow{AF}$ at A; thus m is one line parallel to ℓ through A.

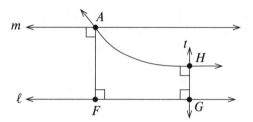

Fig. 17.13. Proof of the all-or-nothing theorem.

To find another one, choose any point $G \in \ell$ other than F, and let t be the line perpendicular to ℓ at G (Fig. 17.13). Then t and $\overleftrightarrow{AF}$ are parallel because they have ℓ as a common perpendicular, so $A \notin t$. Let H be the foot of the perpendicular from A to t. The common perpendicular theorem implies that $\overline{AH} \parallel \overline{FG}$ and $\overline{AF} \parallel \overline{GH}$, so $AFGH$ is a parallelogram by the parallelogram lemma, which has three right angles by construction. If $\overleftrightarrow{AH}$ were equal to m, then the fourth angle of $AFGH$ would be a right angle because $m \perp \overleftrightarrow{AF}$; thus $AFGH$ would be a rectangle, which is impossible. Therefore, $\overleftrightarrow{AH}$ is a line distinct from m that is parallel to ℓ through A. $\qquad\square$

Corollary 17.13. *In every model of neutral geometry, either the Euclidean parallel postulate or the hyperbolic parallel postulate holds.*

Proof. The preceding theorem shows that if the Euclidean parallel postulate is false, then the hyperbolic parallel postulate is true. $\qquad\square$

Exercises

17A. Prove that Playfair's postulate is equivalent to the EPP.

17B. Prove that the alternate interior angle postulate is equivalent to the EPP.

17C. Prove that Proclus's postulate is equivalent to the EPP.

17D. Prove that the transitivity postulate is equivalent to the EPP.

17E. Prove that in neutral geometry, the following postulate is equivalent to the EPP: *If ℓ and m are parallel lines, then any line perpendicular to one of them is perpendicular to the other.*

17F. Prove Corollary 17.9 (the Saccheri–Legendre theorem for convex polygons). [Hint: Use induction on the number of vertices.]

17G. The PYTHAGOREAN POSTULATE is the following statement: *if a right triangle has legs of lengths a and b and hypotenuse of length c, then $a^2 + b^2 = c^2$*. Prove that in neutral geometry, the Pythagorean postulate is equivalent to the EPP. [Hint: Assuming the Pythagorean postulate is true, construct an isosceles right triangle, draw an altitude to the hypotenuse, and prove that you have constructed a triangle with zero defect.]

17H. THALES' POSTULATE is the following statement: *any angle inscribed in a semicircle is a right angle*. (The definition of an angle inscribed in a semicircle is the same in neutral geometry as it is in Euclidean geometry: an angle whose vertex lies on the circle and whose sides both contain interior points that are endpoints of a diameter.) Prove that in neutral geometry, Thales' postulate is equivalent to the EPP. [Hint: Use Corollary 9.20.]

17I. THE CIRCUMCIRCLE POSTULATE is the following statement: *for every triangle, there is a circle that contains all of its vertices*. Prove that in neutral geometry, the circumcircle postulate is equivalent to the EPP. [Hint: Assuming the circumcircle postulate is true, let ℓ be a line and let A be a point not on ℓ. Let t be the line perpendicular to ℓ through A, and let m be any line through A and distinct from t. It suffices to show that if m is not perpendicular to t, then m intersects ℓ (why?). Let F be the foot of the perpendicular from A to ℓ, and let M be the midpoint of $\overline{AF}$. Let P and Q be the reflections of M through ℓ and m, respectively (see Theorem 7.10). Show that there is a circle containing M, P, and Q, and apply Theorem 7.9 to the center of the circle.]

17J. THE SAS SIMILARITY POSTULATE is the following statement: *if $\triangle ABC$ and $\triangle DEF$ are triangles such that $\angle A \cong \angle D$ and $AB/DE = AC/DF$, then $\triangle ABC \sim \triangle DEF$*. Prove that in neutral geometry, the SAS similarity postulate is equivalent to the EPP. [Remark: This postulate is essentially identical to one that Birkhoff included in his list of proposed postulates for Euclidean geometry (Appendix B). Note that the SAS similarity postulate also implies the SAS postulate, by applying it in the special case in which $AB/DE = AC/DF = 1$. Thus Birkhoff used this postulate as a substitute for both the Euclidean parallel postulate and SAS; this is one way in which he was able to keep the number of his postulates down to four.]

Chapter 18

Introduction to Hyperbolic Geometry

In this chapter, we begin our study of hyperbolic geometry. This is the axiomatic system consisting of the primitive terms and postulates of neutral geometry together with the hyperbolic parallel postulate. It gets its name primarily from the hyperboloid model, which is the model of hyperbolic geometry most closely analogous to the sphere. Some authors call it ***Lobachevskian geometry*** in honor of Nikolai Lobachevsky, one of its discoverers. We know it is consistent because Beltrami [**Bel68**] constructed models of it based on the Beltrami–Klein model, the Poincaré disk, and the Poincaré half-plane.

The Hyperbolic Parallel Postulate

In addition to the nine postulates of neutral geometry, throughout this chapter and the next we assume the following postulate.

> **Postulate 10H (The Hyperbolic Parallel Postulate).** *For each line ℓ and each point A that does not lie on ℓ, there are at least two distinct lines that contain A and are parallel to ℓ.*

Every theorem in neutral geometry is also a theorem in hyperbolic geometry, including the Saccheri–Legendre and defect addition theorems from the preceding chapter. In addition, the work we did in the preceding chapter quickly yields several other theorems in hyperbolic geometry: because the Euclidean parallel postulate is false in hyperbolic geometry, each of the postulates that we showed to be equivalent to the EPP is also false. In particular, we have the following important theorem.

Theorem 18.1. *In hyperbolic geometry, there does not exist a rectangle.*

Proof. This follows from Theorem 17.10, which shows that the existence of a rectangle is equivalent to the EPP. $\qquad\square$

The next theorem expresses one of the most fundamental features of hyperbolic geometry.

Theorem 18.2 (Hyperbolic Angle-Sum Theorem). *In hyperbolic geometry, every convex polygon has positive defect.*

Proof. By the Saccheri–Legendre theorem, every triangle has nonnegative defect. Because the EPP is false, Theorem 17.11 shows that the weak angle-sum postulate is also false, so no triangle has zero defect. The only possibility, therefore, is that every triangle has positive defect. The result for convex polygons with more than three vertices is then proved by induction on the number of vertices, just as in the proof of Corollary 17.9.   □

One of the oddest features of hyperbolic geometry follows from the fact that Wallis's postulate is false: we cannot construct similar triangles with arbitrary scale factors. In fact, the next theorem shows that we cannot construct similar triangles at all in hyperbolic geometry (except for congruent triangles, which are similar for silly reasons).

Theorem 18.3 (AAA Congruence Theorem). *In hyperbolic geometry, if there is a correspondence between the vertices of two triangles such that all three pairs of corresponding angles are congruent, then the triangles are congruent under that correspondence.*

Proof. Suppose $\triangle ABC$ and $\triangle DEF$ are triangles in which $\angle A \cong \angle D$, $\angle B \cong \angle E$, and $\angle C \cong \angle F$. If any side of $\triangle ABC$ is congruent to the corresponding side of $\triangle DEF$, then the two triangles are congruent by ASA. So suppose for the sake of contradiction that all three pairs of corresponding sides are noncongruent: $AB \neq DE$, $AC \neq DF$, and $BC \neq EF$. Since there are three inequalities and each one must be either "greater than" or "less than," at least two of them must go in the same direction. Without loss of generality, let us assume that $AB < DE$ and $AC < DF$.

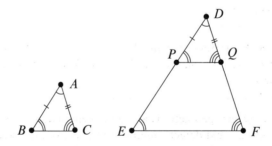

Fig. 18.1. Proof of the AAA congruence theorem.

Choose points $P \in \text{Int}\,\overline{DE}$ and $Q \in \text{Int}\,\overline{DF}$ such that $\overline{DP} \cong \overline{AB}$ and $\overline{DQ} \cong \overline{AC}$ (Fig. 18.1). It follows from SAS that $\triangle DPQ \cong \triangle ABC$, and therefore $\angle DPQ \cong \angle B \cong \angle E$ and $\angle DQP \cong \angle C \cong \angle F$. By the polygon splitting theorem, $PQFE$ is a convex quadrilateral (in fact, a trapezoid by the corresponding angles theorem). Since $\angle EPQ$ is supplementary to $\angle DPQ$ and therefore also to $\angle E$, and since $\angle FQP$ is supplementary to $\angle DQP$ and thus to $\angle F$, it follows that $PQFE$ has angle sum equal to 360° and defect zero. This contradicts the hyperbolic angle-sum theorem.   □

This theorem is remarkable because it implies that it is impossible to construct a scale model or a scaled image of anything—in a hyperbolic world, a photograph would necessarily distort the shape of its subject unless it were actual size.

Another way of interpreting the AAA congruence theorem is that it guarantees the existence of an "absolute" standard of length, independent of any arbitrary conventions. In Euclidean geometry, the distances given to us by the distance postulate are in a certain sense arbitrary: we could multiply all distances by a positive constant c and all areas by c^2, and the theorems of Euclidean geometry would still be true with this new distance scale. (This just corresponds to changing the units of our measuring scale, as from feet to meters.) Thus if two remotely separated civilizations in a Euclidean universe wanted to agree on a unit of length, they would have to rely on some physical phenomenon having the same properties in both locations, such as the wavelength of light emitted from a certain kind of atom.

However, in hyperbolic geometry things are very different. If our hypothetical civilizations lived in a hyperbolic world, they could simply agree that the universal unit of length is, say, the side length of a 45°-45°-45° triangle. (There always exists such a triangle in hyperbolic geometry, although this is not easy to prove.) Because the AAA theorem guarantees that all such triangles are congruent, there would be no ambiguity about what this length is. (There is no ambiguity about what a 45° angle is either, since it can be described as an angle created by the bisector of a right angle, and a right angle in turn can be described as any angle that forms a linear pair with a congruent angle.)

Here is one more consequence of the AAA congruence theorem; it shows another way in which hyperbolic geometry differs markedly from Euclidean geometry.

Theorem 18.4. *The postulates of hyperbolic geometry are not categorical.*

Proof. Let $\mathcal{M}$ be a model of hyperbolic geometry (such as the Poincaré disk). Choose any positive real number $c \neq 1$, and create a new model $\mathcal{M}_c$ by defining the points, lines, and angle measures to be the same as those of $\mathcal{M}$ but defining distances in $\mathcal{M}_c$ to be c times the corresponding distances in $\mathcal{M}$.

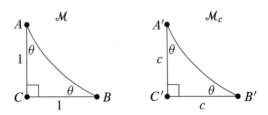

Fig. 18.2. The hyperbolic postulates are not categorical.

To see that the models $\mathcal{M}$ and $\mathcal{M}_c$ cannot be isomorphic, begin by constructing an isosceles right triangle $\triangle ABC$ in $\mathcal{M}$ with legs of length 1 and right angle at C. (See Fig. 18.2. Such a triangle is easily constructed by drawing two perpendicular rays and marking off a segment of length 1 along each of the rays.) We have no way of knowing what the measures of its acute angles are, but we do know that they are equal (by the isosceles triangle theorem) and less than 45° (by the hyperbolic angle-sum theorem). Let θ denote the common measure of $\angle A$ and $\angle B$. Thus $\mathcal{M}$ contains an isosceles right triangle with angle measures θ-θ-$90°$ and legs of length 1.

Let $\triangle A'B'C'$ denote the corresponding triangle in $\mathcal{M}_c$. (It is actually the same triangle, but with different side lengths.) Then our definition of $\mathcal{M}_c$ guarantees that $\triangle A'B'C'$ has angle measures θ-θ-$90°$ and legs of length c. Assume for contradiction that there is an isomorphism F between $\mathcal{M}$ and $\mathcal{M}_c$. If we let $A'' = F(A)$, $B'' = F(B)$, and $C'' = F(C)$, then by definition of an isomorphism, $\triangle A''B''C''$ is an isosceles right triangle in $\mathcal{M}_c$ with angle measures θ-θ-$90°$ and legs of length 1. Then $\triangle A'B'C'$ and $\triangle A''B''C''$ are two triangles in $\mathcal{M}_c$ that satisfy the AAA hypotheses but are not congruent, thus contradicting the AAA congruence theorem.   □

Although the preceding theorem shows that there are nonisomorphic models of hyperbolic geometry, that is only part of the story. It is a fact (which we will not prove) that every model of hyperbolic geometry is isomorphic to one of the models $\mathcal{M}_c$ that we constructed above. Thus, in a sense, the axioms of hyperbolic geometry are categorical except for a more or less arbitrary "scale factor."

It is worth thinking about why this argument doesn't work in the Euclidean case. Starting with the Cartesian model $\mathcal{M}$, we can certainly create a new model $\mathcal{M}_c$ by multiplying all of the distances by a positive constant c. But these two models turn out to be isomorphic: an isomorphism $F : \mathcal{M} \to \mathcal{M}_c$ can be defined by letting $F(x, y) = (cx, cy)$ for every point (x, y) in the model $\mathcal{M}$. If we started with a different model $\mathcal{M}'$ and multiplied its distances by c to create a new model $\mathcal{M}_c'$, this particular technique might not be available, but Theorem 13.19 guarantees that there is *some* isomorphism from $\mathcal{M}'$ to $\mathcal{M}_c'$. A way to visualize the situation is to imagine the first model $\mathcal{M}$ as a large life-size painting and $\mathcal{M}_c$ as a reproduction of that painting at a smaller scale. When you look at $\mathcal{M}_c$, it's impossible to tell (without relying on external clues) whether you're looking at the original painting from far away or looking at the reduced reproduction from nearby. In hyperbolic geometry, you would see the difference, because the reduction would necessarily distort angles and therefore shapes of objects.

Saccheri and Lambert Quadrilaterals

After triangles, rectangles are probably the second most important figures in Euclidean geometry. Since rectangles are nonexistent in hyperbolic geometry, it is natural to look for other figures that might play analogous roles.

There are two obvious ways that one might attempt to construct a rectangle without benefit of the Euclidean parallel postulate, each of which leads to a useful class of quadrilaterals in hyperbolic geometry. One way is to start with a segment $\overline{AB}$, construct congruent perpendicular segments $\overline{AD}$ and $\overline{BC}$ on the same side of $\overleftrightarrow{AB}$ (Fig. 18.3), and join C and D to form a quadrilateral $ABCD$. The other is to start with two segments $\overline{AB}$ and $\overline{BC}$ meeting perpendicularly at B, construct perpendiculars to them at A and C, and let D be the point (if there is one) where the two perpendiculars meet (Fig. 18.4).

In Euclidean geometry, Theorem 10.27 would imply that the quadrilateral pictured in Fig. 18.3 must be a rectangle, and the angle-sum theorem for quadrilaterals would imply that the one pictured in Fig. 18.4 must be a rectangle. But in hyperbolic geometry, the remaining angles cannot be right angles, because that would violate the hyperbolic angle-sum theorem. Thus we make the following definitions. A **Saccheri quadrilateral** is a

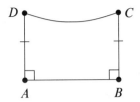

Fig. 18.3. A Saccheri quadrilateral.

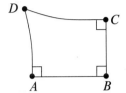

Fig. 18.4. A Lambert quadrilateral.

quadrilateral with a pair of congruent opposite sides that are both perpendicular to a third side, and a *Lambert quadrilateral* is a quadrilateral with three right angles.

Saccheri quadrilaterals are named after Giovanni Saccheri (whom we met in Chapter 1 and heard from again in the preceding chapter), and Lambert quadrilaterals are named after the Swiss mathematician Johann Heinrich Lambert (1728–1777). Although the nomenclature is well established in Western tradition, it is not historically accurate: both types of quadrilaterals were studied many centuries earlier by Islamic mathematicians such as Omar Khayyam.

When we draw figures in hyperbolic geometry, it is generally impossible to make all the line segments appear straight while making all distances and angles appear correct, simply because the plane in which we are drawing our pictures is a Euclidean plane (at least to a very close approximation). We will generally attempt to make the angles look correct and to make the lines look straight whenever possible. But sometimes lines have to be drawn as curves in order to make the angles appear correct, and it is usually impossible to make all distances appear correct. A useful way to visualize hyperbolic figures is in the Poincaré disk model (see Examples 2.16 and 6.12, in which all angles are correct but most lines look like arcs of circles. Fig. 18.5 illustrates Saccheri and Lambert quadrilaterals in the Poincaré disk model. In these figures, the hyperbolic "lines" containing the sides are shown as dashed lines or curves, to make it clear that each side is part of a diameter or a circle intersecting the unit circle perpendicularly.

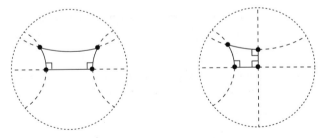

Fig. 18.5. Saccheri and Lambert quadrilaterals in the Poincaré disk model.

For a Saccheri quadrilateral, we make the following definitions:

- The two congruent opposite sides are called the *legs*.
- The side perpendicular to the legs is called the *base*.
- The side opposite the base is called the *summit*.
- The two angles adjacent to the summit are called the *summit angles*.

- The segment joining the midpoints of the summit and the base is called the **midsegment**.

In a Lambert quadrilateral:

- The angle that is not a right angle is called the **fourth angle**.
- The vertex of the fourth angle is called the **fourth vertex**.

Here are a few elementary properties, whose proofs are left for you to carry out.

Theorem 18.5 (Properties of Saccheri Quadrilaterals). *Every Saccheri quadrilateral has the following properties:*

(a) *It is a convex quadrilateral.*

(b) *Its diagonals are congruent.*

(c) *Its summit angles are congruent and acute.*

(d) *Its midsegment is perpendicular to both the base and the summit.*

(e) *It is a parallelogram.*

Proof. Exercise 18B. ☐

Analogously, for Lambert quadrilaterals, we have the following properties.

Theorem 18.6 (Properties of Lambert Quadrilaterals). *Every Lambert quadrilateral has the following properties:*

(a) *It is a convex quadrilateral.*

(b) *Its fourth angle is acute.*

(c) *It is a parallelogram.*

Proof. Exercise 18C. ☐

There are also some important properties of Saccheri and Lambert quadrilaterals involving comparisons between side lengths, but they are somewhat less elementary to prove. The key is the following lemma.

Lemma 18.7 (Scalene Inequality for Quadrilaterals). *Suppose $ABCD$ is a quadrilateral in which $\angle A$ and $\angle B$ are right angles. Then $AD > BC$ if and only if $m\angle D < m\angle C$.*

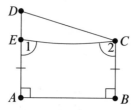

Fig. 18.6. Proof of Lemma 18.7.

Proof. First assume that $AD > BC$ (Fig. 18.6). Let E be the point in Int $\overline{AD}$ such that $AE = BC$. Then $ABCE$ is a Saccheri quadrilateral, so its summit angles (labeled $\angle 1$ and $\angle 2$ in the diagram) are congruent. By the exterior angle inequality, $m\angle D < m\angle 1$, and therefore $m\angle D < m\angle 2$ by substitution. If we can show that $\overrightarrow{CB} * \overrightarrow{CE} * \overrightarrow{CD}$, then it follows that $m\angle 2 < m\angle BCD$ (the whole angle is greater than the part), so by transitivity $m\angle D < m\angle BCD$ and we are done.

To show that $\overrightarrow{CB} * \overrightarrow{CE} * \overrightarrow{CD}$, note first that $ABCD$ is a trapezoid by the common perpendicular theorem, so it is convex. That means that E lies in the closed half-planes $\text{CHP}\left(\overleftrightarrow{CB}, D\right)$ and $\text{CHP}\left(\overleftrightarrow{CD}, B\right)$. Now, E is not on $\overleftrightarrow{CD}$ because consecutive sides of a quadrilateral are noncollinear, and it is not on $\overleftrightarrow{CB}$ because $\overleftrightarrow{DA} \parallel \overleftrightarrow{CB}$. Thus E is actually in the intersection of the *open* half-planes $\text{OHP}\left(\overleftrightarrow{CB}, D\right)$ and $\text{OHP}\left(\overleftrightarrow{CD}, B\right)$, which is the interior of $\angle DCB$. It follows from the interior lemma that $\overrightarrow{CB} * \overrightarrow{CE} * \overrightarrow{CD}$. This completes the proof that $AD > BC$ implies $m\angle D < m\angle C$.

To prove the converse, assume that $m\angle D < m\angle C$. By trichotomy, we must have $AD < BC$, $AD = BC$, or $AD > BC$. If $AD < BC$, then the preceding argument implies $m\angle D > m\angle C$, which is a contradiction. Similarly, if $AD = BC$, then $ABCD$ is a Saccheri quadrilateral, and Theorem 18.5 implies that $\angle D \cong \angle C$, again a contradiction. The only remaining possibility is that $AD > BC$. $\qquad\square$

Theorem 18.8. *In a Lambert quadrilateral, either side between two right angles is strictly shorter than its opposite side.*

Proof. Suppose $ABCD$ is a Lambert quadrilateral with fourth vertex D. We need to show that $AD > BC$ and $CD > AB$. Since $\angle D$ is acute by Theorem 18.6, it follows that $m\angle D < m\angle C$, so Lemma 18.7 implies that $AD > BC$. The same argument with A and C reversed shows that $CD > AB$. $\qquad\square$

There are also corresponding inequalities for Saccheri quadrilaterals, which can be proved by using the fact that the midsegment of a Saccheri quadrilateral divides it into two Lambert quadrilaterals.

Theorem 18.9. *In a Saccheri quadrilateral, the base is strictly shorter than the summit, and the midsegment is strictly shorter than either leg.*

Proof. Exercise 18D. $\qquad\square$

Asymptotic Rays

In hyperbolic geometry, because there are two or more lines through a point that are parallel to a given line, it is natural to ask exactly how many parallels there are and how to distinguish them from each other.

To get an idea of how to identify the different kinds of parallel lines, suppose that ℓ is a line and A is a point not on ℓ and that m and m' are two distinct lines through A and parallel to ℓ. Now imagine a third line n going through A but free to rotate around A (Fig. 18.7). What will happen as the direction of n changes? Most people would guess correctly that as n rotates from m to m', it remains parallel to ℓ. What happens when we rotate it past m'? At some point in the rotation, the line has to start intersecting ℓ. As we will see below, as n rotates toward ℓ, there is one last direction at which it is parallel to ℓ, and any further

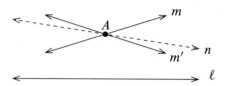

Fig. 18.7. Imagining parallel lines through A.

rotation results in a line that intersects ℓ. Our main goal in this section is to understand exactly what happens in that direction. It is useful to focus first on *rays* starting at A.

If $\overrightarrow{AB}$ and $\overrightarrow{CD}$ are two rays, we say that $\overrightarrow{AB}$ ***is asymptotic to*** $\overrightarrow{CD}$ if all of the following conditions are satisfied (see Fig. 18.8):

(i) $\overrightarrow{AB}$ and $\overrightarrow{CD}$ lie on the same side of the line $\overleftrightarrow{AC}$ connecting their endpoints.

(ii) $\overrightarrow{AB}$ does not intersect $\overrightarrow{CD}$.

(iii) Every ray between $\overrightarrow{AB}$ and $\overrightarrow{AC}$ does intersect $\overrightarrow{CD}$.

The idea is that as a ray rotates around A toward $\overrightarrow{CD}$, the direction of $\overrightarrow{AB}$ is the last direction in which the ray does not intersect $\overrightarrow{CD}$. (We are not claiming yet that there *is* such a ray; we are just giving a name to such a ray in case it does exist.) We use the notation $\overrightarrow{AB} \mid \overrightarrow{CD}$ to mean that $\overrightarrow{AB}$ is asymptotic to $\overrightarrow{CD}$.

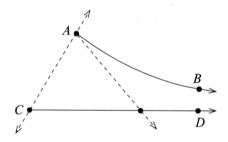

Fig. 18.8. Asymptotic rays.

Our first observation is that the lines determined by asymptotic rays are parallel to each other.

Theorem 18.10. *If* $\overrightarrow{AB} \mid \overrightarrow{CD}$, *then* $\overleftrightarrow{AB} \parallel \overleftrightarrow{CD}$.

Proof. Assume for the sake of contradiction that $\overrightarrow{AB} \mid \overrightarrow{CD}$ but $\overleftrightarrow{AB}$ and $\overleftrightarrow{CD}$ are not parallel. Thus there is a point X where the two lines intersect (Fig. 18.9). Property (i) in the definition of asymptotic rays implies that these rays intersect $\overleftrightarrow{AC}$ only at their two distinct endpoints, so X cannot be on $\overleftrightarrow{AC}$. If X is on the same side of $\overleftrightarrow{AC}$ as D and B, then by Theorem 3.46 it must lie on both rays $\overrightarrow{AB}$ and $\overrightarrow{CD}$; but this is ruled out by property (ii).

The only remaining possibility is that X is on the opposite side of $\overleftrightarrow{AC}$ from B and D. It follows that X is not in either ray $\overrightarrow{AB}$ or $\overrightarrow{CD}$, so $X * A * B$ and $X * C * D$. The exterior angle inequality shows that $m\angle CAB > m\angle ACX$. Let $\overrightarrow{AE}$ be the ray on the same side of $\overleftrightarrow{AC}$ as B and D such that $\angle CAE \cong \angle ACX$. Then $\overleftrightarrow{AE} \parallel \overleftrightarrow{CD}$ by the alternate interior

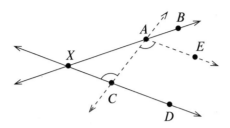

Fig. 18.9. Asymptotic rays are parallel.

angles theorem. By the ordering lemma for rays, the fact that $m\angle CAE < m\angle CAB$ implies that $\overrightarrow{AC} * \overrightarrow{AE} * \overrightarrow{AB}$. Therefore, property (iii) in the definition of asymptotic rays ensures that $\overrightarrow{AE}$ must intersect $\overrightarrow{CD}$, which is a contradiction. □

The next proof is rated R.

Theorem 18.11 (Existence and Uniqueness of Asymptotic Rays). *Suppose $\overrightarrow{CD}$ is a ray and A is a point not on $\overleftrightarrow{CD}$. Then there exists a unique ray starting at A and asymptotic to $\overrightarrow{CD}$.*

Proof. First we prove existence. We start by defining a set S of real numbers, consisting roughly of angle measures of rays from A that meet $\overrightarrow{CD}$. More precisely, let S be the set of all real numbers x strictly between 0 and 180 such that the ray starting at A, lying on the same side of $\overleftrightarrow{AC}$ as D, and making angle measure x with $\overrightarrow{AC}$ has nonempty intersection with $\overrightarrow{CD}$.

We will prove the following two crucial properties of the set S:

(a) If $y \in S$ and $0 < x < y$, then $x \in S$.

(b) If $y \in S$, there exists $z \in S$ such that $z > y$.

To prove (a), assume $y \in S$ and $0 < x < y$. Let $\overrightarrow{AX}$ and $\overrightarrow{AY}$ be the rays on the same side of $\overleftrightarrow{AC}$ as D with $m\angle CAX = x$ and $m\angle CAY = y$ (Fig. 18.10). Then $\overrightarrow{AC} * \overrightarrow{AX} * \overrightarrow{AY}$ by the ordering lemma for rays. The fact that $y \in S$ means that $\overrightarrow{AY}$ intersects $\overrightarrow{CD}$ at a point P, and then the crossbar theorem applied to $\triangle ACP$ implies that $\overrightarrow{AX}$ intersects $\overrightarrow{CD}$ as well. Thus $x \in S$ as claimed.

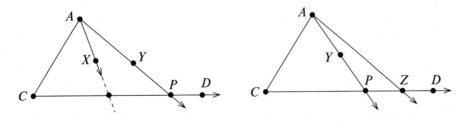

Fig. 18.10. Proof of property (a).   **Fig. 18.11.** Proof of property (b).

To prove (b), let y be any number in S and let $\overrightarrow{AY}$ be the ray on the same side of $\overleftrightarrow{AC}$ as D with $m\angle CAY = y$ (Fig. 18.11). The fact that $y \in S$ means that $\overrightarrow{AY}$ intersects $\overrightarrow{CD}$

at a point P. If Z is any point such that $C * P * Z$, then $\overrightarrow{AC} * \overrightarrow{AP} * \overrightarrow{AZ}$ by betweenness vs. betweenness, and therefore $m\angle CAZ > m\angle CAP$. Writing $z = m\angle CAZ$, we see that z is an element of $\mathcal{S}$ greater than y.

Next we will show that $\mathcal{S}$ is nonempty and bounded above. If P is any interior point of $\overline{CD}$, then P lies on the same side of $\overleftrightarrow{AC}$ as D by the Y-lemma applied to $\overrightarrow{CD}$, and thus the ray $\overrightarrow{AP}$ lies on that side; and $\overrightarrow{AP}$ obviously has nonempty intersection with $\overrightarrow{CD}$ (see Fig. 18.12). If we set $\alpha = m\angle CAP$, then $\alpha \in \mathcal{S}$ by definition, so $\mathcal{S}$ is nonempty. To show that $\mathcal{S}$ is bounded above, let $\beta = 180° - m\angle ACD$, and suppose x is any number in $\mathcal{S}$. Then there is a point $Q \in \text{Int}\,\overrightarrow{CD}$ such that $m\angle CAQ = x$. Because $\angle ACD$ and $\angle CAQ$ are two angles of $\triangle ACQ$, Corollary 5.14 shows that the sum of their measures is less than $180°$, so it follows that $x = m\angle CAQ < 180° - m\angle ACD = \beta$. Thus β is an upper bound for $\mathcal{S}$.

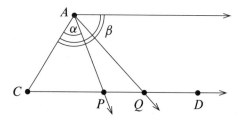

Fig. 18.12. Proof that $\mathcal{S}$ is nonempty and bounded above.

The least upper bound property of the real numbers guarantees that $\mathcal{S}$ has a unique least upper bound; call it θ. Because $\alpha \in \mathcal{S}$ and θ is an upper bound for $\mathcal{S}$, we have $\theta \geq \alpha > 0$; on the other hand, because β is an upper bound for $\mathcal{S}$ and θ is the *least* upper bound, we also have $\theta \leq \beta < 180$. Thus $0 < \theta < 180$. Let $\overrightarrow{AB}$ be the ray on the same side of $\overleftrightarrow{AC}$ as D such that $m\angle CAB = \theta$. We will complete the proof of existence by showing that $\overrightarrow{AB} \mid \overrightarrow{CD}$.

By construction, $\overrightarrow{AB}$ satisfies condition (i) in the definition of asymptotic rays. To prove (ii), assume for the sake of contradiction that $\overrightarrow{AB}$ intersects $\overrightarrow{CD}$. This means that $\theta \in \mathcal{S}$, and therefore by property (b) above there is a number $z \in \mathcal{S}$ such that $z > \theta$; but this contradicts the fact that θ is an upper bound for $\mathcal{S}$. To prove that $\overrightarrow{AB}$ satisfies (iii), suppose $\overrightarrow{AE}$ is any ray between $\overrightarrow{AB}$ and $\overrightarrow{AC}$, and let $x = m\angle EAC$. Then $x < \theta$ by Theorem 4.11(e) (the whole angle is greater than the part). Thus x cannot be an upper bound for $\mathcal{S}$ (because θ is the *least* upper bound), so there is some $y \in \mathcal{S}$ such that $y > x$. Property (a) then implies $x \in \mathcal{S}$, which means that $\overrightarrow{AE}$ intersects $\overrightarrow{CD}$. This completes the proof that $\overrightarrow{AB} \mid \overrightarrow{CD}$.

Finally, we have to prove uniqueness. Suppose for the sake of contradiction that $\overrightarrow{AB}$ and $\overrightarrow{AB'}$ are two different rays that start at A and are asymptotic to $\overrightarrow{CD}$, and let $\theta = m\angle CAB$ and $\theta' = m\angle CAB'$. We may assume $\theta < \theta'$. Because both rays are on the same side of $\overleftrightarrow{AC}$ by definition of asymptotic rays, the ordering lemma for rays shows that $\overrightarrow{AC} * \overrightarrow{AB} * \overrightarrow{AB'}$. Then the assumption that $\overrightarrow{AB'} \mid \overrightarrow{CD}$ implies that $\overrightarrow{AB}$ intersects $\overrightarrow{CD}$, which contradicts the assumption that $\overrightarrow{AB}$ is also asymptotic to $\overrightarrow{CD}$. $\square$

Next we need to develop three important properties of asymptotic rays that we will use later in our analysis of parallel lines: *symmetry, endpoint independence*, and *transitivity*. The proofs of these three properties use only the definition, existence, and uniqueness of asymptotic rays, together with some familiar facts from neutral geometry and the small angle lemma; but they are rather detailed and technical, so on first reading you might want to read the theorem statements carefully but skim the proofs.

The first property we need is a symmetry relation. Notice that the *definition* of asymptotic rays is asymmetric: in property (iii) of the definition of asymptotic rays, $\overrightarrow{AB}$ and $\overrightarrow{CD}$ play very different roles. Nonetheless, as the next theorem shows, the relation "asymptotic to" is reversible. This proof is rated PG.

Theorem 18.12 (Symmetry Property of Asymptotic Rays). *Suppose $\overrightarrow{AB}$ and $\overrightarrow{CD}$ are two distinct rays. Then $\overrightarrow{AB} \mid \overrightarrow{CD}$ if and only if $\overrightarrow{CD} \mid \overrightarrow{AB}$.*

Proof. It suffices to prove that $\overrightarrow{AB} \mid \overrightarrow{CD}$ implies $\overrightarrow{CD} \mid \overrightarrow{AB}$, because the argument is the same in both directions. Assume $\overrightarrow{AB} \mid \overrightarrow{CD}$. It is immediate from our hypothesis that the first two conditions in the definition of $\overrightarrow{CD} \mid \overrightarrow{AB}$ are satisfied: $\overrightarrow{CD}$ and $\overrightarrow{AB}$ lie on the same side of $\overleftrightarrow{AC}$ and do not intersect.

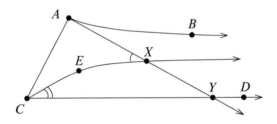

Fig. 18.13. Proof of the symmetry of asymptotic rays.

To show that the third condition is satisfied, let $\overrightarrow{CE}$ be any ray between $\overrightarrow{CA}$ and $\overrightarrow{CD}$, and assume for the sake of contradiction that $\overrightarrow{CE}$ does not intersect $\overrightarrow{AB}$ (Fig. 18.13). Since $m\angle ECD$ is a positive number, the small angle lemma (Lemma 17.5) guarantees that there is a point $X \in \overrightarrow{CE}$ such that $m\angle AXC < m\angle ECY$. We need to verify that X is in the interior of $\angle BAC$. On the one hand, the fact that $\overrightarrow{CE}$ lies in the interior of $\angle ACD$ implies that X is on the same side of $\overleftrightarrow{AC}$ as D and hence also as B. On the other hand, $\overrightarrow{CE}$ does not intersect $\overrightarrow{AB}$ by hypothesis, and it cannot intersect the ray opposite $\overrightarrow{AB}$ because that ray is on the wrong side of $\overleftrightarrow{AC}$; thus X is on the same side of $\overleftrightarrow{AB}$ as C. It follows that X is in the interior of $\angle BAC$, so the assumption $\overrightarrow{AB} \mid \overrightarrow{CD}$ implies that $\overrightarrow{AX}$ intersects $\overrightarrow{CD}$ at a point Y. It follows from betweenness vs. betweenness that $A * X * Y$, and therefore $\angle AXC$ is an exterior angle for $\triangle CXY$. But by construction, $\angle AXC$ is smaller than the remote interior angle $\angle ECY$, which contradicts the exterior angle inequality. $\square$

The next theorem shows that if we start with a pair of asymptotic rays, then changing the endpoint of one of them to another point on the same line does not change the fact that they are asymptotic. (By the symmetry property we just proved, the same is true if we change the endpoint of the other ray.) This proof is also rated PG.

Theorem 18.13 (Endpoint Independence of Asymptotic Rays). *Suppose $\vec{r}$, $\vec{s}$, and $\vec{s}\,'$ are rays such that $\vec{s}\,' \subseteq \vec{s}$. Then $\vec{r} \mid \vec{s}$ if and only if $\vec{r} \mid \vec{s}\,'$.*

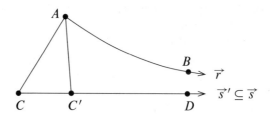

Fig. 18.14. Changing the endpoint of an asymptotic ray.

Proof. If $\vec{s} = \vec{s}\,'$, then the theorem is trivially true, so assume that $\vec{s}$ and $\vec{s}\,'$ are not the same ray. Write $\vec{r} = \overrightarrow{AB}$. Let C and C' be the endpoints of $\vec{s}$ and $\vec{s}\,'$, respectively, and let D be a point in the interior of $\vec{s}\,'$. Then we can write $\vec{s} = \overrightarrow{CD}$ and $\vec{s}\,' = \overrightarrow{C'D}$ with $C * C' * D$ (Fig. 18.14).

Assume first that $\vec{r} \mid \vec{s}$. The first step of the proof is to show that $\overrightarrow{AC} * \overrightarrow{AC'} * \overrightarrow{AB}$. The hypothesis implies that $\overrightarrow{CD}$ and $\overrightarrow{AB}$ are on the same side of $\overleftrightarrow{AC}$, so in particular C' and B lie on the same side of that line. On the other hand, the fact that $\overrightarrow{AB}$ and $\overleftrightarrow{CD}$ are parallel implies that C' and C are on the same side of $\overleftrightarrow{AB}$. Thus C' is in the interior of $\angle BAC$, which implies $\overrightarrow{AC} * \overrightarrow{AC'} * \overrightarrow{AB}$.

Now we need to show that the three defining properties of asymptotic rays are satisfied by $\overrightarrow{AB}$ and $\overrightarrow{C'D}$. To prove (i), note that $\overrightarrow{AC} * \overrightarrow{AC'} * \overrightarrow{AB}$ implies that B and C are on opposite sides of $\overleftrightarrow{AC'}$ (Theorem 4.12). Also, D and C are on opposite sides of the same line because $C * C' * D$. Therefore B and D are on the same side of $\overleftrightarrow{AC'}$, and the Y-lemma then implies that $\overrightarrow{AB}$ and $\overrightarrow{C'D}$ are on that side as well. Next, (ii) follows from the fact that $\overrightarrow{AB}$ and $\overrightarrow{C'D}$ are contained in the parallel lines $\overleftrightarrow{AB}$ and $\overleftrightarrow{CD}$. To prove (iii), suppose $\overrightarrow{AX}$ is any ray between $\overrightarrow{AB}$ and $\overrightarrow{AC'}$. Then X is on the same side of $\overleftrightarrow{AB}$ as C', and the fact that $m\angle BAX < m\angle BAC' < m\angle BAC$ implies that $\overrightarrow{AB} * \overrightarrow{AX} * \overrightarrow{AC}$ by the ordering lemma for rays. Therefore $\overrightarrow{AX}$ must intersect $\overrightarrow{CD}$ because $\overrightarrow{AB} \mid \overrightarrow{CD}$. Since $\overrightarrow{AX}$ is on the same side of $\overleftrightarrow{AC'}$ as B and thus also as D, the intersection point must in fact be in $\overrightarrow{C'D}$. This completes the proof of (iii) and shows that $r \mid s'$.

Conversely, assume $\vec{r} \mid \vec{s}\,'$. Once again, we have to show first that $\overrightarrow{AC} * \overrightarrow{AC'} * \overrightarrow{AB}$. The hypothesis implies that B and D are on the same side of $\overleftrightarrow{AC'}$, and $C * C' * D$ implies that C and D are on opposite sides; thus B and C are on opposite sides of that line, which means that $\angle CAC'$ and $\angle C'AB$ are adjacent angles. Since all three rays $\overrightarrow{AC}$, $\overrightarrow{AC'}$, and $\overrightarrow{AB}$ lie in $\text{HR}\big(\overleftrightarrow{AB}, C\big)$, the adjacency lemma implies that $\overrightarrow{AC} * \overrightarrow{AC'} * \overrightarrow{AB}$.

To show that $\overrightarrow{AB} \mid \overrightarrow{CD}$, note first that $\overrightarrow{AC} * \overrightarrow{AC'} * \overrightarrow{AB}$ implies that B and C' are on the same side of $\overleftrightarrow{AC}$, so (i) holds. Next, as in the previous case, $\overrightarrow{AB}$ and $\overrightarrow{CD}$ are disjoint because they are contained in parallel lines, so (ii) holds. Finally, to prove (iii), let $\overrightarrow{AX}$ be a ray between $\overrightarrow{AC}$ and $\overrightarrow{AB}$. This means X is on the same side of $\overleftrightarrow{AB}$ as C and on the same side of $\overleftrightarrow{AC}$ as B. Now we have three cases, depending on where X lies with respect to $\overleftrightarrow{AC'}$. If X is on the same side of $\overleftrightarrow{AC'}$ as B, then $\overrightarrow{AB} * \overrightarrow{AX} * \overrightarrow{AC'}$ by the interior lemma, so $\overrightarrow{AX}$ intersects $\overrightarrow{C'D} \subseteq \overrightarrow{CD}$ by the fact that $\overrightarrow{AB} \mid \overrightarrow{C'D}$. If X lies on $\overleftrightarrow{AC'}$, then $\overrightarrow{AX} = \overrightarrow{AC'}$, which certainly intersects $\overrightarrow{CD}$. Finally, if X is on the opposite side of $\overleftrightarrow{AC'}$

from B, then it is on the same side of $\overleftrightarrow{AC'}$ as C and on the same side of $\overleftrightarrow{AC}$ as C', so $\overrightarrow{AC} * \overrightarrow{AX} * \overrightarrow{AC'}$. Then the crossbar theorem shows that $\overrightarrow{AX}$ intersects $\overline{CC'} \subseteq \overrightarrow{CD}$. □

Finally, we prove a transitive property for asymptotic rays. It is important to remember that transitivity does not hold for parallel *lines* in hyperbolic geometry. In fact, as we showed in Chapter 17, transitivity of parallelism is equivalent to the Euclidean parallel postulate. This is yet another PG-rated proof.

Theorem 18.14 (Transitive Property of Asymptotic Rays). *Suppose $\overrightarrow{AB}$, $\overrightarrow{CD}$, and $\overrightarrow{EF}$ are three rays such that $\overrightarrow{AB} \mid \overrightarrow{EF}$ and $\overrightarrow{CD} \mid \overrightarrow{EF}$, and assume that $\overrightarrow{AB}$ and $\overrightarrow{CD}$ are noncollinear. Then $\overrightarrow{AB} \mid \overrightarrow{CD}$.*

Proof. The proof is divided into two main cases.

CASE 1. *A and C lie on opposite sides of $\overleftrightarrow{EF}$.* Then there is a point E' where $\overline{AC}$ meets $\overleftrightarrow{EF}$. We can choose a point F' such that the ray $\overrightarrow{E'F'}$ either contains or is contained in $\overrightarrow{EF}$, and then the endpoint independence theorem shows that $\overrightarrow{AB} \mid \overrightarrow{E'F'}$ and $\overrightarrow{CD} \mid \overrightarrow{E'F'}$ (Fig. 18.15).

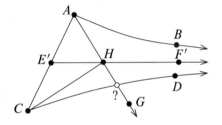

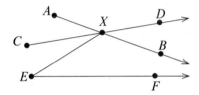

Fig. 18.15. Transitivity: Case 1.

Fig. 18.16. Proof that $\overrightarrow{AB}$ and $\overrightarrow{CD}$ are disjoint.

We will prove that $\overrightarrow{AB} \mid \overrightarrow{CD}$. Property (i) in the definition of asymptotic rays is immediate, because $\overrightarrow{AB}$ and $\overrightarrow{CD}$ both lie on the same side of $\overleftrightarrow{AC}$ as $\overrightarrow{E'F'}$. Property (ii) also holds, because $\overrightarrow{AB}$ and $\overrightarrow{CD}$ are on opposite sides of $\overleftrightarrow{E'F'}$, so they cannot intersect. To prove (iii), suppose $\overrightarrow{AG}$ is any ray between $\overrightarrow{AB}$ and $\overrightarrow{AC}$. The fact that $\overrightarrow{AB} \mid \overrightarrow{E'F'}$ implies that $\overrightarrow{AG}$ intersects $\overrightarrow{E'F'}$ at a point H. Without loss of generality, let us assume that $A * H * G$ and $E' * H * F'$ (which we can achieve by replacing G and F' by different points on the same rays if necessary). Now endpoint independence implies that $\overrightarrow{HF'} \mid \overrightarrow{CD}$. Because $E' * H * F'$, we see that E' and F' are on opposite sides of $\overleftrightarrow{HG}$, and the Y-lemma shows that E' and C are on the same side of that line; thus C and F' are on opposite sides, so $\angle CHG$ and $\angle GHF'$ are adjacent angles. Because the rays $\overrightarrow{HC}$, $\overrightarrow{HG}$, and $\overrightarrow{HF'}$ are all in $\mathrm{HR}\big(\overrightarrow{HF'}, C\big)$, it follows from the adjacency lemma that $\overrightarrow{HC} * \overrightarrow{HG} * \overrightarrow{HF'}$. Since $\overrightarrow{HF'} \mid \overrightarrow{CD}$, this implies that $\overrightarrow{AG}$ intersects $\overrightarrow{CD}$ as needed.

CASE 2: *A and C lie on the same side of $\overleftrightarrow{EF}$.* First we need to rule out the possibility that $\overrightarrow{AB}$ and $\overrightarrow{CD}$ might intersect. Assume for contradiction that X is a point in $\overrightarrow{AB} \cap \overrightarrow{CD}$ (Fig. 18.16). Then (after we replace B and D by other points on the same rays if necessary) endpoint independence implies that $\overrightarrow{XB} \mid \overrightarrow{EF}$ and $\overrightarrow{XD} \mid \overrightarrow{EF}$. The uniqueness part of Theorem 18.11 then implies that $\overrightarrow{XB} = \overrightarrow{XD}$, which contradicts our assumption that $\overrightarrow{AB}$ and $\overrightarrow{CD}$ were noncollinear. Thus $\overrightarrow{AB}$ and $\overrightarrow{CD}$ are disjoint.

To prove the other properties of asymptotic rays, we consider two possibilities, which need to be treated slightly differently. We continue to assume that A and C lie on the same side of $\overleftrightarrow{EF}$.

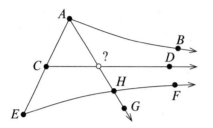

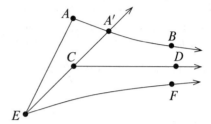

Fig. 18.17. Transitivity: Case 2a.   **Fig. 18.18.** Transitivity: Case 2b.

CASE 2A: *A, C, and E are collinear.* In this case, either $A * C * E$ or $C * A * E$; since the argument is the same in both cases, we can assume it is the former (Fig. 18.17). We will show again that $\overrightarrow{AB} \mid \overrightarrow{CD}$. As before, property (i) is automatic, because $\overrightarrow{AB}, \overrightarrow{CD}$, and $\overrightarrow{EF}$ all lie on the same side of $\overleftrightarrow{AE}$; and we proved (ii) in the preceding paragraph. To prove (iii), let $\overrightarrow{AG}$ be a ray between $\overrightarrow{AB}$ and $\overrightarrow{AC}$. Then $\overrightarrow{AB} \mid \overrightarrow{EF}$ implies that $\overrightarrow{AG}$ intersects $\overrightarrow{EF}$ at a point H. Because A and H are on opposite sides of $\overleftrightarrow{CD}$, there must be a point where $\overrightarrow{AH}$ intersects $\overleftrightarrow{CD}$. Since $\overrightarrow{AH}$ is on the same side of $\overleftrightarrow{AC}$ as $\overrightarrow{CD}$, the intersection point must be on $\overrightarrow{CD}$, not on its opposite ray.

CASE 2B: *A, C, and E are not collinear.* Then the rays $\overrightarrow{EA}, \overrightarrow{EC}$, and $\overrightarrow{EF}$ all lie in $\mathrm{HR}(\overleftrightarrow{EF}, A)$, so the ordering lemma for rays shows that either $\overrightarrow{EF} * \overrightarrow{EC} * \overrightarrow{EA}$ or $\overrightarrow{EF} * \overrightarrow{EA} * \overrightarrow{EC}$, depending on whether $\angle FEA$ or $\angle FEC$ is larger; without loss of generality, say it is the former (Fig. 18.18). Since $\overrightarrow{EF} \mid \overrightarrow{AB}$ and $\overrightarrow{EC}$ is between $\overrightarrow{EF}$ and $\overrightarrow{EA}$, it follows that $\overrightarrow{EC}$ intersects $\overrightarrow{AB}$ at a point A'. We may as well assume (replacing B by a different point on the same ray if necessary) that $A * A' * B$, and then endpoint independence guarantees that $\overrightarrow{A'B} \mid \overrightarrow{EF}$. Now, A', C, and E are collinear (we have drawn the picture as if $A' * C * E$, but we could also have $C * A' * E$), so we are back in the situation of Case 2a, with A replaced by A'. $\qquad\square$

Asymptotic Triangles

There is a type of configuration formed from two rays and a segment that is useful in many situations in hyperbolic geometry. We define an ***asymptotic triangle*** to be any set of points consisting of the union of two asymptotic rays together with the segment connecting their endpoints. Thus if $\overrightarrow{AB} \mid \overrightarrow{CD}$, the union $\overrightarrow{AB} \cup \overline{BC} \cup \overrightarrow{CD}$ is an asymptotic triangle, which we denote by $\sqsubset BACD$ (Fig. 18.19).

The name *asymptotic triangle* is chosen because it is natural to think of this figure as being like an ordinary triangle, one of whose vertices has moved off "to infinity." In the Poincaré disk, for example, the two rays of an asymptotic triangle appear to meet on the boundary of the disk (Fig. 18.20). (Since *triangle* means "three angles," the name is really a misnomer. Probably *asymptotic trigon* or *asymptotic biangle* would be more appropriate; but the name *asymptotic triangle* is well established in this context, probably because it is suggestive of the appearance of the figure.)

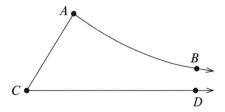

Fig. 18.19. An asymptotic triangle.

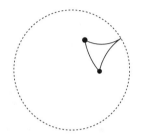

Fig. 18.20. Asymptotic triangle in the Poincaré disk.

In keeping with the analogy between asymptotic triangles and ordinary triangles, we make the following definitions. If $\sqsubset BACD$ is an asymptotic triangle, the segment $\overline{AC}$ is called its *finite edge*, the angles $\angle BAC$ and $\angle ACD$ are called its *angles* or *interior angles*, and the points A and C are called its *vertices*. Two asymptotic triangles are said to be *congruent* if there is a correspondence between their vertices such that the corresponding angles are congruent and the finite edges both have the same length. Thus $\sqsubset BACD \cong \sqsubset B'A'C'D'$ means that $\angle A \cong \angle A'$, $\angle C \cong \angle C'$, and $\overline{AC} \cong \overline{A'C'}$.

If $\sqsubset BACD$ is an asymptotic triangle, we define the *angle sum* and *defect of* $\sqsubset BACD$, denoted respectively by $\sigma(\sqsubset BACD)$ and $\delta(\sqsubset BACD)$, to be the numbers

$$\sigma(\sqsubset BACD) = m\angle A + m\angle C,$$
$$\delta(\sqsubset BACD) = 180 - \sigma(\sqsubset BACD).$$

An *exterior angle of* $\sqsubset BACD$ is any angle that forms a linear pair with either $\angle A$ or $\angle C$ (see Fig. 18.21). Given an exterior angle, the interior angle that is not adjacent to it is called the *remote interior angle* for that exterior angle.

There are many striking ways in which asymptotic triangles behave similarly to ordinary triangles. In this section we will explore several of them. If we think of the missing "angle" at infinity as being a zero-degree angle, then the statements of these theorems should not be too surprising.

Theorem 18.15 (SA Congruence Theorem for Asymptotic Triangles). *Let $\sqsubset BACD$ and $\sqsubset B'A'C'D'$ be asymptotic triangles such that $\overline{AC} \cong \overline{A'C'}$ and $\angle C \cong \angle C'$. Then $\sqsubset BACD \cong \sqsubset B'A'C'D'$.*

Proof. We need only show that $\angle A \cong \angle A'$. If not, then one of these angles is larger; without loss of generality, assume that $m\angle A' < m\angle A$. Let $\overrightarrow{AE}$ be the ray on the same side of $\overleftrightarrow{AC}$ as B and D such that $\angle CAE \cong \angle A'$. Then $\overrightarrow{AE}$ is between $\overrightarrow{AB}$ and $\overrightarrow{AC}$ by the

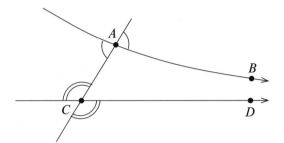

Fig. 18.21. Exterior angles of an asymptotic triangle.

ordering lemma for rays, so the definition of asymptotic rays implies that $\overrightarrow{AE}$ intersects $\overrightarrow{CD}$ at a point F (see Fig. 18.22).

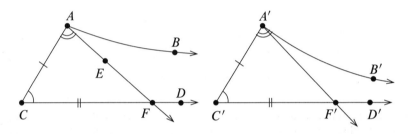

Fig. 18.22. The SA congruence theorem.

Let F' be the point on the ray $\overrightarrow{C'D'}$ such that $\overline{C'F'} \cong \overline{CF}$. It follows from SAS that $\triangle A'C'F' \cong \triangle ACF$, and therefore $\angle C'A'F' \cong \angle CAF$, which in turn is congruent to $\angle C'A'B'$. Thus $\overrightarrow{A'F'} = \overrightarrow{A'B'}$ by the unique ray theorem. This is a contradiction, because $\overrightarrow{A'B'}$ does not intersect $\overrightarrow{C'D'}$. □

Theorem 18.16 (Asymptotic Triangle Copying Theorem). *Suppose $\sqsubset BACD$ is an asymptotic triangle and $\overrightarrow{C'D'}$ is a ray. On each side of $\overleftrightarrow{C'D'}$, there is a ray $\overrightarrow{A'B'}$ asymptotic to $\overrightarrow{C'D'}$ such that $\sqsubset B'A'C'D' \cong \sqsubset BACD$.*

Proof. Exercise 18G. □

Theorem 18.17 (Angle-Sum Theorem for Asymptotic Triangles). *In any asymptotic triangle, the sum of the measures of the two interior angles is strictly less than 180°.*

Proof. Let $\sqsubset BACD$ be an asymptotic triangle, and assume for the sake of contradiction that $m\angle A + m\angle C \geq 180°$.

Suppose first that $m\angle A + m\angle C > 180°$ (Fig. 18.23). Let $\overrightarrow{AE}$ be the ray on the same side of $\overleftrightarrow{AC}$ as B and D such that $m\angle CAE = 180° - m\angle C$. Because $m\angle CAB > 180° - m\angle C$, the ordering lemma for rays implies that $\overrightarrow{AE}$ is between $\overrightarrow{AC}$ and $\overrightarrow{AB}$, so it must intersect $\overrightarrow{CD}$ because $\overrightarrow{AB} \mid \overrightarrow{CD}$. However, the consecutive interior angles theorem implies that $\overleftrightarrow{AE} \parallel \overleftrightarrow{CD}$, so this is a contradiction.

On the other hand, if the angle sum is exactly 180° (Fig. 18.24), let Y be a point such that $Y * C * D$, and let $\overrightarrow{AX}$ be the ray starting at A and asymptotic to $\overrightarrow{CY}$. Then

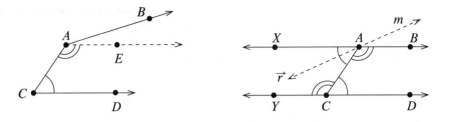

Fig. 18.23. Angle sum greater than 180°. **Fig. 18.24.** Angle sum equal to 180°.

the hypothesis together with the linear pair theorem implies that $\angle ACY \cong \angle CAB$, so the SA congruence theorem implies that $\angle XAC \cong \angle ACD$. By substitution, $m\angle XAC + m\angle CAB = 180°$, so the partial converse to the linear pair theorem implies that X, A, and B are collinear.

Now let m be a line through A and parallel to $\overleftrightarrow{CD}$, different from $\overleftrightarrow{AB}$. One of the rays of m must lie on the same side of $\overleftrightarrow{AB}$ as C; call that ray $\vec{r}$. Then $\vec{r}$ cannot lie on $\overrightarrow{AC}$ (because $\overleftrightarrow{AC}$ intersects $\overleftrightarrow{CD}$), so it must lie in either Int $\angle XAC$ or Int $\angle BAC$, depending on which side of $\overleftrightarrow{AC}$ it lies on. In either case, the definition of asymptotic rays ensures that $\vec{r}$ intersects $\overleftrightarrow{CD}$, contradicting our assumption that $m \parallel \overleftrightarrow{CD}$. Thus the angle sum cannot be equal to 180°. ☐

Theorem 18.18 (Exterior Angle Inequality for Asymptotic Triangles). *The measure of an exterior angle of an asymptotic triangle is strictly greater than the measure of its remote interior angle.*

Proof. Exercise 18H. ☐

Theorem 18.19 (AA Congruence Theorem for Asymptotic Triangles). *Let $\sqsubset BACD$ and $\sqsubset B'A'C'D'$ be asymptotic triangles. If $\angle A \cong \angle A'$ and $\angle C \cong \angle C'$, then $\sqsubset BACD \cong \sqsubset B'A'C'D'$.*

Proof. Exercise 18I. ☐

Theorem 18.20 (SA Inequality). *Suppose $\sqsubset BACD$ and $\sqsubset B'A'C'D'$ are asymptotic triangles with $\angle C \cong \angle C'$. Then $AC > A'C'$ if and only if $m\angle A < m\angle A'$.*

Proof. Exercise 18J. ☐

Theorem 18.21 (Defect Addition for Asymptotic Triangles). *Suppose $\sqsubset BACD$ is an asymptotic triangle, E is a point in Int $\overline{AC}$, and $\overrightarrow{EF} \mid \overrightarrow{AB}$. Then*

$$\delta(\sqsubset BAEF) + \delta(\sqsubset FECD) = \delta(\sqsubset BACD).$$

Proof. Exercise 18K. ☐

Theorem 18.22 (Pasch's Theorem for Asymptotic Triangles). *Suppose $\sqsubset BACD$ is an asymptotic triangle and ℓ is a line that does not contain A or C. If ℓ intersects $\overrightarrow{AB}$, then ℓ also intersects either $\overrightarrow{CD}$ or $\overline{AC}$.*

Proof. Exercise 18L. ☐

Exercises

18A. Given any positive number ε, prove that there exists a triangle with defect less than $\varepsilon°$. [Hint: Use the defect addition theorem.]

18B. Prove Theorem 18.5 (properties of Saccheri quadrilaterals).

18C. Prove Theorem 18.6 (properties of Lambert quadrilaterals).

18D. Prove Theorem 18.9 (side inequalities in Saccheri quadrilaterals).

18E. Prove that if two Saccheri quadrilaterals have congruent bases and congruent summits, then they are congruent. [Hint: If not, make a copy of the smaller quadrilateral inside the larger, and connect the midpoints of their summits.]

18F. Prove that AAASA is a congruence theorem for convex quadrilaterals in hyperbolic geometry.

18G. Prove Theorem 18.16 (the asymptotic triangle copying theorem).

18H. Prove Theorem 18.18 (the exterior angle inequality for asymptotic triangles).

18I. Prove Theorem 18.19 (the AA congruence theorem for asymptotic triangles).

18J. Prove Theorem 18.20 (the SA inequality for asymptotic triangles).

18K. Prove Theorem 18.21 (defect addition for asymptotic triangles).

18L. Prove Theorem 18.22 (Pasch's theorem for asymptotic triangles).

18M. Given any asymptotic triangle $\sqsubset BACD$ and any point $E \in \text{Int}\,\overline{AC}$, prove that there is a unique line through E that does not meet either $\overrightarrow{AB}$ or $\overrightarrow{CD}$. [This shows that the analogue of Pasch's theorem does not apply to a line through the finite side of an asymptotic triangle.]

18N. Given any positive number ε, prove that there exists an asymptotic triangle with defect less than $\varepsilon°$.

Chapter 19

Parallel Lines in Hyperbolic Geometry

In this chapter, we undertake a detailed analysis of parallel lines in hyperbolic geometry. The key observation is that pairs of parallel lines fall into two distinct classes: two lines are said to be ***asymptotically parallel*** if they are parallel and one of them contains a ray that is asymptotic to a ray in the other; they are said to be ***ultraparallel*** if they are parallel but not asymptotically parallel. If ℓ and m are lines, we will use the notation $\ell \mid m$ to mean that ℓ is asymptotically parallel to m. (There is no standard notation for ultraparallel lines.) The notation $\ell \parallel m$ means simply that ℓ and m are parallel, so they could be either asymptotically parallel or ultraparallel. Thanks to Theorem 18.10, any two lines that contain asymptotic rays are automatically asymptotically parallel.

We will see that among all parallels to a given line through a given point, there are exactly two lines that are asymptotically parallel to the given line, and these are "closest" to the given line in a certain sense. All other parallel lines through the given point are ultraparallel. In the Poincaré disk model, asymptotically parallel lines are those that would meet at a point on the boundary of the circle (if the boundary were included in the model), while all other pairs of lines that don't meet are ultraparallel (see Fig. 19.1).

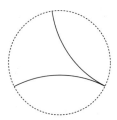
Asymptotically parallel lines.

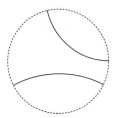
Ultraparallel lines.

Fig. 19.1. Parallel lines in the Poincaré disk model.

Our main task in this chapter is to show that the two classes of parallel lines can be distinguished in three different ways. First, we will classify parallels to a given line ℓ through a given point A not on ℓ in terms of the angles they make with the perpendicular from A to ℓ. Second, we will classify parallel lines according to whether they admit a common perpendicular or not: ultraparallel lines have a common perpendicular, while asymptotically parallel lines do not. Finally, we will classify parallel lines in terms of the distance between them: asymptotically parallel lines approach each other arbitrarily closely in one direction (hence the designation "asymptotic") and diverge in the other direction, while ultraparallel lines diverge from each other in both directions.

Let us begin with a couple of lemmas that will simplify some of the arguments to come. First we record a basic property of asymptotically parallel lines that follows from endpoint independence of asymptotic rays.

Lemma 19.1. *Suppose ℓ and m are asymptotically parallel lines. If A is any point on ℓ and C is any point on m, then one of the rays in ℓ starting at A is asymptotic to the ray in m starting at C and lying on the same side of $\overleftrightarrow{AC}$ (Fig. 19.2).*

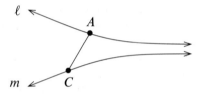

Fig. 19.2. Asymptotically parallel lines.

Proof. The assumption that $\ell \mid m$ means that there are rays $\overrightarrow{PQ} \subseteq \ell$ and $\overrightarrow{XY} \subseteq m$ such that $\overrightarrow{PQ} \mid \overrightarrow{XY}$. First we will show that there exists $B \in \ell$ such that $\overrightarrow{AB} \mid \overrightarrow{XY}$. There are three cases: $A = P$, $A \notin \overrightarrow{PQ}$, or $A \in \operatorname{Int} \overrightarrow{PQ}$. If $A = P$, then we can just take $B = Q$. If $A \notin \overrightarrow{PQ}$, then $\overrightarrow{AQ}$ is a ray that contains $\overrightarrow{PQ}$, and therefore it is asymptotic to $\overrightarrow{XY}$ by endpoint independence, so again we can take $B = Q$. If $A \in \operatorname{Int} \overrightarrow{PQ}$, then we can let B be a point such that $P * A * B$, and it follows that $\overrightarrow{AB} \subseteq \overrightarrow{PQ}$, so $\overrightarrow{AB} \mid \overrightarrow{XY}$ by endpoint independence. A similar argument shows that we can find $D \in m$ such that $\overrightarrow{AB} \mid \overrightarrow{CD}$. By the definition of asymptotic rays, B and D must lie on the same side of $\overleftrightarrow{AC}$. □

The next lemma is a simple consequence of the definition of asymptotic rays. We will use it several times below.

Lemma 19.2. *Suppose $\overrightarrow{AB} \mid \overrightarrow{CD}$ and E is a point on the same side of $\overleftrightarrow{AC}$ as B and D such that $\overrightarrow{AE} \neq \overrightarrow{AB}$ and $\overrightarrow{AE} \cap \overrightarrow{CD} = \varnothing$ (Fig. 19.3). Then $\overrightarrow{AE} * \overrightarrow{AB} * \overrightarrow{AC}$.*

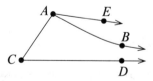

Fig. 19.3. Proof of Lemma 19.2.

Proof. Because $\overrightarrow{AE}$ and $\overrightarrow{AB}$ are both in HR $\left(\overleftrightarrow{AC}, B\right)$ and $\overrightarrow{AE} \neq \overrightarrow{AB}$, it follows that $m\angle EAC \neq m\angle BAC$. If $m\angle EAC < m\angle BAC$, then the ordering lemma for rays implies that $\overrightarrow{AC} * \overrightarrow{AE} * \overrightarrow{AB}$, and then the fact that $\overrightarrow{AB} \mid \overrightarrow{CD}$ implies that $\overrightarrow{AE}$ intersects $\overrightarrow{CD}$, contradicting the hypothesis. The only other possibility is $m\angle EAC > m\angle BAC$, and then the ordering lemma for rays implies $\overrightarrow{AE} * \overrightarrow{AB} * \overrightarrow{AC}$. $\qquad\square$

Parallel Lines and Angles

One way to classify parallels to a given line through a given point is in terms of the angles they make with a perpendicular. To state this classification theorem concisely, let us introduce another definition. If ℓ and m are two lines that intersect at a point A, then there are four angles at A, each made by a ray in ℓ together with a ray in m. In this situation, we define the ***angle measure between the lines*** ℓ ***and*** m to be the smallest of the measures of the angles made by a ray in ℓ and a ray in m. Thus ℓ is perpendicular to m if and only if the angle measure between them is $90°$.

The next proof completely describes the set of parallel lines through a given point not on a line. The proof is rated PG.

Theorem 19.3 (Classification of Parallels through a Point). *Suppose ℓ is a line and A is a point not on ℓ. Let F be the foot of the perpendicular from A to ℓ. There are exactly two lines through A that are asymptotically parallel to ℓ, and they make equal acute angle measures with $\overleftrightarrow{AF}$. A line through A is ultraparallel to ℓ if and only if it makes a larger angle measure with $\overleftrightarrow{AF}$ than the two asymptotically parallel lines, and it intersects ℓ if and only if it makes a smaller angle measure. (See Fig. 19.4.)*

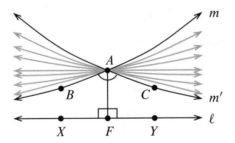

Fig. 19.4. Classification of parallels through a point.

Proof. Choose points $X, Y \in \ell$ such that $X * F * Y$. By the existence and uniqueness of asymptotic rays, there are unique rays $\overrightarrow{AB}$ and $\overrightarrow{AC}$ starting at A such that $\overrightarrow{AB} \mid \overrightarrow{FX}$ and $\overrightarrow{AC} \mid \overrightarrow{FY}$. The lines m and m' containing these rays are asymptotically parallel to ℓ. If there were any other line asymptotically parallel to ℓ through A, then Lemma 19.1 would guarantee that it contained a ray starting at A and asymptotic to one of the rays in ℓ starting at F. But the uniqueness part of Theorem 18.11 guarantees that there are no other such rays, so m and m' are the only lines through A and asymptotically parallel to ℓ.

Let $\theta = m\angle CAF$. The angle-sum theorem for asymptotic triangles (Theorem 18.17) applied to $\sqsubset CAFY$ shows that $\theta < 90°$. Then the SA congruence theorem shows that

$m\angle BAF = \theta$ as well. Thus m and m' are distinct (because otherwise they would form a linear pair with two acute angles), and they make equal acute angle measures with $\overleftrightarrow{AF}$.

Suppose t is a line through A that makes an angle measure smaller than θ with $\overleftrightarrow{AF}$. If $t = \overleftrightarrow{AF}$, then it intersects ℓ at F. Otherwise, t must contain a ray on one side of $\overleftrightarrow{AF}$ or the other that makes an angle measure smaller than θ with $\overrightarrow{AF}$; and then the definition of asymptotic rays ensures that this ray must intersect ℓ.

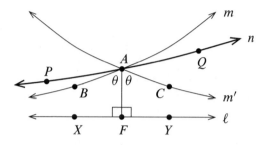

Fig. 19.5. An ultraparallel line through A.

On the other hand, suppose n is any line through A that makes an angle measure with $\overleftrightarrow{AF}$ larger than θ, and choose points P and Q on n such that $P * A * Q$ and P is on the same side of $\overleftrightarrow{AF}$ as X (Fig. 19.5). By definition of the angle measure between two lines, this means that both angles $\angle PAF$ and $\angle QAF$ have measures larger than θ, so the ordering lemma for rays shows that $\overrightarrow{AP} * \overrightarrow{AB} * \overrightarrow{AF}$ and $\overrightarrow{AQ} * \overrightarrow{AC} * \overrightarrow{AF}$. If either $\overrightarrow{AP}$ or $\overrightarrow{AQ}$ were to intersect ℓ, the crossbar theorem would imply that either $\overrightarrow{AB}$ or $\overrightarrow{AC}$ also intersects ℓ, which is a contradiction. Thus $n \parallel \ell$, and because n is not equal to one of the two asymptotically parallel lines, it is ultraparallel to ℓ. $\square$

It is also useful to visualize the parallels through a point in the Poincaré disk model (Fig. 19.6).

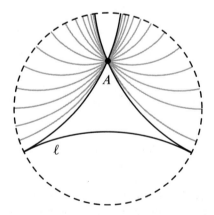

Fig. 19.6. The parallels through a point in the Poincaré disk model.

It is worth noting that even though there is a transitive property of asymptotic rays, the preceding theorem implies that there is no such property of asymptotically parallel lines.

In the situation of the theorem, there are two lines through A that are both asymptotically parallel to ℓ, but they are not asymptotically parallel to each other because they intersect at A. The problem is that they contain rays that are asymptotic to *different* rays in ℓ.

Parallel Lines and Common Perpendiculars

Our next main result in this chapter is a classification of parallel lines according to whether they admit a common perpendicular or not. A familiar feature of parallel lines in Euclidean geometry is that they always have infinitely many common perpendiculars, because any line perpendicular to one of two parallel lines is also perpendicular to the second (Theorem 10.5). In hyperbolic geometry, common perpendiculars are much harder to come by. In this section, we develop the tools for finding them.

Theorem 19.4 (Uniqueness of Common Perpendiculars). *If ℓ and m are parallel lines that admit a common perpendicular, then the common perpendicular is unique.*

Proof. Assume for the sake of contradiction that ℓ and m are parallel lines with two distinct common perpendiculars t and t' (Fig. 19.7). If t and t' meet ℓ at A and A', respectively, and they meet m at B and B', then $AA'B'B$ is a rectangle, which is impossible. $\qquad\square$

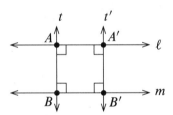

Fig. 19.7. Proof that a common perpendicular is unique.

In fact, as we will see below, there exist parallel lines that admit *no* common perpendiculars. The next theorem gives a useful condition that is sufficient to guarantee that a common perpendicular exists.

Theorem 19.5. *Suppose ℓ and m are parallel lines. If there are two distinct points on ℓ that are equidistant from m, then ℓ and m have a common perpendicular.*

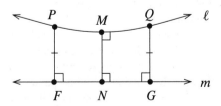

Fig. 19.8. Proof of Theorem 19.5.

Proof. Suppose $\ell \parallel m$, and P and Q are distinct points on ℓ such that $d(P,m) = d(Q,m)$ (Fig. 19.8). Because ℓ and m are parallel, P and Q lie on the same side of m. Let F and G be the feet of the perpendiculars to m from P and Q, respectively. Then $PQGF$ is a trapezoid by the trapezoid lemma, and the assumption that $d(P,m) = d(Q,m)$ implies that it is a Saccheri quadrilateral. If $\overline{MN}$ is the midsegment of $PQGF$, then $\overleftrightarrow{MN}$ is perpendicular to both ℓ and m by Theorem 18.5(d). $\square$

The next theorem and its proof are due to David Hilbert. The proof, which is rated R, draws on most of the hyperbolic results we have proved so far.

Theorem 19.6 (Ultraparallel Theorem). *Two distinct lines are ultraparallel if and only if they admit a common perpendicular.*

Proof. Let ℓ and m be two distinct lines. Assume first that they admit a common perpendicular t. Then they are parallel by the common perpendicular theorem, so they are either ultraparallel or asymptotically parallel. Suppose they are asymptotically parallel. Let A and C be the points where t meets m and ℓ, respectively. By Lemma 19.1, there are points $B \in m$ and $D \in \ell$ such that $\overrightarrow{AB} \mid \overrightarrow{CD}$, and therefore $\square BACD$ is an asymptotic triangle. But $m\angle A + m\angle C = 180°$, which contradicts the angle-sum theorem for asymptotic triangles. The only remaining possibility is that ℓ and m are ultraparallel.

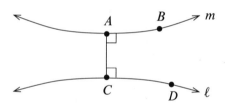

Fig. 19.9. Lines with a common perpendicular must be ultraparallel.

Conversely, assume ℓ and m are ultraparallel. Let A and B be two distinct points on m. Then A and B are on the same side of ℓ by Lemma 7.18. If A and B are equidistant from ℓ, then Theorem 19.5 shows that ℓ and m have a common perpendicular, and we are done. Assume therefore that they are not equidistant; without loss of generality, we may assume that $d(A,\ell) > d(B,\ell)$. We will show that there are two other points on m that are equidistant from ℓ.

Let F and G be the feet of the perpendiculars from A to ℓ and from B to ℓ, respectively, and choose points X and Y such that $A * B * X$ and $F * G * Y$ (Fig. 19.10). Because $AF > BG$, we can choose a point C in the interior of $\overline{AF}$ such that $\overline{CF} \cong \overline{BG}$. Let $\vec{r}$ be the ray starting at C, on the same side of $\overleftrightarrow{AF}$ as B, and making an angle with $\overrightarrow{CF}$ that is congruent to $\angle GBX$. We will show below that $\vec{r}$ intersects $\overrightarrow{AX}$ at a point D. Accepting this for the moment, we continue the proof as follows. Let E be the point on $\overrightarrow{BX}$ such that $\overline{BE} \cong \overline{CD}$, and let K and L be the feet of the perpendiculars from D to ℓ and from E to ℓ, respectively. Since $\overleftrightarrow{EL} \parallel \overleftrightarrow{BG}$ by the common perpendicular theorem, L is on the same side of $\overleftrightarrow{BG}$ as E, which is the opposite side from $\overleftrightarrow{AF}$. In particular, this implies that $F * G * L$. (We are not making any claims about which side of $\overleftrightarrow{BG}$ the points D and K are on; we have pictured them on the same side as $\overleftrightarrow{AF}$, but they might be on the other side or even on $\overleftrightarrow{BG}$. The proof works exactly the same in all three cases.)

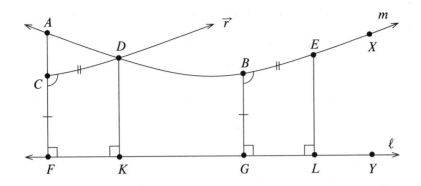

Fig. 19.10. Finding a common perpendicular for ultraparallel lines.

Now, $KFCD$ and $LGBE$ are both trapezoids by the trapezoid lemma, and they are congruent by AASAS. This implies that $DK = EL$, so D and E are equidistant from ℓ. We just need to check that they are distinct. If they were the same point, then K and L would also be equal, and then the congruence $\overline{FK} \cong \overline{GL}$ would imply that L is equidistant from F and G and thus is equal to the midpoint of $\overline{FG}$ by Lemma 3.26. This in turn would imply $F * L * G$, contradicting the conclusion in the previous paragraph. Thus $D \neq E$, so Theorem 19.5 shows that ℓ and m admit a common perpendicular.

It remains only to show that $\overrightarrow{r}$ intersects $\overrightarrow{AX}$. This is the trickiest part of the proof. Introduce three new rays $\overrightarrow{s}, \overrightarrow{u}, \overrightarrow{v}$ as follows: $\overrightarrow{s}$ is the ray starting at F and asymptotic to $\overrightarrow{r}$; $\overrightarrow{u}$ is the ray starting at F and asymptotic to $\overrightarrow{AX}$; and $\overrightarrow{v}$ is the ray starting at G and asymptotic to $\overrightarrow{BX}$. Let R, S, U, V be interior points of $\overrightarrow{r}, \overrightarrow{s}, \overrightarrow{u}$, and $\overrightarrow{v}$, respectively (Fig. 19.11).

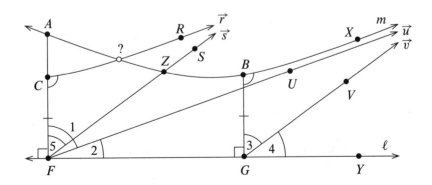

Fig. 19.11. Proving that $\overrightarrow{r}$ intersects $\overrightarrow{AX}$.

The fact that ℓ and m are ultraparallel means that $\overrightarrow{GY}$ is not asymptotic to $\overrightarrow{BX}$, so $\overrightarrow{GY} \neq \overrightarrow{GV}$. Therefore, since $\overrightarrow{GY}$ does not meet $\overrightarrow{BX}$, Lemma 19.2 implies that $\overrightarrow{GB} * \overrightarrow{GV} * \overrightarrow{GY}$. Similar reasoning shows that $\overrightarrow{FA} * \overrightarrow{FU} * \overrightarrow{FY}$. Label angles $\angle 1$ through $\angle 5$ as shown in Fig. 19.11. Then $\angle 1$ and $\angle 2$ are complementary, as are $\angle 3$ and $\angle 4$; but we do not yet know anything about $\angle 5$. We wish to show that $m\angle 5 < m\angle 1$.

Endpoint independence of asymptotic rays implies that $\vec{v} \mid \overrightarrow{AX}$, and then the transitive property of asymptotic rays implies that $\vec{u} \mid \vec{v}$, so $\sqsubset UFGV$ is an asymptotic triangle. The exterior angle inequality for asymptotic triangles implies that $m\angle 4 > m\angle 2$. It follows algebraically that the complements of these two angles satisfy the opposite inequality: $m\angle 3 < m\angle 1$.

Because $\overline{CF} \cong \overline{BG}$ and $\angle C \cong \angle B$, it follows from the SA congruence theorem that $\sqsubset RCFS \cong \sqsubset XBGV$. Thus $m\angle 5 = m\angle 3$, and by substitution, $m\angle 5 < m\angle 1$ as claimed.

From the ordering lemma for rays, we conclude that $\vec{s}$ is between $\vec{u}$ and $\overrightarrow{FA}$. Because $\overrightarrow{FU} \mid \overrightarrow{AX}$, we conclude that $\vec{s}$ intersects $\overrightarrow{AX}$ at a point Z. Consider now the intersection of the line $\overleftrightarrow{r}$ with $\triangle AFZ$: because $\overleftrightarrow{r}$ is parallel to $\overrightarrow{FZ}$ and meets $\overleftrightarrow{AF}$ only at C, it does not contain any of the vertices A, F, or Z and does not meet $\overline{FZ}$. Thus Pasch's theorem shows that $\overleftrightarrow{r}$ intersects $\overline{AZ}$. Because the ray $\vec{r}$ is the part of $\overleftrightarrow{r}$ on the same side of $\overleftrightarrow{AF}$ as Z, the intersection must in fact be in the ray $\vec{r}$. This proves that $\vec{r}$ intersects $\overrightarrow{AX}$ and thus completes the proof of the ultraparallel theorem. ☐

Distances between Parallel Lines

Our final way of classifying parallel lines is in terms of the distances between them. One essential feature of parallel lines in Euclidean geometry is that they are equidistant. In fact, many people think of equidistance as a defining characteristic of parallel lines. In hyperbolic geometry, things are very different, as we will see. In this section, we will show that parallel lines are never equidistant and that asymptotically parallel lines and ultraparallel lines behave very differently with respect to distance.

Our first theorem shows that no two lines can be equidistant in hyperbolic geometry.

Theorem 19.7. *Suppose ℓ and m are two different lines. No three distinct points on ℓ are equidistant from m.*

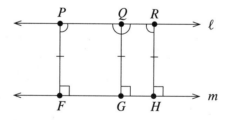

Fig. 19.12. Two lines cannot have three equidistant points.

Proof. Let ℓ and m be distinct lines, and suppose P, Q, and R are three distinct points on ℓ such that $d(P,m) = d(Q,m) = d(R,m)$. Without loss of generality, we may assume that $P * Q * R$. Since ℓ and m are not the same line, these distances cannot all be zero, so none of the points P, Q, R lies on m. At least two of them have to be on the same side of m, so Theorem 7.23 implies that $\ell \parallel m$. Therefore all three points are on the same side of m (because otherwise ℓ and m would have to intersect).

Let F, G, and H be the feet of the perpendiculars to m from P, Q, and R, respectively (Fig. 19.12). Then $PQGF$, $QRHG$, and $PRHF$ are all Saccheri quadrilaterals with bases

$\overline{FG}$, $\overline{GH}$, and $\overline{FH}$, respectively. It follows that the two summit angles of each quadrilateral are congruent: $\angle FPQ \cong \angle PQG$, $\angle GQR \cong \angle QRH$, and $\angle FPQ \cong \angle QRH$. By transitivity, therefore, $\angle PQG \cong \angle GQR$. Since these two angles form a linear pair, they are both right angles. But they are also summit angles of Saccheri quadrilaterals, so they are acute. This is a contradiction. □

Next we prove a simple lemma about distances between *nonparallel* lines: it says that if two rays form an acute angle, then there are points on one ray that are arbitrarily far from the other. This lemma is named after the Greek philosopher Aristotle (384–322 BCE), because he implicitly used it as part of an argument that the universe must have finite size.[1] The following proof is for the case of hyperbolic geometry; but this lemma also holds in Euclidean geometry, with an easier proof (see Exercise 19B).

Lemma 19.8 (Aristotle's Lemma). *Suppose $\vec{r}$ and $\vec{s}$ are rays with the same endpoint such that $\angle rs$ is acute, and c is any positive real number. Then there is a point $P \in \vec{r}$ such that $d\left(P, \overleftrightarrow{s}\right) > c$.*

Proof. Let O be the common endpoint of $\vec{r}$ and $\vec{s}$, and let A be an arbitrary point in the interior of $\vec{r}$ (Fig. 19.13). Let B be the point on $\vec{r}$ such that $O * A * B$ and $AB = OA$. Let F and G be the feet of the perpendiculars from A and B to $\overleftrightarrow{s}$, respectively.

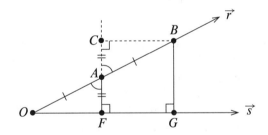

Fig. 19.13. Proof of Aristotle's lemma.

Drop a perpendicular from B to $\overleftrightarrow{AF}$, and call the foot C. Then C cannot be equal to A, because then $\triangle OAF$ would have two right angles; and C cannot lie in the interior of $\overrightarrow{AF}$, because then the right triangle BAC would have an obtuse angle at A. It follows that $C * A * F$ and that $\triangle ABC$ is a right triangle with right angle at C. Because $\angle BAC$ and $\angle OAF$ are vertical angles, they are congruent, and therefore $\triangle OAF \cong \triangle BAC$ by AAS. In particular, this means that $CA = AF$, and therefore $CF = 2AF$. Now $CFGB$ is a Lambert quadrilateral with its fourth vertex at B, and thus Theorem 18.8 implies that $BG > CF = 2AF$. This shows that given any interior point $A \in \vec{r}$, we can find another point $B \in \vec{r}$ whose distance from $\overleftrightarrow{s}$ is more than twice as large as that of A.

---

[1] Aristotle's argument goes roughly like this [**Ari39**, Book 1, Part 5]: Assume the universe is infinitely large. Then any two rays emanating from the center of the earth must have infinite length. Because the distance between the rays becomes arbitrarily large as you go farther out, the distance between their "ends" must be infinite. Because the entire universe obviously(!) rotates about the earth once each day, this means that the "ends" of the rays have to traverse an infinite distance in a finite amount of time, which is impossible.

Repeating the above construction, we can find another point $B' \in \vec{r}$ whose distance is more than twice that of B. After finitely many such constructions, we eventually obtain a point $P \in \vec{r}$ whose distance from $\overleftrightarrow{s}$ is greater than c as claimed. ☐

Aristotle's lemma is the key ingredient in the proof of the following analogous lemma about distances between parallel lines. It is rated PG.

Lemma 19.9. *Suppose ℓ and m are parallel lines and $\vec{r} \subseteq \ell$ is a ray that is not asymptotic to any ray in m. For any positive real number c, there is a point $P \in \vec{r}$ such that $d(P, m) > c$.*

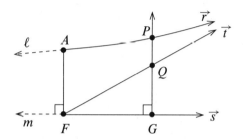

Fig. 19.14. Proof of Lemma 19.9

Proof. Let A be the endpoint of $\vec{r}$, let F be the foot of the perpendicular from A to m, and let $\vec{s}$ be the ray in m starting at F and on the same side of $\overleftrightarrow{AF}$ as $\vec{r}$ (Fig. 19.14). Let $\vec{t}$ be the ray starting at F and asymptotic to $\vec{r}$. If $\vec{t}$ were equal to $\vec{s}$, it would follow that $\vec{s} \mid \vec{r}$ and therefore (by the symmetry property of asymptotic rays) that $\vec{r} \mid \vec{s}$, contradicting the hypothesis. Thus $\vec{t} \neq \vec{s}$, and Lemma 19.2 ensures that $\vec{s} * \vec{t} * \vec{FA}$, so $m\angle st < 90°$.

By Aristotle's lemma, there is a point $Q \in \vec{t}$ such that $d(Q, m) > c$. Let G be the foot of the perpendicular from Q to m. By Pasch's theorem for asymptotic triangles (applied to the asymptotic triangle formed by $\vec{r}$, $\vec{t}$, and $\overline{AF}$), $\overleftrightarrow{QG}$ must intersect either $\overline{AF}$ or $\vec{r}$. Since $\overleftrightarrow{AF} \parallel \overleftrightarrow{GQ}$ by the common perpendicular theorem, the intersection point must lie in $\vec{r}$. Call this point P. Now, G and A are on opposite sides of $\overleftrightarrow{t}$ by Theorem 4.12, and P and A are on the same side of that line because ℓ and $\overleftrightarrow{t}$ are parallel, so it follows that P and G are on opposite sides of $\overleftrightarrow{t}$. Thus $P * Q * G$, and it follows that $PG > QG > c$, so P is the point we seek. ☐

The following theorem shows that the distance between ultraparallel lines is smallest where they meet the common perpendicular and increases without bound in both directions from there.

Theorem 19.10. *Suppose ℓ and m are ultraparallel lines and A is the point where ℓ meets their common perpendicular.*

(a) *If B is any point on ℓ other than A, then $d(B, m) > d(A, m)$.*

(b) *If B and C are points on ℓ such that $A * B * C$, then $d(C, m) > d(B, m)$.*

(c) *Given any positive real number c, on each side of the common perpendicular there exists a point $P \in \ell$ such that $d(P, m) > c$.*

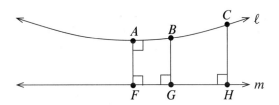

Fig. 19.15. Distances between ultraparallel lines.

Proof. Let F be the foot of the perpendicular from A to m. To prove (a), let B be an arbitrary point on ℓ that is distinct from A, and let G be the foot of the perpendicular from B to m (Fig. 19.15). Then $ABGF$ is a Lambert quadrilateral, so $BG > AF$ by Theorem 18.8. This implies $d(B,m) > d(A,m)$.

Next, to prove (b), suppose B and C are any two points on ℓ such that $A * B * C$, and let G and H be the feet of the perpendiculars from ℓ to m. Because $ABGF$ and $ACHF$ are both Lambert quadrilaterals, their fourth angles $\angle ABG$ and $\angle BCH$ are both acute. Because $\angle CBG$ forms a linear pair with $\angle ABG$, it is obtuse. Therefore, the scalene inequality for quadrilaterals (Lemma 18.7) applied to $BCHG$ implies that $CH > BG$, which means that $d(C,m) > d(B,m)$.

Finally, (c) follows immediately from Lemma 19.9 applied to either of the rays in ℓ starting at A. □

Our final theorem is analogous to the preceding one, but this time for asymptotically parallel lines. Among other things, it explains the geometric significance of the term "asymptotically parallel." It shows that if ℓ and m are asymptotically parallel lines, then the distance from ℓ to m decreases in the direction of their asymptotic rays and becomes arbitrarily small, while it increases in the opposite direction and becomes arbitrarily large. The proof is rated PG.

Theorem 19.11 (Saccheri's Repugnant Theorem). *Suppose ℓ and m are asymptotically parallel lines.*

(a) *If A and B are distinct points of ℓ such that $\overrightarrow{AB}$ is asymptotic to a ray in m, then $d(A,m) > d(B,m)$.*

(b) *If p and q are any positive real numbers, then there are points $P, Q \in \ell$ such that $d(P,m) > p$ and $d(Q,m) < q$.*

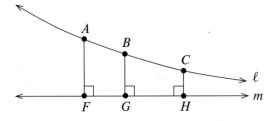

Fig. 19.16. Distances between asymptotically parallel lines.

Proof. To prove (a), suppose A and B are distinct points on ℓ such that $\overrightarrow{AB}$ is asymptotic to a ray in m. Let C be a point such that $A * B * C$, and let F, G, and H be the feet of the perpendiculars to m from A, B, and C, respectively (Fig. 19.16). It follows from endpoint independence that $\overrightarrow{AB} \mid \overrightarrow{FG}$ and $\overrightarrow{BC} \mid \overrightarrow{GH}$, so both $\sqsubset BAFG$ and $\sqsubset CBGH$ are asymptotic triangles. The angle-sum theorem for asymptotic triangles then implies that $\angle BAF$ and $\angle CBG$ are both acute. Because $\angle ABG$ forms a linear pair with $\angle CBG$, it is obtuse. Therefore, $m\angle ABG > m\angle BAF$, and it follows from the scalene inequality for quadrilaterals that $AF > BG$. This proves (a).

To prove (b), suppose p and q are positive real numbers, and let A, B, C, F, G, H be as in the preceding paragraph. The ray $\overrightarrow{BA}$ cannot be asymptotic to any ray in m: if it were, then it would have to be asymptotic to $\overrightarrow{GF}$ by endpoint independence, but then the facts that $\angle ABG$ is obtuse and $\angle BGF$ is right would imply that $\sqsubset ABGF$ violates the angle-sum inequality for asymptotic triangles. Thus the ray $\vec{r} = \overrightarrow{BA}$ satisfies the hypothesis of Lemma 19.9, so that lemma implies that there is a point $P \in \overrightarrow{BA}$ such that $d(P, m) > p$. This proves the first claim of (b).

It remains only to prove the second claim. If $q > BG$, then we can take $Q = B$ and the claim is proved; so assume henceforth that $q \leq BG$. Let c be a real number such that $0 < c < q$, and let X be the point in the interior of $\overline{BG}$ such that $XG = c$. Let $\overrightarrow{XY}$ be the ray starting at X and asymptotic to $\overrightarrow{GF}$, and let W be a point such that $W * X * Y$ (Fig. 19.17).

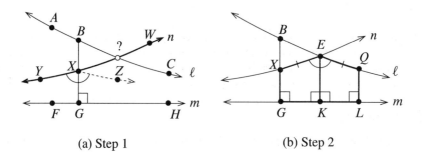

(a) Step 1   (b) Step 2

Fig. 19.17. Proof of Theorem 19.11.

STEP 1: $\overrightarrow{XW}$ *intersects* $\overrightarrow{BC}$. To see this, let $\overrightarrow{XZ}$ be the ray starting at X and asymptotic to $\overrightarrow{GH}$ (Fig. 19.17(a)). It follows from Theorem 19.3 that $\angle YXG \cong \angle ZXG$ and both of these angles are acute. Since $\angle BXZ$ forms a linear pair with $\angle ZXG$, it is obtuse. Therefore, since $\angle BXW$ is acute (by the vertical angles theorem) and on the same side of $\overleftrightarrow{BG}$ as $\angle BXZ$, the ordering lemma for rays guarantees that $\overrightarrow{XB} * \overrightarrow{XW} * \overrightarrow{XZ}$. Now, $\overrightarrow{XZ}$ is also asymptotic to $\overrightarrow{BC}$ by transitivity of asymptotic rays, so $\overrightarrow{XW}$ intersects $\overrightarrow{BC}$ as claimed. Let E be the point of intersection.

STEP 2: *There is a point* $Q \in \ell$ *such that* $d(Q, m) = c$. Let Q be the point on ℓ such that $B * E * Q$ and $EQ = EX$ (Fig. 19.17(b)), and let K and L be the feet of the perpendiculars to m from E and Q, respectively. Then endpoint independence of asymptotic rays implies that $\overrightarrow{EX} \mid \overrightarrow{KG}$ and $\overrightarrow{EQ} \mid \overrightarrow{KL}$, and SA congruence implies that $\angle XEK \cong \angle QEK$. Therefore, the quadrilaterals $GKEX$ and $LKEQ$ are congruent by AASAS. It follows that $QL = XG$, and therefore $d(Q, m) = c < q$. $\qquad\square$

As was mentioned in Chapter 1, in the early 1700s, the Italian mathematician Giovanni Saccheri set out to prove by contradiction that Euclid's fifth postulate follows from the other four. To do so, he assumed that the Euclidean parallel postulate was false and proceeded to prove a great many consequences of that assumption, which we now recognize as theorems in hyperbolic geometry. At the end of his work, he arrived at a result essentially equivalent to Theorem 19.11 above. He then wrote that this is "repugnant to the nature of the straight line," and concluded that therefore his original assumption must have been false. This, of course, does not qualify as a contradiction, because he had not given mathematical definitions for "repugnant" or "the nature of a straight line." In 1733, he published his work, triumphantly titled *Euclides ab Omni Naevo Vindicatus* ("Euclid Freed of Every Flaw") [**Sac86**]. If he had been prepared to recognize the significance of his own discovery instead of falling back at the last minute on his intuition regarding parallel lines, he might have been credited as the originator of non-Euclidean geometry a century before Bolyai, Lobachevsky, and Gauss.

Exercises

19A. Suppose ℓ and m are two lines cut by a transversal that makes a pair of congruent alternate interior angles. Prove the ℓ and m are ultraparallel.

19B. Use similar triangles to prove that Aristotle's lemma (Lemma 19.8) is valid in Euclidean geometry.

19C. Prove that in hyperbolic geometry, there exists a proper angle whose interior contains an entire line. [Hint: Look at Fig. 19.4 for inspiration.]

19D. Prove that in hyperbolic geometry, for every line ℓ, there exists a triangle whose sides are all parallel to ℓ. [Hint: Start with a point $A \notin \ell$; let $\vec{r}$ and $\vec{s}$ be the rays starting at A and asymptotic to rays in ℓ; and let $B \in \vec{r}$ and $C \in \vec{s}$ be two points that are equidistant from A. Prove that all three sides of $\triangle ABC$ are parallel to ℓ.]

19E. Suppose ℓ is a line and A is a point not on ℓ. Theorem 19.3 shows that in hyperbolic geometry, there are exactly two lines through A and asymptotically parallel to ℓ, and they both make the same angle measure with the perpendicular from A to ℓ. This angle measure is called the ***angle of parallelism*** for A and ℓ.

    (a) Show that the angle of parallelism depends only on the distance from A to ℓ: if ℓ' is a line and A' is a point such that $d(A', \ell') = d(A, \ell)$, then the angle of parallelism for A' and ℓ' is the same as that for A and ℓ.

    (b) Show that the angle of parallelism is a decreasing function of distance: if ℓ' is a line and A' is a point such that $d(A', \ell') > d(A, \ell)$, then the angle of parallelism for A' and ℓ' is strictly less than that for A and ℓ. [Hint: Use Theorem 18.20.]

Epilogue: Where Do We Go from Here?

In the preceding chapters, we have explored the roots of axiomatic geometry and have seen some of the ways in which our knowledge of Euclidean and non-Euclidean geometries can be placed on a firm axiomatic foundation. In this final chapter, we take a step back to look around at some of the other ways that geometry can be studied and used. Modern geometry is a vast field of study, with applications in almost every other field of mathematics as well as in countless areas of science and practical endeavors. Now that we have solid ground to stand on, let's take some time to see what we can see from here.

Solid Geometry

In addition to the plane geometry that we have studied in this book, most geometry courses since the time of Euclid have also included a study of three-dimensional figures (traditionally called "solid geometry"). Among the SMSG postulates, there are seven that are specifically tailored for three-dimensional geometry (Postulates 5, 6, 7, 8, 10, 21, and 22 in Appendix C), and most high-school geometry courses include at least a short treatment of three-dimensional geometry. Typical topics include intersection properties of lines and planes in space and volumes and surface areas of geometric solids such as polyhedra, cones, cylinders, and spheres.

We have focused our attention only on plane geometry, because all of the central ideas and techniques of axiomatic geometry are already well illustrated in the study of geometry in two dimensions. However, three-dimensional geometry has obvious practical applications, and now that you have become proficient with the axiomatic approach to geometry, you should find it easy to apply these ideas to solid geometry.

Analytic Geometry and Calculus

Although the axiomatic method presented in this book is extremely efficient for proving most of the basic properties of lines, angles, polygons, and circles that we have discussed, at some point it is necessary to go beyond these simple shapes. For physics, engineering, computer graphics, art, and most mathematical applications, one needs to be able to analyze more complicated geometric figures, starting with ellipses, parabolas, and hyperbolas and continuing with such specialized curves as logarithmic spirals, leminscates, and catenaries. (If you don't know what these are, they're easy to look up online.) Although some of these geometric figures can be studied within the axiomatic framework, the complications grow rapidly and it soon ceases to be fun.

The invention of coordinate geometry by Descartes in the seventeenth century brought powerful new tools to bear on geometric problems. From this point of view, all kinds of curves can be treated in a unified way as solution sets to equations in two variables, so arbitrary geometric shapes can be studied just as easily as lines, angles, polygons, and circles. The crowning achievement of the mathematicians who pursued this point of view was the development of calculus, which provided systematic techniques for analyzing areas, arc lengths, angle measures, shapes of curves, and much more. Many of these techniques extend almost effortlessly to three dimensions or even more.

Because the postulates of Euclidean geometry are categorical, every model of Euclidean geometry is isomorphic to the Cartesian model (see Theorem 13.19). Thus no generality is lost by focusing on the Cartesian model—any statements involving only points, lines, distances, and angle measures that are true in the Cartesian model will be true in every model of Euclidean geometry, and vice versa. For this reason, and because the tools offered by the Cartesian model are so powerful, when mathematicians study any Euclidean phenomena beyond the theorems discussed in this book, they almost invariably do so in the context of the Cartesian model. (Axiomatic Euclidean geometry provides the underlying motivation for the definitions of the Cartesian model and remains the simplest way to understand elementary facts about polygons and circles, so the work you have done in studying this book has by no means been wasted!)

Spherical Geometry

This book has concentrated on Euclidean geometry and hyperbolic geometry, because those are the geometries that can be studied efficiently using some variant of Euclid's axioms. We mentioned in Chapter 1 that spherical geometry can be viewed (if you squint) as an interpretation of the axiom system comprising Euclid's first four postulates plus the elliptic parallel postulate. But we have not pursued spherical geometry axiomatically, because many of the neutral postulates (most notably the ruler postulate) would have to be radically modified in order to hold on the sphere.

This is not to say, however, that spherical geometry is unimportant—after all, it is a much better description of the large-scale geometry of the surface of the planet we live on than either Euclidean or hyperbolic geometry, and the very word *geometry* descends from the Greek word *geometrein*, meaning "to measure the earth." Spherical geometry has vital practical applications (such as navigation, mapmaking, astronomy, and global positioning systems), as well as being a rich and fascinating area of study from a purely mathematical point of view.

It is possible to revise some of the neutral postulates to obtain an axiomatic system that describes the geometry of the sphere (one way to do this is described in [**Gre08**]). But it is much more efficient, and much more powerful, to approach spherical geometry analytically. Because a sphere can be modeled as a certain subset of $\mathbb{R}^3$, all of the tools of three-dimensional analytic geometry and calculus are available. Using these techniques, it is possible to prove many theorems analogous to the ones we have proved in the Euclidean setting, such as results about distances and angles, spherical triangles (all of which have angle sums greater than 180°), areas, and arc lengths. There is an entire branch of trigonometry called *spherical trigonometry*, which studies the relationships between side lengths and angle measures in spherical triangles.

Further Topics in Hyperbolic Geometry

In hyperbolic geometry, in contrast with spherical geometry, the axiomatic approach can take us somewhat farther. As you have seen in Chapters 18–19, it is possible to carry out a detailed qualitative analysis of the theory of parallel lines, for example. However, as soon as one attempts to address more quantitative questions, such as angle sums of specific triangles or circumferences of circles or areas or trigonometric relationships, the axiomatic approach quickly becomes bogged down.

Much more can be done than we have attempted. In particular, one hyperbolic phenomenon that is of central importance but that we have not treated is *area*. To treat area axiomatically in the hyperbolic setting, one need only introduce an area postulate analogous to our Euclidean area postulate. The definition of an area function makes perfectly good sense in hyperbolic geometry. However, the requirement that unit squares have area 1 obviously does not make sense, because there are no squares in hyperbolic geometry. So what should take the place of this property? We could pick some arbitrary figure and assign area 1 to it, such as a 45°-45°-45° triangle; but why that figure and not some other?

This turns out to be a rather deep question, and studying it leads to new insights into hyperbolic geometry. It turns out, surprisingly, that we have already seen an area function in hyperbolic geometry. Recall from Chapter 17 that the *defect* of a polygon is the difference between the expected angle sum in Euclidean geometry and its actual angle sum. It is immediate from the definitions that congruent polygons have the same defect, and the defect addition theorem shows that defects add like areas, at least in the special case of a convex simple region cut into two by a chord. From here, it is relatively straightforward to extend the definition of defect to general polygonal regions and prove that it actually satisfies the two defining properties of an area function.

So defect is an area function, but are there any others? Well, yes: it is easy to check that any constant multiple of the defect is also an area function. (Remember, so far we have not insisted that the area function must give a particular result for any specific region.)

Here is the amazing fact: *in hyperbolic geometry, every area function is a constant multiple of the defect.* This is quite messy to prove from the axioms, which is why we have not done so; but if you are interested, you can find a complete proof in [**MP91**].

Thus, in each model of hyperbolic geometry, there are lots of area functions, but they are just constant multiples of each other. That still leaves open the question of which constant is appropriate. Given that hyperbolic geometry possesses a natural length scale (as described in Chapter 18 after the proof of the AAA congruence theorem), one might expect that there should be a natural choice of units for measuring area too. There is, but it is not completely straightforward to describe. It can be shown that very small hyperbolic triangles have very small defects and therefore look very much like Euclidean triangles. (This is a familiar phenomenon in spherical geometry as well: on the scale of a city block, for example, triangles on the surface of the earth have angle sums that are practically indistinguishable from 180°.) Thus as the side lengths of triangles get smaller, their defects approach zero. A reasonable requirement for an area function, therefore, is that the areas of small triangles should be very close to the the area that the Euclidean area formula would give, $\frac{1}{2}$(base)(height).

In any model of hyperbolic geometry, for each positive real number s there exists an isosceles right triangle with legs of length s. (Just choose points on two perpendicular lines at a distance s from the intersection point.) All such triangles with the same leg lengths are congruent by SAS, so they have the same defect; let $\delta(s)$ denote the defect of any such right triangle. In Euclidean geometry, such a triangle would have area equal to $\frac{1}{2}s^2$. This is not the case in hyperbolic geometry, but it can be shown that in every model of hyperbolic geometry, the following limit exists:

$$k = \lim_{s \to 0} \frac{\frac{1}{2}s^2}{\delta(s)}. \tag{20.1}$$

With this choice of k, a natural definition of area in hyperbolic geometry is given by the *defect formula*:

$$\alpha(\mathcal{P}) = k \cdot \delta(\mathcal{P}).$$

Then (20.1) guarantees that for small isosceles right triangles, the area is better and better approximated by $\frac{1}{2}s^2$. Note that different models of hyperbolic geometry (corresponding to different constant multiples of the distance) will give different values of k.

This formula has lots of interesting consequences. One of the most striking is that in any model of hyperbolic geometry, there is a universal upper bound on the areas of triangles: since defects of triangles are always less than 180° by definition, it follows from the defect formula that the area of every triangle is less than $180k$.

Transformational Geometry

To the ancient Greeks, there was but one geometry: it was the science of shape, and it described the absolute, eternal, true relationships among shapes in the real world. As our understanding of the foundations of geometry (and mathematics) has deepened, we have seen our conception of geometry evolve to encompass ever more varied kinds of geometry: the Cartesian plane, hyperbolic geometry, spherical geometry, finite incidence geometries, to name a few. The variety is fascinating, but it makes one yearn for a unified framework in which to understand them all.

One approach to such a unification is based on the notion of *transformations* of the plane, which are just bijective maps from the plane to itself. These include as an important special case the *rigid motions* described in Appendix I: bijective maps from the plane to itself that preserve all relevant geometric properties (angle measures, betweenness, collinearity, and distances). As Appendix I shows, in neutral geometry one can postulate the existence of certain rigid motions as an alternative to the SAS postulate.

Whichever approach is taken to the axioms, rigid motions form a fascinating subject of study in their own right. They also lead naturally to a study of symmetry: in any geometry, one can define a *symmetry* of a geometric figure to be a rigid motion that takes the figure to itself. For example, a Euclidean square has rotational symmetries about its center and reflection symmetries across its diagonals. The set of all symmetries of a given figure forms a structure known as a *group* and is called the *symmetry group* of the object. In general, a group is a set of unspecified objects endowed with a "multiplication" operation that is associative but not necessarily commutative, that has an identity, and such that each object has an inverse. Groups are one of the main topics in the mathematical field known as *abstract algebra*. For symmetry groups, the "multiplication" is actually composition of symmetries, the identity is the transformation that takes every point to itself, and the inverse of any symmetry is the transformation that "undoes" the symmetry by taking every point back to where it started.

The study of transformations has become quite popular in geometry courses in North American secondary schools, and most such courses include some treatment of rigid motions and symmetries regardless of whether the axiom system is based on the SAS postulate or the reflection postulate.

In the late nineteenth century, shortly after the discovery of hyperbolic geometry, the German mathematician Felix Klein (whose name adorns the Beltrami–Klein model of hyperbolic geometry) suggested a unifying framework for understanding all of the geometries that were known at the time (Euclidean, hyperbolic, and spherical geometries, among others), based on a generalization of rigid motions. He observed that for each of the known geometries, the set of all rigid motions forms a group, called the *symmetry group of the geometry*, and the symmetry groups of the different geometries have very different algebraic properties. In an influential paper originally published in 1872 [**Kle93**], Klein proposed using the symmetry groups themselves as a unifying principle for classifying geometries. Thus we now define a *Klein geometry* to be a set of "points" together with a group of bijective maps from that set to itself, called *symmetries of the geometry*, with the property that given any two points in the set, there is a symmetry that takes one to the other. Euclidean, hyperbolic, and spherical geometries can all be viewed as Klein geometries, as can many others. This approach to unifying geometries is now known as the *Erlangen program*, because Erlangen, Germany, is where Klein originally published his proposal.

The Erlangen program did not turn out to be the basis for the grand unification of geometry that Klein had hoped it would be; that honor went to *differential geometry*, described below. However, the mathematical philosophy that Klein espoused—characterizing a mathematical subject in terms of the properties preserved by a group of symmetries—has become enormously influential and pervades most fields of mathematics today.

Differential Geometry

Klein's proposal did not become the universal organizing principle for geometry as he had hoped, primarily because mathematicians soon realized that there were even more general notions of geometry that did not fit into the Klein program because they were not sufficiently uniform. For example, although spherical geometry provides a good approximate model for the surface of the earth, it breaks down at a certain level of detail because the earth is not perfectly round. There is another more general framework that subsumes Klein geometries and opens the door to much more general geometries.

To understand what it is, it is worth going back to the time of Gauss, one of the original discoverers of non-Euclidean geometry. Around the same time as he was exploring non-Euclidean geometry, in a completely separate series of investigations, Gauss had been using the tools of calculus to study the geometric properties of curves and surfaces and had developed some highly sophisticated techniques for the purpose. (It apparently didn't occur to him to make the connection between his surface investigations and non-Euclidean geometry.)

Gauss's techniques were refined and generalized by successive generations of mathematicians and gradually coalesced into a systematic framework for studying geometry in just about any guise in which it might manifest itself. That framework is known as *differential geometry*. It uses the tools of calculus to study shapes of curves, surfaces, and their higher-dimensional analogues. It is possible to use differential geometry to construct models of Euclidean geometry, hyperbolic geometry, spherical geometry, as well as every Klein geometry, and to compare them and classify them. Most of the theorems of hyperbolic geometry that we have mentioned but not proved, such as the defect formula for area, can be proved straightforwardly using the tools of differential geometry.

Forty years after the discovery of non-Euclidean geometry, Beltrami used differential geometry to construct models of hyperbolic geometry and to show that his models were all isomorphic to each other. Using these tools, one can also show that, although the axioms for hyperbolic geometry are not categorical, the only flexibility is the one we have already discussed: given any two models of hyperbolic geometry, it is always possible to find a positive constant c such that if the distances in one model are all multiplied by c, then the models are isomorphic. This is very similar to spherical geometry—two spheres of different radii are nonisomorphic geometries, but it is always possible to multiply the distances in one model by a constant so that they become isomorphic.

Thus a reasonable next step for those who want to continue their study of geometry toward the frontiers of modern research is to look into differential geometry. A good place to start at the undergraduate level is the book [**dC76**], which treats the differential geometry of curves and surfaces. There you will find the tools you need to make a deep study of spherical geometry, to construct all of the various models of hyperbolic geometry and understand the relationships among them, and much more. Don't stop here—a vast world of geometry awaits your exploration!

Hilbert's Axioms

David Hilbert first published his axiomatic system for Euclidean geometry in 1899, and then he repeatedly revised it thereafter. The list below is based on the tenth German edition of Hilbert's book, using the 1971 translation by Leo Unger [**Hil71**] (slightly edited to use more familiar notation and terminology).

In Hilbert's axiomatic system, the primitive objects are ***point***, ***line***, and ***plane***; and the primitive relations are ***lies on***, ***between***, and ***congruent***.

Group I: Axioms of Incidence

I.1 *For every two points A, B there exists a line that contains each of the points A, B.*

I.2 *For every two points A, B there exists no more than one line that contains each of the points A, B.*

I.3 *There exist at least two points on each line. There exist at least three points that do not lie on any line.*

I.4 *For any three points A, B, C that do not lie on the same line, there exists a plane P that contains each of the points A, B, C. For every plane, there exists a point which it contains.*

I.5 *For any three points A, B, C that do not line on one and the same line, there exists no more than one plane that contains each of the three points A, B, C.*

I.6 *If two points of a line ℓ lie in a plane P, then every point of ℓ lies in the plane P.*

I.7 *If two planes P, Q have a point A in common, then they have at least one more point B in common.*

I.8 *There exist at least four points which do not lie in a plane.*

Group II: Axioms of Order

II.1 *If a point B lies between a point A and a point C, then the points A, B, C are three distinct points of a line, and B then also lies between C and A.*

II.2 *For two points A and C, there always exists at least one point B on the line $\overleftrightarrow{AC}$ such that C lies between A and B.*

II.3 *Of any three points on a line, there exists no more than one that lies between the other two.*

II.4 *Let A, B, C be three points that do not lie on a line, and let ℓ be a line in the plane ABC which does not meet any of the points A, B, C. If the line ℓ passes through a point of the segment $\overline{AB}$, it also passes through a point of the segment $\overline{AC}$ or through a point of the segment $\overline{BC}$.*

Group III: Axioms of Congruence

III.1 *If A, B are two points on a line ℓ and A' is a point on the same line or on another line ℓ', then it is always possible to find a point B' on a given ray of ℓ' starting at A' such that $\overline{AB} \cong \overline{A'B'}$.*

III.2 *If segments $\overline{A'B'}$ and $\overline{A''B''}$ are congruent to the same segment $\overline{AB}$, then $\overline{A'B'}$ and $\overline{A''B''}$ are congruent to each other.*

III.3 *On a line ℓ, let $\overline{AB}$ and $\overline{BC}$ be two segments which, except for B, have no points in common. Furthermore, on the same line or another line ℓ', let $\overline{A'B'}$ and $\overline{B'C'}$ be two segments which, except for B', have no points in common. In that case, if $\overline{AB} \cong \overline{A'B'}$ and $\overline{BC} \cong \overline{B'C'}$, then $\overline{AC} \cong \overline{A'C'}$.*

III.4 *Let $\angle rs$ be an angle in a plane P and ℓ' a line in a plane P', and let a definite side of ℓ' in P' be given. Let $\overrightarrow{r'}$ be a ray on ℓ' starting at a point O'. Then there exists in the plane P' one and only one ray $\overrightarrow{s'}$ such that $\angle r's' \cong \angle rs$ and at the same time all the interior points of $\angle r's'$ lie on the given side of ℓ'.*

III.5 *(SAS) If for two triangles $\triangle ABC$ and $\triangle A'B'C'$ the congruences $\overline{AB} \cong \overline{A'B'}$, $\overline{AC} \cong \overline{A'C'}$, and $\angle BAC \cong \angle B'A'C'$ hold, then the congruence $\angle ABC \cong \angle A'B'C'$ is also satisfied.*

Group IV: Axiom of Parallels

IV.1 *(EUCLID'S AXIOM) Let ℓ be any line and A a point not on it. Then there is at most one line in the plane determined by ℓ and A that passes through A and does not intersect ℓ.*

Group V: Axioms of Continuity

V.1 *(ARCHIMEDES' AXIOM) If $\overline{AB}$ and $\overline{CD}$ are any segments, then there exists a number n such that n copies of $\overline{CD}$ constructed contiguously from A along the ray $\overrightarrow{AB}$ will pass beyond the point B.*

V.2 *(LINE COMPLETENESS) An extension of a set of points on a line with its order and congruence relations that would preserve the relations existing among the original elements as well as the fundamental properties of line order and congruence that follow from Axiom Groups I–III and from Axiom V.1 is impossible.*

Birkhoff's Postulates

These postulates are taken from the 1932 article *A set of postulates for plane geometry, based on scale and protractor*, by George D. Birkhoff [**Bir32**]. In Birkhoff's system, the primitive terms are ***point***, ***line***, ***distance***, and ***angle***.

Postulate I (The Postulate of Line Measure). *The points $A, B, \ldots$ of any line ℓ can be put into one-to-one correspondence with the real numbers x so that $|x_B - x_A| = d(A, B)$ for all points A, B.*

Postulate II (The Point-Line Postulate). *One and only one straight line ℓ contains two given points P, Q ($P \neq Q$).*

Postulate III (The Postulate of Angle Measure). *The half-lines $\ell, m, \ldots$ through any point O can be put into one-to-one correspondence with the real numbers $a \pmod{2\pi}$ so that, if $A \neq O$ and $B \neq O$ are points on ℓ and m, respectively, the difference $a_m - a_l \pmod{2\pi}$ is $\angle AOB$. Furthermore, if the point B on m varies continuously in a line r not containing the vertex O, the number a_m varies continuously also.*

Postulate IV (The Similarity Postulate). *If in two triangles, $\triangle ABC$, $\triangle A'B'C'$, and for some constant $k > 0$, $d(A', B') = k\,d(A, B)$, $d(A', C') = k\,d(A, C)$, and also $\angle B'A'C' = \pm\angle BAC$, then also $d(B', C') = k\,d(B, C)$, $\angle C'B'A' = \pm\angle CBA$, $\angle A'C'B' = \pm\angle ACB$.*

The SMSG Postulates

These axioms, published in the 1961 textbook [**SMSG**] by the School Mathematics Study Group, form the basis for most axiom systems used in U.S. high-school geometry texts today. In the SMSG axiom system, the primitive terms are ***point***, ***line***, and ***plane***.

Postulate 1. *Given any two distinct points there is exactly one line that contains them.*

Postulate 2 (Distance Postulate). *To every pair of distinct points there corresponds a unique positive number. This number is called the distance between the two points.*

Postulate 3 (Ruler Postulate). *The points of a line can be placed in a correspondence with the real numbers such that:*

1. *To every point of the line there corresponds exactly one real number.*
2. *To every real number there corresponds exactly one point of the line.*
3. *The distance between two distinct points is the absolute value of the difference of the corresponding real numbers.*

Postulate 4 (Ruler Placement Postulate). *Given two points P and Q of a line, the co-ordinate system can be chosen in such a way that the coordinate of P is zero and the coordinate of Q is positive.*

Postulate 5.

(a) *Every plane contains at least three noncollinear points.*

(b) *Space contains at least four noncoplanar points.*

Postulate 6. *If two points lie in a plane, then the line containing these points lies in the same plane.*

Postulate 7. *Any three points lie in at least one plane, and any three noncollinear points lie in exactly one plane.*

Postulate 8. *If two planes intersect, then that intersection is a line.*

Postulate 9 (Plane Separation Postulate). *Given a line and a plane containing it, the points of the plane that do not lie on the line form two sets such that:*

1. *Each of the sets is convex.*
2. *If P is in one set and Q is in the other, then segment $\overline{PQ}$ intersects the line.*

Postulate 10 (Space Separation Postulate). *The points of space that do not lie in a given plane form two sets such that:*

1. *Each of the sets is convex.*
2. *If P is in one set and Q is in the other, then segment $\overline{PQ}$ intersects the plane.*

Postulate 11 (Angle Measurement Postulate). *To every angle there corresponds a real number between $0°$ and $180°$.*

Postulate 12 (Angle Construction Postulate). *Let $A\overrightarrow{B}$ be a ray on the edge of the half-plane H. For every r between $0°$ and $180°$, there is exactly one ray $\overrightarrow{AP}$ with P in H such that $m\angle PAB = r$.*

Postulate 13 (Angle Addition Postulate). *If D is a point in the interior of $\angle BAC$, then $m\angle BAC = m\angle BAD + m\angle DAC$.*

Postulate 14 (Supplement Postulate). *If two angles form a linear pair, then they are supplementary.*

Postulate 15 (SAS Postulate). *Given a one-to-one correspondence between two triangles (or between a triangle and itself). If two sides and the included angle of the first triangle are congruent to the corresponding parts of the second triangle, then the correspondence is a congruence.*

Postulate 16 (Parallel Postulate). *Through a given external point there is at most one line parallel to a given line.*

Postulate 17. *To every polygonal region there corresponds a unique positive real number called the area.*

Postulate 18. *If two triangles are congruent, then the triangular regions have the same area.*

Postulate 19. *Suppose that the region R is the union of two regions R_1 and R_2. If R_1 and R_2 intersect at most in a finite number of segments and points, then the area of R is the sum of the areas of R_1 and R_2.*

Postulate 20. *The area of a rectangle is the product of the length of its base and the length of its altitude.*

Postulate 21. *The volume of a rectangular parallelepiped is equal to the product of the length of its altitude and the area of its base.*

Postulate 22 (Cavalieri's Principle). *Given two solids and a plane. If for every plane that intersects the solids and is parallel to the given plane the two intersections determine regions that have the same area, then the two solids have the same volume.*

The Postulates Used in This Book

Postulates of Neutral Geometry

The primitive terms in neutral geometry are ***point***, ***line***, ***distance*** (between points), and ***measure*** (of an angle).

Postulate 1 (The Set Postulate). *Every line is a set of points, and there is a set of all points called **the plane**.*

Postulate 2 (The Existence Postulate). *There exist at least three distinct noncollinear points.*

Postulate 3 (The Unique Line Postulate). *Given any two distinct points, there is a unique line that contains both of them.*

Postulate 4 (The Distance Postulate). *For every pair of points A and B, the distance from A to B is a nonnegative real number determined by A and B.*

Postulate 5 (The Ruler Postulate). *For every line ℓ, there is a bijective function $f : \ell \to \mathbb{R}$ with the property that for any two points $A, B \in \ell$, we have*

$$AB = |f(B) - f(A)|.$$

Postulate 6 (The Plane Separation Postulate). *For any line ℓ, the set of all points not on ℓ is the union of two disjoint subsets called the **sides of ℓ**. If A and B are distinct points not on ℓ, then A and B are on the same side of ℓ if and only if $\overline{AB} \cap \ell = \varnothing$.*

Postulate 7 (The Angle Measure Postulate). *For every angle $\angle ab$, the measure of $\angle ab$ is a real number in the closed interval $[0, 180]$ determined by $\angle ab$.*

Postulate 8 (The Protractor Postulate). *For every ray $\overrightarrow{r}$ and every point P not on $\overleftrightarrow{r}$, there is a bijective function $g : \mathrm{HR}(\overrightarrow{r}, P) \to [0, 180]$ that assigns the number 0 to $\overrightarrow{r}$ and the number 180 to the ray opposite $\overrightarrow{r}$ and such that if $\overrightarrow{a}$ and $\overrightarrow{b}$ are any two rays in $\mathrm{HR}(\overrightarrow{r}, P)$,* *then*

$$m\angle ab = \left| g\left(\vec{b}\right) - g\left(\vec{a}\right) \right|.$$

Postulate 9 (The SAS Postulate). *If there is a correspondence between the vertices of two triangles such that two sides and the included angle of one triangle are congruent to the corresponding sides and angle of the other triangle, then the triangles are congruent under that correspondence.*

Postulates of Euclidean Geometry

Postulates 1–9 of neutral geometry, plus the following:

Postulate 10E (The Euclidean Parallel Postulate). *For each line ℓ and each point A that does not lie on ℓ, there is a unique line that contains A and is parallel to ℓ.*

Postulate 11E (The Euclidean Area Postulate). *There exists a unique area function α with the property that $\alpha(\mathcal{R}) = 1$ whenever $\mathcal{R}$ is a square region with sides of length 1.*

Postulates of Hyperbolic Geometry

Postulates 1–9 of neutral geometry, plus the following:

Postulate 10H (The Hyperbolic Parallel Postulate). *For each line ℓ and each point A that does not lie on ℓ, there are at least two distinct lines that contain A and are parallel to ℓ.*

The Language
of Mathematics

In this appendix, we describe in some detail the language mathematicians use to express theorems and their proofs. After the discovery of non-Euclidean geometry prompted a reevaluation of the way mathematical terminology was used by Euclid and other mathematical writers (see Chapter 1), mathematicians realized that the language we use to state theorems and construct their proofs must have a precise, unambiguous meaning. Over the past hundred years or so, mathematicians have developed a system of conventions for expressing mathematical ideas, which is designed to minimize the possibility of ambiguity.

The purpose of this appendix is to describe the "grammar" for constructing and interpreting mathematical arguments. As we do so, it is good to bear in mind that our discussion about mathematical statements and their logic is not itself a rigorous mathematical discussion—in particular, we are not going to try to treat mathematical logic itself as an axiomatic system. (It can be done, but that's another subject entirely.) This means that some of the terms we use to describe statements will not get rigorous definitions and the claims we make about the validity of mathematical deductions will not get rigorous proofs. Instead, we will describe the rules for constructing and interpreting mathematical statements and appeal to common sense and everyday experience with logical reasoning to justify them.

Simple Statements

The basic building blocks of mathematical proofs are ***mathematical statements***. These are assertions of facts about "mathematical objects"—things like numbers, points, lines, vectors, sets, functions, angles, triangles—the usual flora and fauna of mathematical discourse.

Most importantly, a mathematical statement must be *unambiguous*, and it must be *either true or false, but not both*. (For any particular statement, we might not *know* whether it is true or false, but it must in principle be one or the other.) Thus an English sentence such as "This statement is false" cannot be considered as a mathematical statement, because if it were true, we would have to conclude that it was false; and if it were false, we would have to conclude that it was true. Similarly, "The number 42 is interesting" cannot be considered as a mathematical statement unless we are prepared to give an unambiguous definition of the concept of an "interesting number."

Mathematical statements can be wonderfully complex, but the complicated ones are all built up in a systematic way from simpler ones. The basic building blocks are ***simple statements*** (also called ***atomic statements***), which are ones that cannot be expressed as combinations of smaller statements. They can be written using mathematical symbols, or words in English (or some other natural language), or some combination of the two. Here are some examples of simple statements (in these examples, we assume that a, b, and c represent specific real numbers; A, B, and C represent points; and ℓ and m represent lines):

- $2 + 2 = 4$.

- $0 < 1$.

- $\sin \pi = 0$.

- $\pi = 3$.

- $a \in \mathbb{R}$.

- $\ell \perp m$.

- A lies on ℓ.

- $\dfrac{-b + \sqrt{b^2 - 4ac}}{2a} = 5$.

- a is positive.

- $\angle ABC$ is acute.

- $\angle ABC$ is a right angle.

Simple mathematical statements typically come in two basic types. The first type expresses a mathematical relationship between two mathematical objects. Some of the mathematical objects that appear in the sentences above are $2 + 2$, 4, 0, 1, π, $\sin \pi$, a, $\mathbb{R}$, ℓ, m, A, and $\angle ABC$. The "verb" in such a statement is a *mathematical relation*, such as "equals" ($=$), "is less than" ($<$), "is an element of" ($\in$), "is perpendicular to" ($\perp$), or "lies on" (no symbol). All the sentences above except the last three are of this type.

The other type of simple statement has "is" as its verb and expresses the fact that a particular mathematical object has a particular property (such as "acute" or "positive") or is a particular type of object (such as a "right angle"). The last three sentences are of this type.

Notice that not all of the statements in the list above are true—for example, the statement "$\pi = 3$" is patently false; while we do not know if the last seven statements are true until we are told what the letters $a, b, c, \ell, m, A, B, C$ stand for. However, once particular values for all of these letters are chosen, they are all perfectly good mathematical statements regardless of whether they are true or false.

Logical Connectives

To express more complicated mathematical thoughts, we can combine simple statements together to create **compound statements**: these are statements built up from simple statements using **logical connectives**, which are words that function somewhat like grammatical conjunctions in the structure of mathematical statements. There are five basic logical connectives: *and*, *or*, *not*, *implies*, and *if and only if*.

Conjunctions

The easiest logical connective to understand is **and**. If P and Q are any mathematical statements whatsoever, then "P and Q" is another mathematical statement, called the **conjunction of P and Q**. It is sometimes symbolized by $P \wedge Q$. For example, we might combine the first two statements in our list above to form the statement "$2 + 2 = 4$ and $0 < 1$."

Whether a conjunction is true or not depends only on whether the individual statements that are being combined are true. As the wording suggests, the conjunction "P and Q" is true when P and Q are both true, and it is false otherwise. This rule is summarized in a display called a *truth table*. Such a table lists every possible combination of truth values of the individual statements P and Q, together with the resulting truth value of the compound statement "P and Q." Here is the truth table for "and":

P	Q	$P \wedge Q$
T	T	T
T	F	F
F	T	F
F	F	F

(E.1)

Although logical symbols like "$\wedge$" are useful for analyzing the structure of mathematical logic, when mathematicians write mathematical statements in English (as we ordinarily do when writing up proofs for human consumption), it is customary to avoid logical symbols and write out the logical connectives using English words. Conjunctions are usually written using the word "and"; but there are a few other ways they can be expressed, some of which might not at first appear to be conjunctions. All of the following sentences express the conjunction of the two statements "a is a positive number" and "$a < 10$"; in each case, the italicized word(s) convey the same mathematical meaning as *and*:

- a is a positive number *and* $a < 10$.

- a is a positive number *in addition to* being less than 10.

- a is a positive number *such that* $a < 10$.

- a is a positive number *but* $a < 10$.

- a is a positive number less than 10.

(As the last example shows, sometimes there is not any word at all that explicitly marks a sentence as a conjunction.)

Disjunctions

The next logical connective is **or**. If P and Q are mathematical statements, then "P or Q" is a mathematical statement, called the **disjunction of P and Q**. It is symbolized by $P \vee Q$.

Before we describe the mathematical meaning of a disjunction, let's look at the way "or" is used in everyday speech. In ordinary English, "or" might be interpreted in different ways, depending on the context: most often it is interpreted as *exclusive or*, which means "one or the other, but not both"; while other times it is interpreted as *inclusive or*, meaning "one or the other or both." For example, if you're eating at Joe's diner and Joe tells you, "The sandwich comes with soup or salad," you will understand immediately that you can have one or the other, but not both. On the other hand, if you're applying for a job and Joe says, "You need to have a high-school diploma or two years' experience," you will understand equally well that it is fine if you have both.

For mathematical statements, this kind of ambiguity and dependence on context will not do. In order to interpret a sentence like "a is an integer or $a < 10$" as a mathematical statement, we must be able to agree on an unambiguous, well-defined meaning that depends only on the truth or falsity of the simple statements that make it up ("a is an integer" and "$a < 10$"), not on the context in which it's written.

For these reasons, mathematicians have agreed on the convention that the word "or" in a mathematical statement always means *inclusive or*. Thus if P and Q are statements, then "P or Q" is a statement that is true if P is true or if Q is true or if both are true; and otherwise it is false. Its precise meaning is captured in the following truth table:

P	Q	$P \vee Q$
T	T	T
T	F	T
F	T	T
F	F	F

There are only two common ways of expressing disjunctions in English sentences:

- a is an integer or $a < 10$.

- Either a is an integer or $a < 10$.

Note that the addition of "either" does not change the fact that the "or" in a mathematical statement is an *inclusive or*. If you want to express *exclusive or* in a mathematical context, the only way to do so is to say explicitly what you mean: "a is an integer or $a < 10$, but not both."

In the strict grammar of mathematical statements, logical connectives such as "and" and "or" can be used only to connect *statements*, not to connect mathematical objects, properties, or relations. In an English sentence, we might say "a is positive or zero," but to analyze that sentence we have to recognize it as a disjunction in which two complete statements are joined by "or": "either a is positive or a is zero." Similarly, equations involving the "plus or minus" symbol are secretly disjunctions: the statement $x = \pm 1$ really means "either $x = +1$ or $x = -1$."

Negation

The third logical connective is the word ***not***; it is in some ways the simplest, but it can present surprising challenges. If P is a mathematical statement, then "not P," symbolized by $\neg P$ (or $\sim P$ in some books), is also a mathematical statement, called the ***negation of P***. It is false when P is true and true when P is false, as summarized in this truth table:

P	$\neg P$
T	F
F	T

When writing mathematical statements in English, we usually express the negation of a statement by inserting the word "not" at a grammatically appropriate place in the sentence; but when necessary, it can also be expressed by preceding the sentence with a negating phrase such as "it is not the case that ... ." The negation of a symbolic relation (like $=$, $<$, or $\in$) is usually expressed by placing a slash through the relation symbol (as in $\neq$, $\not<$, or $\notin$). Each of the following sentences expresses the negation of "a is equal to 0":

- a is not equal to 0.
- $a \neq 0$.
- It is not the case that $a = 0$.
- It is not true that $a = 0$.
- It is false that $a = 0$.

Negation is easy to understand when it is applied to a simple statement; but when it is applied to a compound statement, things can get complicated. It is important to know how to simplify the negations of compound statements.

To negate a conjunction $P \wedge Q$, we just look at the truth table (E.1) to see that $P \wedge Q$ is false whenever P is false or Q is false or both are false. This suggests that the negation of "P and Q" should have the same truth values as "not P or not Q." The following truth table shows that this is indeed the case:

P	Q	$P \wedge Q$	$\neg(P \wedge Q)$	$\neg P$	$\neg Q$	$\neg P \vee \neg Q$
T	T	T	F	F	F	F
T	F	F	T	F	T	T
F	T	F	T	T	F	T
F	F	F	T	T	T	T

Thus wherever the negation of a statement such as "P and Q" occurs, we can freely replace it with the statement "not P or not Q." A similar analysis shows that the negation of $P \vee Q$ has the same truth value as $\neg P \wedge \neg Q$:

P	Q	$P \vee Q$	$\neg(P \vee Q)$	$\neg P$	$\neg Q$	$\neg P \wedge \neg Q$
T	T	T	F	F	F	F
T	F	T	F	F	T	F
F	T	T	F	T	F	F
F	F	F	T	T	T	T

In brief, *the negation of a conjunction is the disjunction of the negations*, and *the negation of a disjunction is the conjunction of the negations*.

Of course, we can also negate a negation. Because every statement is either true or false but not both, the negation of a negated statement has the same truth value as the original statement. That is, $\neg(\neg P)$ has the same truth value as P, as the following truth table shows:

P	$\neg P$	$\neg(\neg P)$
T	F	**T**
F	T	**F**

Conditional Statements

The most important logical connective in mathematics is **implies**. If P and Q are mathematical statements, then "P implies Q" is a mathematical statement, symbolized by $P \Rightarrow Q$ (or, in some books, $P \rightarrow Q$). The expression "If P, then Q" means exactly the same thing. Any such compound statement is called an **implication** or a **conditional statement**. In the conditional statement $P \Rightarrow Q$, the statement P is called the **hypothesis**, and Q is called the **conclusion**.

In everyday English, the words "implies" and "if/then" carry many shades of meaning, such as causation or temporal proximity, that would be difficult or impossible to interpret in the bright unambiguous light of mathematics. When the veterinarian says, "If your cat has eaten poison, then she will die," you will probably understand that the vet is expressing a causal relationship (as well as a short time span) between the hypothesis and the conclusion. Contrast this with the sentence "If your cat is white, then she will die." A (perhaps slightly pedantic) case can be made that this sentence is true, because if your cat does happen to be white, then it certainly is the case that she will die, just because all cats die. However, under ordinary circumstances, most people would probably consider this statement to be nonsense, because there is no relationship between the hypothesis and the conclusion; and some might even consider it to be false for this reason.

In formal mathematical discourse, if we want to interpret "implies" as a logical connective, then we must be prepared to give a definite truth value to the statement $P \Rightarrow Q$ whenever P and Q are two mathematical statements, and that truth value must depend *only* on the truth values of P and Q. The clearest way to see what that truth value should be is to consider the circumstances under which we would consider a conditional statement to be *false*. Suppose a teacher says to a class, "If you scored higher than 95% on the exam, then you got an A." When they look at their papers, here is what the first four students find:

- Alice scored 99% on the exam and got an A.
- Bob scored 85% on the exam and got a B.
- Carmela scored 94% on the exam and got an A.
- Duane scored 100% on the exam and got a B.

Alice and Bob would certainly believe that the teacher was telling the truth: in Alice's case, the hypothesis is true and the conclusion is true, while in Bob's case they are both false. In Carmela's case, the hypothesis is false, but the conclusion is true nonetheless; Carmela might be pleasantly surprised, but she would have no reason to think the teacher was not telling the truth, because the teacher said nothing about what would happen to someone who did not score higher than 95%. But Duane's case is different: the hypothesis is true (he scored higher than 95%), but the conclusion is false (he did not get an A), and he will indignantly conclude that the teacher's statement is false. The inescapable inference is this: the only time a conditional statement is clearly false is when the hypothesis is true but the conclusion is false.

To ensure that every mathematical statement has an unambiguous meaning, mathematicians have agreed to interpret every conditional statement in this way: if P and Q are statements, then the conditional statement $P \Rightarrow Q$ is false in the case that P is true and Q is false; in all other cases, the conditional statement is true. This is summarized in the following truth table:

$$
\begin{array}{cc|c}
P & Q & P \Rightarrow Q \\
\hline
T & T & T \\
T & F & F \\
F & T & T \\
F & F & T \\
\end{array}
\tag{E.2}
$$

This can lead to some unexpected interpretations. For example, the following mathematical statements are silly, but they are all true:

- If $\sqrt{2} = 1$, then $\sqrt{2} > 100$.
- If $\sqrt{2} > 100$, then $\pi > 3$.
- If $\pi > 3$, then $\sqrt{2} \neq 1$.

(You should convince yourself that each of these statements is true by using the truth table above.) Of course, it would be hard to conceive of a situation in which a mathematician would write statements like these (other than as illustrations of how to interpret conditional statements); but the fact that each of them has an unambiguous interpretation as a true statement turns out to be of great importance in ensuring that our logic is sound. We will revisit this issue later in this appendix.

In English sentences, there is a wide variety of ways of expressing conditional statements. All of the following English sentences mean the same thing:

- If a is negative, then $a < 1$.
- a being negative implies $a < 1$.
- Assuming a is negative, it follows that $a < 1$.
- Given that a is negative, $a < 1$.
- Suppose a is negative; then $a < 1$.
- Whenever a is negative, $a < 1$.
- Provided that a is negative, $a < 1$.
- a being negative is a sufficient condition for $a < 1$.
- For a to be negative, a necessary condition is that $a < 1$.

In every one of these statements, the hypothesis is "a is negative," and the conclusion is $a < 1$. Pay special attention to the last two sentences: in sentences like these, the ***sufficient condition*** is always the hypothesis (the hypothesis is sufficient to ensure that the conclusion is true), while the ***necessary condition*** is the conclusion (if the hypothesis is true, it is necessary that the conclusion also be true).

The hypothesis does not need to come first. All of the statements above can be rephrased so that the conclusion is stated first; it is the key words such as "if " and "given" that signal the hypothesis, not the word order. All of the following sentences also mean "If a is negative, then $a < 1$":

- $a < 1$ if a is negative.
- $a < 1$ is implied by a being negative.
- $a < 1$, assuming that a is negative.
- $a < 1$ given that a is negative.
- $a < 1$, supposing that a is negative.
- $a < 1$ whenever a is negative.
- $a < 1$ provided that a is negative.
- For $a < 1$, a sufficient condition is that a is negative.
- $a < 1$ is a necessary condition for a to be negative.

Because implications form such an important part of mathematical discourse, it is very important to know how to negate them. The key is the truth table (E.2): it shows that the implication $P \Rightarrow Q$ is false exactly when P is true and Q is false, and in no other circumstances. Thus $\neg(P \Rightarrow Q)$ has the same truth values as $P \wedge \neg Q$. The following truth table confirms this:

P	Q	$P \Rightarrow Q$	$\neg(P \Rightarrow Q)$	P	$\neg Q$	$P \wedge \neg Q$
T	T	T	F	T	F	F
T	F	F	T	T	T	T
F	T	T	F	F	F	F
F	F	T	F	F	T	F

Beginning students frequently lose sight of this fact, and when faced with the task of writing the negation of an implication like $P \Rightarrow Q$, they want to write another implication (such as $\neg P \Rightarrow \neg Q$ or $\neg P \Rightarrow Q$ or $P \Rightarrow \neg Q$ or even $Q \Rightarrow P$). If you remember the following mantra, you will avoid falling into this trap:

The negation of an implication is never an implication.

Instead, it is a conjunction: the negation of $P \Rightarrow Q$ is $P \wedge \neg Q$. For example:

STATEMENT: If $a^2 = 4$, then $a = 2$.
NEGATION: $a^2 = 4$ and $a \neq 2$.

Starting with the implication $P \Rightarrow Q$ and negating it once, we get the conjunction $P \wedge \neg Q$, as we have seen. Then if we negate again, following the rule for negating a conjunction discussed above, we obtain $\neg P \vee Q$. How can this be, given that the negation of a negation is supposed to have the same meaning as the original statement? A glance at the truth tables for $P \Rightarrow Q$ and $\neg P \vee Q$ reveals the answer:

P	Q	$P \Rightarrow Q$	$\neg P$	Q	$\neg P \vee Q$
T	T	T	F	T	T
T	F	F	F	F	F
F	T	T	T	T	T
F	F	T	T	F	T

This shows that the statements $P \Rightarrow Q$ and $\neg P \vee Q$ are always equivalent. Thus if it is convenient, we can always replace an implication by the corresponding disjunction.

The Contrapositive and Converse of an Implication

Given an implication $P \Rightarrow Q$, we can form two other closely related implications. The first one, called the ***contrapositive*** of the original implication, is the statement $\neg Q \Rightarrow \neg P$. The contrapositive turns out to have exactly the same truth value as the original implication. In everyday English, this is usually easy to recognize. For example, think back to the teacher's announcement that we discussed earlier: "If you scored higher than 95% on the exam, then you got an A." If this statement were true, then any student who did not get an A would be justified in concluding that he did not score higher than 95% on the exam, so the contrapositive is also true: "if you did not get an A, then you did not score higher than 95% on the exam."

The following truth table shows that the contrapositive has the same truth value as the original implication in every case:

P	Q	$P \Rightarrow Q$	$\neg Q$	$\neg P$	$\neg Q \Rightarrow \neg P$
T	T	**T**	F	F	**T**
T	F	**F**	T	F	**F**
F	T	**T**	F	T	**T**
F	F	**T**	T	T	**T**

The second important related implication is called the ***converse*** of the original implication. Given an implication $P \Rightarrow Q$, its converse is the implication $Q \Rightarrow P$. By examining the truth tables for an implication and its converse, we can see that there is no necessary relation between the two statements—the fact that either one is true does not necessarily tell us anything about the truth of the other.

P	Q	$P \Rightarrow Q$	Q	P	$Q \Rightarrow P$
T	T	**T**	T	T	**T**
T	F	**F**	F	T	**T**
F	T	**T**	T	F	**F**
F	F	**T**	F	F	**T**

For example, consider the following mathematical statements about a real number a:

STATEMENT: If $a \neq 0$, then $a^2 > 0$.
CONVERSE: If $a^2 > 0$, then $a \neq 0$.
[Both the original implication and its converse are always true.]

STATEMENT: If $a > 2$, then $a > 0$.
CONVERSE: If $a > 0$, then $a > 2$.
[The original implication is always true, but the converse is false, for example, when $a = 1$.]

STATEMENT: If $a^2 = 4$, then $a = -2$.
CONVERSE: If $a = -2$, then $a^2 = 4$.
[The original implication is false when $a = 2$, but the converse is always true.]

STATEMENT: If $a > 0$, then $a < 2$.
CONVERSE: If $a < 2$, then $a > 0$.
[The original implication is false whenever $a \geq 2$, and the converse is false whenever $a \leq 0$.]

In ordinary discourse, it is very common to confuse statements with their converses. For example, when a teacher says to a class, "If you scored higher than 95% on the exam, then you got an A," the students might reasonably expect that if they got an A, it means they must have scored higher than 95%; this is why Carmela was surprised by her grade in our previous example. But from a strictly logical point of view, this conclusion would not be justified. In mathematical reasoning, it is vital that we strictly observe the distinction between a statement and its converse—knowing that an implication is true tells us *absolutely nothing* about the truth of its converse.

Biconditionals

This brings us naturally to the last of our logical connectives: ***if and only if***. If P and Q are mathematical statements, then "P if and only if Q" is a mathematical statement, called a ***biconditional statement*** or an ***equivalence*** and symbolized by $P \Leftrightarrow Q$ (or sometimes $P \leftrightarrow Q$). This statement is true when P and Q are both true or both false, and it is false otherwise. The truth table below shows that its meaning is exactly the same as that of the conjunction "P implies Q and Q implies P."

P	Q	$P \Leftrightarrow Q$	$P \Rightarrow Q$	$Q \Rightarrow P$	$P \Rightarrow Q \wedge Q \Rightarrow P$
T	T	**T**	T	T	**T**
T	F	**F**	F	T	**F**
F	T	**F**	T	F	**F**
F	F	**T**	T	T	**T**

In an English sentence, a biconditional statement can be phrased in several ways. Here are some ways of expressing the biconditional statement formed from the two statements "$a \neq 0$" and "$a^2 > 0$":

- $a \neq 0$ if and only if $a^2 > 0$.
- $a \neq 0$ is equivalent to $a^2 > 0$.
- $a \neq 0$ is a necessary and sufficient condition for $a^2 > 0$.
- $a \neq 0$ precisely when $a^2 > 0$.

Some writers abbreviate "if and only if" as "iff." This is a handy shortcut when writing on the blackboard or in one's own informal notes; but the best advice from experienced and careful writers is to avoid using this abbreviation in formal writing. When you write proofs to be read by others, it is much easier on the reader if you write out the full phrase.

A few words are in order regarding the meaning of "only if" in the phrase "if and only if." We can break the statement "P if and only if Q" into two parts: "P if Q," and "P only if Q." As we observed above, "P if Q" is one way of expressing the implication $Q \Rightarrow P$. (The statement to which the word *if* is attached is always the hypothesis, whether it comes first or last in the sentence.) On the other hand, "P only if Q" expresses the other half of the biconditional, namely $P \Rightarrow Q$. This can be confusing, because the appearance of the word "if" just before Q seems to suggest that Q is the hypothesis; but when "if" appears in the combination "only if," it always signals the *conclusion*. Thus the following two sentences mean the same thing:

- If $a > 1$, then a is positive.
- $a > 1$ only if a is positive.

Mathematicians seldom use "only if" on its own to express a conditional statement because it can be confusing; but because it appears so frequently in biconditionals, it is important to be clearly aware of which implication it expresses.

Equivalence and Equality

It is worth taking a second look at the term "equivalent." As explained above, the statement "P is equivalent to Q" is a way of saying $P \Leftrightarrow Q$, which means that P and Q have the same truth values. If this is known to be the case (that is, if "$P \Leftrightarrow Q$" is known to be a true statement), then Q can be deduced from P and vice versa, wherever either one appears in a logical argument. We have already seen several examples of equivalent statements: no matter what statements P and Q represent,

$$\neg(P \wedge Q) \quad \text{is always equivalent to} \quad \neg P \vee \neg Q;$$
$$\neg(P \vee Q) \quad \text{is always equivalent to} \quad \neg P \wedge \neg Q;$$
$$\neg(P \Rightarrow Q) \quad \text{is always equivalent to} \quad P \wedge \neg Q;$$
$$P \Rightarrow Q \quad \text{is always equivalent to} \quad \neg Q \Rightarrow \neg P.$$

Note that the word *equivalent* applies only to *statements*, not to mathematical objects (numbers, points, lines, etc.). It does not make sense to say "$\sqrt{4}$ is equivalent to 2." Instead, when you want to say that two mathematical objects are the same, use "equal": "$\sqrt{4}$ is equal to 2." Only statements can be equivalent; only mathematical objects can be equal. (Full disclosure: it should be noted for the record that, in higher mathematics, one can define a type of relation called an *equivalence relation* between mathematical objects, and once such a relation has been defined, one can say that objects that satisfy that relation are "equivalent" to each other. But that always requires a special definition, and we never use the term "equivalent" in that way in this book.)

Although, as explained in Chapter 1, Euclid used the word "equal" to mean "the same size," in modern mathematics the word typically means only one thing: to say that mathematical objects are *equal* is to say that they are the same object. Thus an equation like $a = b$ is a statement that the symbols a and b represent the very same number (or point or line or whatever type of object a and b stand for). When we write $2/4 = 1/2$, we are asserting that these are two different mathematical expressions for the same real number. To say that two mathematical objects are *distinct* or *different* means simply that they are not equal.

This interpretation leads immediately to several important properties that equality has, no matter what kinds of objects it is used with. Suppose a, b, and c represent mathematical objects of any type whatsoever. Equality satisfies the following properties:

- REFLEXIVE PROPERTY OF EQUALITY: $a = a$.
- SYMMETRIC PROPERTY OF EQUALITY: If $a = b$, then $b = a$.
- TRANSITIVE PROPERTY OF EQUALITY: If $a = b$ and $b = c$, then $a = c$.
- SUBSTITUTION PROPERTY OF EQUALITY: If $a = b$, then b may be substituted for some or all occurrences of a in any mathematical statement.

These are not really mathematical properties; rather, they are logical properties that follow immediately from our interpretation of the word *equal*.

Quantifiers

When we first described mathematical statements at the beginning of this appendix, we insisted that every statement must be either true or false, but not both. As a consequence, a sentence like "$a > 0$" cannot actually be considered to be a statement until we assign a specific value to the symbol a, like $a = 2$ or $a = \pi$. But this is far too limiting: in ordinary mathematical discourse, we frequently make statements in which symbols stand for whole classes of objects, and we have no trouble interpreting such sentences as true or false. For example, most people would probably consider the following two sentences to be unambiguous mathematical statements, with the first one true and the second one false:

- If x is a real number, then $x^2 \geq 0$.
- If n is an integer, then $n > 0$.

We will see in this section how to analyze such sentences.

First let us introduce some terminology. A ***variable*** is a mathematical symbol (usually a letter) that has not been assigned a specific value but is used to stand for any value within some specified set of allowable values; this set of allowable values is called the ***domain*** of the variable. A sentence containing one or more variables is called an ***open sentence*** (or sometimes a ***predicate***), provided that it becomes a (true or false) mathematical statement whenever each of the variables is assigned a specific value from its domain.

Thus, for example, suppose we agree that the variable n is to represent an unspecified integer (so the domain of n is the set of all integers). Then "$n > 0$" is not a statement because we cannot know whether it is true or false without further specifying what n is; instead, it is an open sentence. Any variable that appears in an open sentence and has not been assigned a value is called a ***free variable***. For purposes of discussing the logic of open sentences, we will often use a notation such as $P(x)$ to symbolize a generic open sentence that has x as a free variable, with notations such as $Q(x, y)$, $R(x, y, z)$, etc., symbolizing open sentences with more than one free variable. (If some symbols have already been assigned specific values, as in "let $a = 2$," then they are not considered to be free variables.)

To turn an open sentence into a statement, we need a *quantifier*. There are two types of quantifiers in mathematical logic: *universal* and *existential*. Let's examine each in turn.

Universal Quantifiers

The first type of quantifier we will study is called a ***universal quantifier***; it is usually expressed with a phrase such as "for all." The basic syntax works like this. Suppose $P(x)$ is an open sentence with one free variable x, in which the domain of x is some predetermined set D of mathematical objects. (For example, D could be the set of real numbers, denoted by $\mathbb{R}$; or the set of all integers, denoted by $\mathbb{Z}$; or the set of all points in the plane; or the set of all isosceles triangles.) Then the following is a valid mathematical statement, called a ***universal statement***:

$$\text{for all } x,\ P(x).$$

This statement is true if $P(x)$ becomes a true statement no matter what element of the domain D is substituted for x; otherwise it is false. This universal statement can be symbolized as follows:

$$\forall x, P(x).$$

The phrase "for all x" (or $\forall x$) in this statement is called a ***universal quantifier***. It is not a statement in itself but rather is a "prefix" that can be attached to an open sentence to turn it into a statement. For example, consider the open sentence "$x^2 \geq 0$." If we agree that x represents an arbitrary real number, then "for all x, $x^2 \geq 0$" is a universal statement, which happens to be true. On the other hand, "for all x, $x > 0$" is a false universal statement, because there is at least one allowable value we can substitute for x ($x = 0$, for example) that makes it false.

Besides "for all," there are other ways that universal quantifiers can be expressed in English. Each of the following sentences means the same as "$\forall x,\ x^2 \geq 0$":

- For all x, $x^2 \geq 0$.
- For each x, $x^2 \geq 0$.
- For every x, $x^2 \geq 0$.
- Given any x, $x^2 \geq 0$.
- Every x satisfies $x^2 \geq 0$.
- For any x, $x^2 \geq 0$.

You will notice that the word "given" can be interpreted either as a universal quantifier or as the hypothesis of an implication. When it is used in an implication, it usually signals the presence of an implicit universal quantifier, which we will discuss below.

One must be extremely careful about using "any" to express a universal quantifier, however, because it might not always be interpreted as intended. For example, suppose we try to define a new term called a "happy set," with the following definition: "we say that S is a ***happy set*** if any number in S is positive." Does S being a happy set mean that, if we choose any arbitrary number in S, then it will necessarily be positive (in other words, *every* number in S is positive)? Or does it mean just that there is some number in S that is positive? If you use "any" to express a universal quantifier, make sure that it cannot be easily misinterpreted.

In order to correctly interpret a quantified statement, we must know the domain of each quantified variable; otherwise the statement is meaningless. For example, the statement "for all n, $n \geq 1$" is false if we understand the domain of n to be the set of all integers. However, if we change the domain to be the set of all *positive* integers, then it becomes a true statement. And if we take the domain to be the set of all points in the plane, then it is not a statement at all, because it is meaningless to say that a point in the plane is greater than or equal to 1.

There are various ways that the domain of a quantified variable can be specified. One way is to stipulate in advance that certain variables represent elements of a certain domain. This is what we did in the examples above, when we stipulated that x represents an arbitrary real number.

It is usually clearer, however, to specify the domain of a quantifier in the quantified statement itself. This can be done by indicating the domain right in the quantifier:

- For all $x \in \mathbb{R}$, $x^2 \geq 0$.
- For all real numbers x, $x^2 \geq 0$.
- $\forall x \in \mathbb{R}$, $x^2 \geq 0$.

Alternatively, the domain of a universal quantifier can be specified in the body of the statement being quantified, usually as part of the hypothesis of an implication:

- For all x, if $x \in \mathbb{R}$, then $x^2 \geq 0$.
- $\forall x$, $x \in \mathbb{R} \Rightarrow x^2 \geq 0$.

An open sentence with two or more free variables can be turned into a statement by adding universal quantifiers for all of the variables. For example, if $Q(x, y)$ is an open sentence with free variables x and y, both with domain D, then we can form the universal statement "For all x and y in D, $Q(x, y)$." This can be symbolized as $\forall x, y \in D, Q(x, y)$. The following statements all mean the same thing:

- For all real numbers x and y, $x + y = y + x$.
- For all $x, y \in \mathbb{R}$, $x + y = y + x$.
- $\forall x, y \in \mathbb{R}, x + y = y + x$.

If the variables don't all have the same domain, then they can be expressed with multiple quantifiers. For example, the following two statements mean the same thing (here $\mathbb{Z}^+$ denotes the set of positive integers):

- For all real numbers x and y and all positive integers n, $(xy)^n = x^n y^n$.
- $\forall x, y \in \mathbb{R}$, $\forall n \in \mathbb{Z}^+$, $(xy)^n = x^n y^n$.

It is possible that the domain of a quantified variable might actually contain no elements that satisfy the hypothesis. For example, consider the following statement:

- For all real numbers x, if $x^2 < 0$, then $x < 0$.

Whenever x is a real number, the hypothesis of this implication is false (because x^2 is never negative), so according to the truth table for implications, the implication is true. This means that the universal statement is true.

A similar thing happens if the domain contains no elements at all, for example:

- For all integers n strictly between 0 and 1, $n^2 > 0$.

In this case, the domain (integers strictly between 0 and 1) contains no elements. To analyze this statement, we can rephrase it by moving the domain into the hypothesis of the quantified statement:

- For all n, if n is an integer strictly between 0 and 1, then $n^2 > 0$.

Now you can see that the hypothesis is false for every n, so once again the quantified statement is true. (You can think of it this way: the conclusion $n^2 > 0$ is true for all of the integers n between 0 and 1—all none of them!) A universal statement in which the domain has no elements (or no elements that satisfy the hypothesis in the case of a universal implication) is said to be ***vacuously true***.

In practice, universal quantifiers are often not explicitly expressed at all when the statement being quantified is an implication. The following three statements have exactly the same meaning:

- If $x \in \mathbb{R}$, then $x^2 \geq 0$.
- For all x, if $x \in \mathbb{R}$, then $x^2 \geq 0$.
- For all $x \in \mathbb{R}$, $x^2 \geq 0$.

The first of these three statements is an example of a statement with an ***implicit universal quantifier***. This occurs in the following situation: if an implication contains the same free variable(s) in both the hypothesis and the conclusion and there are no explicit quantifiers for those variables, then the sentence is to be understood as if each free variable were governed by a universal quantifier. Thus the following statement should be interpreted as if it were preceded by "For all n":

- If n is a positive integer, then $n \geq 1$.

(Of course, the domain of each variable must be made clear, either in the surrounding context or by being explicitly specified as part of the hypothesis.) Most mathematical theorems are stated in the form of universal implications, with the universal quantifier either implicit or explicit.

We mentioned earlier that we would revisit the issue of why the truth table we have given for $P \Rightarrow Q$ is the appropriate one. Universal implications make this abundantly clear. Consider the following universal statement:

$$\text{for every real number } x, \text{ if } x > 1, \text{ then } x > 0.$$

Everyone would agree that this is a true statement. Let's see what it means for a few specific values of x:

x	$x > 1$	$x > 0$	$x > 1 \Rightarrow x > 0$
0	F	F	T
1	F	T	T
2	T	T	T

If we want to say that the implication "$x > 1 \Rightarrow x > 0$" is true for *every* real number x, then the table above shows that we have to consider it to be true when the hypothesis and conclusion are both false (as when $x = 0$), when the hypothesis is false but the conclusion is true ($x = 1$), and when they are both true ($x = 2$). Since we have already agreed that it is false when the hypothesis is true but the conclusion is false, we have no choice but to accept that the truth table for $P \Rightarrow Q$ must be as we defined it.

Existential Quantifiers

The other kind of quantifier is called an ***existential quantifier*** and is usually expressed with a phrase such as "there exists." If $P(x)$ is an open sentence with one free variable x, then the following is a valid mathematical statement, called an ***existence statement*** or an ***existential statement***:

$$\text{there exists } x \text{ such that } P(x).$$

Symbolically, this is written

$$\exists x, P(x).$$

(Notice that there is no specific logical symbol for the phrase "such that": in an existence statement, that phrase serves only as "connective tissue" between the object x whose existence is being asserted and the statement $P(x)$ that it is supposed to satisfy. In the symbolic version of the statement, its place is taken by a comma.) This existence statement is true if there is *at least one* x in the given domain such that $P(x)$ is a true statement, and it is false otherwise.

Just like variables in universal quantifiers, the variable in an existential quantifier must have a clearly specified domain, so we know what kind of object is being asserted to exist. The domain can be determined by the context, for example by saying something like "throughout this discussion, the variable x represents a real number." But it is usually better to specify the domain in the statement itself. This can be done by including the domain in the quantifier: "there exists $x \in \mathbb{R}$ such that $x^2 = 2$." Alternatively, the domain can be specified by including it in the condition to be satisfied by the object: "there exists x such that $x \in \mathbb{R}$ and $x^2 = 2$." Notice that when the domain is specified in the body of an existential statement, it becomes part of a conjunction; but, as we saw earlier, when it is specified in the body of a universal statement, it becomes part of the hypothesis.

Here are some ways that an existential quantifier can be expressed in an English sentence. All of these sentences mean the same thing.

- There exists $x \in \mathbb{R}$ such that $x^2 = 2$.
- For some $x \in \mathbb{R}$, $x^2 = 2$.
- There is a real number x such that $x^2 = 2$.
- There is at least one $x \in \mathbb{R}$ such that $x^2 = 2$.

A common variation on the theme of existence statements is a statement of *existence and uniqueness*. Instead of "there exists x in D such that $P(x)$," we might say "there exists a *unique* x in D such that $P(x)$." Its meaning should be clear: whereas an ordinary existence statement asserts that there exists *at least one* $x \in D$ satisfying the given condition, an existence and uniqueness statement asserts that there is *one and only one* such x. For example, in Chapter 3 we study the following existence and uniqueness statement: "given any two distinct points, there is a unique line that contains both of them." An existence and uniqueness statement is often symbolized by adding an exclamation point after the symbol for "there exists"; thus "there exists a unique x in D such that $P(x)$" can be symbolized as

$$\exists! x \in D, P(x). \tag{E.3}$$

As always, the domain of the quantifier must be either specified or understood from the context.

"Existence and uniqueness" is not really a new kind of quantifier; rather, it is just shorthand for the conjunction of two separate statements, one asserting existence ("there exists at least one such x") and the other asserting uniqueness ("there exists at most one such x"). We already know what the existence assertion means. The uniqueness assertion is really a short way of saying "if x_1 and x_2 are elements of D that satisfy $P(x_1)$ and $P(x_2)$, then they are equal." (Note the implicit universal quantifier in this statement.) Thus we can expand the existence and uniqueness statement (E.3) as follows:

$$\left(\exists x \in D, P(x)\right) \wedge \left(\forall x_1, x_2 \in D, P(x_1) \wedge P(x_2) \Rightarrow x_1 = x_2\right). \tag{E.4}$$

It is important to understand the distinction between the mathematical terms "unique" and "distinct": to say that something is *unique* is to say that any two objects satisfying the same condition must be equal to each other; while to say that two or more objects are *distinct* is to say that they are *not* equal to each other. Mathematically, these terms are not interchangeable, even though in nonmathematical contexts, they are sometimes used interchangeably. For example, in computer science, one speaks of the number of "unique visitors" to a website during a certain period of time, meaning the number of *different* people who logged on to the site. A mathematician would prefer to call these "distinct visitors."

Mixed Quantifiers

As we mentioned earlier in the section on universal quantifiers, open sentences with multiple variables need multiple quantifiers to turn them into statements. If two or more variables are quantified by quantifiers of the same type (universal or existential), then it does not matter what order we express the quantifiers in. The following two universal statements are equivalent:

- For every $x \in \mathbb{R}$ and every $n \in \mathbb{Z}$, $(x^2)^n = x^{2n}$.
- For every $n \in \mathbb{Z}$ and every $x \in \mathbb{R}$, $(x^2)^n = x^{2n}$.

Similarly, the following existence statements mean the same thing:

- There exist $x \in \mathbb{R}$ and $n \in \mathbb{Z}$ such that $x^n = 1$.
- There exist $n \in \mathbb{Z}$ and $x \in \mathbb{R}$ such that $x^n = 1$.

(Note that when the existence of more than one object is being asserted, the grammatically appropriate phrase is "there exist" in place of "there exists.")

However, if one sentence has two (or more) quantifiers of *different* types, then the order makes a big difference. Consider the following two statements:

- For every $x \in \mathbb{R}$, there exists $y \in \mathbb{R}$ such that $y > x$.
- There exists $y \in \mathbb{R}$ such that for every $x \in \mathbb{R}$, $y > x$.

The first statement says that given any real number x, there must exist a real number y that is larger than x. This is a true statement, because given x, we can take $y = x + 1$, which satisfies $y > x$. Because of the order in which the quantifiers were stated, x is given to us first, and we are allowed to choose y depending on x.

The second statement is very different: it says that there exists some real number y, which has the property that it is greater than *every* real number x. This is certainly not true.

The moral is that when quantifiers of different types are used in the same sentence, order matters.

Negating Quantified Statements

Just as negations of ordinary compound statements can be simplified, we can also simplify negations of quantified statements by paying close attention to their meanings.

Let us begin with universal statements. A universal statement such as "$\forall x \in D$, $P(x)$" is true when every $x \in D$ (if there are any) satisfies the statement $P(x)$. Therefore it is false if and only if there is some x in D that does *not* satisfy $P(x)$. The negation of this universal statement, therefore, is an existence statement:

STATEMENT: $\forall x \in D$, $P(x)$.
NEGATION: $\exists x \in D$, $\neg P(x)$.

Notice that the domain has not changed, because to conclude that the original statement is false we must find some x *in the original domain D* that satisfies $\neg P(x)$.

Now consider existence statements. A negated existence statement is called a ***nonexistence statement***. Often it is simplest to express such a statement simply by replacing "there exists" by "there does not exist." Thus, for example, we could express the negation of "there exists a real number x such that $x^2 = 2$" in the form "there does not exist a real number x such that $x^2 = 2$."

However, for purposes of analyzing negated existence statements, it is also important to be able to expand them the way we did with negated universal statements. In order for the statement "$\exists x \in D, P(x)$" to be false, it must be the case that there is no element of D that satisfies the condition $P(x)$, which is the same as saying all elements of D (if there are any) satisfy $\neg P(x)$. Thus the negation of an existence statement is equivalent to a universal statement:

STATEMENT: $\exists x \in D, \ P(x)$.
NEGATION: $\forall x \in D, \ \neg P(x)$.

These rules can be combined to negate very complicated mathematical statements. To do so efficiently, it is usually easiest if you write the statement in symbolic form and then read the statement from left to right, negating as you go. Each time you see a universal quantifier in the original statement, replace it by an existential quantifier and negate the whole statement that follows it (the part being quantified). When you see an existential quantifier, replace it by a universal quantifier and negate the statement that follows it. When you see a conjunction, disjunction, or implication, negate it according to the rules we discussed earlier in this appendix.

Here is an example of how to negate a complicated statement step-by-step. The statement we will negate is the following:

- For every line ℓ and every point A, if A does not lie on ℓ, then there exists a line m such that A lies on m and m is parallel to ℓ.

We begin by writing the statement in symbolic form. For the purpose of symbolizing it, we let $\mathscr{L}$ denote the set of all lines and $\mathscr{P}$ the set of all points; we use $\in$ to mean both "is an element of" (a set) and "lies on" (a line), and $\parallel$ to mean "is parallel to."

- $\forall \ell \in \mathscr{L}, \ \forall A \in \mathscr{P}, \ A \notin \ell \Rightarrow (\exists m \in \mathscr{L}, \ A \in m \wedge m \parallel \ell)$.

Here is a step-by-step process for negating this statement. (Remember that the negation of an implication is a conjunction and the negation of a conjunction is a disjunction.) Each of the statements in this sequence is equivalent to the previous one:

1. $\neg\big(\forall \ell \in \mathscr{L}, \ \forall A \in \mathscr{P}, \ A \notin \ell \Rightarrow (\exists m \in \mathscr{L}, \ A \in m \wedge m \parallel \ell)\big)$.

2. $\exists \ell \in \mathscr{L}, \ \neg\big(\forall A \in \mathscr{P}, \ A \notin \ell \Rightarrow (\exists m \in \mathscr{L}, \ A \in m \wedge m \parallel \ell)\big)$.

3. $\exists \ell \in \mathscr{L}, \ \exists A \in \mathscr{P}, \ \neg\big(A \notin \ell \Rightarrow (\exists m \in \mathscr{L}, \ A \in m \wedge m \parallel \ell)\big)$.

4. $\exists \ell \in \mathscr{L}, \ \exists A \in \mathscr{P}, \ A \notin \ell \wedge \neg\big(\exists m \in \mathscr{L}, \ A \in m \wedge m \parallel \ell\big)$.

5. $\exists \ell \in \mathscr{L}, \ \exists A \in \mathscr{P}, \ A \notin \ell \wedge \big(\forall m \in \mathscr{L}, \ \neg(A \in m \wedge m \parallel \ell)\big)$.

6. $\exists \ell \in \mathscr{L}, \ \exists A \in \mathscr{P}, \ A \notin \ell \wedge \big(\forall m \in \mathscr{L}, \ A \notin m \vee m \nparallel \ell\big)$.

Now we can rephrase the negation as an English sentence:

- There exist a line ℓ and a point A such that A does not lie on ℓ, and for every line m, either A does not lie on m or m is not parallel to ℓ.

To rephrase it a little more fluently, it is best to go back to statement 4, in which the nonexistence statement has not been expanded:

- There exist a line ℓ and a point A not on ℓ, with the property that no line contains A and is parallel to ℓ.

Special care must be taken when negating an *existence and uniqueness statement*. Recall that such a statement can always be expanded into a conjunction like statement (E.4): the first part is the existence statement, and the second part is a universal implication expressing uniqueness. The negation will therefore always be a disjunction. Simply stated, there are two ways for an existence and uniqueness statement to be false: either existence can fail (if there is no object satisfying the given conditions) or uniqueness can fail (if there are at least two distinct such objects).

To see how this works in practice, consider the following (false) existence and uniqueness statement:

- There exists a unique real number x such that $x^2 = 2$.

Symbolically,

- $\left(\exists x \in \mathbb{R},\ x^2 = 2 \right) \wedge \left(\forall x_1, x_2 \in \mathbb{R},\ (x_1{}^2 = 2\ \wedge\ x_2{}^2 = 2) \Rightarrow x_1 = x_2 \right).$

Following the procedure outlined above, we can negate this statement systematically by working from left to right. For clarity, we leave the negated existence statement unexpanded as before.

1. $\neg\left(\left(\exists x \in \mathbb{R},\ x^2 = 2 \right) \wedge \left(\forall x_1, x_2 \in \mathbb{R},\ (x_1{}^2 = 2\ \wedge\ x_2{}^2 = 2) \Rightarrow x_1 = x_2 \right) \right).$
2. $\neg\left(\exists x \in \mathbb{R},\ x^2 = 2 \right) \vee\ \neg\left(\forall x_1, x_2 \in \mathbb{R},\ (x_1{}^2 = 2\ \wedge\ x_2{}^2 = 2) \Rightarrow x_1 = x_2 \right).$
3. $\neg\left(\exists x \in \mathbb{R},\ x^2 = 2 \right) \vee\ \left(\exists x_1, x_2 \in \mathbb{R},\ x_1{}^2 = 2\ \wedge\ x_2{}^2 = 2\ \wedge\ x_1 \neq x_2 \right).$

In words, this means the following:

- Either there does not exist a real number whose square is equal to 2 or there exist two different real numbers whose squares are both equal to 2.

This, of course, is a true statement (why?).

Mathematical Definitions

We end this appendix with a few words about *mathematical definitions*. In rigorous modern mathematics, to give a term a **mathematical definition** means to specify a precise, unambiguous condition that an object x must satisfy, in the form of an open sentence $P(x)$, in order for x to be considered as an instance of the term being defined.

When stating a definition, it is conventional to introduce the condition with the word "if"; however, in the context of a definition, it is always understood that x satisfies the definition *if and only if* $P(x)$ is true. For example, suppose we wish to define *greebsnitch* as a mathematical term. We might write,

We say x is a **greebsnitch** if x is a real number greater than 42.

In this case, the defining condition is the open sentence "x is a real number greater than 42," and what the definition *means* is the following biconditional:

x is a greebsnitch $\Leftrightarrow$ $x \in \mathbb{R}$ and $x > 42$.

Once the definition is made, the two open sentences "x is a greebsnitch" and "$x \in \mathbb{R}$ and $x > 42$" are thenceforth equivalent, so either one can be deduced from the other.

Exercises

EA. Each of the statements below is an implication. In these statements, you may assume that A, B, and C are points, ℓ and m are lines, and x is a real number, all of which have been previously defined; there are no implicit universal quantifiers in these statements. For each statement, do all of the following:

- Identify the hypothesis and the conclusion.
- Write the contrapositive.
- Write the converse.
- Write the negation.

Your answers should be written as ordinary (unambiguous) English sentences. Expand the negations as explained in this appendix; don't just write "It is not true that ... ."

(a) If A, B, and C all lie on ℓ, then they are collinear.
(b) If ℓ is a line, then it contains at least two distinct points.
(c) A quadrilateral is a parallelogram if it is a rectangle.
(d) For a triangle to be isosceles, it is necessary that it have two equal angles.
(e) x is divisible by 4 only if it is even.
(f) If $2x + 1 = 5$, then $x = 2$ or $x = 3$.
(g) If the 10^{100}th decimal digit of π is even, then $\sqrt{5} = 2$.

EB. For each of the statements below, do the following:

- Write the statement in symbolic form.
- Write its negation in symbolic form, and expand out each negation to eliminate all appearances of the symbol "$\neg$."
- Then rewrite the negation as an ordinary English sentence.

In some of these statements, you will have to introduce names for the quantified variables.

(a) Every point lies either on ℓ or on m.
(b) For any three points A, B, and C, if they are collinear, then there is another point D that is not equal to A, B, or C.
(c) For every line ℓ, if ℓ contains three distinct points, then it has points in common with three distinct lines.
(d) There exist at least two distinct points.
(e) Given any two distinct points, there is at least one line that contains both of them.
(f) Given any two distinct points, there is at most one line that contains both of them.
(g) Given any two distinct points, there is a unique line that contains both of them.
(h) Given any line, there are at least two distinct points that lie on it.
(i) Given any line, there is at least one point that does not lie on it.
(j) There exists a line ℓ such that for every point A, A lies on ℓ.
(k) There exists a point A that does not lie on any line.
(l) There exist three distinct points that do not all lie on any one line.
(m) In any triangle ABC, if $AB = AC$, then $\angle B = \angle C$.
(n) Given a line ℓ and a point A that does not lie on ℓ, there are at least two different lines that contain A and are parallel to ℓ.
(o) In any triangle, the sum of the measures of any two angles is less than $180°$.

EC. As mentioned in the text, most mathematical theorems are universal implications. Often when the theorem statement is written, the universal quantifier is implicit. Each of the following statements contains one or more implicit universal quantifiers. Rewrite each statement (in ordinary English, not symbols), making the quantifier(s) explicit; then negate the quantified statement. (The symbol $\cong$ means "is congruent to"; its meaning is discussed in Chapters 3 and 4, but you don't need to know what it means to do this exercise.)

(a) If ℓ is a line and A is a point not on that line, then there are no lines that contain A and are parallel to ℓ.

(b) If ABC and DEF are triangles such that $AB \cong DE$, $AC \cong DF$, and $BC \cong EF$, then $\angle A \cong \angle D$, $\angle B \cong \angle E$, and $\angle C \cong \angle F$.

(c) If $x^2 = 2$, then x is not a rational number.

(d) If the hypotenuse and one leg of a right triangle are congruent to the hypotenuse and one leg of another, then the triangles are congruent.

(e) The sum of the interior angles of a triangle is $180°$.

Proofs

A *theorem* is a mathematical statement together with a proof of that statement. So what is a proof? Of course, it is meant to be a convincing argument that the theorem follows logically from the axioms and previously proved theorems. But in ordinary discourse, different people might be convinced by different types of arguments. The mathematical community, by contrast, has reached a near-universal consensus about what constitutes a valid and convincing proof. This is what gives modern mathematics its power.

In its most basic form, a mathematical proof is just a sequence of mathematical statements, connected to each other by strict rules that describe what types of statements may be added and in what order. If, by following these rules, you produce a sequence of statements, the last of which is the statement of the theorem you're trying to prove, then you have produced a proof that will convince any mathematically educated reader or listener beyond reasonable doubt that your theorem is correct.

In this appendix, we focus on the underlying structure of a proof as a sequence of statements, each justified according to the conventions of mathematical reasoning. To make that structure as plain as possible, we begin by presenting our proofs in a "two-column" format, similar to the way proofs are introduced in some high-school geometry courses. In the left column is the sequence of statements that constitute the proof; in the right column are the reasons that justify the steps.

The Structure of Mathematical Proofs

There are several different types of proofs, each of which is appropriate in certain circumstances. We will discuss the most important types later in this appendix. But to give us something concrete to work with now, let's start by talking about the simplest type of proof of the simplest type of theorem statement, an implication: $P \Rightarrow Q$. Here is a "template" for a proof of this implication:

Theorem. $P \Rightarrow Q$.

Proof.

	Statement	Reason
	Assume P.	(hypothesis)
	...	(...)
Goal:	Q.	(...)
Conclusion:	$P \Rightarrow Q$	$\square$

The "$\square$" is an end-of-proof symbol, which in modern mathematical writing usually takes the place of the q.e.d. (or q.e.f.) that Euclid used to indicate the end of a proof.

The idea of this kind of proof is this: to verify that $P \Rightarrow Q$ is true, we need only verify that whenever P is true, it follows logically that Q is also true. If P happens to be false, then the implication is automatically true, as you can see by examining the truth table for "implies." This is why we get to carry out the proof under the assumption that P is true.

The proof will be correct provided that each step is justified by one or more of the following six types of reasons:

- by hypothesis,
- by a definition,
- by an axiom,
- by a previously proved theorem,
- by a previous step in the same proof,
- by the laws of logic.

Let us examine each of these types of justification in turn. We will illustrate the way these justifications are used by exhibiting "snippets" of proofs that exemplify the types of moves we can make and how they are justified.

By hypothesis: As we explained above, the proof of an implication is carried out under the assumption that the hypothesis is true. So the first step of most proofs is to assume that the hypothesis is true, and the justification for this step is "by hypothesis." Later, we will see some other kinds of hypotheses that can be assumed in different types of proofs.

By a definition: In Appendix E, we explained that a definition is always a biconditional statement (even though most definitions are worded using only "if"). Thus a definition yields two implications, either of which can be used to justify any step of a proof. For example, if we have defined a *greebsnitch* to be a real number greater than 42, we may make either of the following moves in a proof:

Statement	Reason
...	(...)
x is a greebsnitch.	(...)
$x \in \mathbb{R}$ and $x > 42$.	(definition of *greebsnitch*)
...	(...)

or

Statement	Reason
...	(...)
$y \in \mathbb{R}$ and $y > 42$.	(...)
y is a greebsnitch.	(definition of *greebsnitch*)
...	(...)

"By definition" is also an appropriate justification for a step that merely serves to define a new symbolic name by giving it a specific value. For example:

Statement	Reason
...	(...)
Let $x = \sqrt{2}$.	(defining x)
...	(...)

By an axiom: A mathematical proof is always carried out in the context of a particular axiomatic system. At any point, one or more of the axioms can be used to justify a statement. For example, one of the axioms of incidence geometry in Chapter 2 (Incidence Axiom 4) states that every line contains at least two points. Here is an example of how that axiom can be used in a proof in incidence geometry:

Statement	Reason
...	(...)
m is a line.	(...)
There exist two distinct points A and B that lie on m.	(Incidence Axiom 4)
...	(...)

In order to use an axiom to justify a step, it is essential to have already established that the hypotheses of the axiom are satisfied, which includes verifying that the object you are applying it to is an element of the correct domain. For example, Incidence Axiom 4 applies to every line, so in order to apply it to m in the snippet above, all we need to verify is that m is a line. But Incidence Axiom 2 applies to two *distinct* points, so if we wish to apply it to points A and B, we must first establish in a previous step that A and B are distinct.

By a previously proved theorem: Similarly, once a theorem has been proved within the context of the axiomatic system, the statement of that theorem can be used to justify a step in a later proof, in the same way you would use an axiom. Of course, you have to ensure that you only use *previously* proved theorems, so as to avoid circular reasoning, and, just as in the case of axioms, it is vital to check that the hypotheses of the theorem are satisfied before you can apply it.

By a previous step in the same proof: Usually, the justification for a step in a proof is some combination of reasons, including one or more previous steps in the proof. It is always understood that the immediately preceding step is part of the justification for each statement in a proof, so if your statement is justified by that step together with some other reasons (such as a definition or an axiom), just cite the other reasons. However, if your reasoning depends on one or more steps other than, or in addition to, the immediately preceding one, you should cite all such previous steps (including the immediately preceding one if appropriate) as part of your justification. For example, in the proof snippet above, if the use of Incidence Axiom 4 did not come immediately after the assertion that m is a line, the proof might look like this:

Statement	Reason
. . .	(. . .)
6. m is a line.	(. . .)
. . .	(. . .)
10. There exist distinct points A and B that lie on m.	(Incidence Axiom 4 and Step 6)
. . .	(. . .)

By the laws of logic: The rules for interpreting logical statements discussed in Appendix E lead to many general principles for deducing that if certain types of statements are true, then another related statement must also be true. For example, if P and Q are statements and we know that $P \Rightarrow Q$ is true and also that P is true, then we can conclude logically that Q must be true. Similarly, if we know that $P \vee Q$ is true and that Q is false, then we can conclude that P must be true.

Such principles are called *rules of inference*, and they are studied intensively in logic courses. The two rules of inference mentioned in the preceding paragraph can be summarized in the following tables:

$$\frac{\begin{array}{c} P \Rightarrow Q \\ P \end{array}}{Q} \qquad \frac{\begin{array}{c} P \vee Q \\ \neg Q \end{array}}{P}$$

In each case, the statements above the line are called the *premises*, and the statement below the line is called the *conclusion*. The rule of inference says that if the premises are known to be true, then it can be concluded logically that the conclusion is also true.

Rules of inference can often be seen to be true by common sense; more formally, they are verified by examining truth tables. For example, to justify the first rule of inference we described above, just notice that in the truth table for $P \Rightarrow Q$, there is only one row in which $P \Rightarrow Q$ is true and P is also true, and in that row, Q is true. The second rule can similarly be justified by examining the truth table for $P \vee Q$:

P	Q	$P \Rightarrow Q$
T	T	T
T	F	F
F	T	T
F	F	T

P	Q	$P \vee Q$
T	T	T
T	F	T
F	T	T
F	F	F

Here are some other commonly used rules of inference. They can all be justified by appealing to appropriate truth tables.

$$\frac{\begin{array}{c} P \Rightarrow Q \\ \neg Q \end{array}}{\neg P} \qquad \frac{\begin{array}{c} P \\ Q \end{array}}{P \wedge Q} \qquad \frac{P}{P \vee Q} \qquad \frac{P \wedge Q}{P}$$

There are also rules of inference that apply to quantified statements. First consider universal quantifiers: if we know that the statement "$\forall x \in D, \ Q(x)$" is true and we also know that a is an element of D, then we can conclude logically that $Q(a)$ is true. Thus we have the following rule of inference:

$$\forall x \in D, \ Q(x)$$
$$a \in D$$
$$\overline{\qquad\qquad}$$
$$Q(a)$$

(This is exactly what we did in the last two proof snippets above: Incidence Axiom 4 is the universal statement that for every line ℓ, there exist two distinct points that lie on ℓ, and a previous step says that m is a line, so we can conclude logically that there exist two distinct points that lie on m.)

Existence statements are handled a little differently. If an axiom, previous theorem, or previous step established an existence statement, such as "there exists a line containing A and B," then we can introduce a symbolic name to refer to such an object, as in "let ℓ be a line containing A and B." Note that such a statement must always be justified by a previously proved or assumed existence statement.

If a certain step in our proof is an existence assertion that gives a name to the object that is being asserted to exist, such as "there exists $x \in D$ such that $Q(x)$," then this statement in itself serves to introduce the symbol x into the discussion as an element of the domain D that satisfies $Q(x)$, and we can thenceforth use the fact that x has those properties. This corresponds to the following rule of inference:

$$\exists x \in D, \ Q(x)$$
$$\overline{\qquad\qquad}$$
$$x \in D \ \land \ Q(x)$$

For example, suppose a step in our proof says "there exists a line ℓ containing A." Thereafter, the symbol ℓ is understood to represent a specific line that contains A.

When you use an existence statement to justify the introduction of an object, it's important to bear in mind that you cannot assume anything about the object other than what's guaranteed by the existence statement. Thus, having introduced ℓ with a statement like "there exists a line ℓ containing A," you cannot count on ℓ having any special properties except the fact that it contains A. It's useful to imagine that the actual value to be assigned to the variable is being chosen by a mischievous and annoyingly literal gremlin. The gremlin is required to choose a value that satisfies the stated condition, so if the existence statement says "there exists a real number x between a and b," you know that the gremlin will choose some real number between a and b. But if there's any way the gremlin can choose it so as to complicate or invalidate your proof (say, by choosing zero or a number extremely close to a or b or a very large negative number or an irrational number), then he will probably do so. Your proof has to work no matter what real number between a and b the gremlin chooses.

Another situation in which the laws of logic are used in proofs is in applying the reflexive, symmetric, transitive, and substitution properties of equality. Thus, for example, if $Q(x)$ is an open sentence with one free variable and a, b, c are variables representing any types of objects whatsoever, the following are valid inferences:

$$Q(a) \qquad\qquad a = b$$
$$a = b \qquad\qquad b = c$$
$$\overline{\qquad\qquad}\qquad\quad\overline{\qquad\qquad}$$
$$Q(b) \qquad\qquad a = c$$

In proofs, the typical justification for the left-hand inference is "by substitution" (or "by substitution into equation (n)," if it is clearer), and the justification for the right-hand one is "by transitivity." It is usually not necessary to give an explicit justification when using the reflexive and symmetric properties of equality.

We use the laws of logic all the time in proofs, often in concert with other types of justification. It is usually not necessary to cite "logic" explicitly as a reason, except in one circumstance: if a step is justified *solely* by the laws of logic and one or more previous steps, it can be helpful to mention "logic" as a justification, so it is clear to the reader that no other theorem or axiom is being used. Here is a simple example:

Statement	Reason
...	(...)
2. ℓ and m are either equal or parallel.	(...)
...	(...)
5. $\ell \neq m$.	(...)
6. ℓ and m are parallel.	(Steps 2 and 5, logic)
...	(...)

The remarkable thing about mathematical proofs is that each of these forms of justification is simple, universal, and for the most part noncontroversial. If you write a proof that includes every step and explains why each step is justified in one or more of these ways, then nobody will argue with the correctness of your proof.

Of course, most proofs are not written with this level of detail, because they would be exceedingly long and boring to read. But if you understand the principles of how proofs are constructed and make sure that your proofs *could* in principle be written in this way, you will be able to write convincing arguments and avoid mistakes. You will also be able to read and understand other people's proofs and judge whether they are valid or not.

Direct Proofs

Just as there are several types of mathematical statements, there are several types of argument that can serve as proofs. In the next few sections, we will summarize the most important types and give a template for constructing each type.

The most straightforward type of proof is called a *direct proof*: this is one in which we assume the hypotheses and then, using the rules of deduction that we discussed above, derive the conclusion. It is easiest to set up when applied to a simple implication. This is the type of proof template we introduced as an illustration at the beginning of this appendix. For convenience, here it is again:

Template F.1 (Direct Proof of an Implication).

Theorem. $P \Rightarrow Q$.

Proof.

	Statement	Reason
	Assume P.	(hypothesis)
	...	(...)
Goal:	Q.	(...)
Conclusion:	$P \Rightarrow Q$.	$\square$

Notice that the last step is the statement of the theorem being proved, which follows logically from the preceding steps. This is the point of a proof: to construct a series of statements, each of which follows logically from the preceding ones and such that the last one is the theorem being proved. Euclid always ended his proofs with a restatement of what had been proved. In practice, unlike Euclid, modern mathematicians do not usually restate the theorem being proved as the final step, provided that the structure of the proof makes it obvious that the theorem follows. Thus a direct proof of an implication is more likely to appear as follows:

Statement	Reason
Assume P.	(hypothesis)
$\ldots$	$(\ldots)$
Goal: Q.	$(\ldots)$ $\quad\square$

There are a number of variations on direct proofs. One of the most common is *proof by contrapositive*. Because the contrapositive of an implication is equivalent to the original implication, we can always prove an implication by proving its contrapositive instead, if that is more convenient. Here is the template:

Template F.2 (Proof by Contrapositive).

Theorem. $P \Rightarrow Q$.

Proof.

Statement	Reason
Assume $\neg Q$.	(hypothesis)
$\ldots$	$(\ldots)$
Goal: $\neg P$.	$(\ldots)$ $\quad\square$

Proofs of Universal Statements

The simple implications we have been considering so far are not very realistic; most theorem statements contain variables and (implicit or explicit) universal quantifiers. Here is the basic template for proving a universal statement:

Template F.3 (Universal Statement).

Theorem. $\forall x \in D, \ Q(x)$.

Proof.

Statement	Reason
Let x be an arbitrary element of D.	(hypothesis)
$\ldots$	$(\ldots)$
Goal: $Q(x)$.	$(\ldots)$ $\quad\square$

Pay close attention to the wording of the first step in this template. The word "let" is generally reserved for a special role in mathematical arguments: it should be used only to introduce a new symbolic name for something. One situation in which it occurs frequently is when assigning a specific value to a variable: "let $a = 2$." Another such situation is one we talked about earlier, in connection with the rule of inference for an existence statement: if we have established that there exists an object with a certain property, then we can use the word "let" to introduce a symbolic name for such an object: "let ℓ be a line containing A and B."

A third situation in which "let" is appropriate is at the beginning of a proof of a universal statement, as in the template above, to introduce a symbol that represents a universally quantified variable in the proof. You should avoid using "let" to introduce anything other than a new symbolic name—for example, it would not be appropriate to use "let" to introduce an assumption, as in "let AB and AC be equal." It would be much better to say "assume that AB and AC are equal," or "suppose AB and AC are equal." (Occasionally, mathematicians do carelessly use the word "let" instead of "assume" or "suppose" in sentences like these, but your proofs will be clearer if you learn from the start to use "let" only to introduce new symbolic names.)

The other important word in the first step is "arbitrary." This is reserved for universally quantified objects and indicates that the symbol x can stand for any unspecified element of the domain D. This means that we are not allowed to assume anything special about it except that it is an element of D (unless the subsequent steps in the proof call for such an assumption, an example of which we will see in the next template we consider). As long as we remember this restriction, the steps of our proof will be applicable to any element of D whatsoever, and we will have proved that each element $x \in D$ must satisfy the conclusion $Q(x)$. It is not always necessary to include the word "arbitrary" when introducing a universally quantified variable; you will often see the first step in a universal proof written simply as "let $x \in D$." But including the word "arbitrary" reminds us that x has to stand for any and every element of the specified domain and that we can make no other special assumptions about it. Imagine that the actual value assigned to x has been chosen by the same gremlin we mentioned earlier and that he has worked hard to find any weak spots in your logic and exploit them.

Many of the templates we give here can be combined to obtain more complicated proof structures, for example by embedding one type of proof as a series of steps inside a longer proof of a different type. An important and common instance of this occurs when the theorem statement is a *universal implication*. In this case, the structure of the proof combines the patterns for proving both a universal statement and an implication. The vast majority of all theorems in mathematics are stated as universal implications (often with an implicit universal quantifier), so it is important to become very familiar with the template for such proofs.

Template F.4 (Universal Implication).

Theorem. $\forall x \in D,\ P(x) \Rightarrow Q(x)$.

Proof.

	Statement	Reason
	Let x be an arbitrary element of D.	(hypothesis)
	Assume $P(x)$.	(hypothesis)
	$\ldots$	$(\ldots)$
Goal:	$Q(x)$.	$(\ldots)$ $\square$

Of course, a universal implication can also be proved by proving the contrapositive; we leave it to you to figure out the details of all the possible variations on such combinations. You will see many of them in action in the proofs you encounter throughout this book.

Proofs of Conjunctions

Another type of theorem you will encounter is one in which you must prove a conjunction. Most often, the conjunction occurs as the conclusion of an implication, as in "$P \Rightarrow Q_1 \wedge Q_2$." In this case, the idea is simple: to prove the conclusion, we must prove that Q_1 and Q_2 are both true, so the proof will have two parts, one for each conclusion. Here is what the template looks like:

Template F.5 (Proof of a Conjunction).

Theorem. $P \Rightarrow Q_1 \wedge Q_2$.

Proof.

	Statement	Reason
	Assume P.	(hypothesis)
	Part 1: Proof of Q_1	
	...	(...)
	...	(...)
Goal:	Q_1.	(...)
	Part 2: Proof of Q_2	
	...	(...)
	...	(...)
Goal:	Q_2.	(...)

The hypothesis P is stated before the beginning of Part 1 to make it clear that it can be used throughout both parts of the proof.

Proofs of Equivalence

A common type of theorem is an *equivalence theorem*, one that asserts a biconditional like $P \Leftrightarrow Q$. As we discussed in Appendix E, this statement is equivalent to the conjunction "$P \Rightarrow Q \wedge Q \Rightarrow P$," so a proof of equivalence has to be carried out in two parts like a proof of a conjunction.

Template F.6 (Proof of Equivalence).

Theorem. $P \Leftrightarrow Q$.

Proof.

	Statement	Reason
	Part 1: Proof that $P \Rightarrow Q$	
	Assume P.	(hypothesis)
	...	(...)
Goal:	Q.	(...)
	Part 2: Proof that $Q \Rightarrow P$	
	Assume Q.	(hypothesis)
	...	(...)
Goal:	P.	(...)

A common variation on this pattern is to prove one or the other of the implications by proving its contrapositive. For example, a common way to prove $P \Leftrightarrow Q$ is first to prove $P \Rightarrow Q$ and then to prove $\neg P \Rightarrow \neg Q$.

Proof by Cases

Another important variation on direct proof is *proof by cases*. This is needed whenever you need to prove that two or more different hypotheses lead to the same conclusion. The most common example of this is a theorem whose hypothesis is a disjunction (an "or" statement). For example, suppose we want to prove a statement of the form "$P_1 \vee P_2 \Rightarrow Q$." In this case, the hypothesis tells us that either P_1 or P_2 is true, but we don't know which one; you need to show that either hypothesis leads to the conclusion that Q is true. For such a theorem, it is necessary to carry out two proofs, one for the case in which P_1 is true, and the other for the case in which P_2 is true. The template looks like this:

Template F.7 (Proof by Cases).

Theorem. $P_1 \vee P_2 \Rightarrow Q$.

Proof.

	Statement	**Reason**
	Assume $P_1 \vee P_2$.	(hypothesis)
	Case 1:	
	Assume P_1.	(hypothesis)
	...	(...)
Goal:	Q.	(...)
	Case 2:	
	Assume P_2.	(hypothesis)
	...	(...)
Goal:	Q.	(...)

Note the difference between *cases* and *parts* of a proof: as explained in the preceding section, your proof will have multiple *parts* when you have two or more different conclusions to prove, with the same or different hypotheses. On the other hand, it will have multiple *cases* when there is only one conclusion to prove, but it must be proved under two or more different hypotheses. The typical patterns are summarized in the following diagrams:

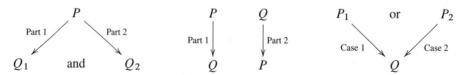

Proofs by cases can be useful even when there is not a disjunction in the hypothesis. If, at any point during a proof, you have deduced a statement of the form $P_1 \vee P_2$, then it is legitimate to divide the remainder of the proof into cases depending on whether P_1 is true or P_2 is true.

In fact, sometimes cases are useful even when no disjunction is evident. If one of two (or more) possibilities must hold and different proofs are needed to handle the different possibilities, then a proof by cases is called for. In effect, we introduce a disjunction of the form $R \vee \neg R$ (which is always true by the laws of logic) and then use those two cases to continue the proof. Consider the following snippet:

Statement	Reason
...	(...)
5. ℓ and m are lines.	(...)
6. Either ℓ and m are parallel or they are not.	(logic)
Case 1:	
7. Assume ℓ and m are parallel.	(hypothesis)
...	(...)
Case 2:	
10. Assume ℓ and m are not parallel.	(hypothesis)
...	(...)

In practice, in cases like this, the explicit statement of the disjunction (Step 6 in the preceding snippet) is frequently omitted from the proof, as long as it will be obvious to the reader that at least one of the specified cases must hold.

Another situation in which proof by cases is useful, albeit in a more subtle way, is when the *conclusion* of the theorem is a disjunction, as in a theorem of the form "$P \Rightarrow Q_1 \vee Q_2$." In this case, we need to prove that Q_1 or Q_2 is true, but we might not know in advance which one. One way to do this is to choose one of the two conclusions, say Q_1, and consider two cases: either Q_1 is true or it is not. If it is true, then the conclusion $Q_1 \vee Q_2$ is true automatically by the laws of logic. In the case that Q_1 is not true, you have to prove that Q_2 is true (using the assumption $\neg Q_1$ if necessary). Here is how the template looks:

Template F.8 (Proof of a Disjunction).

Theorem. $P \Rightarrow Q_1 \vee Q_2$.

Proof.

	Statement	Reason
	Assume P.	(hypothesis)
	Case 1:	
	Assume Q_1.	(hypothesis)
	$Q_1 \vee Q_2$.	(logic)
	Case 2:	
	Assume $\neg Q_1$.	(hypothesis)
	...	(...)
Goal:	Q_2.	(...)
Conclusion:	$Q_1 \vee Q_2$.	$\square$

Be sure you don't fall into the trap of trying to use "assume Q_1" and "assume Q_2" as your two cases: a proof by cases works only if you already know that one of two (or more) alternatives must be true, either by hypothesis or by an axiom or by the rules of logic or because you have already proved it. When we are trying to prove $Q_1 \vee Q_2$, we do not know that Q_1 or Q_2 must be true until *after* we prove it, so assuming that one or the other must be true amounts to circular reasoning. However, we do know that either Q_1 or $\neg Q_1$ must be true (by logic), which is why the proof template given above works.

Indirect Proofs

All of the types of proofs we have discussed so far are variants of direct proofs: we assume one or more hypotheses and reason directly until we reach the desired conclusion(s). Now we will discuss a very different type of proof, called an *indirect proof*, in which we assume that the theorem statement is false and derive a contradiction. For this reason, an indirect proof is also called a *proof by contradiction*. One example is the proof given in Chapter 1 that $\sqrt{2}$ is irrational (see Theorem 1.1).

An indirect proof can be used for any type of theorem. Here's the general template:

Template F.9 (Indirect Proof).

Theorem. Q.

Proof.

	Statement	**Reason**
	Assume $\neg Q$.	(hypothesis for contradiction)
	...	(...)
Goal:	Contradiction.	(...)
Conclusion:	Q.	□

The "contradiction" can be any statement that is known to be false, but usually it is a statement of the form $R \wedge \neg R$, where R is any statement whatsoever. Typically, one of the previous steps in the proof asserts some mathematical statement R, and then a later step asserts its negation, $\neg R$. The contradiction step then refers to the two contradictory steps as reasons.

As with other forms of proof, the final step restating the theorem that has been proved can be omitted if it will be clear to the intended readers that the proof is complete. But in indirect proofs, especially long ones, the logic can get a little complicated, so it is often worth restating the result anyway just to remind the reader why we were deriving a contradiction.

The idea of an indirect proof is this: the statement Q, being a mathematical statement, is either true or false. The proof shows that if Q were false, it would lead to a proof of something we know to be false. Therefore, the only remaining possibility is that Q is true.

More formally, indirect proofs are justified by the following rule of inference:

$$\frac{\neg Q \Rightarrow (R \wedge \neg R)}{Q}$$

In words, if the falsity of Q implies a contradiction, then Q must be true. This can be verified by looking at the truth table for the statement $\neg Q \Rightarrow (R \wedge \neg R)$.

Because the first step in an indirect proof is the assumption of the negation of what is to be proved, it is vitally important that you be able to precisely formulate the negation of any mathematical statement. This is why we focused so much attention on negations in Appendix E.

The most common use of indirect proof is to prove an implication, such as $P \Rightarrow Q$. In that case, the hypothesis for contradiction is the negation of $P \Rightarrow Q$, which is $P \wedge \neg Q$. Thus we get to assume both that P is true and that Q is false. Since P is what we would normally assume for a direct proof, only the $\neg Q$ assumption needs to be labeled as a hypothesis for contradiction. The template looks like this:

Template F.10 (Indirect Proof of an Implication).

Theorem. $P \Rightarrow Q$.

Proof.

	Statement	**Reason**
	Assume P.	(hypothesis)
	Assume $\neg Q$.	(hypothesis for contradiction)
	...	(...)
Goal:	Contradiction.	(...)
Conclusion:	$P \Rightarrow Q$.	(logic)

Existence Proofs

Many theorems in geometry are existence statements. Existence proofs do not really constitute a separate kind of proof, because existence is typically proved by some combination of the direct or indirect methods we have already introduced; but because existence proofs display some special features of their own, it is useful to treat them separately.

A simple existence statement is one of the form "$\exists y \in E, \; Q(y)$." To prove such a statement, we need only show that there is one element y in E that satisfies the condition $Q(y)$. We get to choose or construct the element in any way that is convenient. (In this case, it is the proof writer who chooses it, not the gremlin!) Once we have chosen it, we then need to prove that it satisfies the condition $Q(y)$. Thus a proof of existence will typically have two parts: the first part describes how to construct or choose an appropriate element $y \in E$, and the second part proves that y has the desired properties.

This kind of proof is often called a *constructive proof*, because the object y whose existence is being asserted is sometimes "constructed" in the way one constructs geometric figures. Even if y is not really constructed but is just chosen from among some already-existing set of objects, this type of proof is still called a constructive proof as long as it gives a definite formula, rule, or algorithm for how to choose y. Here is a template:

Template F.11 (Constructive Existence Proof).

Theorem. $\exists y \in E, \; Q(y)$.

Proof.

Statement	Reason
Part 1: Choosing y	
...	(...)
...	(...)
Let $y = \ldots$.	(...)
Part 2: Proof that $y \in E$ and $Q(y)$	
...	(...)
...	(...)
Goal: $y \in E \wedge Q(y)$.	(...) □

The first part of a constructive existence proof might require a little more ingenuity than other kinds of proofs, because there are no general guidelines about how to construct or choose an object that satisfies the desired conditions; you have to look at each particular situation and figure out how to use the given information to find an object of the right type.

In practice, the fact that y is an element of the set E is often obvious from the way y is defined, so that part of the conclusion might not be mentioned explicitly. Moreover, the proof that y satisfies the condition $Q(y)$ is sometimes deeply embedded in the choice of y, so the two parts of an existence proof might not be easily separable; in such cases, it is permissible to present the entire proof in one part, as long as it is easy for the reader to see that you have proved that $Q(y)$ holds for the chosen value of y.

Most often, an existence statement occurs as the conclusion of a different type of statement, such as a universal statement. Templates for such proofs are simply combinations of the template for an existence proof with the appropriate template for the enclosing statement. Here is a template for one of the most common types of existence theorems:

Template F.12 (Universal Existence Proof).

Theorem. $\forall x \in D,\ \exists y \in E,\ Q(x, y)$.

Proof.

Statement	Reason
Let x be an arbitrary element of D.	(hypothesis)
Part 1: Choosing y	
...	(...)
Let $y = \ldots$.	(...)
Part 2: Proof that $y \in E$ and $Q(x, y)$	
...	(...)
Goal: $y \in E \wedge Q(x, y)$.	(...) □

The important feature of this template is that, due to the order in which the quantifiers appear, the universally quantified variable x is introduced first, and then you get to choose or construct y in a way that might depend on x; typically, different choices of x will result in different choices of y. For example, to prove the statement $\forall x \in \mathbb{R},\ \exists y \in \mathbb{R},\ y > x$, we would start by letting x be an arbitrary real number and then defining y by some formula involving x, such as $y = x + 1$.

Existence statements often appear as conclusions in other types of statements as well, such as universal implications, multiple quantifiers, etc. For example, consider the statement "for every pair of real numbers a, b, if $a < b$, then there is a real number y such that $a < y < b$." This is a universal implication with two universal variables and with an existence statement as its conclusion. To write a constructive proof of this theorem, you get to start by assuming all of the hypotheses:

Statement	Reason
Let a and b be real numbers.	(hypothesis)
Assume $a < b$.	(hypothesis)
$\ldots$	$(\ldots)$

After that, you construct an appropriate y (for example, you could set $y = (a+b)/2$) and then show that it satisfies the required conclusion. All of the propositions of Euclid that solve geometric construction problems are universal existence theorems of this type, and there are many more throughout this book.

Existence theorems frequently come paired with uniqueness theorems. As explained in Appendix E, an existence and uniqueness statement like $\exists! y \in E,\ Q(y)$ is really a conjunction of two statements: the existence statement "$\exists y \in E,\ Q(y)$" and the uniqueness statement "$\forall y_1, y_2 \in E,\ Q(y_1) \wedge Q(y_2) \Rightarrow y_1 = y_2$." Thus an existence and uniqueness proof will have two parts, one part for existence and one for uniqueness. Here's a template:

Template F.13 (Universal Existence and Uniqueness Proof).

Theorem. $\forall x \in D,\ \exists! y \in E,\ Q(x, y)$.

Proof.

Statement	Reason
Let $x \in D$.	(hypothesis)
Part 1: Existence	
Part 1a: Choosing y	
$\ldots$	$(\ldots)$
Let $y = \ldots$.	$(\ldots)$
Part 1b: Proof that $y \in E$ and $Q(x, y)$	
$\ldots$	$(\ldots)$
Goal: $y \in E \wedge Q(y)$.	$(\ldots)$
Part 2: Uniqueness	
Let $y_1, y_2 \in E$.	(hypothesis)
Assume $Q(x, y_1)$ and $Q(x, y_2)$ are true.	(hypothesis)
$\ldots$	$(\ldots)$
Goal: $y_1 = y_2$	$(\ldots)$ $\square$

There are three common ways to structure a proof of uniqueness:

- Assume y_1 and y_2 are arbitrary elements of E satisfying $Q(x, y_1)$ and $Q(x, y_2)$, and prove that $y_1 = y_2$. (This is the direct approach followed in the template above.)

- Assume that y_1 is an arbitrary element of E satisfying $Q(x, y_1)$, and prove that y_1 is equal to the element $y \in E$ that was produced in the first part of the proof. (To be logically complete, one would then argue that if y_1 and y_2 are two elements of E satisfying $Q(x, y_1)$ and $Q(x, y_2)$, then they are both equal to y and thus equal to each other by transitivity. But it is usually unnecessary to write out this step explicitly.)

- Alternatively, you can use an indirect proof of uniqueness: assume that y_1 and y_2 are *distinct* elements of E satisfying $Q(x, y_1)$ and $Q(x, y_2)$, and derive a contradiction.

All other things being equal, direct proofs are generally preferable to indirect ones; but for uniqueness proofs, sometimes the extra information provided by the assumption $y_1 \neq y_2$ is needed to get the proof going, and in such cases an indirect proof of uniqueness might be the only kind available.

It is also possible to prove *existence theorems* indirectly, by assuming that there is no object satisfying the desired conditions and deriving a contradiction. But constructive existence proofs are always preferable when they can be found, because they come with explicit algorithms for producing the objects in question, which can be extremely useful in applications. All of the existence proofs in this book are constructive.

Nonexistence Proofs

Closely related to existence proofs are proofs of *nonexistence statements*. For example, having defined certain points A, B, and C, we might wish to prove that there is no line that contains them all. Of course, as explained in Appendix E, a negated existence statement is equivalent to a universal statement: "for every line ℓ, it is not true that ℓ contains A, B, and C," so we could attempt to prove this universal statement. But it is almost always easier to prove nonexistence indirectly, by assuming that an object exists with the given properties and deriving a contradiction. Here is a template for the simplest case:

Template F.14 (Nonexistence Proof).

Theorem. $\neg(\exists y \in E,\ Q(y))$.

Proof.

Statement	Reason
Assume $\exists y \in E,\ Q(y)$.	(hypothesis for contradiction)
$\ldots$	$(\ldots)$
Goal: Contradiction.	$(\ldots)$

Proof by Mathematical Induction

The last type of proof we will consider is *mathematical induction*. This is a very specialized type of proof for proving universal statements about integers. We use it only a few times in this book, but it is indispensable in those instances.

Here is an intuitive description of how it works: suppose we wish to prove that a statement $Q(n)$ is true whenever n is a positive integer. We could start checking individual values—showing that $Q(1)$ is true, $Q(2)$ is true, $Q(3)$ is true, etc. But of course, that would never lead to a proof that the statement is true for *every* positive integer, because some very large n might be a counterexample, and we might never get far enough to discover it. But suppose, while checking values, we discover a systematic argument that allows us to prove that whenever $Q(n)$ is true, it follows logically that $Q(n + 1)$ must necessarily also be true. We could prove $Q(1)$ by hand and then program a computer, using our systematic argument, to print out a proof that $Q(2)$ follows from $Q(1)$, then that $Q(3)$ follows from $Q(2)$, etc. Although this would still only produce proofs of finitely many cases, if our systematic argument is sound, then we can be confident that the statement could eventually be verified for any specific positive integer. Mathematical induction is a formalized way of making this intuition rigorous.

The general pattern of a proof by mathematical induction, therefore, has two parts, called the *base case* and the *inductive step*. In the base case, you simply prove by hand that $Q(1)$ is true (this is usually easy); in the inductive step, you prove that the implication $Q(n) \Rightarrow Q(n+1)$ holds for every positive integer n.

Mathematical induction is not restricted only to statements about all *positive* integers; it works just as well for any statement about all integers greater than or equal to some specified base case, say $n = n_0$. The idea is the same: for the base case, you prove that $Q(n_0)$ is true; and for the inductive step, you prove that $Q(n) \Rightarrow Q(n + 1)$ whenever $n \geq n_0$.

The following theorem gives a formal justification for why mathematical induction is a valid method of proof.

Theorem F.15 (The Principle of Mathematical Induction). *Let n_0 be an integer, and let $Q(n)$ be an open sentence in which n is a free variable representing an arbitrary integer greater than or equal to n_0. Suppose the following:*

(a) BASE CASE: *$Q(n_0)$ is true.*

(b) INDUCTIVE STEP: *For each integer $n \geq n_0$, $Q(n) \Rightarrow Q(n+1)$.*

Then $Q(n)$ is true for every integer $n \geq n_0$.

Proof. Assume for the sake of contradiction that there is some integer $n \geq n_0$ such that $Q(n)$ is false. Then there must be a *smallest* such integer; call it $n_{\min}$. Now $n_{\min}$ cannot be equal to n_0, because hypothesis (a) shows that $Q(n_0)$ is true. It follows that $n_{\min} > n_0$, so if we let $n_{\text{smaller}} = n_{\min} - 1$, then we have $n_{\text{smaller}} \geq n_0$. Since n_{smaller} is less than $n_{\min}$ and $n_{\min}$ is the smallest integer greater than or equal to n_0 for which $Q(n_{\min})$ is false, it must be the case that $Q(n_{\text{smaller}})$ is true. But (b) shows that $Q(n_{\text{smaller}}) \Rightarrow Q(n_{\text{smaller}} + 1)$, so $Q(n_{\text{smaller}} + 1)$ is also true. Because we arranged our definitions so that $n_{\text{smaller}} + 1 = n_{\min}$, this means $Q(n_{\min})$ is true, which is a contradiction. Thus our assumption was false, and we conclude that $Q(n)$ is true for every integer $n \geq n_0$. $\square$

Here is a template for an induction proof:

Template F.16 (Proof by Mathematical Induction).

Theorem. *For every integer $n \geq n_0$, $Q(n)$.*

Proof.

	Statement	Reason
	Base case	
	...	(...)
	...	(...)
Goal:	$Q(n_0)$.	(...)
	Inductive step	
	Let n be an integer such that $n \geq n_0$.	(hypothesis)
	Assume $Q(n)$.	(inductive hypothesis)
	...	(...)
Goal:	$Q(n+1)$.	(...)
Conclusion:	For every integer $n \geq n_0$, $Q(n)$.	(induction principle) □

Sets and Functions

The axioms for plane geometry presented in this book use set theory as an underlying foundation. In our description of the axioms beginning in Chapter 3, we assume that you are familiar with the basic terminology and properties of set theory and functions. In this appendix, we give a brief review of the parts of the theory that we will use. For a somewhat more complete introduction, consult any introductory book on mathematical reasoning, such as [**Ecc97**] or [**Vel06**]. If you want a more systematic introduction to set theory as a subject in its own right, you can consult an introductory set theory text such as [**Dev93**] or [**Hal74**]. Throughout this appendix, we use the concepts and terminology of mathematical logic described in Appendices E and F.

Basic Concepts

A *set* is a collection of objects, considered as a whole. The objects that make up the set are called its *elements* or its *members*. The elements of a set may be any objects whatsoever, but for our purposes, they will usually be *mathematical objects*, as described in Appendix E. The notation $x \in X$ means that the object x is an element of the set X. The words *collection* and *family* are synonyms for set.

In rigorous axiomatic developments of set theory, the words *set* and *element* are taken as primitive undefined terms. (It would be very difficult to *define* the word "set" without using some word such as "collection," which is essentially a synonym for "set.") Instead of giving a general mathematical definition of what it means to be a set or for an object to be an element of a set, mathematicians characterize each particular set by giving a precise definition of what it means for an object to be a element of *that* set—this is called the set's *membership criterion*. The membership criterion for a set X is a statement of the form "$x \in X \Leftrightarrow P(x)$," where $P(x)$ is some open sentence that is true precisely for those objects x that are elements of X, and no others. For example, if $\mathbb{Q}$ is the set of all rational numbers, then the membership criterion for $\mathbb{Q}$ could be expressed as follows:

$$x \in \mathbb{Q} \quad \Leftrightarrow \quad x = p/q \text{ for some integers } p \text{ and } q \text{ with } q \neq 0.$$

The essential characteristic of sets is that two sets are equal if and only if they have the same elements. Thus if X and Y are sets, then $X = Y$ if and only if every element of X is an element of Y and every element of Y is an element of X. Symbolically,

$$X = Y \quad \text{if and only if} \quad \forall x, \; x \in X \Leftrightarrow x \in Y.$$

If X and Y are sets such that every element of X is also an element of Y, then we say X is a **subset of Y**, written $X \subseteq Y$. Thus

$$X \subseteq Y \quad \text{if and only if} \quad \forall x, \; x \in X \Rightarrow x \in Y.$$

The notation $Y \supseteq X$ ("Y is a **superset of X**") means the same as $X \subseteq Y$. Using the concept of subsets, we can restate the criterion for two sets to be equal as follows:

$$X = Y \quad \text{if and only if} \quad X \subseteq Y \text{ and } Y \subseteq X.$$

If $X \subseteq Y$ but $X \neq Y$, we say that X is a **proper subset of Y** (or Y is a **proper superset of X**). Some authors use the notations $X \subset Y$ and $Y \supset X$ to mean that X is a proper subset of Y; however, since other authors use the symbol "$\subset$" to mean *any* subset, not necessarily proper, we generally avoid using this notation and instead say explicitly when a subset is proper.

Defining Sets

We begin with a few sets whose existence and basic properties we take for granted.

- THE EMPTY SET: The set containing no elements is called the **empty set**; it is denoted by $\varnothing$. It is the only set that satisfies the following equivalent properties:

$$\neg(\exists x, \; x \in \varnothing) \qquad \text{or} \qquad \forall x, \; x \notin \varnothing.$$

- THE SET OF REAL NUMBERS: The set of all real numbers is denoted by $\mathbb{R}$. The properties of this set are summarized in Appendix H.

- THE SET OF INTEGERS: The integers (or whole numbers), denoted by $\mathbb{Z}$, form a subset of $\mathbb{R}$. The properties of the integers are also summarized in Appendix H.

Beyond these basic sets, there are essentially two ways to define new sets. You will see that in each case, the set is completely determined by its membership criterion.

- DEFINING A SET BY LISTING ELEMENTS: Given any list of objects that can be explicitly named, the set containing those objects and no others is denoted by listing the objects between braces: $\{c_1, c_2, \ldots, c_n\}$. The membership criterion is easy to express:

$$a \in \{c_1, c_2, \ldots, c_n\} \qquad \Leftrightarrow \qquad a = c_1 \text{ or } a = c_2 \text{ or } \ldots \text{ or } a = c_n.$$

For example, the set $\{0, 1, 2\}$ contains the numbers 0, 1, and 2, and nothing else. Because a set is completely determined by which elements it contains, it does not matter what order the elements are listed in or whether they are repeated; the notations $\{0, 1, 2\}$, $\{2, 1, 0\}$, and $\{0, 0, 1, 2, 1, 1\}$ all denote the same set. A set containing exactly one element is called a **singleton**.

- DEFINING A SET BY SPECIFICATION: Given a set D and an open sentence $P(x)$ in which x represents an element of D, there is a set whose elements are precisely those $x \in D$ for which $P(x)$ is true. This set is denoted by either of the notations

$\{x \in D : P(x)\}$ or $\{x \in D \mid P(x)\}$. (This notation is often called ***set-builder notation***.) Here is the membership criterion for this set:

$$a \in \{x \in D : P(x)\} \quad \Leftrightarrow \quad a \in D \text{ and } P(a).$$

If the domain of x is understood or is implicit in the condition $P(x)$, the same set can be denoted by $\{x : P(x)\}$. For example, the set of all positive real numbers can be described by either of the following notations:

$$\{x \in \mathbb{R} : x > 0\} \quad \text{or} \quad \{x : x \in \mathbb{R} \text{ and } x > 0\}.$$

For some sets, there is a formula that represents a typical element of the set, as some variable or variables run through all elements of some predetermined domain. For example, the set of perfect squares can be described as the set of all numbers of the form n^2 as n runs through the integers. In this case, we often use the following variant of set-builder notation:

$$\{n^2 : n \in \mathbb{Z}\}.$$

This is shorthand notation for $\{x : x = n^2 \text{ for some } n \in \mathbb{Z}\}$.

Operations on Sets

There are three important operations that can be used to combine sets to obtain other sets.

- UNION: Given any sets X and Y, their ***union***, denoted by $X \cup Y$, is the set whose elements are all the objects that are elements of X or elements of Y (or both). The membership criterion is

$$x \in X \cup Y \qquad \text{if and only if} \qquad x \in X \text{ or } x \in Y.$$

Unions of more than two sets are defined similarly:

$$x \in X_1 \cup \cdots \cup X_n \qquad \text{if and only if} \qquad x \in X_1 \text{ or } \ldots \text{ or } x \in X_n.$$

- INTERSECTION: Given sets X and Y, their ***intersection***, denoted by $X \cap Y$, is the set whose elements are all the objects that are elements of both X and Y; thus

$$x \in X \cap Y \qquad \text{if and only if} \qquad x \in X \text{ and } x \in Y.$$

Just as for unions, we can define intersections of more than two sets:

$$x \in X_1 \cap \cdots \cap X_n \qquad \text{if and only if} \qquad x \in X_1 \text{ and } \ldots \text{ and } x \in X_n.$$

Given two sets X and Y, we say that ***X and Y intersect*** if $X \cap Y \neq \varnothing$, meaning that they have at least one element in common. We say that ***X and Y are disjoint*** if $X \cap Y = \varnothing$ (i.e., if they do not intersect), meaning that they have no elements in common. To say that more than two sets are disjoint means that each pair of sets is disjoint; in other words, there is no element that lies in more than one of the sets.

- SET DIFFERENCE: If X and Y are sets, their ***difference***, denoted by $X \smallsetminus Y$, is the set of all elements in X that are not in Y:

$$x \in X \smallsetminus Y \qquad \text{if and only if} \qquad x \in X \text{ and } x \notin Y.$$

Ordered Pairs and Cartesian Products

An ***ordered pair*** is a choice of two objects (which could be the same or different), called the ***components*** of the ordered pair, together with a specification of which is the first component and which is the second. The notation (a,b) means the ordered pair in which a is the first component and b is the second. The defining characteristic is that two ordered pairs are equal if and only if their first components are equal and their second components are equal:

$$(a,b) = (a',b') \qquad \text{if and only if} \qquad a = a' \text{ and } b = b'.$$

Notice that the ordered pair (a,b) is not the same as the *set* $\{a,b\}$, because the order of components matters in the former but not in the latter. Thus $(1,2)$ and $(2,1)$ are different ordered pairs, but $\{1,2\}$ and $\{2,1\}$ are the same set.

Given two sets X and Y, the set of all ordered pairs of the form (x,y) with $x \in X$ and $y \in Y$ is called the ***Cartesian product of X and Y*** and is denoted by $X \times Y$. Thus

$$X \times Y = \{(x,y) : x \in X \text{ and } y \in Y\}.$$

More generally, for any positive integer n, an ***ordered n-tuple*** is a choice of n objects arranged in a sequence, denoted by $(a_1,\dots,a_n)$. The criterion for equality of ordered n-tuples is the obvious generalization of that for ordered pairs: two ordered n-tuples $(a_1,\dots,a_n)$ and $(a'_1,\dots,a'_n)$ are equal if and only if all of their corresponding components are equal:

$$(a_1,\dots,a_n) = (a'_1,\dots,a'_n) \qquad \text{if and only if} \qquad a_1 = a'_1, \ a_2 = a'_2, \ \dots, \text{ and } a_n = a'_n.$$

If $X_1,\dots,X_n$ are sets, the notation $X_1 \times \cdots \times X_n$ denotes the set of all ordered n-tuples of the form $(x_1,\dots,x_n)$, in which $x_1 \in X_1$, $x_2 \in X_2$, $\dots$, and $x_n \in X_n$. In particular, if all of the sets are the same set X, the n-fold Cartesian product $X \times \cdots \times X$ is usually denoted by X^n; it is the set of all ordered n-tuples of the form $(x_1,\dots,x_n)$ in which $x_1,\dots,x_n$ are all elements of X. For example, $\mathbb{R}^n$ denotes the n-fold Cartesian product of $\mathbb{R}$ with itself; it is just the set of all ordered n-tuples $(x_1,\dots,x_n)$ in which each x_i is a real number.

Proofs in Set Theory

Many theorems in geometry (as well as in most other fields of mathematics) are actually statements about sets. Here are some guidelines for proving such statements.

- PROVING THAT SOMETHING IS AN ELEMENT OF A SET: If X is a specific set and you wish to prove that some object a is an element of X, then you just have to prove that a satisfies the membership criterion for that set. Thus, for example, if X is defined by specification, as in $X = \{x \in D : P(x)\}$, a proof that $a \in X$ needs to show the following:

$$a \in D \text{ and } P(a).$$

- PROVING THAT ONE SET IS A SUBSET OF ANOTHER: Suppose X and Y are sets. To prove that $X \subseteq Y$, just prove the following universal implication:

$$\forall x, \ x \in X \Rightarrow x \in Y.$$

- PROVING THAT TWO SETS ARE EQUAL: Suppose X and Y are sets. Because the statement "$X = Y$" is equivalent to "$X \subseteq Y$ and $Y \subseteq X$," a proof that the sets X and

Y are equal to each other will typically have two parts:

$$\text{Part 1: } \forall x, \ x \in X \Rightarrow x \in Y.$$
$$\text{Part 2: } \forall x, \ x \in Y \Rightarrow x \in X.$$

- PROVING THAT A SET IS NONEMPTY: To prove that a set X is not equal to the empty set, you just need to prove that it contains at least one element. This is an existence statement:

$$\exists x, \ x \in X.$$

- PROVING THAT A SET IS EMPTY: If X is a set, to prove that $X = \varnothing$ is to prove that there does not exist any element of X:

$$\neg(\exists x, \ x \in X).$$

Like any nonexistence statement, this is usually easiest to do by an indirect proof: assume that there exists an element $x \in X$ and derive a contradiction.

Here is an example of a typical proof of set equality.

Theorem G.1. *If X and Y are any sets, then $(X \smallsetminus Y) \cup Y = X \cup Y$.*

Proof. First we will show that $(X \smallsetminus Y) \cup Y \subseteq X \cup Y$. Suppose x is an arbitrary element of $(X \smallsetminus Y) \cup Y$. By definition of union, this means that $x \in X \smallsetminus Y$ or $x \in Y$. In the case that $x \in X \smallsetminus Y$, this means by definition that $x \in X$ and $x \notin Y$; in particular, $x \in X$, so $x \in X \cup Y$. In the second case, $x \in Y$, which also implies $x \in X \cup Y$.

Conversely, we will show that $X \cup Y \subseteq (X \smallsetminus Y) \cup Y$. Suppose $x \in X \cup Y$. Then either $x \in X$ or $x \in Y$. If $x \in Y$, then $x \in (X \smallsetminus Y) \cup Y$ by definition of union. On the other hand, if $x \in X$, then there are two subcases: either $x \in Y$ or not. If $x \in Y$, then $x \in (X \smallsetminus Y) \cup Y$ as before; if $x \notin Y$, then we have $x \in X \smallsetminus Y$, which also implies $x \in (X \smallsetminus Y) \cup Y$. $\qquad\square$

The following general facts about sets can sometimes be used to streamline proofs of set equality.

Theorem G.2 (Distributive Laws for Sets). *For all sets X, Y, and Z,*

(a) $X \cap (Y \cup Z) = (X \cap Y) \cup (X \cap Z)$,

(b) $X \cup (Y \cap Z) = (X \cup Y) \cap (X \cup Z)$.

Proof. Exercise GA. $\qquad\square$

Theorem G.3 (DeMorgan's Laws). *For all sets X, Y, and Z,*

(a) $X \smallsetminus (Y \cap Z) = (X \smallsetminus Y) \cup (X \smallsetminus Z)$,

(b) $X \smallsetminus (Y \cup Z) = (X \smallsetminus Y) \cap (X \smallsetminus Z)$.

Proof. Exercise GB. $\qquad\square$

Functions

Suppose X and Y are sets. A *function from X to Y* is a rule that assigns to each element $x \in X$ one and only one element $y \in Y$, called the *value of f at x*. If f is a function and $x \in X$, the value of f at x is usually denoted by $f(x)$, so the equation $y = f(x)$ means "y is the value of f at x." The set X is called the *domain of f*, and the set Y is called its *codomain* (or sometimes its *range*). The notation $f: X \to Y$ means "f is a function from X to Y" (or, depending on how it is used in a sentence, "the function f from X to Y" or "f, a function from X to Y" or "f, from X to Y"). The words *map* and *mapping* are synonyms for function.

To say that two functions are *equal* is to say that they have the same domain, same codomain, and the same rule of assignment. Thus, if $f: X \to Y$ and $g: X' \to Y'$ are functions, then $f = g$ if and only if $X = X'$, $Y = Y'$, and $f(x) = g(x)$ for every $x \in X$.

To *define* a function $f: X \to Y$, you have to specify the domain X and the codomain Y and then give a rule (or a formula or an algorithm) that specifies unambiguously, for every element $x \in X$, what should be the value $f(x) \in Y$. Typically the rule is expressed by writing a formula, as in "$f(x) = \ldots$"; this is always understood to mean that $f(x)$ is given by the indicated formula for *every* x in the domain.

To ensure that the given rule actually defines a function, you have to make sure that it gives one and only one value for every element of the domain. The following two quantified statements express the requirements for f to be a well-defined function:

$$\forall x \in X, \ \exists y \in Y, \ f(x) = y; \tag{G.1}$$

$$\forall x_1, x_2 \in X, \ x_1 = x_2 \Rightarrow f(x_1) = f(x_2). \tag{G.2}$$

The first condition says that for each x in the domain, f produces some value in the codomain. The second condition says that f yields *only one* value for each element of the domain; in other words, if you give f the same input twice, it will produce the same output both times. The first condition is often expressed by saying that f is *everywhere defined*, and the second by saying that it is *uniquely defined*. If both of these conditions hold, we say that f is *well defined*. (This is just a fancy way of saying that f is a function.)

For example, we could define the "squaring function" from $\mathbb{R}$ to $\mathbb{R}$ by saying "let $f: \mathbb{R} \to \mathbb{R}$ be the function defined by $f(x) = x^2$." This means that the domain of f is $\mathbb{R}$, the codomain is also $\mathbb{R}$, and for every $x \in \mathbb{R}$, the value of f at x is x^2. Because x^2 always yields a single real number whenever x is a real number, this is well defined.

We could also define the "absolute value function" $g: \mathbb{R} \to \mathbb{R}$ by

$$g(x) = \begin{cases} x, & x \geq 0, \\ -x, & x \leq 0. \end{cases}$$

To check that g is well defined, we first note that there is at least one formula for $g(x)$ whenever x is any real number, so g is everywhere defined. If $x \neq 0$, only one of the formulas applies, so $g(x)$ is a uniquely defined real number; however, we have given two formulas for $g(0)$, since both $x \geq 0$ and $x \leq 0$ apply when $x = 0$. Since both formulas yield the same value for $g(0)$, namely $g(0) = 0$, we can conclude that g is also uniquely defined, and thus we have a well-defined function.

Notice that when we define the rule of assignment for a function, there is actually an implied universal quantifier: in the definition of the function $f: \mathbb{R} \to \mathbb{R}$ above, the statement "$f(x) = x^2$" actually means "for all $x \in \mathbb{R}$, $f(x) = x^2$." Thereafter, we should only use the notation $f(x)$ when x refers to some *specific* element of the domain. The notation $f(x)$ refers to the *value of f at a specific element x*, which is an element of Y; while the notation f refers to the entire function (including its domain, its codomain, and its rule of assignment). Thus, in careful mathematical writing, it is not appropriate to refer to "the function $f(x)$."

Surjective, Injective, and Bijective Functions

Suppose $f: X \to Y$ is a function. Every value of f is an element of the codomain Y; but the definition does not require that every element of Y is a value of the function. For example, consider the squaring function $f: \mathbb{R} \to \mathbb{R}$ that we mentioned above, defined by $f(x) = x^2$ for all $x \in \mathbb{R}$. Not all elements of $\mathbb{R}$ occur as values of this function, because x^2 never takes on negative values. Functions that *do* take on all values of the codomain have a special name: a function $f: X \to Y$ is said to be ***surjective***, or to map X ***onto*** Y, if for every $y \in Y$, there is some $x \in X$ such that $f(x) = y$. Symbolically, f is surjective if and only if it satisfies the following condition:

$$\forall y \in Y, \ \exists x \in X, \ f(x) = y. \tag{G.3}$$

Another important aspect of the definition of a function is that it does not prevent different elements of the domain from being mapped to the same element of the codomain. The squaring function, for example, maps two different real numbers to each positive real number, because $f(x) = f(-x)$. Functions for which this never occurs also have a special name: a function $f: X \to Y$ is said to be ***injective*** or ***one-to-one*** if $f(x_1) = f(x_2)$ implies $x_1 = x_2$ whenever x_1 and x_2 are elements of X. Symbolically, f is injective if and only if

$$\forall x_1, x_2 \in X, \ f(x_1) = f(x_2) \Rightarrow x_1 = x_2. \tag{G.4}$$

This is often simpler to understand if phrased in terms of the contrapositive: if $x_1 \neq x_2$, then $f(x_1) \neq f(x_2)$. In words, this means that f takes different elements of the domain to different elements of the codomain. Be sure you understand the difference between condition (G.1) for f to be everywhere defined (part of what it means to be a function) and condition (G.3) for it to be surjective; and likewise, be sure you understand the difference between (G.2) for f to be uniquely defined (also part of the definition of a function) and condition (G.4) for it to be injective.

A function that is both surjective and injective is said to be ***bijective***. Putting together the definitions above, we see that to say $f: X \to Y$ is bijective means that for every $x \in X$, there is one and only one $y \in Y$ such that $f(x) = y$ (this says that f is a well-defined function); and for every $y \in Y$, there is one and only one $x \in X$ such that $f(x) = y$ (there exists at least one such x by surjectivity and at most one by injectivity). For this reason, the term ***one-to-one correspondence*** is often used as a synonym for a bijective function; but you have to take care to remember that a "one-to-one correspondence" (i.e., a bijective function) is different from a "one-to-one function" (i.e., an injective function). Because of this potential ambiguity and also because "onto" is awkward to use as an adjective, in this book we use only the terms *surjective*, *injective*, and *bijective* for these concepts.

Proofs about Functions

Here are some common types of proofs involving functions.

- PROVING THAT A FUNCTION IS WELL DEFINED: Suppose we are given two sets X and Y and a rule of assignment that is supposed to represent a function $f: X \to Y$. To prove that f is well defined, you have to prove the following existence and uniqueness statement:

  for every $x \in X$, there is one and only one $y \in Y$ such that $f(x) = y$.

- PROVING THAT TWO FUNCTIONS ARE EQUAL: As we mentioned above, to prove that two functions are equal, you have to show that they have the same domain, same codomain, and same rule of assignment. Since the domain and codomain are usually stated as part of the definition of a function, it is the rule of assignment that typically has to be checked. Thus if $f: X \to Y$ and $g: X \to Y$ are two functions from X to Y, proving that $f = g$ means proving the following universal statement:

  for every $x \in X$, $f(x) = g(x)$.

- PROVING THAT A FUNCTION IS SURJECTIVE: To prove that $f: X \to Y$ is surjective, you must prove the universal statement (G.3). Thus you would begin by letting y be an arbitrary element of Y, and then somehow you would produce an element $x \in X$ that satisfies the equation $f(x) = y$. Often, an appropriate x can be found by writing down the equation $f(x) = y$ and solving for x. The process of finding such an x is not actually part of the proof (although it might be helpful to include it so that the reader sees where the value of x came from); the actual *proof* of surjectivity is simply the demonstration that the chosen value of x satisfies the equation $f(x) = y$.

- PROVING THAT A FUNCTION IS INJECTIVE: To prove that $f: X \to Y$ is injective, you need to prove the universal implication (G.4). Typically, such a proof begins by letting x_1 and x_2 be arbitrary elements of X and assuming that $f(x_1) = f(x_2)$; then you have to prove that this equation implies $x_1 = x_2$. (It is almost always easier to prove injectivity using (G.4) instead of its contrapositive, even though the contrapositive gives a better intuitive feeling for what injectivity means.)

- PROVING THAT A FUNCTION IS BIJECTIVE: Finally, to prove that f is bijective, you have to prove both of the statements above. Thus a proof of bijectivity will have two parts: one part to prove surjectivity and another to prove injectivity.

It is important to observe that injectivity, surjectivity, and bijectivity depend critically on the choice of domain and codomain, not only on the rule of assignment. For example, the following theorem shows how the same rule of assignment can yield functions with different properties, just by changing the domain and codomain. It will also serve as an illustration of how to prove surjectivity, injectivity, and bijectivity.

Theorem G.4. *Let* $W = \{x \in \mathbb{R} : x \geq 0\}$ *denote the set of all nonnegative real numbers, and define four functions as follows:*

$$f_1 : \mathbb{R} \to \mathbb{R}, \quad \text{defined by } f_1(x) = x^2,$$
$$f_2 : \mathbb{R} \to W, \quad \text{defined by } f_2(x) = x^2,$$
$$f_3 : W \to \mathbb{R}, \quad \text{defined by } f_3(x) = x^2,$$
$$f_4 : W \to W, \quad \text{defined by } f_4(x) = x^2.$$

Then f_1 *is neither injective nor surjective;* f_2 *is surjective but not injective;* f_3 *is injective but not surjective; and* f_4 *is bijective.*

Proof. First we consider f_1. To prove that f_1 is not surjective, we need to prove the negation of the statement "for all $y \in \mathbb{R}$, there exists $x \in \mathbb{R}$ satisfying $x^2 = y$." This negation is equivalent to the following existence statement: "there exists $y \in \mathbb{R}$ such that there is no $x \in \mathbb{R}$ satisfying $x^2 = y$." In other words, we need to find real number y that is not the square of another real number. Any negative number will do: for example, -1 is in the codomain but there is no real number x such that $f_1(x) = -1$." Thus f_1 is not surjective.

To show that f_1 is not injective, we need to prove the negation of "for all $x_1, x_2 \in \mathbb{R}$, if $(x_1)^2 = (x_2)^2$, then $x_1 = x_2$." Thus again we need to prove an existence statement: "there exist $x_1, x_2 \in \mathbb{R}$ such that $(x_1)^2 = (x_2)^2$ but $x_1 \neq x_2$." We can take, for example, $x_1 = 1$ and $x_2 = -1$, because $1^2 = (-1)^2$ while $1 \neq -1$.

Next consider f_2. This function is not injective for the same reason as f_1: the numbers 1 and -1 are two distinct elements of the domain of f_2 such that $1^2 = (-1)^2$. To show that f_2 is surjective, we need to prove the following statement: "for every $y \in W$, there exists $x \in \mathbb{R}$ such that $x^2 = y$." Let $y \in W$ be arbitrary (or, in other words, let y be an arbitrary nonnegative real number). Theorem H.10(h) in Appendix H shows that y has a nonnegative square root $x = \sqrt{y}$, and this number satisfies $f_2(x) = \left(\sqrt{y}\right)^2 = y$.

Now consider f_3. It is not surjective because it never produces negative values: for example, -1 is element of the codomain, but there is no $x \in W$ such that $x^2 = -1$. To prove that f_3 is injective, suppose x_1 and x_2 are elements of W (i.e., nonnegative real numbers) such that $f_3(x_1) = f_3(x_2)$, which means $(x_1)^2 = (x_2)^2$. Then Theorem H.10(l) implies that $x_1 = \pm x_2$. Thus there are two cases: either $x_1 = -x_2$ or $x_1 = x_2$. In the first case, since $x_2 \geq 0$, it follows by algebra that $x_1 = -x_2 \leq 0$; but since our hypothesis that $x_1 \in W$ implies $x_1 \geq 0$, it follows from Theorem H.8(s) that $x_1 = 0$, and therefore $x_2 = 0$ as well. Thus $x_1 = x_2$ in this case. In the second case, the same result holds by hypothesis.

Finally, consider f_4. To show that f_4 is surjective, we just carry out the same argument that we used to prove surjectivity of f_2. Similarly, the argument that f_3 is injective also applies to f_4. Alternatively, we can just note that Theorem H.10(h) shows that for every $y \in W$ there exists one and only one $x \in W$ such that $x^2 = y$, namely $x = \sqrt{y}$. $\quad\square$

Exercises

GA. Prove Theorem G.2 (the distributive laws for sets).

GB. Prove Theorem G.3 (De Morgan's laws).

GC. Decide which of the following statements are true for all sets X, Y, and Z. For each one that is true, give a proof; for each one that is false, give an example of *specific* set(s) for which the statement is untrue.

(a) $\varnothing \in X$.

(b) $\varnothing \subseteq X$.

(c) $X \subseteq X$.

(d) $X \subseteq (X \cup Y)$.

(e) $X \subseteq (X \cap Y)$.

(f) $(X \cup Y) \subseteq X$.

(g) $(X \cap Y) \subseteq X$.

(h) $X \cup (X \cap Y) = X$.

(i) $X \cap (X \cup Y) = X$.

(j) $X \cap Y = (X \smallsetminus Y) \cap Y$.

(k) $X \smallsetminus Y = (X \cup Y) \smallsetminus Y$.

(l) $X \smallsetminus (X \smallsetminus Y) = Y$.

(m) $X \smallsetminus (Y \smallsetminus X) = X$.

(n) $X \subseteq Y$ and $X \subseteq Z \Leftrightarrow X \subseteq (Y \cup Z)$.

(o) $X \subseteq Y$ or $X \subseteq Z \Leftrightarrow X \subseteq (Y \cup Z)$.

(p) $X \subseteq Y$ and $X \subseteq Z \Leftrightarrow X \subseteq (Y \cap Z)$.

(q) $X \subseteq Y$ or $X \subseteq Z \Leftrightarrow X \subseteq (Y \cap Z)$.

GD. For each of the following functions, decide whether it is injective, surjective, neither, or both and prove your answer correct. In these definitions, $\mathbb{R}^+$ denotes the set of all positive real numbers: $\mathbb{R}^+ = \{x \in \mathbb{R} : x > 0\}$.

(a) $g_1 : \mathbb{R}^+ \to \mathbb{R}^+$; $g_1(x) = 1/x$.

(b) $g_2 : \mathbb{R} \to \mathbb{R}$; $g_2(x) = 2x - 3$.

(c) $g_3 : \mathbb{R}^2 \to \mathbb{R}$; $g_3(x, y) = xy$.

(d) $g_4 : \mathbb{R} \to \mathbb{R}^2$; $g_4(x) = (x, x^2)$.

Properties of
the Real Numbers

Because our axioms for plane geometry are predicated on an understanding of the real number system, it is important to establish clearly what properties of the real numbers we are taking for granted. In this appendix, we summarize those properties.

Primitive Terms

Just as in geometry, some terms must remain undefined to avoid circularity. You know intuitively what these terms mean already; these descriptions are just meant to ensure that we're all thinking about the same things when we use the terms. Everything we need to know about these terms is contained in the properties listed below.

- ***Real number:*** Intuitively, a real number represents a point on the number line or a (signed) distance left or right from the origin or any quantity that has a finite or infinite decimal representation. Real numbers include integers, positive and negative fractions, and irrational numbers like $\sqrt{2}$, π, and e.

- ***Integer:*** An integer is a whole number (positive, negative, or zero).

- ***Zero:*** The number zero is denoted by 0.

- ***One:*** The number one is denoted by 1.

- ***Addition:*** The result of adding two real numbers x and y is denoted by $x + y$ and is called the ***sum of x and y***.

- ***Multiplication:*** The result of multiplying two real numbers x and y is denoted by xy or $x \cdot y$ or $x \times y$ and is called the ***product of x and y***.

- ***Less than:*** To say that x is less than y, denoted by $x < y$, means intuitively that x is to the left of y on the number line.

Definitions

In all the definitions below, x and y represent arbitrary real numbers.

- The set of all real numbers is denoted by $\mathbb{R}$, and the set of all integers is denoted by $\mathbb{Z}$.

- The numbers **2** through **10** are defined by $2 = 1 + 1$, $3 = 2 + 1$, etc. The decimal representations for other numbers are defined by the usual rules of decimal notation: for example, 23 is defined to be $2 \cdot 10 + 3$, etc.

- The *additive inverse* (or *negative*) *of x* is the number $-x$ that satisfies $x + (-x) = 0$ and whose existence and uniqueness are guaranteed by Theorem H.4(a) below.

- The *difference between x and y*, denoted by $x - y$, is the real number defined by $x - y = x + (-y)$ and is said to be obtained by *subtracting y from x*.

- If $x \neq 0$, the *multiplicative inverse*, or *reciprocal, of x* is the number x^{-1} that satisfies $x \cdot x^{-1} = 1$, whose existence and uniqueness are guaranteed by Theorem H.6(a) below.

- If $y \neq 0$, the *quotient of x and y*, denoted by x/y, is the real number defined by $x/y = xy^{-1}$ and is said to be obtained by *dividing x by y*.

- A real number is said to be *rational* if it is equal to p/q for some integers p and q with $q \neq 0$. The set of all rational numbers is denoted by $\mathbb{Q}$.

- A real number is said to be *irrational* if it is not rational.

- The statement *x is less than or equal to y*, denoted by $x \leq y$, means $x < y$ or $x = y$.

- The statement *x is greater than y*, denoted by $x > y$, means $y < x$.

- The statement *x is greater than or equal to y*, denoted by $x \geq y$, means $x > y$ or $x = y$.

- A real number x is said to be *positive* if $x > 0$ and *negative* if $x < 0$.

- A real number x is said to be *nonnegative* if $x \geq 0$ and *nonpositive* if $x \leq 0$.

- A real number x is said to be *nonzero* if $x \neq 0$.

- If a and b are real numbers such that $a < b$, the *open interval* (a,b) and the *closed interval* $[a,b]$ are the following sets:
$$(a,b) = \{x \in \mathbb{R} : a < x < b\}; \qquad [a,b] = \{x \in \mathbb{R} : a \leq x \leq b\}.$$

- If m and n are integers with $n \neq 0$, then the statements *m is divisible by n* and *n divides m* both mean that m/n is an integer.

- An integer is said to be *even* if it is divisible by 2.

- An integer is said to be *odd* if it is not even.

- An integer p is said to be *prime* if $p > 1$ and p is divisible by no positive integers other than 1 and p.

- For any real number x, the *absolute value of x*, denoted by $|x|$, is defined by
$$|x| = \begin{cases} x & \text{if } x \geq 0, \\ -x & \text{if } x < 0. \end{cases}$$

- If x is a real number and n is a positive integer, the *nth power of x*, denoted by x^n, is the product of n factors of x. The *square* of x is the number $x^2 = x \cdot x$.

- If x is a nonnegative real number, the ***square root of x***, denoted by $\sqrt{x}$, is the unique nonnegative real number whose square is x (see Theorem H.10(h) below).
- If S is a set of real numbers, a real number b is said to be an ***upper bound for S*** if $b \geq x$ for every x in S. It is said to be a ***least upper bound for S*** if every other upper bound b' for S satisfies $b' \geq b$. The terms ***lower bound*** and ***greatest lower bound*** are defined similarly.

Properties

If our purpose were to develop a rigorous theory of real numbers and integers, we would choose a few of these properties as axioms and use them to prove the others. For the purposes of this book, though, you can treat all of these as if they were axioms. In all of the properties below, the letters w, x, y, z represent arbitrary real numbers unless otherwise specified.

Theorem H.1 (Properties of Equality).

(a) *If $x = y$, then $-x = -y$.*

(b) *If $x = y$ and x and y are nonzero, then $x^{-1} = y^{-1}$.*

(c) *If $x = y$ and $z = w$, then $x + z = y + w$, $xz = yw$, and $x - z = y - w$.*

(d) *If $x = y$ and $z = w$ and z and w are both nonzero, then $x/z = y/w$.*

(e) *If $x = y$, then $x^2 = y^2$.*

Theorem H.2 (Closure Properties).

(a) *Every integer is a real number.*

(b) *If x and y are integers, then so are $x + y$ and xy.*

(c) *If x and y are real numbers, then so are $x + y$ and xy.*

Theorem H.3 (Properties of Zero and One).

(a) *0 and 1 are integers.*

(b) *$0 < 1$.*

(c) *$0 + x = x$.*

(d) *$0x = 0$.*

(e) *$1x = x$.*

(f) *If $xy = 0$, then $x = 0$ or $y = 0$.*

Theorem H.4 (Properties of Negatives).

(a) *If x is a real number, there is a unique real number $-x$ such that $x + (-x) = 0$.*

(b) *If x is an integer, then so is $-x$.*

(c) *$-0 = 0$.*

(d) *$-(-x) = x$.*

(e) *$-x = (-1)x$.*

(f) *$(-x)y = -(xy) = x(-y)$.*

(g) *$(-x)(-y) = xy$.*

(h) $-(x+y) = (-x) + (-y) = -x - y.$

(i) $-(x-y) = y - x.$

(j) $-(-x-y) = x + y.$

Theorem H.5 (Commutative, Associative, and Distributive Laws).

(a) $x + y = y + x.$

(b) $xy = yx.$

(c) $(x+y) + z = x + (y+z).$

(d) $(xy)z = x(yz).$

(e) $x(y+z) = xy + xz.$

(f) $(x+y)z - xz + yz.$

(g) $x + x = 2x.$

(h) $x(y-z) = xy - xz.$

(i) $(x-y)z = xz - yz.$

(j) $(x+y)(z+w) = xz + xw + yz + yw.$

(k) $(x+y)(z-w) = xz - xw + yz - yw.$

(l) $(x-y)(z-w) = xz - xw - yz + yw.$

Theorem H.6 (Properties of Inverses).

(a) *If x is any nonzero real number, there is a unique real number x^{-1} such that $x \cdot x^{-1} = 1$.*

(b) $1^{-1} = 1.$

(c) $(x^{-1})^{-1} = x$ *if x is nonzero.*

(d) $(-x)^{-1} = -(x^{-1})$ *if x is nonzero.*

(e) $(xy)^{-1} = x^{-1}y^{-1}$ *if x and y are nonzero.*

(f) $(x/y)^{-1} = y/x$ *if x and y are nonzero.*

Theorem H.7 (Properties of Quotients).

(a) $x/1 = x.$

(b) $1/x = x^{-1}$ *if x is nonzero.*

(c) $(x/y)(z/w) = (xz)/(yw)$ *if y and w are nonzero.*

(d) $(x/y)/(z/w) = (xw)/(yz)$ *if y, z, and w are nonzero.*

(e) $(xz)/(yz) = x/y$ *if y and z are nonzero.*

(f) $(-x)/y = -(x/y) = x/(-y)$ *if y is nonzero.*

(g) $(-x)/(-y) = x/y$ *if y is nonzero.*

(h) $x/y + z/w = (xw + yz)/(yw)$ *if y and w are nonzero.*

(i) $x/y - z/w = (xw - yz)/(yw)$ *if y and w are nonzero.*

Theorem H.8 (Properties of Inequalities).

(a) (TRICHOTOMY LAW) *If x and y are real numbers, then exactly one of the following three possibilities must hold: $x < y$, $x = y$, or $x > y$.*

(b) *If $x < y$, then $-x > -y$.*

(c) *If $x < y$ and x and y are both positive or both negative, then $x^{-1} > y^{-1}$.*

(d) *If $x < y$ and $y < z$, then $x < z$.*

(e) *If $x \le y$ and $y \le z$, then $x \le z$.*

(f) *If $x \le y$ and $y < z$, then $x < z$.*

(g) *If $x < y$ and $y \le z$, then $x < z$.*

(h) *If $x < y$ and $z < w$, then $x + z < y + w$.*

(i) *If $x \le y$ and $z < w$, then $x + z < y + w$.*

(j) *If $x \le y$ and $z \le w$, then $x + z \le y + w$.*

(k) *If $x < y$ and $z > 0$, then $xz < yz$.*

(l) *If $x < y$ and $z < 0$, then $xz > yz$.*

(m) *If $x \le y$ and $z \ge 0$, then $xz \le yz$.*

(n) *If $x \le y$ and $z \le 0$, then $xz \ge yz$.*

(o) *If $x < y$ and $z \le w$ and x, y, z, w are positive, then $xz < yw$.*

(p) *If $x \le y$ and $z \le w$ and x, y, z, w are positive, then $xz \le yw$.*

(q) *$xy > 0$ if and only if x and y are both positive or both negative.*

(r) *$xy < 0$ if and only if one is positive and the other is negative.*

(s) *If $x \le y$ and $y \le x$, then $x = y$.*

Theorem H.9 (Properties of Absolute Values).

(a) *If x is any real number, then $|x| \ge 0$.*

(b) *$|x| = 0$ if and only if $x = 0$.*

(c) *$|-x| = |x|$.*

(d) *$|x^{-1}| = 1/|x|$ if $x \ne 0$.*

(e) *$|xy| = |x||y|$.*

(f) *$|x/y| = |x|/|y|$ if $y \ne 0$.*

(g) (THE TRIANGLE INEQUALITY FOR REAL NUMBERS) *$|x + y| \le |x| + |y|$.*

(h) *If x and y are both nonnegative, then $|x| \ge |y|$ if and only if $x \ge y$.*

(i) *If x and y are both negative, then $|x| \ge |y|$ if and only if $x \le y$.*

Theorem H.10 (Properties of Squares and Square Roots).

(a) *If x is any real number, then $x^2 \ge 0$.*

(b) *$x^2 = 0$ if and only if $x = 0$.*

(c) *$x^2 > 0$ if and only if $x \ne 0$.*

(d) *$(-x)^2 = x^2$.*

(e) *$(x^{-1})^2 = 1/x^2$.*

(f) *If x and y are both nonnegative, then $x < y \Rightarrow x^2 < y^2$.*

(g) *If x and y are both nonpositive, then $x < y \Rightarrow x^2 > y^2$.*

(h) *If x is any nonnegative real number, there exists a unique nonnegative real number $\sqrt{x}$ such that $\left(\sqrt{x}\right)^2 = x$.*

(i) *If x and y are both nonnegative, then $x < y \Rightarrow \sqrt{x} < \sqrt{y}$.*

(j) *$\sqrt{x^2} = |x|$.*

(k) *If $x^2 = y$, then $x = \pm\sqrt{y}$.*

(l) *If $x^2 = y^2$, then $x = \pm y$.*

Theorem H.11 (Density and Completeness Properties).

(a) (LEAST UPPER BOUND PROPERTY) *If S is any nonempty set of real numbers and S has an upper bound, then S has a unique least upper bound.*

(b) (DENSITY) *If x and y are real numbers such that $x < y$, then there exist a rational number q such that $x < q < y$ and an irrational number r such that $x < r < y$.*

(c) *There does not exist a smallest positive real number or a smallest positive rational number.*

Theorem H.12 (Properties of Integers).

(a) *If n is a positive integer, then $n \geq 1$.*

(b) *If m and n are integers such that $m > n$, then $m \geq n + 1$.*

(c) *There does not exist a largest integer.*

(d) (THE ARCHIMEDEAN PROPERTY OF REAL NUMBERS) *If M and ε are positive real numbers, there exists a positive integer n such that $n\varepsilon > M$.*

(e) (THE WELL-ORDERING PRINCIPLE) *Every nonempty set of positive integers contains a smallest number.*

Theorem H.13 (Properties of Divisibility).   *In each of the following statements, m, n, and p are assumed to be integers.*

(a) *If mn is divisible by p and p is prime, then m is divisible by p or n is divisible by p.*

(b) *If p is prime and m^2 is divisible by p, then m is divisible by p.*

(c) *m is even if and only if $m = 2k$ for some integer k, and it is odd if and only if $m = 2k + 1$ for some integer k.*

(d) *$m + n$ is even if and only if m and n are both odd or both even.*

(e) *$m + n$ is odd if and only if one of the summands is even and the other is odd.*

(f) *mn is even if and only if m or n is even.*

(g) *mn is odd if and only if m and n are both odd.*

(h) *n^2 is even if and only if n is even, and it is odd if and only if n is odd.*

(i) *Every rational number has a unique representation as a quotient of the form $x = m/n$, in which $n > 0$ and there is no prime that divides both m and n. Such a representation is said to be in **lowest terms**.*

Theorem H.14 (Properties of Exponents). *In these statements, m and n are positive integers.*

(a) $x^n y^n = (xy)^n$.

(b) $x^{m+n} = x^m x^n$.

(c) $(x^m)^n = x^{mn}$.

(d) $x^n / y^n = (x/y)^n$ *if y is nonzero.*

Rigid Motions: Another Approach

In Chapter 5, we mentioned that it is possible to substitute some sort of motion postulate in place of the SAS postulate and thereby to vindicate Euclid's "method of superposition." In this appendix we describe that approach.

The intuitive notion of "moving" geometric figures around is formalized in the following concept. A **transformation of the plane** is a bijective map F from the plane to itself. A **rigid motion of the plane** is a transformation such that for any points A, B, C, if $A' = F(A)$, $B' = F(B)$, and $C' = F(C)$, then the following properties hold:

A. ANGLE MEASURE IS PRESERVED: If $\angle ABC$ is an angle, then so is $\angle A'B'C'$, and $m\angle A'B'C' = m\angle ABC$.

B. BETWEENNESS OF POINTS IS PRESERVED: If $A * B * C$, then $A' * B' * C'$.

C. COLLINEARITY IS PRESERVED: If A, B, C are collinear, then so are A', B', C'.

D. DISTANCE IS PRESERVED: $AB = A'B'$.

Technically, the requirement that rigid motions preserve betweenness is redundant, because if F is a bijective map from the plane to itself that preserves collinearity and distances, it follows from the partial converse to the betweenness theorem for points (Theorem 3.11, whose proof did not rely on the SAS postulate) that F automatically preserves betweenness as well. However, including betweenness and arranging the properties in alphabetical order makes the list easy to remember—this is often referred to as the **ABCD property** of rigid motions.

Without trying to prove anything yet, let us think intuitively about a few examples of rigid motions in Euclidean geometry. If you imagine a portion of the plane as a piece of paper sitting on a tabletop, you can visualize rigid motions as the results of sliding the paper along the table or picking the paper up and turning it over, without changing any of the geometric relationships among figures on the paper. Here are some familiar types of rigid motions (see Fig. I.1):

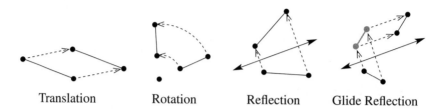

Translation   Rotation   Reflection   Glide Reflection

Fig. I.1. Some rigid motions of the plane.

- TRANSLATION: Moving every point a fixed distance in the same direction.
- ROTATION: Moving every point through a certain angular measure around a fixed center.
- REFLECTION: Moving every point to its "mirror image" across a fixed line.

In addition, there is a fourth type of rigid motion that may be less familiar but is equally important:

- GLIDE REFLECTION: Reflecting across a line ℓ and then translating parallel to ℓ.

Some properties of these types of rigid motions are developed in the exercises at the end of this appendix.

Reflection Implies SAS

In this section, we show that it is possible to replace the SAS postulate with a postulate about the existence of reflections. For this purpose, we assume only Postulates 1–8 of neutral geometry, but not the SAS postulate. Thus we can use all of the results that we proved in Chapters 3 and 4, as well as the material that was covered in Chapter 5 before the introduction of the SAS postulate, including the definition of triangles and the transitive property of congruence of triangles. We can also use the unique triangle theorem (Theorem 5.11), because its proof uses only results from Chapters 3 and 4, as you can check. But we cannot use the SAS postulate or any of the results that follow from it.

The next lemma shows that a rigid motion transforms each triangle into one that is congruent to the original.

Lemma I.1. *Suppose F is a rigid motion and A, B, C are noncollinear points. Let* $A' = F(A)$, $B' = F(B)$, *and* $C' = F(C)$. *Then* A', B', C' *are also noncollinear, and* $\triangle A'B'C' \cong \triangle ABC$.

Proof. The main part of the proof is to show that A', B', C' are noncollinear. We should note that this does not follow directly from the fact that rigid motions preserve collinearity: the definition of rigid motions says that collinearity of A, B, C implies collinearity of A', B', C'; it does not say that *noncollinearity* of A, B, C implies noncollinearity of A', B', C'.

To prove that A', B', C' are noncollinear, assume for the sake of contradiction that there is a line ℓ that contains all three of them. By Hilbert's betweenness axiom (Theorem

3.9), exactly one of these three points is between the other two; for definiteness, let us say that $A' * B' * C'$. Because $A'B' < A'C'$, which in turn is equal to AC, Euclid's segment cutoff theorem (Corollary 3.37) shows that there is a unique point $P \in \text{Int} \overline{AC}$ such that $AP = A'B'$. Then because $AP = A'B'$ and $AC = A'C'$, it follows from the segment subtraction theorem (Theorem 3.22(c)) that $PC = B'C'$.

Let $P' = F(P)$. Because A, P, C are collinear and F preserves collinearity, it follows that A', P', C' are collinear; and because F preserves distances and betweenness, it follows that $A' * P' * C'$ and $A'P' = AP = A'B'$. However, we are assuming that B' satisfies $A' * B' * C'$; thus P' and B' are both points on $\overrightarrow{A'C'}$ at the same distance from A, so the unique point theorem (Corollary 3.36) implies that $P' = B'$, or in other words $F(P) = F(B)$. Because P lies on $\overleftrightarrow{AB}$ but B does not, P and B are distinct, so this contradicts the fact that F is injective. This completes the proof that A', B', C' are noncollinear.

Now, because F preserves distances, we have $AB = A'B'$, $BC = B'C'$, and $AC = A'C'$; and because it preserves angle measures, we also have $m\angle ABC = m\angle A'B'C'$, $m\angle ACB = m\angle A'C'B'$, and $m\angle BAC = m\angle B'A'C'$. Thus $\triangle A'B'C' \cong \triangle ABC$. $\square$

If we wish to use rigid motions in place of the SAS postulate, we need a postulate that ensures there are enough rigid motions to superimpose one triangle upon another if they satisfy the SAS hypotheses. It turns out that we don't need to assume very much, because all of the needed rigid motions can be built up from successive reflections. To state the postulate concisely, we introduce the following definition. If F is a rigid motion and A is a point, F is said to *fix A*, and A is said to be a *fixed point of F*, if $F(A) = A$.

As we will see below, the following postulate serves as an adequate substitute for the SAS postulate.

> **Postulate 9′ (The Reflection Postulate).** *For every line ℓ, there is a rigid motion R_ℓ, called **reflection across ℓ**, that fixes every point of ℓ but no other points.*

Before proving that the reflection postulate implies SAS, let us develop some basic properties of reflections.

Lemma I.2 (Properties of Reflections). *Suppose ℓ is a line and A is a point not on ℓ.*

(a) *A and $R_\ell(A)$ lie on opposite sides of ℓ.*

(b) *A point A' is equal to $R_\ell(A)$ if and only if ℓ is the perpendicular bisector of $\overline{AA'}$.*

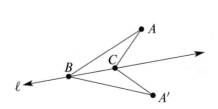

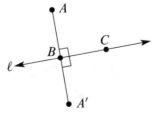

Fig. I.2. A and A' are on opposite sides of ℓ.   **Fig. I.3.** ℓ is the perpendicular bisector of $\overline{AA'}$.

Proof. To prove (a), let $A' = R_\ell(A)$, and let B and C be any two distinct points on ℓ, so $R_\ell(B) = B$ and $R_\ell(C) = C$ (Fig. I.2). Since we are assuming $A \notin \ell$, the points A, B, C are noncollinear, so Lemma I.1 implies that A', B, C are noncollinear and $\triangle A'BC \cong \triangle ABC$. In particular, this means that $A' \notin \ell$. If A' were on the same side of ℓ as A, the unique triangle theorem would imply $A = A'$, which is a contradiction because $A \notin \ell$ and therefore A is not a fixed point of R_ℓ. Thus A and A' lie on opposite sides of ℓ.

Next we prove (b). Assume first that $A' = R_\ell(A)$ (Fig. I.3). By the result of part (a), there is a point B where ℓ meets the interior of $\overline{AA'}$. Let C be any point on ℓ different from B. Since $R_\ell(B) = B$ and $R_\ell(C) = C$, it follows that $\angle ABC \cong \angle A'BC$. These two angles are thus congruent and adjacent, so they are both right by Corollary 4.15. Since $AB = A'B$ (because R_ℓ preserves distances), it follows that B is the midpoint of $\overline{AA'}$, and thus ℓ is the perpendicular bisector as claimed.

To prove the converse, assume A' is a point such that ℓ is the perpendicular bisector of $\overline{AA'}$. Let B be the midpoint of $\overline{AA'}$, which is on ℓ, and let C be any point on ℓ different from B. Then $R_\ell(B) = B$ and $R_\ell(C) = C$. If we set $A'' = R_\ell(A)$, then $m\angle A''BC = m\angle ABC = 90°$. Part (a) implies that A'' is on the opposite side of ℓ from A and thus on the same side of ℓ as A'. Since $m\angle A'BC$ is also equal to $90°$, it follows from the unique ray theorem (Corollary 4.6) that $\overrightarrow{CA'} = \overrightarrow{CA''}$; and since $CA' = CA = CA''$, it follows from the unique point theorem (Corollary 3.36) that $A' = A'' = R_\ell(A)$. □

The next theorem is the main result of this section.

Theorem I.3 (Reflection Implies SAS). *If Postulates 1–8 and the reflection postulate are true, then the SAS postulate is also true.*

Proof. Assume that Postulates 1–8 and the reflection postulate are true. To prove that the SAS postulate is also true, assume that its hypotheses are satisfied: thus we assume that $\triangle ABC$ and $\triangle DEF$ are triangles in which $\overline{AB} \cong \overline{DE}$, $\angle B \cong \angle E$, and $\overline{BC} \cong \overline{EF}$. Using reflections, we will define three new triangles $\triangle A'B'C'$, $\triangle A''B''C''$, and $\triangle A'''B'''C'''$, each of which is congruent to $\triangle ABC$ and such that $\triangle A'''B'''C'''$ is actually equal to $\triangle DEF$.

STEP 1: *Find $\triangle A'B'C' \cong \triangle ABC$ with $A' = D$.* For this step, we have two cases. If $D \neq A$, let ℓ_1 be the perpendicular bisector of $\overline{AD}$ (Fig. I.4), and define $A' = R_{\ell_1}(A) = D$, $B' = R_{\ell_1}(B)$, and $C' = R_{\ell_1}(C)$. Otherwise, if $D = A$, then we simply let $\triangle A'B'C' = \triangle ABC$. In either case, we have $A' = D$ and $\triangle A'B'C' \cong \triangle ABC$.

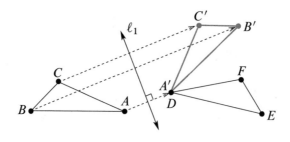

Fig. I.4. Proof that reflection implies SAS: Step 1.

STEP 2: *Find $\triangle A''B''C'' \cong \triangle ABC$ with $A'' = D$ and $B'' = E$.* Note that A' is distinct from both B' and E, so $\angle B'A'E$ is an angle. Now we have three possibilities: $\angle B'A'E$ is a proper angle, a straight angle, or a zero angle. First suppose $\angle B'A'E$ is a proper angle (Fig. I.5). In this case, we let $\overrightarrow{A'Q}$ be the bisector of $\angle B'A'E$, let $\ell_2 = \overleftrightarrow{A'Q}$, and write $A'' = R_{\ell_2}(A')$, $B'' = R_{\ell_2}(B')$, and $C'' = R_{\ell_2}(C')$. Then $A'' = A' = D$ because A' lies on ℓ_2. We wish to show that $B'' = E$. Now, B'' is on the opposite side of ℓ_2 by Lemma I.2 above, and E is on the opposite side of ℓ_2 by Theorem 4.12; so B'' and E are on the same side of ℓ_2. Also, $m\angle QA''B'' = m\angle QA''B'$ because F preserves angle measures, and $m\angle QA''B' = m\angle QA''E$ by definition of angle bisector, so $m\angle QA''B'' = m\angle QA''E$ by transitivity, and thus $\overrightarrow{A''B''} = \overrightarrow{A''E}$ by the unique ray theorem. Since $A''B'' = A'B' = A'E = A''E$, it follows from the unique point theorem that $B'' = E$ as claimed.

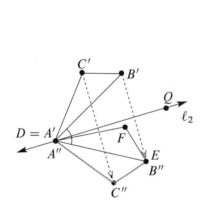

Fig. I.5. Step 2: proper angle case.

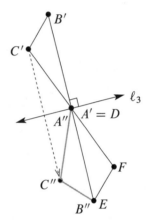

Fig. I.6. Step 2: straight angle case.

The second possibility is that $\angle B'A'E$ is a straight angle (Fig. I.6). Then $B' * A' * E$ and $B'A' = A'E$, so A' is the midpoint of $\overline{B'E}$. If ℓ_3 is the perpendicular bisector of this segment, then Lemma I.2 shows that $R_{\ell_3}(A') = A'$ and $R_{\ell_3}(B') = E$. In this case we set $A'' = R_{\ell_3}(A') = A'$, $B'' = R_{\ell_3}(B') = E$, and $C'' = R_{\ell_3}(C')$.

Finally, if $\angle B'A'E$ is a zero angle, then $\overrightarrow{A'B'}$ and $\overrightarrow{A'E}$ are the same ray, and since $A'B' = A'E$, it follows from the unique point theorem that $B' = E$. In this case, we simply set $\triangle A''B''C'' = \triangle A'B'C'$.

In all three cases above, $\triangle A''B''C''$ is congruent to $\triangle A'B'C'$ and thus also (by transitivity) to $\triangle ABC$, and it shares two vertices with $\triangle DEF$: $A'' = D$ and $B'' = E$.

STEP 3: *Find $\triangle A'''B'''C''' \cong \triangle ABC$ with $A''' = D$, $B''' = E$, and $C''' = F$.* Again there are two cases. If $C'' = F$, then we can just set $\triangle A'''B'''C''' = \triangle A''B''C''$ and be done with it. On the other hand, if $C'' \neq F$, then C'' and F must be on opposite sides of $\overleftrightarrow{A''B''}$, because otherwise the unique triangle theorem would guarantee that they are the same point. Let $\ell_4 = \overleftrightarrow{A''B''}$, and let $A''' = R_{\ell_4}(A'') = A'' = D$, $B''' = R_{\ell_4}(B'') = B'' = E$, and $C''' = R_{\ell_4}(C'')$. Then Lemma I.2 implies that C''' is on the same side of ℓ_4 as F and $\triangle A'''B'''C''' \cong \triangle A''B''C''$, and the unique triangle theorem guarantees that $C''' = F$ as desired. Thus by transitivity of congruence, we have $\triangle ABC \cong \triangle A'B'C' \cong \triangle A''B''C'' \cong \triangle A'''B'''C''' = \triangle DEF$. $\qquad\square$

This theorem shows that we could have chosen the reflection postulate in place of the SAS postulate, and we would have been able to prove all of the same theorems in neutral, Euclidean, and hyperbolic geometries. In addition, there are some other advantages to basing an axiomatic system on the theory of rigid motions.

One of the most useful features of rigid motions is that they allow one to give a uniform definition of *congruence* that applies to figures of all kinds, instead of having to give separate definitions as we did for congruence of segments, angles, triangles, convex polygons, etc. In this approach, one says that two sets of points S_1 and S_2 are **congruent** if there is a rigid motion F such that $F(S_1) = S_2$, where $F(S_1) = \{F(A) : A \in S_1\}$ represents the set of images of points in S_1. Of course, if this definition is adopted, then for each different type of figure, one must prove a theorem that shows congruence is equivalent to having equal corresponding measurements—for example, there would be a theorem for triangles that says two triangles are congruent if and only if there is a correspondence between their vertices such that corresponding side lengths and corresponding angle measures are equal. Thus the main advantage of using this general definition of congruence is one of conceptual unity rather than efficiency.

Another advantage of basing the axiomatic system on rigid motions is that it leads naturally to a deeper study of rigid motions for their own sake. Since moving things around is part of our everyday experience with geometric objects, this is a natural, useful, and satisfying field of study. Some of the types of theorems that can be proved are indicated in the exercises at the end of this appendix.

SAS Implies Reflection

Thanks to Theorem I.3, we now know that the reflection postulate is strong enough to allow us to prove all the theorems we proved using the SAS postulate. One might wonder whether it is even stronger than SAS—might the reflection postulate allow us to prove theorems that cannot be derived from SAS? In this section, we prove that it is not stronger, because the reflection postulate can also be derived from SAS.

Theorem I.4 (SAS Implies Reflection). *If Postulates 1–8 and the SAS postulate are true, then the reflection postulate is also true.*

Proof. Assume that Postulates 1–8 and the SAS postulate are all true. Because of this assumption, we may use all of the theorems of neutral geometry that were proved in Chapters 3–9.

Let ℓ be a line. We need to prove that there is a rigid motion R_ℓ that fixes points on ℓ but no others. Here is how to define R_ℓ. If A is a point on ℓ, we just let $R_\ell(A) = A$. If $A \notin \ell$, we let $R_\ell(A) = A'$, where A' is the unique point such that ℓ is the perpendicular bisector of $\overline{AA'}$; the existence and uniqueness of such a point is guaranteed by Theorem 7.10.

Before we show that R_ℓ is an isometry, we will show first that applying R_ℓ twice brings every point back to its initial position:

$$R_\ell(R_\ell(A)) = A \quad \text{for all } A. \tag{I.1}$$

To prove this, we consider two cases. First, if $A \in \ell$, then $R_\ell(A) = A$, so $R_\ell(R_\ell(A)) = R_\ell(A) = A$. On the other hand, if $A \notin \ell$, let $A' = R_\ell(A)$; by definition, this means that ℓ

is the perpendicular bisector of $\overline{AA'}$. Since $\overline{AA'} = \overline{A'A}$, this also means that $A = R_\ell(A')$, so $R_\ell(R_\ell(A)) = R_\ell(A') = A$.

Next we need to show that R_ℓ is bijective. To prove injectivity, suppose A_1 and A_2 are points such that $R_\ell(A_1) = R_\ell(A_2)$. Applying R_ℓ to both sides and using (I.1), we obtain

$$A_1 = R_\ell(R_\ell(A_1)) = R_\ell(R_\ell(A_2)) = A_2,$$

so R_ℓ is injective. To prove surjectivity, let A' be an arbitrary point, and let $A = R_\ell(A')$. Then (I.1) implies

$$R_\ell(A) = R_\ell(R_\ell(A')) = A',$$

so R_ℓ is surjective.

For the rest of the proof, for brevity, let us adopt the convention that if A is any point, then A' denotes the point $R_\ell(A)$.

Our next task is to show that R_ℓ preserves distances. Let A and B be arbitrary points. We need to show that $A'B' = AB$. There are six cases, illustrated in Fig. I.7.

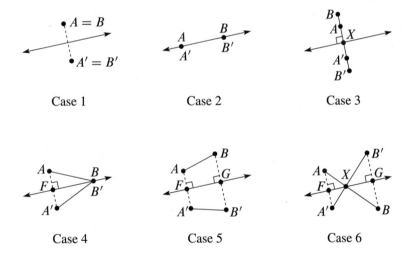

Case 1   Case 2   Case 3

Case 4   Case 5   Case 6

Fig. I.7. Proof that reflections are rigid motions.

CASE 1: $A = B$. In this case, $A' = B'$ and therefore $A'B' = 0 = AB$. In all of the remaining cases, we assume $A \neq B$.

CASE 2: *A and B both lie on* ℓ. In this case $A' = A$ and $B' = B$, so it is immediate that $A'B' = AB$.

CASE 3: $\overleftrightarrow{AB} \perp \ell$. Let X be the point where $\overleftrightarrow{AB}$ meets ℓ, and let $f : \overleftrightarrow{AB} \to \mathbb{R}$ be a coordinate function for $\overleftrightarrow{AB}$ such that $f(X) = 0$. If $A \notin \ell$, then A' is another point on $\overleftrightarrow{AB}$ such that $XA = XA'$. Since $XA = |f(A) - f(X)| = |f(A)|$ and $XA' = |f(A') = f(X)| = |f(A')|$, it follows that $f(A') = \pm f(A)$. Since $A \neq A'$ and f is injective, it cannot be the case that $f(A') = f(A)$, so we must have $f(A') = -f(A)$. On the other hand, if $A \in \ell$, then $A = A' = X$, and $f(A') = f(A) = 0$, which implies again that $f(A') = -f(A)$. A similar analysis shows that $f(B') = -f(B)$. Therefore

$$A'B' = |f(B') - f(A')| = |(-f(B)) - (-f(A))| = |f(B) - f(A)| = AB.$$

CASE 4: $\overleftrightarrow{AB} \not\perp \ell$, and one of the points lies on ℓ but the other does not. We may assume $A \notin \ell$. Let F be the midpoint of $\overline{AA'}$, which lies on ℓ. Then $\triangle AFB \cong \triangle A'FB$ by SAS, so $AB = A'B = A'B'$.

CASE 5: $\overleftrightarrow{AB} \not\perp \ell$, and A and B lie on the same side of ℓ. Let F be the midpoint of $\overline{AA'}$ and let G be the midpoint of $\overline{BB'}$, so $F, G \in \ell$. Then $AF \parallel BF$ by the common perpendicular theorem, so $ABGF$ and $A'B'GF$ are both trapezoids by the trapezoid lemma. It then follows from SASAS that $ABGF \cong A'B'GF$, so $AB = A'B'$.

CASE 6: $\overleftrightarrow{AB} \not\perp \ell$, and A and B lie on opposite sides of ℓ. Let F and G be the midpoints of $\overline{AA'}$ and $\overline{BB'}$ as before, and let X be the point where $\overline{AB}$ meets ℓ. It follows that $A * X * B$, and therefore $\angle AXF \cong \angle BXG$ by the vertical angles theorem. The diagram suggests that $A' * X * B'$ as well, but we do not yet know that because we have not proved that reflections preserve betweenness or collinearity. Note that the argument in Case 5 shows that $A'X = AX$ and $B'X = BX$, so $\triangle A'XF \cong \triangle AXF$ and $\triangle B'XG \cong \triangle BXG$ by SAS. Therefore, $\angle A'XF \cong \angle AXF$ and $\angle BXG \cong \angle B'XG$, so $\angle A'XF \cong \angle B'XG$ by transitivity. Now it follows from the partial converse to the vertical angles theorem (Theorem 4.18) that $\overrightarrow{XA'}$ and $\overrightarrow{XB'}$ are opposite rays, and thus it is indeed the case that $A' * X * B'$. Therefore,

$$A'B' = A'X + XB' = AX + XB = AB.$$

This completes the proof that R_ℓ preserves distances.

The rest of the properties now follow easily. First consider betweenness. Suppose $A * B * C$. Then $AB + BC = AC$ by the betweenness theorem for points, and because F preserves distances it follows that $A'B' + B'C' = A'C'$. Now it follows from the converse to the betweenness theorem (Corollary 5.20) that $A' * B' * C'$, so R_ℓ preserves betweenness.

Next we prove that R_ℓ preserves collinearity. Suppose A, B, C are collinear points. If the points are not all distinct, then the set $\{A', B', C'\}$ contains at most two points, so there is certainly a line containing them. If the points A, B, C are distinct, then one of them is between the other two by Hilbert's betweenness axiom (Theorem 3.9), say $A * B * C$. The fact that R_ℓ preserves betweenness implies $A' * B' * C'$, which means in particular that A', B', C' are collinear.

Finally, to see that R_ℓ preserves angle measures, suppose $\angle ABC$ is an angle. Because R_ℓ is injective, it follows that neither A' nor C' is equal to B', so $\angle A'B'C'$ is also an angle. If $m\angle ABC = 180°$, then $A * B * C$, which implies $A' * B' * C'$ and thus $m\angle A'B'C' = 180°$ as well. If $m\angle ABC = 0°$, then $C \in \text{Int } \overrightarrow{AB}$, which means $A * C * B$, $C = B$, or $A * B * C$. It follows that $A' * C' * B'$, $C' = B'$, or $A' * B' * C'$, respectively, each of which implies $\overrightarrow{A'C'} = \overrightarrow{A'B'}$, so $m\angle A'B'C' = 0°$. Finally, if $\angle ABC$ is a proper angle, then $\triangle ABC$ is a triangle. If A', B', C' were collinear, then A, B, C would also be collinear because $R_\ell(A') = A$, $R_\ell(B') = B$, and $R_\ell(C') = C$, so $\triangle A'B'C'$ is also a triangle. Because R_ℓ preserves distances, it follows that $\triangle A'B'C' \cong \triangle ABC$ by SSS, and therefore $m\angle A'B'C' = m\angle ABC$. Thus R_ℓ preserves angle measures. $\qquad\square$

Exercises

The following exercises should all be done in the context of Euclidean geometry. You may assume either the SAS postulate or the reflection postulate; since each one implies the other, they both are true in Euclidean geometry. Some of these exercises are more elaborate than most of the ones in this book, so they might serve well as starting points for independent projects.

IA. Given rigid motions F and G, the ***composition of F with G*** is the map $G \circ F$ from the plane to itself defined by $G \circ F(A) = G(F(A))$ for every point A (i.e., first apply F, then G). Prove that every composition of rigid motions is a rigid motion.

IB. Suppose A, B, C are noncollinear points and F and G are two rigid motions such that $F(A) = G(A)$, $F(B) = G(B)$, and $F(C) = G(C)$. Prove that $F(X) = G(X)$ for every point X.

IC. Prove that every rigid motion of the plane can be expressed as a composition of one, two, or three reflections. [Hint: If F is an arbitrary rigid motion, argue as in the proof of Theorem I.3 to show that there is another rigid motion F' composed of at most three reflections that agrees with F at three noncollinear points.]

ID. Suppose X and Y are two distinct points. A rigid motion F is called a ***translation along $\overline{XY}$*** if it satisfies the following property: if A is any point and $A' = F(A)$, then $AA' = XY$ and $\overleftrightarrow{AA'}$ is either parallel to or equal to $\overleftrightarrow{XX'}$. Prove that a rigid motion is a translation if and only if it is a composition of two reflections across parallel lines.

IE. Suppose O is a point and θ is a real number such that $0 \le \theta \le 180$. A rigid motion F is called a ***rotation about O through angle θ*** if it satisfies the following property: $F(O) = O$, and if A is any point distinct from O and $A' = F(A)$, then $m\angle AOA' = \theta°$. Prove that a rigid motion is a rotation if and only if it is a composition of two reflections across nonparallel lines.

IF. A rigid motion F is called a ***glide reflection*** if it is a composition of a reflection across a line ℓ followed by a translation along a segment parallel to ℓ. Prove that a rigid motion is a glide reflection if and only if it is a composition of reflections across three distinct lines, the first of which is perpendicular to the other two. [Remark: It can be shown that every rigid motion is either a reflection, a translation, a rotation, or a glide reflection. See [**Ven05**, Chapter 12] for a proof.]

References

[Ari39] Aristotle, *On the Heavens*, Harvard University Press, 1939. With an English translation by W. K. C. Guthrie.

[Arc97] Archimedes, *The Works of Archimedes*, translated by T. L. Heath, C. J. Clay and Sons, London, 1897. Available at www.archive.org/details/worksofarchimede029517mbp/.

[Arc02] ———, *The Works of Archimedes*, translated by T. L. Heath, Dover, Mineola, NY, 2002.

[ADM09] Jeremy Avigad, Edward Dean, and John Mumma, *A formal system for Euclid's Elements*, Rev. Symb. Log. **2** (2009), no. 4, 700–768.

[Bal87] W. W. Rouse Ball, *Mathematical Recreations and Essays*, 13th ed., Dover, Mineola, NY, 1987.

[Bel68] Eugenio Beltrami, *Saggio di interpretazione della geometria non-euclidea*, Giornale di Mathematiche **6** (1868), 285–315. English translation by John Stillwell: "Essay on the interpretation of noneuclidean geometry," *Sources of Hyperbolic Geometry*, Amer. Math. Soc., Providence, RI, 1996.

[Bir32] George D. Birkhoff, *A Set of Postulates for Plane Geometry, Based on Scale and Protractor*, Ann. Math. **33** (1932), no. 2, 329–345.

[BB41] George D. Birkhoff and Ralph Beatley, *Basic Geometry*, Scott, Foresman, Chicago, 1941.

[Con10] John B. Conway, *Mathematical Connections: A Capstone Course*, Amer. Math. Soc., Providence, RI, 2010.

[CG67] H. S. M. Coxeter and S. L. Greitzer, *Geometry Revisited*, Math. Assoc. of America, Washington, DC, 1967.

[Des54] René Descartes, *The Geometry of René Descartes*, Dover, 1954. With an English translation by David Eugene Smith and Marcia L. Latham.

[Dev93] Keith Devlin, *The Joy of Sets*, 2nd ed., Springer, New York, 1993.

[dC76] Manfredo P. do Carmo, *Differential Geometry of Curves and Surfaces*, Prentice-Hall Inc., Englewood Cliffs, N.J., 1976. Translated from the Portuguese.

[Dud94] Underwood Dudley, *The Trisectors*, revised edition, Math. Assoc. of America, Washington, DC, 1994.

[Ecc97] Peter J. Eccles, *An Introduction to Mathematical Reasoning: Numbers, Sets and Func-tions*, Cambridge University Press, Cambridge, 1997.

[Euc98] Euclid, *Euclid's Elements*, translated by Dominic E. Joyce, 1998. Available at `aleph0.clarku.edu/~djoyce/java/elements/elements.html`.

[Euc02] _____, *Euclid's Elements* (Dana Densmore, ed.), translated by Thomas L. Heath, Green Lion Press, Santa Fe, NM, 2002.

[Gre08] Marvin J. Greenberg, *Euclidean and non-Euclidean geometries*, 4th ed., W. H. Freeman, New York, 2008.

[Hal74] Paul R. Halmos, *Naive Set Theory*, Springer-Verlag, New York, 1974.

[Hog05] Jan P. Hogendijk, *Al-Mu'taman ibn Hud, 11th century king of Saragossa and brilliant mathematician*, Historia Mathematica **22** (1005), no. 1, 1–18.

[HW08] G. H. Hardy and E. M. Wright, *An Introduction to the Theory of Numbers*, 6th ed., Oxford University Press, Oxford, 2008.

[Har12] Michael Hartl, *The Tau Manifesto* (2012). Available at `tauday.com/tau-manifesto.pdf`.

[Har00] Robin Hartshorne, *Geometry: Euclid and Beyond*, Springer-Verlag, New York, 2000.

[Her99] Israel N. Herstein, *Topics in Algebra*, 3rd ed., John Wiley & Sons, Inc., New York, 1999.

[Hil71] D. Hilbert, *The Foundations of Geometry*, Open Court Publishing Co., Chicago, 1971. Translated from the tenth German edition by Leo Unger.

[Jac03] Harold R. Jacobs, *Geometry: Seeing, Doing, Understanding*, 3rd ed., W. H. Freeman, New York, 2003.

[Kle93] Felix Klein, *A comparative review of recent researches in geometry*, translated by M. W. Haskell, Bull. Amer. Math. Soc. **2** (1893), 215–249, available at `arXiv:0807.3161v1`.

[Liv02] Mario Livio, *The Golden Ratio*, Broadway Books, New York, 2002.

[Loo68] Elisha Scott Loomis, *The Pythagorean Proposition*, National Council of Teachers of Mathematics, Washington, 1968. Reprint of the 2d ed., 1940.

[Ma99] Liping Ma, *Knowing and Teaching Elementary Mathematics*, Routledge, Mahwah, NJ, 1999.

[Mar96] George E. Martin, *The Foundations of Geometry and the Non-Euclidean Plane*, Springer-Verlag, New York, 1996. Corrected third printing of the 1975 original.

[MP91] Richard S. Millman and George D. Parker, *Geometry: A Metric Approach with Models*, 2nd ed., Springer-Verlag, New York, 1991.

[Moi77] Edwin E. Moise, *Geometric Topology in Dimensions 2 and 3*, Springer-Verlag, New York, 1977.

[Moi90] _____, *Elementary Geometry from an Advanced Standpoint*, 3rd ed., Addison–Wesley, Reading, MA, 1990.

[Pla95] John Playfair, *Elements of Geometry: Containing the First Six Books of Euclid, with Two Books on the Geometry of Solids. To Which Are Added, Elements of Plane and Spherical Trigonometry*, Bell & Bradfute and G. G. & J. Robinson, London, 1795.

[Pro70] Proclus, *A Commentary on the First Book of Euclid's Elements*, translated by Glenn R. Morrow, Princeton University Press, Princeton, NJ, 1970.

[Sac86] Girolamo Saccheri, *Euclides ab Omni Naevo Vindicatus*, translated by George Bruce Halsted, Chelsea, New York, 1986.

[SMSG] School Mathematics Study Group, *Geometry*, Yale University Press, New Haven, 1961.

[UC02] University of Chicago School Mathematics Project, *Geometry, Parts I and II*, 2nd ed., Prentice Hall, Glenview, IL, 2002.

[Vel06] Daniel J. Velleman, *How to Prove It*: *A structured approach*, 2nd ed., Cambridge University Press, Cambridge, 2006.

[Ven05] Gerard A. Venema, *Foundations of Geometry*, Pearson Prentice-Hall, Upper Saddle River, NJ, 2005.

Index

图字：01-2023-0928号

公理化几何学

Gonglihua Jihexue

图书在版编目 (CIP) 数据

公理化几何学 = Axiomatic Geometry：英文 /
(美) 约翰·李 (John M. Lee) 著. -- 影印本
. -- 北京：高等教育出版社, 2024.2
ISBN 978-7-04-061256-1

Ⅰ.①公… Ⅱ.①约… Ⅲ.①几何学—英文

Ⅳ.①O18

中国国家版本馆CIP 数据核字(2023) 第189813 号

策划编辑　李华英　　　责任编辑　李华英
封面设计　张申申　　　责任印制　耿　轩

出版发行　高等教育出版社　　　　　开本　787mm×1092mm　1/16
社址　北京市西城区德外大街4号　　印张　31.25
邮政编码　100120　　　　　　　　　字数　780千字
购书热线　010-58581118　　　　　　版次　2024 年 2 月第 1 版
咨询电话　400-810-0598　　　　　　印次　2024 年 2 月第 1 次印刷
网址　http://www.hep.edu.cn　　　　定价　199.00元
　　　http://www.hep.com.cn
网上订购　http://www.hepmall.com.cn　本书如有缺页、倒页、脱页等质量问题，
　　　http://www.hepmall.com　　　请到所购图书销售部门联系调换
　　　http://www.hepmall.cn　　　　版权所有　侵权必究
印刷　山东临沂新华印刷物流集团有限责任公司　　　[物 料 号 61256-00]

郑重声明